UK OIL AND GAS LAW

VOLUME I

UK OIL AND GAS LAW
CURRENT PRACTICE AND EMERGING TRENDS

3rd edition

VOLUME I
RESOURCE MANAGEMENT AND
REGULATORY LAW

Edited by
Greg Gordon, LL.B., Dip.L.P., LL.M., Ph.D.
Senior Lecturer and Head of School of Law,
University of Aberdeen

John Paterson, LL.B., Dip.L.P., LL.M., Ph.D.
Professor of Law and Vice-Principal,
University of Aberdeen

Emre Üşenmez, B.Sc., B.A., LL.M., M.Sc., Ph.D.
Lecturer in Law, University of Aberdeen

Editorial Assistant
James Cowie, LL.B.

EDINBURGH
University Press

Note: all chapters and sections are prefixed "I" to denote Volume I; cross-references that are prefixed "II" refer to items in Volume II.

Edinburgh University Press is one of the leading university presses in the UK. We publish academic books and journals in our selected subject areas across the humanities and social sciences, combining cutting-edge scholarship with high editorial and production values to produce academic works of lasting importance. For more information visit our website: edinburghuniversitypress.com

First edition published in 2007 by Dundee University Press
Second edition 2011

Edinburgh University Press Ltd
The Tun – Holyrood Road
12 (2f) Jackson's Entry
Edinburgh EH8 8PJ

Typeset in Sabon by Fakenham Prepress Solutions, Fakenham, Norfolk NR21 8NN, and printed and bound in Great Britain

A CIP record for this book is available from the British Library

ISBN 978 1 4744 2018 1 (paperback)
ISBN 978 1 4744 2019 8 (webready PDF)
ISBN 978 1 4744 2020 4 (epub)

Published with the support of the University of Edinburgh Scholarly Publishing Initiatives Fund.

CONTENTS

List of Figures and Tables   vii

List of Contributors   ix

List of Abbreviations and Acronyms   x

Foreword to the First Edition   xvi

Preface to the First Edition   xviii

Preface to the Second Edition   xx

Preface to the Third Edition   xxi

Table of Cases   xxiii

Table of Statutes   xxxix

Table of Statutory Instruments   lvii

Table of European Legislation   lxxvii

Table of International Instruments   lxxxiii

Introduction and Context

I-1   Oil and Gas Law on the United Kingdom Continental
      Shelf: Current Practice and Emerging Trends in Public
      and Regulatory Law   3
      Greg Gordon, John Paterson, Emre Üşenmez and
      James Cowie

I-2   Rejuvenating Activity in the North Sea Oil and Gas
      Industry: The Role of Tax Incentives in Context   14
      Alexander Kemp and Linda Stephen

I-3   The UK's Energy Security   42
      Emre Üşenmez, James Cowie and Greg Gordon

Resource Management

I-4   Petroleum Licensing   81
      Greg Gordon

I-5   The Wood Review and Maximising Economic Recovery
      upon the UKCS   132
      Greg Gordon, John Paterson and Uisdean Vass

I-6   Access to Infrastructure   170
      Uisdean Vass

I-7   The UK's Oil and Gas Fiscal Regime: A Radical
      Evolution   201
      Claire Ralph

I-8   Continental Shelf Boundaries in the North Sea and the
      North Atlantic   221
      Constantinos Yiallourides

Regulatory Law

I-9   Current Practice and Emerging Trends in Regulating
      Onshore Exploration and Production in Great Britain   245
      Tina Hunter, Steven Latta and Greg Gordon

I-10 Health and Safety at Work Offshore   284
      John Paterson

I-11 Environmental Law and Regulation on the UKCS   339
      Luke Havemann and Tina Hunter

I-12 Decommissioning of Offshore Oil and Gas Installations   391
      John Paterson

I-13 Decommissioning Security   435
      Judith Aldersey-Williams

Appendix I-A: Mature Province Initiatives   450

Index   477

FIGURES AND TABLES

FIGURES

I-2.1   Historic UKCS Oil Production by Production Start
        Date   16
I-2.2   Historic UKCS Gas Production by Production Start
        Date   17
I-2.3   Production Efficiency – UK Continental Shelf (Actual
        and Forecast)   18
I-2.4   Potential Total Hydrocarbon Production $70/bbl and
        45p/therm   20
I-2.5   Potential Total Hydrocarbon Production $70/bbl
        and 45p/therm (Production Efficiency Problem Partly
        Resolved)   20
I-2.6   Potential Total Hydrocarbon Production $70/bbl and
        45p/therm (Production Efficiency Problem Resolved)   21
I-2.7   Change in Potential Hydrocarbon Production   22
I-2.8   Change in Potential Development Expenditure   22
I-2.9   Change in Potential Operating Expenditure   22
I-2.10 Undeveloped Discoveries   24
I-2.11 CNS Oil 10 Mboe; Real Post-tax NPV @ 10% / Real
        Devex @ 10%; Oil Price $50/bbl Gas Price 40p/therm   27
I-2-12 CNS Oil 10 Mboe; Real Post-tax NPV @ 10% / Real
        Devex @ 10%; Oil Price $60/bbl Gas Price 45p/therm   28
I-2.13 CNS Oil 20 Mboe; Real Post-tax NPV @ 10% / Real
        Devex @ 10%; Oil Price $50/bbl Gas Price 40p/therm   29
I-2.14 CNS Oil 20 Mboe; Real Post-tax NPV @ 10% / Real
        Devex @ 10%; Oil Price $60/bbl Gas Price 45p/therm   29
I-2.15 CNS Oil 100 Mboe; Real Post-tax NPV @ 10% /
        Real Devex @ 10%; Oil Price $50/bbl Gas Price 40p/
        therm   30
I-2-16 CNS Oil 100 Mboe; Real Post-tax NPV @ 10% / Real
        Devex @ 10%; Oil Price $60/bbl Gas Price 45p/therm   30
I-2.17 CNS Gas 20 Mboe; Real Post-tax NPV @ 10% / Real
        Devex @ 10%; Oil Price $60/bbl Gas Price 45p/therm   31
I-2.18 WoS Oil 100 Mboe; Real Post-tax NPV @ 10% / Real
        Devex @ 10%; Oil Price $50/bbl Gas Price 40p/therm   32
I-2.19 WoS Oil 100 Mboe; Real Post-tax NPV @ 10% / Real
        Devex @ 10%; Oil Price $60/bbl Gas Price 45p/therm   32

I-2.20 NNS Oil 50 Mboe; Real Post-tax NPV @ 10% / Real
Devex @ 10%; Oil Price $60/bbl Gas Price 45p/therm 33

I-2.21 SNS Gas 50 Mboe; Real Post-tax NPV @ 10% / Real
Devex @ 10%; Oil Price $60/bbl Gas Price 45p/therm 34

I-2.22 CNS Project Fast Limited IA Initial Price $55 p/b 40p/
therm Reduced Costs 37

I-2.23 NNS Project Fast Limited IA Initial Price $55 p/b and
40p/therm Reduced Costs 37

I-2.24 SNS Project Fast Limited IA Initial Price $55 p/b and
40p/therm Reduced Costs 38

I-2.25 WoS Project Fast Limited IA Initial Price $55 p/b and
40p/therm Reduced Costs 39

I-2.26 R and D in the UK Energy Sector 40

I-2.27 Scottish Oil and Gas Supply Chain. International and
UK Market Sales 1997-2014, £m (MoD) (including
overseas sales of Scottish subsidiaries) 41

I-3.1 UK Net Energy Export and Crude Oil Production
Levels 1970–2008 46

I-7.1 Graph Illustrating Headline Tax Rates over Time 203

TABLES

I-2.1 UK Oil and Gas Reserves and Resources (bnboe) 23

I-2.2 Rates of Tax on Income and Rates of Relief for
Investment in the UKCS 26

I-2.3 Assumptions for Monte Carlo Modelling by Region
After Cost Reductions 36

CONTRIBUTORS

Judith Aldersey-Williams, B.A., LL.M., Solicitor
Partner, CMS Cameron McKenna Nabarro Olswang LLP

James Cowie, LL.B.
Trainee Solicitor, Jones Day

Greg Gordon, LL.B., Dip.L.P., LL.M., Ph.D.
Senior Lecturer and Head of School of Law, University of Aberdeen

Luke Havemann, B.A., LL.B., Ph.D., Attorney-at-Law
Senior Associate, Bowmans

Tina Hunter, LL.B., Ph.D.
Professor of Law, University of Aberdeen

Alexander Kemp, O.B.E., F.R.S.E.
Schlumberger Professor of Petroleum Economics, Business School,
University of Aberdeen

Steven Latta, B.Sc., Pg. Dip., LL.B., LL.M., Dip.L.P., M.R.I.C.S.,
Solicitor
Assistant Head of Transnational Education, School of Engineering
and the Built Environment, Glasgow Caledonian University

John Paterson, LL.B., Dip.L.P., LL.M., Ph.D.
Professor of Law and Vice-Principal for Internationalisation,
University of Aberdeen

Claire Ralph, ATT
Head of Tax, Falklands Island Government; formerly of Oil & Gas
UK and HM Treasury

Linda Stephen, M.A.
Research Fellow, University of Aberdeen

Uisdean Vass, LL.B., LL.M., Solicitor
Partner, Ledingham Chalmers LLP

Emre Üşenmez, B.Sc., B.A., LL.M., M.Sc., Ph.D.
Lecturer in Law, University of Aberdeen

Constantinos Yiallourides, LL.B., LL.M., Ph.D.
Arthur Watts Research Fellow, British Institute of International and
Comparative Law Teaching Fellow, University of Aberdeen

ABBREVIATIONS AND ACRONYMS

AA	appropriate assessment
AAA/ICDR	American Arbitration Association/ International Court of Dispute Resolution
AAPL	American Association of Professional Landmen
ADR	alternative dispute resolution
AFE	authorisation for expenditure
AIPN	Association of International Petroleum Negotiators
ALARP	as low as reasonably practicable
AMI	Area of Mutual Interest Agreement
API	American Petroleum Institute
ARN	automatic referral notice
ASCOBANS	Agreement on Small Cetaceans of the Baltic and North Seas
BAT	Best Available Technique
BATNA	best alternative to a negotiated agreement
BEIS	Department for Business, Energy and Industrial Strategy
BEP	Best Environmental Practice
BNOC	British National Oil Corporation
boe	barrels of oil equivalent
BPEO	best practicable environmental option
CAEM	Center for the Advancement of Energy Markets
CAR	Construction All Risk
CCS	carbon capture and storage
CCW	Countryside Council for Wales
CEDR	Centre for Effective Dispute Resolution
CEFAS	Centre for Environment, Fisheries and Aquaculture Science
CERM	Co-ordinated Emergency Response Measures
CGT	Capital Gains Tax
CIMAH	Control of Industrial Major Accident Hazard Regulations (1984)
CMR	Convention on the Contract for the International Carriage of Goods by Road

COMAH	Control of Major Accident Hazard Regulations (1999)
CPA 1949	Coast Protection Act 1949
CPC	central product classification
CPR	Civil Procedure Rules (1998)
CRINE	Cost Reduction Initiative for the New Era
CSIS	Center for Strategic & International Studies
CT	Corporation Tax
CTA 2010	Corporation Tax Act 2010
CVA	company voluntary arrangement
DBERR	Department for Business, Enterprise and Regulatory Reform
DEAL	Digital Energy Atlas and Library
DECC	Department of Energy and Climate Change
DEFRA	Department of Environment, Food and Rural Affairs
DEn	Department of Energy
DNV	Det Norske Veritas
DOPWTS	Dispersed Oil in Produced Water Trading Scheme
DSA	decommissioning security agreement
DTI	Department of Trade and Industry
EA	environmental assessment
E&P	exploration and production
EAT	Employment Appeal Tribunal
EC	European Community
ECJ	European Court of Justice
ECT	Energy Charter Treaty
EEA	European Economic Area
EIA	environmental impact assessment
EMT	Environmental Management Team
EMV	expected monetary value
EPC Regulations	Offshore Installations (Emergency Pollution Control) Regulations 2002
ERA	Employment Rights Act 1996
ES	environmental statement
EU	European Union
FEPA 1985	Food and Environment Protection Act 1985
FPAL	First Point Assessment Ltd
FPSO	floating production, storage and offloading
FRS	Fisheries Research Services
FSA	Formal Safety Assessment
FY	financial year
GAAP	generally accepted accounting practice
GATT	General Agreement on Tariffs and Trade

GDP	gross domestic product
GFU	Norwegian Gas Negotiation Committee
GHG	greenhouse gas
GLA	General Lighthouse Authority
H_2S	hydrogen sulphide
HMRC	Her Majesty's Revenue and Customs
HP/HT	high pressure/high temperature
HSC	Health and Safety Commission
HSE	Health and Safety Executive
HSWA 1974	Health and Safety at Work, etc Act 1974
IAPP Certificate	International Air Pollution Prevention Certificate
IATA	International Air Transport Association
ICC	International Chamber of Commerce
ICOP	Infrastructure Code of Practice
ICSID	International Centre for Settlement of Investment Disputes
IEA	International Energy Agency
IEP Agreement	Agreement on an International Energy Program
IGIP	initial gas in place
IMCA	International Maritime Contractors Association
IMHH	Industry Mutual Hold Harmless Deed (strictly, the Mutual Indemnity and Hold Harmless Deed)
IMO	International Maritime Organization
IP	intellectual property
IRR	internal rate of return
IT	income tax
ITF	Industry Technology Facilitator
IUK	Interconnector UK Ltd
JBA	joint bidding agreement
JNCC	Joint Nature Conservancy Council
JOA	joint operating agreement
JOC	Joint Operating Committee
JV	joint venture
KP3	Key Programme 3
LCIA	London Court of International Arbitration
LCP	large combustion plant
LCPD	Large Combustion Plants Directive
LNG	liquefied natural gas
LOC	letter of credit
LOGIC	Leading Oil and Gas Industry Competitiveness

MC	Model Clause
MCA	Maritime and Coastguard Agency
Merchant Shipping (OPRC) Regulations	Merchant Shipping (Oil Pollution Preparedness, Response and Co-operation Convention) Regulations 1998
mmb/d	million barrels of oil per day
MOOIP	moveable oil originally in place
NARUC	National Association of Regulatory Utility Commissioners
NEC Regulations	National Emission Ceilings Regulations 2002
NERC	Natural Environment Research Council
NH_3	ammonia
NO_x	nitrogen oxide
NPI	net profit interest
NPV	net present value
NPV/I	net present value to investment ratio
NSRI	National Subsea Research Institute
NTS	National Transmission System or non-technical summary (in ES)
OC Regulations	Offshore Chemical Regulations 2002
OCA	Offshore Contractors Association
OECD	Organization for Economic Co-operation and Development
OED	Offshore Environment and Decommissioning Unit
OFT	Office of Fair Trading
OGA	Oil and Gas Authority
OGIA	Oil and Gas Independents' Association
OGITF	Oil and Gas Industry Task Force
OGUK	Oil & Gas UK
ONS	Office for National Statistics
OPA Regulations	Offshore Petroleum Activities (Oil Pollution Prevention and Control) Regulations 2005
Opcom	Joint Operating Committee
OPEC	Organization of the Petroleum Exporting Countries
OPOL	Offshore Pollution Liability Agreement
OSPAR	Convention for the Protection of the Marine Environment of the North-East Atlantic 1992
OSPRAG	Offshore Spill Prevention and Response Advisory Group
OTA 1975	Oil Taxation Act 1975
OTA 1983	Oil Taxation Act 1983

PAPS Regulations	Merchant Shipping (Prevention of Air Pollution from Ships) Regulations 2008
PCG	parent company guarantee
PED	Petroleum Engineering Division (of the Department of Energy)
PEDL	Petroleum Exploration and Development Licence
PILOT	successor to the Oil and Gas Industry Task Force
PON	Petroleum Operation Notice
PPSGS Regulations	Merchant Shipping (Prevention of Pollution by Sewage and Garbage from Ships) Regulations 2008
PPWG	Progressing Partnership Work Group
PRT	Petroleum Revenue Tax
PSPA	Petroleum and Submarine Pipelines Act 1975
PTW	permit to work
QCI	qualifying combustion installation
QRA	Quantified Risk Assessment
RFCT	Ring Fence Corporation Tax
ROV	remotely operated vehicle
RPGA	Rules and Procedures Governing Access to Offshore Infrastructure
SAC	Special Area of Conservation
SC	Supplementary Charge
SEA	strategic environmental assessment
SEAM	Senior Executive Appraisal Mediation
SECA	SO_x emission control area
SEPA	Scottish Environment Protection Agency
SGERAD	Scottish Government Environment and Rural Affairs Department
SMS	Safety Management System
SNH	Scottish Natural Heritage
SO_2	sulphur dioxide
SOAEFD	Scottish Office Agriculture, Environment and Fisheries Department
SPA	Special Protection Area
STOOIP	stock tank oil originally in place
t	tonnes (metric)
TDM	Transnational Dispute Management
TFEU	Treaty on the Functioning of the European Union
toe	ton oil equivalent
TPA	transport and processing agreement
TWJA 1878	Territorial Waters Jurisdiction Act 1878

UCTA	Unfair Contract Terms Act (1977)
UK LIFT	United Kingdom Licence Information for Trading
UKAPP Certificate	United Kingdom Air Pollution Prevention Certificate
UKCS	United Kingdom Continental Shelf
UKOOA	United Kingdom Offshore Operators Association (now Oil & Gas UK Ltd)
UNCITRAL	United Nations Commission on International Trade Law
UNCLOS	United Nations Convention on the Law of the Sea
UOA	unit operating agreement
UUOA	unitisation and unit operating agreement
VOC	volatile organic compound
WSCA	Well Services Contractors Association
WTO	World Trade Organization

FOREWORD TO THE FIRST EDITION

I write this Foreword at a time when crude oil prices have jumped to a record high of over US $80 a barrel (West Texas Intermediate). At the same time, there is a world credit crunch, and it remains to be seen what impact this will have upon the oil and gas sector.

At a recent major conference of the Association for the Study of Peak Oil, Lord Oxburgh (the former chairman of Shell) gave a stark warning that the price of oil could hit US $150 per barrel and that oil production could peak within the next 20 years. The rapid increase in the price of oil seems inevitable as demand continues to outstrip supply. However, it is also going to become very expensive indeed to extract oil from the ground. We already see that in our maturing province in the UKCS, with a considerable increase in costs for operating and developing oil and gas fields. This is an industry in a state of flux, and there is a great responsibility on industry lawyers and commercial negotiators to come up with innovative business models and flexible, streamlined legal agreements and processes to facilitate the maximum recovery of remaining reserves in the UKCS. This we must do by working closely with our technical colleagues who are charged with developing increasingly innovative and cost-effective technical solutions to reserves recovery. It is also the responsibility of lawyers, along with our commercial, tax and finance colleagues, to be effective advocates for appropriate changes to UK oil and gas legislation to ensure a successful future for the UKCS. To meet this responsibility, the industry needs dynamic and competent advice at a time when we are experiencing an extreme shortage of experienced oil and gas lawyers. It is all the more important, then, that lawyers coming into our industry have access to reliable and up-to-date reference books on oil and gas law. If we are to meet the challenges ahead, we must pass on the knowledge we already have to a new generation of lawyers; this book helps enormously in that task.

Often our oil and gas industry leaders decry lawyers as those who simply "paper" the deals and arrangements put in place by technical and commercial people. This book goes a long way towards dispelling that myth. It shows the complexity and sophistication of oil and gas law, and its breadth. UK oil and gas law is formed by a layering of statute, commercial agreements, EU and UK competition and procurement law, industry voluntary codes (such as CCOP and

ICOP) and DBERR Guidance. Oil and gas law is a very important field of law and yet there are very few reference sources. This volume is long overdue and very welcome. It describes, in depth, most of the recent developments in this very broad and diverse field. Most importantly, it captures with great clarity the many joint industry and government initiatives since 2000 which impact the legal and commercial arrangements in our sector, for instance those on fallow acreage, stewardship, CCOP and ICOP. It is also the first time the legal basis for these initiatives has been analysed in detail.

There is an enormous challenge ahead. In a time of high oil prices, owners of infrastructure inevitably wish to protect their own production and fair allocation of risk remains difficult to achieve. The "mutual hold harmless" principle is being pushed to its limits, with creeping practices of uncapped liability and indemnity clauses on third-party infrastructure users.

As an industry lawyer for the past 15 years, I have been passionate about improving the way the industry conducts its business to take duplication and waste out of legal processes. This began in 1995 when I worked on the setting up of First Point Assessment Ltd ("FPAL") and the development of the Memorandum and Articles for the new entity. It is with pride that I note that FPAL celebrated its 10th anniversary at Offshore Europe this month.

It has been an enormous privilege for me to have played a part in many of the industry legal working groups which have brought about streamlined agreements (IMHH, Standard Contracts, ICOP, DSA, SPA and Master Deed). We can be proud of what has been achieved and the contribution made by industry and private practice lawyers alike to such progress. This book is an excellent consolidated source on all of these important initiatives and is testimony to the considerable progress made. May it foster even greater academic enquiry and innovation among oil and gas lawyers.

In summary, never has oil and gas law been more complex, never have the expectations of government and industry leaders on lawyers and commercial advisers been higher – and all this at a time of uncertainty as to how the oil and gas market will play out. The industry requires highly competent future lawyers, great clarity of legal thinking and drafting and – above all else – swift close-out of transactions. This excellent book stands us in good stead for meeting the exciting challenges ahead.

Jacquelynn F Craw
Legal Manager, Director and Company Secretary
Talisman Energy (UK) Ltd
September 2007

PREFACE TO THE FIRST EDITION

This book arose as a result of several inter-related stimuli. In developing the LL.M. in Oil and Gas Law, upon which course the editors and several of the book's contributors teach, it became increasingly apparent that while there has been a constant throughput of primary materials in the form of statutes, statutory instruments, guidance and codes, surprisingly little in the way of secondary comment has been published in the area of UK oil and gas law over the years. Moreover, although much of the work published is of a very high standard, there are some noticeable gaps in coverage – and some of the works which are available, and which continue to be of great value, are beginning now to show their age as the UKCS develops and new issues become increasingly relevant.[1] In addition, many of the materials which are available assume a considerable degree of industry knowledge and experience. It can be difficult for a student, or indeed a qualified lawyer making his or her way into the industry, to find a book which will provide a clear but concise account. Finally, many of the books which are available are so highly priced that they are prohibitively expensive to students, and indeed many libraries. The decision to write this book was taken by the editors over coffee while discussing these matters. Much coffee has been drunk by the editors since.

The editors have many people to thank. Our most obvious debt is to the contributors. The book could not have been produced within a reasonable timescale if the editors had had to write it all themselves, and some of the chapters here could not have been written at all. In addition to writing chapters within the book, Margaret Ross, Roderick Paisley, Norman Wisely, Judith Aldersey-Williams and Uisdean Vass read and offered useful comments upon other chapters. Valuable comments have also been received from Lorna Hingston of CMS Cameron McKenna, Bob Ruddiman of McGrigors and Angus Campbell of the University of Aberdeen. The editors are very grateful to all of them for taking the time and trouble to assist. The editors are also very grateful to Carole Dalgleish for commissioning

---

[1] This observation does not apply to Daintith, Willoughby and Hill's excellent and regularly updated *UK Oil and Gas Law*.

the work, and to her and all involved at Dundee University Press for their unfailing commitment and encouragement.

This book is not intended to supplant existing materials, but to supplement them, and hopefully to bring them to the attention of a wider readership. Nor is it intended to be a comprehensive exposition of all legal issues facing the oil and gas industry in the UKCS. There is more that could usefully be said in relation to many of the areas which have been covered, and many other topics could have been selected were it not for the constraints of space and time.[2] Finally, it is hoped that this book will go some way towards stimulating more writing about, and more debate in, what is a fascinating and important area of (or perhaps more properly, *context for*) the law. Towards that end, the editors invoke the spirit of Sir John Skene's dedication to the reader:

"Quhatever I have done, I did it nocht to offend thee or to displease anie man, bot to provoke uthers to doe better."[3]

Greg Gordon
John Paterson
August 2007

---

[2] Environmental law, for instance, is dealt with at several points, but considerations of environmental law as relative to the oil and gas industry could very easily form the subject of a large book on their own

[3] Sir John Skene, *De Verborum Significatione* (1597).

PREFACE TO THE SECOND EDITION

It is very gratifying to see the book go into a second edition. The editors' aim in producing the first edition of this work was to provide a clear, reasonably concise and affordable account of contemporary oil and gas practice in the UKCS. That aim is unchanged. The book attempts to describe the law as it stood in January 2011, but it has been possible to incorporate at proof stage passing reference to some later developments.

As before, the editors have many people to thank. First, the new contributors (Martin Ewan, Luke Havemann and Emre Üşenmez) who have allowed us to expand the scope of the book by authoring chapters on technology in the oilfield, environmental regulation, energy security and taxation. The inclusion of these topics is of great benefit to the book. Second, we must thank all of the original contributors who kindly agreed to update their chapters. Law and practice have certainly not stood still in the 4 years since the first edition of this book was published and in many cases this has involved a significant amount of work. Thanks are also due to David Roper for his preparatory work in the chapter on technology in the oilfield. The editors are also grateful to Christine Gane for allowing us to use her index for the first edition as the basis for the second and to Karen Howatson at Dundee University Press for updating the index. We would also like to thank Carole Dalgleish and all involved at Dundee University Press for their ongoing commitment and support.

Finally, the original editors are delighted to welcome Emre Üşenmez to both the editorial team of this work and the lecturing staff at the University of Aberdeen. Emre has undertaken a significant amount of the editing work for the new edition as well as contributing two new chapters to the book. He also makes a mean cup of coffee. This is not something we say lightly; readers of the preface to the first edition will know the importance which that beverage has played since the very inception of this book. But we should also emphasise that Emre was recruited on the basis of his legal and analytical skills alone.

Greg Gordon
John Paterson
Emre Üşenmez
April 2011

PREFACE TO THE THIRD EDITION

The third edition of this book comes seven years after the second. Those years can fairly be described as turbulent. Oil price, which spent much of the period 2011–2014 at or above $100 per barrel, has since then never risen above $60, falling as low as $30 per barrel in early 2016. As we write this Preface in the summer of 2017, it stands at around $48 per barrel, a price at which much of the production on the United Kingdom Continental Shelf (UKCS) is marginal at best.

The impact upon the industry of the new economic reality can hardly be overstated. There has been an intense focus upon cost-cutting and rationalisation. Thousands of jobs have been lost, within both the service sector and the oil companies themselves. Projects and exploration have been postponed and contracts renegotiated. Companies have merged and insolvencies – previously as rare as hen's teeth – have become prevalent. And the first great wave of decommissioning projects – long anticipated, but previously kept at bay by a combination of technological innovation and high oil price – is upon us.

It is not yet wholly clear what all of these changes portend. Some would argue that they signal the advent of the industry's twilight years. Others contend that – painful as they have been – the industry will emerge leaner, fitter and better placed to ensure that production continues for decades to come.

Oil and gas law and practice has not emerged unchanged from this turmoil. Contracting practice and the fiscal system have both undergone significant change. The Wood Review, with its focus on maximising economic recovery of oil and gas from the UKCS, commenced in a high-price environment but rapidly became a central plank of the Government's attempts to respond to the new low-price environment. And all the while, attempts have been made to launch an onshore unconventionals industry in the face of considerable public opposition and at a time of great constitutional uncertainty. Keeping up with this has been no small task. As a result, the third edition of this book is much changed from the second. It has increased in size to such an extent that it has had to be split into two volumes, the first focusing on issues of resource management and regulation and the second on commercial and contractual issues.

As always, the editors have many people to thank. We are grateful to the authors of the new chapters, to the contributing authors who have had to make painstaking changes to their original work and to the new authors who have joined the writing team in order to update the work of original authors unable, due to pressure of time, to complete the updating task. We are grateful, too, to James Cowie, without whose outstanding editorial assistance the book would have been (even more) seriously delayed. Finally, we are particularly grateful to John Watson and Laura Williamson for commissioning this third edition, and to all at Edinburgh University Press involved in the editing, production and marketing process of the book for all of their diligence, professionalism, enthusiasm and encouragement.

The law is stated as at April 2017. We have, however, been able to take in some subsequent changes to the law at proof stage.

Greg Gordon
John Paterson
Emre Üşenmez
July 2017

TABLE OF CASES

A Turtle Offshore SA v Superior Trading Inc [2008] EWHC
  3034 (Admlty), [2008] 2 CLC 953 .... II–6.05
ACF Chemiefarma NV v Commission (C-41/69) [1970]
  ECR 661 .... II–11.12
Adam v Newbiggin (1888) 13 App Cas 308 .... II–2.15
Advocate (Lord) v Scotsman Publications Ltd 1989 SC
  (HL) 122 .... II–12.29
Advocate (Lord) v Wemyss [1896] 24 R 216, (1899) 2 F
  (HL) 1 .... II–13.01
Ailsa Craig Fishing Co Ltd v Malvern Fishing Co Ltd
  [1983] 1 WLR 964 .... II–6.43
Ainsworth v Inland Revenue Commissioners [2009] UKHL
  31; (2009) 4 All ER 1205; [2009] ICR 985; [2009]
  IRLR 677 .... II–14.71
Aird v Prime Meridian Ltd [2006] EWCA Civ 1866 .... II–15.53
Akzo Chemie BV v Commission (C-62/86) [1991] ECR
  I-3359 .... II–11.19
Aldred Mcalpine Capital Projects Ltd v Tilebox Ltd [2005]
  EWHC 288 (TCC) II–2.67
Alliance Pipeline Ltd v Seibert [2003] 25 Alta LR (4th)
  365 .... II–13.14
Allonby v Accrington & Rossendale College [2004] ICR
  1328 .... II–14.26
Allseas UK Ltd v Greenpeace 2001 SC 844 .... II–15.17
Almelo v Energiebefriff Ijsselmij (C-393/92) [1994] ECR
  I-1477 .... II–11.11
Amoco Production Co v Wilson, 976 P 2d 941 .... II–2.43
Amoco (UK) Exploration Co v Amerada Hess Ltd [1994]
  1 Lloyd's Rep 330 .... II–3.39
Anchor Line (Henderson Brothers) Ltd (No 2), Re [1937]
  Ch 483 .... II–10.69
Anderson v Brattisanni 1978 SLT (Notes) 42 .... II–13.03
Anderson v Stena Drilling Pte Ltd 2006 WL 2524780 .... II–14.37,
                              II–14.38, II–14.40, II–14.41
Arnold v Britton [2015] UKSC 36 .... II–1.01, II–1.08, II–2.80,
                              II–6.34, II–6.36–7, II–6.41, II–6.44
Ashborder BV v Green Gas Power Ltd [2004] EWHC 1517
  (Ch) II–10.83

Aslam v Uber BV [2017] IRLR 4 .... II–14.07
Assessor for Strathclyde Region v BP Refinery
    Grangemouth Ltd 1983 SC 18 .... II–13.14
Atherton (HM Inspector of Taxes) v British Insulated and
    Helsby Cables Ltd [1925] KB 421, 10 TC 155 (CA) I–7.38
Attorney-General of Southern Nigeria v John Holt & Co
    Ltd [1915] AC 599 (PC (S Nigeria)) II–13.22
Auquhirie Land Co Ltd v Scottish Hydro Electric
    Transmission plc, unreported, 10 August 2016 .... II–13.31
Autoclenz Ltd v Belcher [2011] ICR 1157 .... II–14.07, II–14.08

Bahler v Shell Pipe Line Corp (1940, DC Mo) 34 F Supp
    10 .... II–13.17
Bank of Nova Scotia v Société Générale (Canada) (1998)
    87 AR 133, 58 Alta LR (2d) 193 (Alberta CA) II–2.43
Barber v RJB Mining (UK) Ltd [1999] 2 CMLR 833 .... II–14.72
Bates van Winkelhof v Clyde & Co LLP (Public Concern at
    Work intervening) [2014] ICR 730 .... II–14.22, II–14.26,
                                                                              II–14.27
Bear Scotland Ltd v Fulton [2015] 1 CMLR 40; [2015]
    IRLR 15 EAT II–14.64
Berdur Properties (Pty) Ltd v 76 Commercial Road (Pty)
    Ltd 1998 (4) SA 62 (D) II–13.23
Besser v Buckeye Pipe Line Co (1937) 57 Ohio App 341,
    13 NE 2d 927 .... II–13.17
BHP Petroleum Ltd v British Steel plc and Dalmine SpA
    [2000] 2 Lloyd's Rep 277 .... II–6.77, II–6.80
BICC plc v Burndy Corp [1985] Ch 232 .... II–2.64
Blantyre (Lord) v Waterworks Commissioners of
    Dumbarton (1886) 15 R (HL) 56 .... II–13.23
Bleuse v MBT Transport Ltd [2008] ICR 488; [2008] IRLR
    264 .... II–14.43
Bombay Official Assignee v Shroff (1932) 48 TLR 443 .... II–2.86
Borland's Trustee v Steel Bros & Co Ltd [1901] 1 Ch
    279 .... II–2.86
Boss Projects v Bragg 2013 WL 6536645 .... II–14.09
Botham v Ministry of Defence [2006] ICR 250 .... II–14.33,
                                                                       II–14.36, II–14.37
BP Exploration Operating Company Ltd v Dolphin Drilling
    Ltd [2009] EWHC 3319 .... II–5.22–4
Bristol and West Building Society v Mothew [1998] Ch 1 .... II–2.40
British Eagle International Airline Ltd v Cie Nationale Air
    France [1975] 1 WLR 758 .... II–2.87
British Gas Trading v Eastern Electricity plc [1996] EWCA
    Civ 1239 .... II–9.54

British Sugar plc v NEI Power Projects Ltd (1997) 87 BLR
    42 .... II–6.78
Brown v Rice, Patel and the ADR Group [2007] WL
    763674 .... II–15.53
Brown v Voss 105 Wash 2d 366, 715 P 2d 514
    (1986) II–13.23
Bruce v Dalrymple (1731) Elch Serv No 2; 5 Brown's Supp
    220 .... II–13.22
Buchan v Cockburn (1739) Elch 'Clause' 2; M6528 .... II–13.07
Buchan v Hunter, *Unreported Property Cases* p 311 .... II–13.22
Burmah Oil Co (Burma Trading) Ltd v Lord Advocate
    1964 SC (HL) 117 .... II–13.04
Byrne Bros (Formwork) Ltd v Baird [2002] ICR 667 .... II–14.03,
                                            II–14.23, II–14.27

Cable and Wireless v IBM United Kindom Ltd [2002] 2 All
    ER (Comm) 1041 .... II–15.34, II–15.36
Cable and Wireless v Muscat [2006] IRLR 354 .... II–14.13,
                                            II–14.14, II–14.15
Cairn Energy plc v Greenpeace Ltd [2013] CSOH 50 .... II–15.17
Caledonia North Sea Ltd v British Telecommunications plc
    2002 SC (HL) II–6.13
Caledonia North Sea Ltd v London Bridge Engineering Ltd
    2000 SLT 1123; 2002 SC (HL) 117, [2002] 1 All ER
    (Comm) II–1.08, II–6.09, II–6.13, II–6.18, II–6.19–20,
            II–6.24, II–6.28–9, II–6.34, II–6.38, II–6.41–2, II–6.45,
                    II–6.50, II–6.77, II–6.79, II–6.80
Callan v McAvinue, unreported, Irish High Court, 11 May
    1973 .... II–13.22
Campbell v Conoco (UK) Ltd [2003] 1 All ER (Comm)
    35 .... II–4.143, II–6.18, II–6.47
Campbell Discount Co Ltd v Bridge [1962] AC 600 .... II–2.62
Canada Steamship Lines Ltd v The King [1952] AC 192 ..... II–6.38,
                                            II–6.40, II–6.42–3
Canadian Western Natural Gas Co v Empire Trucking Parts
    (1985) Ltd [1998] 61 Alta LR (3rd) II–13.17
Caparo Industries plc v Dickman [1990] 2 AC 605 .... I–12.77,
                                                        II–6.30

Capita Translation and Interpreting Ltd v Siacunas
    (Debarred), Ministry of Justice, Appeal No
    UKEAT/0181/16/RN; 2017 WL 00737371 .... II–14.28
Carlile v Douglas (1731) M 14524 .... II–13.22
Carmichael v National Power plc [2000] IRLR 43 (HL) ... II–14.07,
                                                        II–14.08
Castellain v Preston [1883] 11 QBD 380 .... II–6.04

Cavendish Square Holdings BV v El-Makdessi [2015]
    UKSC 67; [2015] 3 WLR 1373.............II–2.59, II–7.83, II–8.40
CD Robinson Steel v RD Retail Services Ltd [2006] IRLR
    386 ....II–14.64
Central RC v Ferns 1979 SC 136..II–13.17
Chartbrook Ltd v Persimmons Homes Ltd [2009] UKHL
    38; [2009] 1 AC 1101...II–6.36
Cheever v Jefferson Properties Ltd (1995) II–13.03
Christie v Wemyss (1842) 5 D 242 ....II–13.22
Clark v Craig, unreported, Stonehaven Sheriff Court, 12
    February 1993 ....II–13.22
Clarke v Oxfordshire Health Authority [1998] IRLR
    125 ....II–14.08
Clydebank Engineering & Shipbuilding Co Ltd v Don
    Jose Ramos Yzquierdo y Castaneda [1905] AC
    6 ....II–2.61, II–2.74
Coco v A N Clark (Engineers) Ltd [1969] RPC 41 ....II–12.29
Cofts v Vetal Ltd [2006] ICR 250........................II–14.33, II–14.37
Commissioner of Public Works v Hills [1906] AC
    368 ....II–2.59, II–2.74
Commune de Mesquer v Total France SA and Total
    International Ltd ECR 2008 I-04501............................I–11–105
Consafe Engineering (UK) Ltd v Emtunga UK Ltd [1999]
    RPC II–12.16
Consistent Group Ltd v Kalwak [2008] IRLR 505 ....II–14.07
Cooper v Strathclyde RC 1993 GWD 31-2013...................II–13.11
Cornelis v Fernando (1962) 65 NLR 93 ....II–13.07
Cornwall CC v Prater [2006] 2 All ER 1013 ....II–14.08
Cotswold Developments Construction Ltd v Williams
    [2006] IRLR 181 ....II–14.27
CRA v NZ Goldfields Investments [1989] VR 873 (Victoria
    Supreme Court) II–2.64
Craigie v London Borough of Haringey [2006] UKEAT
    0556/06/JOJ..II–14.14, II–14.15
Crehan v Inntrepreneur Pub Co [2007] 1 AC 333 ....II–11.01
Crofts v Veta Ltd [2006] ICR 250 ....II–14.33, II–14.35,
                    II–14.37, II–14.38, II–14.40, II–14.42, II–14.43

Dacas v Brook Street Bureau (UK) Ltd [2004] EWCA Civ
    217; [2004] ICR 1437 .....II–14.11, II–14.12, II–14.14, II–14.15
Daks Simpson Group plc v Kuiper 1994 SLT 689 ....II–15.53
Dalkia Utilities Services plc v Celtech International Ltd
    [2006] EWHC 63 (Comm) II–8.12
Deepak Fertilisers and Petrochemicals Corporation v ICI
    Chemicals & Polymers Ltd (1999) 1 TCLR 2000..........II–6.78

Delimitation of the Continental Shelf (UK/France) (Arbitral
    Award of 30 June 1977) I–8.12
Dewhurst v Citysprint (UK) Ltd (Case No
    2202512/16) II–14.07
Dhuma v Creditsights Ltd UKEAT/02416/12 .... II–14.43
Director General of Telecommunications v Mercury
    Communications Ltd, CA, unreported II–15.60
Dispute Concerning Delimitation of the Maritime
    Boundary in the Bay of Bengal (Bangladesh/Myanmar)
    (ITLOS Judgment) 51 (2012) ILM 840 .... I–8.23–4, I–8.29
Dobie v Burns International Security Services (UK) Ltd
    1984 ICR 812, CA II–14.83
Dole Food Company Inc and Dole Fresh Fruit Europe v
    Commission (C-286/13P) II–11.61
Donoghue v Stevenson [1932] AC 562 .... II–6.30
Duncombe v Secretary of State for Children, Schools and
    Families [2011] 4 All ER 1020 .... II–14.39
Dunlop Pneumatic Tyre Co Ltd v New Garage & Motor
    Co Ltd [1915] AC 79 .... II–2.59, II–2.61, II–2.68, II–2.74
Dunnett v Railtrack plc [2002] 1 WLR 2434 .... II–15.43
Dyce v Hay (1852) 305 .... II–13.22

EC Gransden & Co Ltd v Secretary of State for the
    Environment [1987] 54 P & CR 361 .... I–9.78
ECC Quarries Ltd v Watkins (Inspector of Taxes) [1975] 3
    All ER 843, (1977) 1 WLR 1386 .... I–7.38
El-Makdessi v Cavendish Holdings BV [2015] UKSC 67;
    [2016] AC 1172; [2015] 3 WLR 1373; [2016] 2 All ER
    519; [2016] 2 All ER (Comm) 1; [2016] 1 Lloyd's Rep
    55; [2015] 2 CLC 686; [2016] BLR 1; 162 Con LR 1;
    [2016] RTR 8; [2016] CILL 3769 .... II–1.04, II–2.63,
                                    II–2.69–74, II–2.76, II–2.79
El Paso Field Service Inc v Stephen Minvielle, 867 So 2d
    120 (La App 3d Cir, 2004) II–13.17
Elf Enterprise Caledonia Ltd v Orbit Valve Co Europe
    [1995] 1 All ER 174 .... II–1.08, II–4.144, II–6.11, II–6.27,
                      II–6.34, II–6.38–42, II–6.45, II–6.47
Ellenborough Park, Re [1956] 1 Ch 131 .... II–13.27
Emirates Trading Agency LLC v Prime Mineral Exports
    Private Ltd [2014] EWHC 2104 (Comm) II–15.34
Enviroco v Farstadt [2009] EWCA Civ 1399 .... II–5.58
Erewhon Exploration Ltd v Northstar Energy Corp [1993]
    147 AR 1, 15 Alta LR (3d) 200 9 (Alberta QB) II–2.43
Euro London Appointments v Claessens International
    [2006] EWCA Civ 385 .... II–8.41

Express & Echo Publications Ltd v Tanton [1999] ICR
   693 ....II–14.09, II–14.24
Exxon Corporation v Exxon Life Insurance Consultants
   International Ltd [2982] Ch 119 ....II–12.18
Exxon Pipeline Co v LeBlanc 763 So 2d 128 (La App 1 Cir
   2000) II–13.09

Faccenda Chicken Ltd v Fowler [1986] 1 All ER
   617 ....II–12.30, II–12.33
Farm Assist Ltd (in liquidation) v Secretary of State for the
   Environment, Food and Rural Affairs (No 2) [2009]
   EWHC 1102 (TCC).. II–15.53
Farstad Supply A/S v Enviroco Ltd [2010] UKSC 18, 2010
   SCLR 379 ....II–4.140, II–6.05, II–6.09, II–6.14, II–6.15,
                     II–6.22, II–6.28, II–6.29, II–6.51, II–6.61
Farstad Supply A/S v Enviroco Ltd [2011] UKSC
   16 ....II–6.52, II–10.88
Ferguson v John Dawson & Partners (1976) 1 WLR
   1213 .... II–14.07
Firma C-Trade SA v Newcastle Protection and Indemnity
   Association (The Fanti) (No 2) [1991] 2 AC 1
   (HL)...II–6.03, II–6.05
Fisher v California Cake & Cookie Ltd [1997] IRLR
   212 .... II–14.87
Fontenot v Mesa Petroleum 791 F 2d 1207....................... II–6.18
Forder v Great Western Railway Co [1905] 2 KB
   532 ....II–2.28, II–2.30
Fourie v Marandellas Town Council 1972 Rhodesian Law
   Reports 164..II–13.22
Franks v Reuters Ltd [2003] EWCA 417 ....II–14.12
Fraser v Oystertec plc [2004] BCC 233 .... II–2.89
Friedman v Murray [1952] OWN 295, [1952] 3 DLR 159
   (HC), affirmed [1953] OWN 486; [1953] 3 DLR 313
   (CA)... II–13.23

Gairlton v Stevenson (1677) M 12769.............................. II–13.22
Gallagher v Alpha Catering Services Ltd (t/a Alpha Flight
   Services) [2005] ICR 673.. II–14.52
Gateshead MBC v Secretary of State for the Environment
   [1995] Env LR 37... I–9.74, I–9.79
George Wimpey East Scotland Ltd v Fleming 2006 SLT
   (Lands Tr) 2... II–13.20
Gerald Metals SA v the Trustees of Timis Trust [2016]
   EWHC 2327.. II–15.65
Gibb v United Steel Companies Ltd [1957] 2 All ER 110... II–14.06

Gillespie Bros & Co Ltd v Roy Bowles Transport Ltd
[1973] QB 4000..II–6.38
Glasgow Corporation v McEwan (1899) 2 F (HL) 25........II–13.22
Glencore Energy UK Ltd v Cirrus Oil Services Ltd [2014]
EWHC 87 (Comm)..II–6.78
Glen's Trs v Lancashire and Yorkshire Accident Insurance
Co Ltd (1906) 8 F 915...II–6.33
Goldacre (Offices) Ltd v Nortel UK Ltd [2009] EWHC
3389 (Ch)..II–10.102
Graham v Belfast and Northern Counties Railway Co
[1901] 2 IR 13...II–2.28
Graham v Teesdale (1981) 81 LGR 117................................II–2.29
Grange v Abellio London Ltd [2017] IRLR 108.................II–14.52
Gray v Maxwell (1762) M 12800...II–13.22
Groupement des cartes bancaires (CB) v Commission
(C-67/713P)...II–11.61
Gulf Pipe Line Co v Kaderli (1927, Tex Civ App) 299 SW
534...II–13.14
Gulf Pipe Line Co v Thomason(1927, Tex Civ App) 299
SW 532...II–13.14
Gunlegal Ltd, Re [2003] EWHC 1844 (Ch)..........................II–2.13

Hadley v Baxendale (1854) 9 Ex 341.........II–6.78, II–6.79, II–6.80
Halliburton Energy Services Inc v Smith International
(North Sea) Ltd [2006] RPC 2..II–12.16
Halsey v Milton Keynes General NHS Trust [2004] 1 WLR
3002...II–15.43
Hamilton-Gray v Sherwood, Sheriff Court, 27 August
2002...II–13.17
Hamlyn & Co v Talisker Distillery (1894) 21 R (HL) 21...II–15.65
Harlow v O'Mahony, Appeal No UKEAT/0144/07/LA;
2007 WL 3001900..II–14.17
Hashami v OMV Maurice Energy Ltd [2015] WCA Civ
1171; [2015] 2 CLC 80...II–2.19
Hashwami v Jivraj (London Court of International
Arbitration intervening) [2011] ICR 1004.....................II–14.26
Hayns v Secretary of State for the Environment (1978) 36
P & CR 317..II–13.30
Heatherwood and Wexham Park Hospitals NHS Trust v
Kulubowila 2007 WL919521 (EAT)..................................II–14.15
Henderson v Merrett Syndicates Ltd [1995] 2 AC 145.........II–2.41
Higginbotham v Holme (1812) 19 Ves 88............................II–2.83
High Court in the Office of Fair Trading v Abbey National
plc [2008] EWHC 875 (Comm) and Civ 116 [2010]
1 AC 696...II–8.41

HIH Casualty & General Insurance Ltd v Chase
  Manhattan Bank [2003] UKHL 6, [2003] 2 Lloyd's
  Rep 297 .... II–6.43–5
HJ Banks and Co Ltd v Shell Chemicals UK Ltd CA11/05
  [2005] CSOH 123 .... II–13.04
Holland House Property Investments Ltd v Crabbe 2008
  SLT 777 .... II–15.60
Hollier v Rambler Motors (AMC) Ltd [1972] 2 QB 71 .... II–6.43
Holloway v Chancery Mead [2008] 1 All ER (Comm)
  653 .... II–15.34, II–15.36
Horobin's case [1952] 2 Lloyd's Rep 460 .... II–2.30
Hospital Products Ltd v United States Surgical Corp (1984)
  55 ALR 417 .... II–2.39
Hotel Services Ltd v Hilton International Ltd [2000] 1 All
  ER (Comm) 750 .... II–6.78
Huckvale v Aegean Hotels Ltd (1989) 58 P & CR 163 ...... II–13.27
Hurst v Leeming [2003] 1 Lloyd's Rep 279 .... II–15.43

ICI v Commission (Dyestuffs) (C-48/69) [1972] ECR
  619 .... II–11.12
Industrial Gas Co v Jones (1939) 62 Ohio App 553, 24 NE
  2d 830 .... II–13.17
Investors Compensation Scheme Ltd v West Bromwich
  Building Society [1997] UKHL 28 .... II–6.33, II–6.36
Irvine Knitters Ltd v North Ayrshire Co-operative Society
  Ltd 1978 SC 109 .... II–13.23
Ithaca v NSE [2012] EWHC 1793 (QB) II–2.19, II–2.52–3
ITP SA v Coflexip Stena Offshore Ltd 2004 SLT 1285 .... II–12.16

Jackson v Hughes Dowdall 2008 SC 637 .... II–15.75
James v Greenwich Council [2006] UKEAT
  0006/06/1812 .... II–14.14–16
James v Greenwich Council [2008] ICR 545 .... II–14.14
James v Redcats 2007 WL504779 (EAT); [2007] ICR
  1006 .... II–14.24, II–14.28, II–14.30
James v Redcats (Brands) Ltd [2007] ICR 1006 .... II–14.09
Jay, Ex p; In re Harrison (1880) 14 Ch D 19 .... II–2.83,
                                                      II–2.85, II–2.86
Jeffery v British Council [2016] IRLR 935 .... II–14.42
Jengle v Keetch (1992) 89 DLR (4th) 15 .... II–13.28
Jobson v Johnson [1989] 1 All ER 621 (CA); [1989] 1
  WLR 1026 CA II–2.59, II–2.64, II–2.65
Jodrell, Re (1890) 44 Ch D 590 .... II–6.33
Johnson, Thomas and Thomas (a Firm) v Smith 2016
  GWD 25-456 .... II–13.22

Jones v Tower Hamlets [2001] RPC 23 .... II–12.22

KCA Drilling Ltd v Robert S Breeds 2000 WL 824099
(EAT) II–14.84, II–14.89–93
Kelly v Cooper [1993] AC 205 .... II–2.39
Kelly v Ohio Oil Co, 49 NE 399 (Ohio, 1897) II–3.04
Kerr v Brown 1939 SC 140 .... II–13.03
Knight v Fairway and Kenwood Car Service
UKEAT/0075/12/LA II–14.07

Labinski Ltd v BP Oil Development Ltd, 24 January 2003,
IH, 18 December 2001, OH II–13.14
Lackey v Joule, App, 577 SW 2d 114 .... II–13.22
Landeshauptstadt Kiel v Norbert Jaeger [2003] ECR
1-8389 ....II–14.50, II–14.51, II–14.52, II–14.53,
II–14.55, II–14.58, II–14.66, II–14.74, II–14.75, II–14.78,
II–14.79, II–14.80
Lawson v Serco Ltd, Botham v Ministry of Defence and
Crofts v Veta Ltd [2006] ICR 250 .... II–14.33, II–14.35,
II–14.36, II–14.37, II–14.39, II–14.40, II–14.42, II–14.43
Lean v Hunter 1050 SLT (Notes) 32..................................II–13.25
Leisure (Norwich) (II) Ltd v Luminar Lava Ignite Ltd
[2012] EWHC 951 .... II–10.102
Liscombe v Maughan (1928) 62 OLR 328, [1928] 3 DLR
397 (CA)... II–13.23
Lloyd v McMahon [1987] AC 625 .... II–2.29
Lock v British Gast Trading Ltd [2013] CJEU Case
C-539/12..II–14.64
Lock v British Gast Trading Ltd [2016] EWCA Civ 983;
[2017] 1 CMLR 25..II–14.64
London & Blenheim Estates Ltd v Ladbroke Retail Parks
Ltd [1993] 4 All ER 157...II–13.25
Lundy Granite Co ex p Heavan (1870-71) LR 6 Ch App
462 ....II–10.104
Luscar Ltd v Pembina Resources Ltd (1995) 24 Alta LR
(3d) 305, [1995] 2 WWR 153 (Alberta CA) II–2.43
Lyddon v Englefield Brickwork Ltd [2008] IRLR 198 ....II–14.64

McCain Foods GB Ltd v Eco-Tec (Europe) Ltd [2011]
EWHC 66 (TCC)... II–6.78
MacCartney v Oversley House Management [2006] ICR
510 ....II–14.52, II–14.75, II–14.77, II–14.80
McCosh v Brown & Co's Trs (1889) 1 F (HL) 86 .... II–2.15
MacDonald Estates plc v National Car Parks Ltd 2010 SC
250, 2010 SLT 36 ....II–15.59, II–15.60, II–15.61

Macepark (Wittlebury) Ltd v Sargeant [2003] 2 P & CR
    12 ....II–13.23, II–13.24
MacFarlane v Glasgow City Council [2001] IRLR 7,
    EAT II–14.09, II–14.24
McLellan v Hunter 1987 GWD 21-799 ....II–13.22
McMeechan v Secretary of State for Employment [1995]
    ICR 444 .... II–14.10–11
Madden v Coy [1994] VLR 88 .... II–13.01
Mair v Wood 1948 SC 83 .... II–2.07
Manfredi v Lloyd Adriatico Assicurazioni (Joined Cases
    C-298/04 and 299/04) [2006] 5 CMLR 17 .... II–10.105
Mangold v Helm [2006] All ER (EC) 383 .... II–14.43
Maritime Delimitation and Territorial Questions (Qatar v
    Bahrain) [2002] 40 ICJ Rep 847 .... I–8.02
Markerstudy Insurance Co Ltd v Endsleigh Insurance
    Services Ltd [2010] EWHC 281 (Comm) II–6.78
Market Investigations v Minister for Social Security [1969]
    2 QB 173 ....II–14.06
Martha Envoy [1978] AC 1 .... II–5.21
Midgulf International Ltd v Groupe Chimique Tunisien
    [2010] EWCA 66 (Civ) II–15.65
Millar's Machinery Co v David Way & Son [1935] 40 Com
    Cas 204 .... II–6.78
Moncrieff v Jamieson 2005 1 SC 281 ....II–13.22
Money Markets International Stockbrokers Ltd (in
    liquidation) v London Stock Exchange [2002] 1 WLR
    1150 .... II–2.88
Montgomery v Johnson Underwood Ltd [2001] ICR 819 .... II–14.11
Moody v Steggles (1879) 12 Ch 261 ....II–13.07
Mosaic Oil NL v Angaari Pty Ltd [1990] 8 ACLC 780
    (New South Wales Supreme Court) II–2.64, II–2.91
Motours Ltd v Eurobell (West Kent) Ltd [2003] EWHC
    614 (QB) II–6.77
Murray v Mags of Peebles, 8 Dec 1808, FC II–13.28

National Grid Electricity Transmission plc v Wood, Appeal
    No ULEAT/0432/07/DM, 2007 WL 3002010 .... II–14.17
National Semiconductors (UK) Ltd v UPS Ltd [1996] 2 LL
    Rep 212 .... II–2.30
National Westminster Bank plc v Spectrum Plus Ltd [2005]
    UKHL 41 ....II–10.83
Nelson v Atlantic Power and Gas Ltd 1995 SLT 102 .... II–1.08,
                                                II–6.38, II–6.40, II–6.41
Neste Production Ltd v Shell UK Ltd [1994] 1 Lloyd's Rep
    447 .... II–3.39

Nethermere (St Neots) Ltd v Taverna & Gardiner [1984]
IRLR 240.. II–14.08
Newitt, Ex p; In re Garrud (1991) 16 Ch D 522 .... II–2.86
Newport County Borough v Secretary of State for Wales
and Browning and Ferris Environmentaal Services Ltd
[1998] Env LR 174.. I–9.75, I–9.79
Nigel Witham Ltd v Smith (No 2) [2008] EWHC 12
(TCC) II–15.43
Norscot Rig Management PVT Ltd v Essar Oilfields
Services [2010] EWHC 195 (Comm)........................... II–15.69
North British Railway v Park Yard Co Ltd (1898) 25 R
(HL) 47... II–13.25
North Sea Continental Shelf Cases (Germany v Denmark;
Germany v Netherlands) [1969] ICJ Rep 3....................
I–8.07, I–8.11–12, I–8.34, I–8.35, II–3.05
Nutting v Baldwin [1995] 1 WLR................................. II–2.76–9

Oceanbulk Shipping and Trading SA v TMT Asia Ltd (aka
TMT Asia Ltd v Oceanbulk Shipping and Trading SA)
[2010] UKSC 44 .... II–15.53
Olympia and York Canary Wharf Ltd, Re [1993] BCC
154 .... II–10.99

Parker Hannifin Manufacturing and Parker-Hannifin v
Commission (Case T-146/09 RENV) II–11.03
Peacock v Custins [2001] 2 All ER 827, [2001] 13 EG 152,
CA II–13.23
Penney's Trade Mark, Re [1978] OJ L60/19 .... II–11.12
Perpetual Trustee Co Ltd v BNY Corporate Trustee Services
Ltd and Lehman Brothers Special Financing Inc and
Butters v BBC Worldwide Ltd [2009] EWCA Civ 1160;
[2010] 3 WLR 87; [2010] Bus LR 632; [2010]
BCC 59... II–2.90, II–2.91
Petrofac Offshore Management Ltd v Olley 2005
WL3142404 (EAT) II–14.84–8, II–14.89, II–14.92, II–14.93
Philips v Attorney-General of Hong Kong [1993] 61 BLR
41 .... II–2.66
Phillips v First Secretary of State [2004] JPL 613.................. I–9.80
Photo Production Ltd v Securicor Transport Ltd [1980] AC
827 .... II–6.77
Pickard v Somers (1932) 48 Sh Ct Rep 237....................... II–13.22
Pillar Denton Ltd v Jervis [2014] EWCA Civ
180 .... II–10.102–3, II–10.105
Pimlico Plumbers Ltd v Smith [2017] EWCA Civ
51 .... II–14.09, II–14.27

Pine Energy Consultants Ltd v Talisman Energy (UK) Ltd
    [2008] CSOH 10 .... II–12.29
Polkey v AE Dayton Services Ltd [1987] IRLR 503 .... II–14.87
Polypearl Ltd v E.On Energy Solutions Ltd [2014] EWHC
    3045 (QB)... II–6.78
Porter v Magill [2001] UKHL 67; [2002] 2 AC 357
    (HL)... II–2.29, II–2.31
Prenn v Simmonds [1971] 1 WLR 1381 .... II–6.36
Printers & Finishers Ltd v Holloway [1965] 1 WLR 1 .... II–12.30
Proton Energy Group SA v Orlen Lietuva [2013] EWHC
    2782 .... II–8.07

Quashie v Stringfellow Restaurants Ltd [2012] EWCA Civ
    1735; [2013] IRLR 99 .... II–14.28

R v A-G for Northern Ireland ex p Burns [1999] IRLR
    315 .... II–14.56
R v Kite and OLL Ltd [1994] (unreported) I–10.78
R v Secretary of State for Trade and Industry, ex parte
    Greenpeace Ltd (No 2) [2001] Env L R 221.................... I–4.25
R (ex p Cowl) v Plymouth City Council [2002] 1 WLR
    903 .... II–15.43
R (on the application of Copeland) v Tower Hamlets LBC
    [2010] LLR 654.. I–9.66
R (on the application of Hottak) v Secretary of State for
    Foreign and Commonwealth Affairs [2016] 1 WLR
    3791 .... II–14.43
Rainy Sky v Kookmin Bank [2011] UKSC............................ II–2.80
Rattray v Tayport Patent Slip Co (1868) 5 SLR 219........... II–13.22
Ravat v Halliburton Manufacturing & Services Ltd 2012
    SC (UKSC) 265 .... II–14.39–40, II–14.42
Ready Mixed Concrete (South East) Ltd v Minister of
    Pensions and National Insurance [1968] 2 QB
    497 .... II–14.06, II–14.08, II–14.11
Reardon Smith Line Ltd v Hansen-Tangen [1976] 1 WLR
    989 .... II–6.36
Reed Executive plc v Reed Business Information Ltd [2004]
    1 WLR 3026... II–15.53
Regia Autonoma de Electricitate Renel v Gulf Petroleum
    International Ltd [1996] 1 Lloyd's Rep 67 .... II–15.72
Robb v Salamis M & I Ltd 2007 SLT 158............................ II–6.46
Rochon v Charron, 2 May 2002, Cour du Québec, QCCQ
    705-22-003035-001... II–13.22
Rule v Hazlehaw Properties Ltd and Scottish Power UK plc
    [2017] SC GLA 1.. II–13.11

Russell v Transocean International Resources Ltd [2011]
　　UKSC 57; 2012 SC (UKSC) 250 ..... II–14.51, II–14.62, II–14.80

Safeway Food Stores Ltd v Wellington Motor Co (Ayr) Ltd
　　1976 SLT 53 .... II–13.24
Safeway Ltd v Twigger [2010] EWCA Civ 1472 .... II–11.02
Saga Cruises BDF Ltd v Fincantieri SPA [2016] EWHC
　　1875 (Comm) II–6.79
Saint Line v Richardsons (1940) 2 KB 99 .... II–6.78
Saltman Engineering Co Ltd v Campbell Engineering Co
　　Ltd [1948] 65 RPC 203 .... II–12.29
Salvin's Indenture, Re [1938] 2 All ER 498 .... II–13.24
Scandinavian Trading Tanker Co AB v Flota Petrolea
　　Ecuatorina (The Scapade) [1983] QB 529 .... II–2.64
Scott v Bogle, 6 July 1809, FC II–13.23
Scottish Ambulance Service v Truslove UKEATS/0028/11/
　　BI; 2012 WL 2800455 .... II–14.52
Scottish Highland Distillery Co v Reid (1877) 4 R 1118 .... II–13.22
Scottish & Newcastle plc v G D Construction (St Albans)
　　Ltd [2003] EWCA Civ 16 .... II–6.04
Scottish Oil Company Ltd (in liquidation), Re The (2013)
　　CSIH 108 .... II–10.114
Scottish Power UK plc v BP Exploration and Drilling
　　[2015] EWHC 2658 .... II–6.79
Scotto v Petch [2001] BCC 899 .... II–2.13
Shell UK Ltd v Enterprise Oil plc [1999] 2 Lloyd's Rep
　　456 .... II–3.39
Shetlands Islands Council v BP Petroleum Development Ltd
　　1990 SLT 82 .... II–13.12
Shiloph Spinners Ltd v Harding [1973] AC 671 .... II–2.63
SHV Gas Supply & Trading SAS v Naftomar Shipping &
　　Trading Co Ltd [2006] 1 LLR 163 .... II–8.17
Sindicato de Medicos de Assitencia Publica (SIMAP) v
　　Conselleria de Sanidad y Consumo de la Generalidad
　　Valenciana [2000] ECR 1-7963 .... II–14.54, II–14.73,
　　　　　　　　　　　　　　　　　II–14.78, II–14.79, II–14.80
Skull v Glenister (1864) 16 CB (NS) 81 .... II–13.23
Slessor v Vetco Gray, unreported, 7 July 2006, Court of
　　Session, Outer House II–6.07, II–6.18, II–6.27, II–6.34,
　　　　　　　　　　　　　　　II–6.45, II–6.48–9, II–6.72
Smith v Carillon, Case No A2/2014/0395/EATRF, [2015]
　　EWCA Civ 209; [2015] IRLR 467 .... II–14.17
Smith v South Wales Switchgear Co Ltd [1978] 1 WLR
　　165 .... II–6.43, II–6.47
Smith v UMB Chrysler (Scotland) Ltd 1978 SC (HL) 1 .... II–6.38

Société de Vente de Ciments et Bétons de l'Est SA v Kerpen
& Kerpen GmbH und Ko KG (C-319/82) [1983] ECR
4173 .... II–11.13
Soufflet Negoce SA (2009) EWHC 2454 (Comm); [2010] 1
Lloyd's Rep 718; Affd [2010] EWCA Civ 1102; [2011]
1 Lloyd I Rep 531... II–8.12
South China Sea Arbitration (Phillippines v China) (award
of 12 July 2016) I–8.33
South Lanarkshire Council v Taylor 2005 1 SC 182........... II–13.12
Southern Star Central Gas Pipeline Inc v Murray 190 SW
3d 423 Mo App SD, 2006... II–13.17
Sovmots Investments Ltd v Secretary of State for the
Environment [1979] AC 144... II–13.03
Sport International Bussum BV v Inter-Footwear Ltd
[1984] 1 WLR 776 .... II–2.64
Spree Engineering & Testing Ltd v O'Rourke Civil &
Structural Engineering Ltd, Unreported, 1999 WL
33453546, QBD II–2.17, II–2.18, II–2.19, II–2.22
Star Energy Weald Basin Ltd v Bocardo SA [2010] UKSC
35, [2010] 3 WLR 654 .... I–4.08
Starsin case [2003] UKHL 12; [2004] 1 AC 715.................... II–6.36
Stewart v Stewart (1788) Hume 731................................. II–13.28
Stringer v Minister of Health for Housing and Local
Government [1970] WLR 1281 .... I–9.59
Sul America Cia Nacional de Seguros SA v Enesa
Engenhara SA [2001] EWCA Civ 638 .... II–15.34, II–15.36
Sweeney v Lagan Developments [2007] NICA 11................ II–2.02,
II–2.18, II–2.19, II–2.22

TCS Holdings Ltd v Ashtead Plant Hire Co Ltd 2003 SLT
177 .... II–13.12
Tesco Stores v Secretary of State for Environment [1995] 1
WLR 759.. I–9.69
Texas Eastern v EE Caledonia, CA, 1989............................. II–9.52
Texas Eastern v Enterprise Oil plc, CA, 21 July 1989
(unreported)... II–2.13
Thames Valley Power Ltd v Total Gas & Power Ltd [2006]
Lloyd's Rep 441.............. II–15.14, II–15.31, II–15.58, II–15.61
Thompson v T Lohan (Plant Hire) Ltd [1987] 1 WLR
649 .... II–6.29
TNT Global SPA v Denfleet International Ltd [2007]
EWCA Civ 405; [2008] 1 All ER (Comm) 97; [2007] 2
Lloyd's Rep 504.. II–2.30, II–2.31
Todd v Adams [2002] 2 All ER (Comm)............................... II–2.21
Todd v Scoular 1988 GWD 24-1041 .... II–13.22

Todrick v Western National Omnibus Co [1934] Ch
561 .... II–13.07
Transco plc v HMA 2004 SLT 41 .... I–10.78
Transocean Drilling UK Ltd v Providence Resources plc
[2016] EWCA Civ 372 .... II–6.79
Transocean International Resources Ltd v Russell, Case no
S/104056/04 .... II–14.45
Trevett v Secretary of State for Transport, Local
Government and the Regions [2002] EWHC 2696
(Admin) I–9.81
Truslove v Scottish Ambulance Service [2014] ICR
1232 .... II–14.76, II–14.77, II–14.80

Veba Oil Supply and Trading GmbH v Petrograde Inc
[2001] 2 Lloyd's Rep 731 .... II–15.58, II–15.61
Venture North Sea Gas Ltd v Nuon Exploration &
Production UK Ltd [2010] EWHC 204 .... II–2.19
Voice v Bell [1993] EGCS 128, (1993) 68 P & CR 441 ...... II–13.25

Walker v Crystal Palace Football Club Ltd [1910] 1 KB 87,
CA II–14.06
Watteau v Fenwick [1893] 1 QB 346 .... II–2.38
Weiner v Harris [1910] 1 KB 285 .... II–2.16
Wessanen Foods Ltd v Jofson Ltd [2006] EWHC 1325
(TCC) II–6.77
West Midlands Probation Committee v Secretary for the
Environment [1998] 76 P & CR 589 .... I–9.72
WesternGreco Ltd v ATP Oil and Gas (UK) Ltd [2006]
EWHC 1164 (Comm) II–6.84
Westminster City Council v Great Portland Estates plc
[1985] AC 661 .... I–9.63
White & Carter (Councils) Ltd v McGregor [1962] AC
413 .... II–8.41
Whitmore v Mason (1861) 2 J & H 204 .... II–2.83–4
William Tracey Ltd v Scottish Ministers 2016 SLT 1049 .... II–13.31
Williams' Trs v Macandrew and Jenkins 1960 SLT 246 ...... II–13.22
Windle v Secretary of State for Justice [2016] EWCA Civ
459; [2016] ICR 721 .... II–14.22, II–14.27, II–14.28
Wittenberg v Sunset Personnel Services Ltd
UKEAT/0019/13 .... II–14.43
Wong Kwok-chiang v Longo Construction Ltd (1987)
Hong Kong Law Reports 345 .... II–13.22
Wood Group Engineering (North Sea) Ltd v Robertson,
Appeal No UKEATS/0081/06/MT, 2007 WL
2186972 .... II–14.17

Wood v Capita Insurance Services Ltd [2017] UKSC
    24 ....II–2.80, II–4.141
Wright v Logan (1829) 8 s 247.. II–13.22

Yewens v Noakes 6 (1880) QBD 530 .... II–14.06

TABLE OF STATUTES

1890 Partnership Act II–2.20
     s 1.. II–2.20
     s 2.. II–2.20
        (1) II–2.20
        (2) II–2.20
        (3) II–2.20, II–2.21
     ss 5–18.. II–2.22
     s 19.. II–2.22
1891 Stamp Act
     s 55.. II–9.86
1906 Marine Insurance Act
     s 1.. II–6.04
     s 55(2) II–2.29
1909 Housing & Town Planning Act................................. I–9.29
1918 Petroleum (Production) Act I–4.07
     s 1(1) I–4.07
     Representation of the People Act............................... I–5.05
1925 Law of Property Act
     s 52(1) II–10.89
     s 205(1)(ii).. II–10.89
     s 205(1)(ix)... II–10.89
1928 Drainage of Land Act (Australia)........................... II–13.01
1934 Petroleum (Production) Act I–4.08,
                     I–4.10, I–4.36, I–5.05, I–10.05
     s 1.. I–4.09
        (1) I–4.08, II–3.10
     s 2.. I–4.09
     s 3.. I–4.09
     s 4.. I–4.09
     s 6.. I–4.09
     Sch 2, Cl18 .... I–11.06
1938 Coal Act
     s 3.. I–4.01
1940 Law Reform (Miscellaneous Provisions) (Scotland) Act
     s 3... II.6.50
1946 Atomic Energy Act
     s 6.. I–4.01
     s 7.. I–4.01

National Health Act I–5.05
1947 Town and Country Planning Act I–9.31
1949 Coast Protection Act...................I–11.72, I–11.74–8, I–12.58
      s 34... I–11.72, I–11.74
         (1)(a) I–11.74
         (1)(b) I–11.74
         (1)(c) I–11.74
         (2) I–11.75
         (3) I–11.76
      s 36 (1) I–11.76
      s 36A I–11.75
1959 Coastal Protection Act.. I–11.72
1960 Occupiers' Liability (Scotland) Act II–13.11
1961 Companies (Floating Charges) (Scotland) Act.......... II–10.69
1964 Continental Shelf Act..........I–1.16, I–4.10, I–10.03, I–10.04,
                                                               I–10.13, I–11.84
      s 1
         (1) I–4.09
         (2) I–4.09
         (3) I–4.09, I–5.05
         (7) I–11.79
      s 3... I–10.04
      s 4(1) I–11.72, I–11.74
1965 Gas Act.. II–13.09
      s 4...II–13.02, II–13.31
      s 5...II–13.02, II–13.31
      s 12..II–13.02, II–13.09, II–13.31
      s 13..II–13.02, II–13.09, II–13.31
      War Damage Act.. II–13.05
1969 Employers' Liability (Compulsory Insurance) Act.....I–10.42,
                                                                             II–2.36
1970 Conveyancing and Feudal Reform (Scotland) Act
      Pt II.. II–10.89
1971 Mineral Workings (Offshore Installations) Act I–1.16,
         I–10.03, I–10.12–17, I–10.18, I–10.19, I–10.20, I–10.22,
                            I–10.24, I–10.37, I–10.40, I–12.55
1972 Island of Rockall Act... I–8.13
1973 Prescription and Limitation (Scotland) Act
      s 3(2) II–13.07
      Seas and Submerged Lands Act (Australia)
      s 10A I–4.09
1974 Consumer Credit Act
      s 21... I–4.04
      Health and Safety at Work Act I–1.16, I–9.15, I–10.03,
                            I–10.18–30, I–10.22, I–10.86, II–14.70

s 2.. I–10.22, I–10.85

s 33... I–10.85

　(1) I–10.85

s 37... I–10.87

Sch 3A.. I–10.85

1975　Employment Protection Act.......................... I–10.27

Offshore Petroleum Development (Scotland)

Act.. II–13.09

Oil Taxation Act I–5.05

Pt I... I–5.05

s 1 (2) I–7.09

s 3(1)(i)(hh)–(j) I–13.03

s 13... I–7.02

Petroleum and Submarine Pipe-lines Act.......I–5.05, I–10.17,

　　　　　　　　　I–10.40, I–A.7, I–A.20, I–A.29

s 17... I–5.05

s 18.. I–4.17, I–4.51

Sch 2, Pt II, Model Cl 11 ....I-4.02

Sch 4.. I–5.05

1976　Energy Act I–3.34, I–3.67

s 6... I–3.34

　(6) I–3.34

Restrictive Trade Practices Act............................. II–11.39

1977　Patents Act

Pt II.. II–12.15

s 1

　(1)(a) II–12.13

　(1)(b) II–12.13

　(1)(c) II–12.13

　(1)(d) II–12.13

s 3... II–12.13

s 25(1) II–12.12

s 30

　(2) II–12.14

　(4) II–12.14

s 31

　(3) II–12.14

　(4) II–12.14

s 39(1)(b) II–12.36

s 40... II–12.37

s 60... II–12.12

Unfair Contract Terms Act.............. II–2.38, II–6.05, II–6.77,

　　　　　　　　　　　II–6.83, II–6.84

s 2

　(1) II–6.28, II–6.29

(2) II–6.29
(4) II–6.05, II–6.28
s 3 .... II–6.29
s 4 .... II–6.28
s 12 .... II–6.28
s 16(1)(a) II–6.28
s 16(1)(b) II–6.29
s 17 .... II–6.29
s 18 .... II–6.28
s 26
(3) II–8.02
(4) II–8.02
1979 Sale of Goods Act II–8.02
s 8
(1) II–8.18
(2) II–8.18, II–8.42
s 12 .... II–8.12, II–8.30
s 13 (1) II–8.08, II–8.51
s 14
(2) II–8.08, II–8.51
(2B) II–8.09, II–8.51
s 15(2) II–8.08, II–8.51
s 15A II–8.08
s 16 .... II–8.30
s 27 .... II–8.17
s 61(1) II–8.02
1982 Civil Jurisdiction and Judgments Act II–14.38
Local Government Finance Act
s 20 .... II–2.29
(1) II–2.29
1983 Finance Act
s 36(2) I–7.27
Petroleum Royalties (Relief) Act
s 1 .... I–7.27
(2)(a) I–7.27
1984 Control of Industrial Major Accident Hazard (CIMAH)
Regulations (SI 1984/1902) I–10.38
National Fishing Enhancement Act (PL 98-623) (USA)
Title II I–12.10
Occupiers' Liability Act II–13.11
Telecommunications Act
Sch 2 paras 2–6 .... II.13.31
1985 Companies Act
s 736 .... II–9.50
Environment Protection Act I–12.58

Food and Environment Protection Act........I–11.37, I–11.44,
I–11.49, I–11.72, I–12.22, I–12.58
s 5 .... I–11.49, I–11.72
(a).. I–11.37
(b) I–11.37
s 7... I–11.72
s 8
(3) I–11.49
(10) I–11.49
s 9
(11)(a) I–11.49
(11)(b) I–11.49
s 11... I–11.49
(b) I–11.49
s 21
(2A)(a).. I–11.49
(2A)(b) I–11.49
(6) I–11.49
1986 Company Directors Disqualification Act
s 9(A–E).. II–11.02
Gas Act.. II–13.09
Sch 4.. II–13.01
Insolvency Act
s 1(3)(a) II–10.119
s 72A II–10.95
s 72B–H II–10.95
s 72E... II–10.24, II–10.95
s 178.. II–10.112, II–10.114
s 214.. II–10.116
Sch 2A II–10.95
Sch B1
para 3(1) II–10.97
para 22.. II–10.116
paras 42–44 .... II–10.98
para 64.. II–10.108
para 71.. II–10.117
para 83.. II–10.110
para 99(3) II–10.100
1987 Petroleum Act I–12.22, I–12.36, I–12.41, II–13.09
s 21... I–12.78
s 22... I–12.78
s 23... I–12.78
s 27... II–13.09
Territorial Sea Act
s 1 .... II–13.01

(1)(b) I–9.125
1988 Copyright, Designs and Patents Act
Ch 6.. I–4.04
s 1(1) II–12.18
s 3
(1) II–12.17
(1)(b) II–12.23
(2) II–12.17
s 4.. II–12.17
s 11(2) II–12.34
s 17(1) II–12.19
s 18.. II–12.19
s 18A... II–12.19
s 50A... II–12.24
s 90(3) II–12.35
s 215(3) II–12.34
s 217(2) II–12.24
Court of Session Act
s 27.. II–15.76
Income and Corporation Taxes Act
s 416
(2) II–9.34
(4) II–9.34
(6) II–9.34
Sch 19B, para 4 .... I–7.31
Road Traffic Act
s 87.. I–4.04
1989 Companies Act
s 144.. II–9.50
Electricity Act
Sch 4
para 6(6)(a) II–13.31
para 6(6)(b) II–13.31
1990 Environment Protection Act................................. I–12.58
Law Reform Miscellaneous Provisions (Scotland) Act
s 66.. II–15.66
Sch 7... II–15.66
Town and Country Planning Act I–9.33
s 55.. I–9.38
s 57.. I–9.37
s 70.. I–9.49
(1)(a) I–9.44
(2) I–9.57
1992 Offshore Safety Act...................... I–1.16, I–9.16, I–10.03,
I–10.40–7

Taxation of Chargeble Gains Act
s 8...II–9.18
Trade Union and Labour Relations (Consolidation)
Act...II–14.19, II–14.31
1993 Finance Act
s 185..I–7.09
Radioactive Substances Act...............................I–12.58
1994 Coal Industry Act
s 1(1)(a) I–4.01
1995 Gas Act...II–13.09
s 12..I–6.25
s 19...II–13.28
Merchant Shipping Act....................................I–11.82
s 128(1) I–11.51
s 277..I–11.84
s 293(2)(za)...I–10.71
1996 Arbitration Act II–15.66
s 1...II–15.65
s 5...II–15.65
Employment Rights Act.................................II–14.31
Pt IVA.............................II–14.19, II–14.20
ss 13–23..II–14.71
s 94 (1) II–14.33, II–14.34, II.14.36, II–14.37,
                                       II–14.38, II–14.39
s 98(4) II.14.90
s 108 (1) II–14.02
s 155..II–14.02
s 196.............................II–14.34, II–14.35
s 201.............................II–14.30, II–14.37
s 230.................II–14.22, II–14.26, II–14.46
    (1) II–14.05
    (2) II–14.05
    (3) II–14.20
    (3)(b)...II–14.26
Employment Tribunals Act
s 7(3AA) II–15.47
Housing Grants, Construction and Regeneration
Act..II–5.19
Pt II...II–15.62
s 105..II–15.62
    (2) II–15.63
    (2)(a)...II–15.63
    (2)(c)...II–15.63
    (3) II–15.63
    (4) II–15.63

(6) II–15.63
(7) II–15.63
Petroleum Activities Act (Norway)
s 1-1... I–4.09
s 5-4... I–12.79
1997 Planning (Listed Buildings and Conservation Areas)
(Scotland) Act.. I–9.34
s 26(1) I–9.38
s 37(1) I–9.43
Town and Country Planning Act
Pt 3 .... I–9.36
s 25... I–9.56
s 28... I–9.37
s 70... I–9.49
(2) I–9.57
Town and Country Planning (Scotland) Act I–9.34
Pt 3 .... I–9.36
1998 Competition Act I–6.16, I–6.19, I–6.44, I–6.64–72,
I–6.75, I–6.122, II–11.21
Ch I I–6.64–5, I–6.69
Ch II I–6.64, I–6.70–2
s 9... II–11.17
ss 25–29... II–11.04
s 47A II–11.01
Finance Act
s 29.. I–7.07
Human Rights Act... I–14.13
National Minimum Wage Act.................... II–14.19, II–14.20
Petroleum Act I–1.18, I–4.10, I–4.14, I–5.19, I–9.15,
I–11.73, I–11.79–80, I–12.06, I–12.41–56, I–12.58, I–12.75,
II–2.94, II–13.09
Pt I.. II–5.06
Pt III I–11.45, II–13.09
Pt IV I–11.45, I–12.41, I–12.58, I–12.78, I–13.27
s 2(1) I–9.08
s 3
(1) I–4.10, I–4.14, I–11.22
(3) I–4.14, I–4.32, I–9.08
s 4... I–9.08, I–9.11
(1)(e) I–4.17, I–4.54, I–4.74
(4) I–4.17
s 4A I–9.28
(3) I–9.28
s 7.. II–13.09
s 9.. I–6.16, I–6.19

s 9A I–6.45, II.6.59
  (1) I–5.33
  (1)(b) II–15.10
  (2) I–5.33
  (3) I–5.34
s 9B I–5.34
  (a)–(e) I–5.34
  (c) I–6.46, I–6.47
s 9C I–5.35, I–5.36, I–6.47, II.3.11
  (2) I–5.36
s 9G I–5.39
s 14
  (1)(a) I–11.79
  (1)(b) I–11.79
  (2) I–11.79
s 15
  (3)(c)(i) I–11.79
  (3)(c)(iii) I–11.79
  (3)(c)(iv) I–11.79
s 17F II–13.28
s 17GA I–6.22
s 21
  (2) I–11.80
  (3) I–11.80
s 22 .... I–11.80
s 23 .... I–11.80
s 26 .... I–12.42
  (1) I–11.79
s 27 .... I–12.44, II–13.28
  (1) I–12.44
s 28A I–12.42
s 29 .... I–12.06, I–12.44–5, I–12.52–3, I–12.59–61,
                     I–12.75, I–13.04–10, II–9.76, II–9.77
  (1) I–12.42, I–12.69
  (1A) I–12.42
  (2) I–12.42, I–12.63
  (2A) I–12.42
  (3) I–12.42, I–12.66
  (4) I–12.42
s 30 .... I–13.07–9
  (1) I–12.43, I–12.45
  (1)(a) I–13.08
  (1)(d) I–12.47, I–13.09
  (1)(e) I–12.47
  (2) I–12.44

(2)(b).. I–12.47
(2)(c).. I–12.47
(3) I–12.45
(4) I–12.45
(5) I–12.43
(5)(b).. I–12.45
(6) I–12.43
(8) I–12.43, I–12.44
(8)–(8D) I–12.43
(9) I–12.43, I–12.44
s 31
(3) I–12.45
(5) I–12.45, I–12.59–60
s 31A.. I–13.06
s 31(A1)–(D1)... I–12.51
s 32
(1) I–12.46
(2) I–12.46
(2A)(a)... I–12.46
(2A)(b) I–12.46
(6) I–12.46
(7) I–12.46
s 33
(1) I–12.46
(2) I–12.46
(3) I–12.46
(3A)... I–12.46
(3B) I–12.46
(4) I–12.46
s 34................... I–12.47, I–12.51, I–12.54, I–12.60, I–12.70,
                                             I–13.06, I–13.10, II–9.77
(1)(a) I–12.47
(1)(b)... I–12.47
(2) I–12.47
(3) I–12.47
(4) I–12.47
(4A)... I–12.47
(4B) I–12.47
(5) I–12.47
(6) I–12.47
(7) I–12.47
(7A)... I–12.47
(7B) I–12.47
s 34A
(1) I–12.48

(2) I–12.48
(3) I–12.48
(4) I–12.48
(5) I–12.48
s 35
(1) I–12.49
(2) I–12.49
(3) I–12.49
s 36.. I–12.50, I–12.53
s 36A
(2) I–12.50
(3) I–12.50
(5) I–12.50
(6) I–12.50
(7) I–12.50
(8) I–12.50
(9) I–12.50
(10) I–12.50
s 37
(1) I–12.51
(3) I–12.51
(4) I–12.51
(5) I–12.51
s 37A.. I–12.51
s 38.. I–12.52, I–13.10
(1)–(1B) I–12.52
(2) I–12.52
(2A)... I–12.53
(3) I–12.52
(4) I–12.53
(4A)... I–12.53
(5) I–12.53
(6) I–12.52, I–12.53
s 38A........................... I–12.54, I–13.25, II–2.95
s 38B.. I–12.54
s 39.. I–12.55
(2)(d) I–12.55
(2)(e) I–12.55
(3) I–12.55
(4) I–12.55
(5) I–12.55
(6) I–12.55
s 44.. I–12.42
s 45.. I–12.42, I–12.44
Sch 1.. I–9.08

Scotland Act I–4.13, I–9.112, I–9.113, I–12.82
s 2... I–9.113
s 29(2)(b).. I–9.113
Sch 5 Pt II Head D.. I–9.117
Sch 5 Pt II Head D1.. I–9.117
Sch 5 Pt II Head D2.. I–9.117
Sch 5 Pt II Head D4.. I–9.117
1999   Contracts (Rights of Third Parties) Act..... I–13.27, II–4.140,
II–6.31, II–6.66
Employment Relations Act
s 10..II–14.19, II–14.20
Pollution Prevention and Control Act...................... I–11.60
s 3... I–11.90
2000   Abolition of Feudal Tenure (Scotland) Act
s 1.. I–9.08
s 67.. II–13.12
Financial Services and Markets Act
s 19(1) II–9.22
Utilities Act
Pt V II–13.09
2001   Capital Allowances Act.. I–7.30
Pt 2.. I–7.30
Pt 5.. I–7.30
Pt 6.. I–7.30
s 4... I–7.38
s 56(2) I–7.29
ss 162–165... I–7.30, I–13.03
s 163... I–7.46
2002   Enterprise Act II–10.95, II–11.01
s 188... II–11.02
s 189... II–11.02
2003   Communications Act
s 363... I–4.04
Land Reform (Scotland) Act II–13.03
Title Conditions (Scotland) Act............................... II–13.30
Pt 1... II–13.30
Pt 9... II–13.20
s 75(3)(b)..II–13.15, II–13.30
s 76(2) II–13.22
s 77..II–13.07, II–13.14
2004   Energy Act
s 84... I–12.82
Planning and Compulsory Purchase Act I–9.33
2006   Companies Act
Pt 21A II–10.88

s 859A II–10.68

   (4) II–10.68

s 859D II–10.68

s 859E II–10.68

s 859F(3) II–10.68

s 859H II–10.68

s 1159 .... II–9.50

Offshore Petroleum and Greenhouse Gas Storage Act
(Australia)

s 6 .... I–11.25

s 16 .... I–4.18

Planning (Scotland) Act I–9.34

2007 Consolidated Act No 889 on the Use of the Danish
Subsoil (Denmark)

s 1 .... I–4.09

s 2 .... I–4.09

Corporate Manslaughter and Corporate Homicide

Act I–10.77, I–10.78–84, I–10.86

s 1 .... I–10.83

   (1) I–10.79

   (3) I–10.79

   (4)(b) I–10.80

   (4)(c) I–10.80

   (5) I–10.78, I–10.79

   (6) I–10.83

s 2 .... I–10.79

   (1)(a) I–10.80

   (1)(c) I–10.80

   (4) I–10.80

s 8

   (2) I–10.81

   (3) I–10.81

   (4) I–10.81

s 18 .... I–10.80

s 28(3)(e) I–10.79

2008 Climate Change Act I–3.72

s 1 .... I–3.60

s 4 .... I–3.60

s 5 .... I–3.60

ss 12–14 .... I–3.60

s 33(3) I–3.60

s 36(1) II–3.60

Employment Act

Pt 1 .... II–15.47

s 4 .... II–15.47

Energy Act I–3.71, I–7.45, I–12.06, I–12.58,
                                        II–2.95, II–7.76
Ch 1 .... I–3.57
Ch 2 .... I–3.57, I–12.45
Pt 2 .... I–3.63
s 1 .... I–12.82
s 30 .... I–12.45
s 72
    (2)(a) I–12.43
    (2)(b) I–12.43
    (3) I–12.43
    (4) I–12.45
    (5) I–12.43
    (7) I–12.51
    (8) I–12.51
s 73
    (1) I–12.52
    (2) I–12.52
    (3) I–12.52
    (4) I–12.53
    (5) I–12.53
s 74 .... I–12.54, I–13.25
s 77 .... I–11.29, II–10.99
    (2) I–4.17
Sch 1
    para 10 .... I–12.45
    para 11 .... I–12.45
Sch 3 .... I–4.17, I–11.29, II–10.99
Sch 5, para 11 .... I–12.44
Health and Safety (Offences) Act I–10.77, I–10.85–8
s 1 .... I–10.85
Sch 3A, para 1 .... I–10.85
Planning Act I–9.33
2009   Corporation Tax Act
       s 931 .... I–7.38
       Marine and Coastal Access Act I–4.29, I–4.30, I–12.58
       s 46 .... I–4.30
       s 77 .... I–4.29
2010   Arbitration (Scotland) Act II–15.66
       s 1 .... II–15.65
       s 10 .... II–15.65
       Bribery Act II–3.13, II.5.65
       Corporation Tax Act I–7.49
       Pt 8 Chapter 7 .... I–7.12, I–7.34
       s 37 .... I–7.43

s 39 .... I–7.43
s 42 .... I–7.47
s 274 .... I–7.02
s 279 .... I–7.02
s 279A I–7.14
s 286 .... I–7.08
s 311 .... I–7.32
s 312 .... I–7.32
s 322A I–7.34
s 330 .... I–7.08
   (1) I–7.08
   (3)(a) I–7.08
s 332B I–7.36, I–7.38
s 349A I–7.12
s 351 .... I–7.12
s 356BA I–3.44
s 356BC I–3.44
s 356C I–3.44
s 356JC I–7.40
s 720 .... I–7.12
Sch 3 .... I–7.12
Equality Act II–14.04, II–14.31
s 83 .... II–14.22
   (2)(a) II–14.26
Marine (Scotland) Act I–4.29, I–12.58
2011 Energy Act I–5.34, I–6.19, I–6.21–44, I–6.120, II–7.76
Pt 2 .... I–6.46
s 82 .... I–6.22, I–6.42, I–6.46, I–6.99
   (1) I–6.22, I–6.38, I–6.42, I–6.43, I–6.44
   (2) I–6.25
   (3) I–6.22
   (4) I–6.27, I–6.28, I–6.33, I–6.35, I–6.38, I–6.41,
                           I–6.94, I–6.99
   (5) I–6.28, I–6.75
   (6) I–6.36
   (6)(a) I–6.28, I–6.99
   (6)(b) I–6.28
   (7) I–6.29, I–6.30, I–6.36, I–6.40, I–6.46, I–6.56,
                     I–6.101, I–6.102, I–6.120
   (7)(a) I–6.102
   (7)(b) I–6.103
   (7)(c) I–6.103
   (7)(d) I–6.102
   (8) I–6.120
   (9) .I–6.31, I–6.32, I–6.36, I–6.40, I–6.56, I–6.57, I–6.101

(10) I–6.22, I–6.31, I–6.32, I–6.36, I–6.40, I–6.56,
                                                I–6.101, I–6.120
(11) I–6.27, I–6.31, I–6.33, I–6.34, I–6.35, I–6.36,
                  I–6.38, I–6.44, I–6.46, I–6.56, I–6.59–61
(12) I–6.33
(17) I–6.34
(e) I–6.104
ss 82–84... I–6.46
ss 82–91.................................... I–6.14, I–6.18, I–6.21
s 83......I–6.21, I–6.22, I–6.35, I–6.37, I–6.42, I–6.46, I–6.99
(3) I–6.35
s 84...I–6.38, I–6.41, I–6.42, I–6.46
(2) I–6.39, I–6.40, I–6.46
(3) I–6.39
s 87... I–6.21, I–6.46
(1) I–6.42
(4) I–6.29
s 89A I–6.46
s 89B... I–6.46
s 90... I–6.25
(1) I–6.23, I–6.24, I–6.25
(2) I–6.24, I–6.25
Finance Act
s 7(1) I–7.08
Historic Environment (Amendment) Scotland Act I–9.34
Localism Act... I–9.33
2012   Finance Act
Sch 22... I–7.12
2013   Energy Act.. I–9.119
Finance Act
s 80(3) I–13.36
2014   Finance Act
s 70... I–3.44
Sch 15... I–3.44
Historic Environment (Scotland) Act I–9.34
2015   Consumer Rights Act.....................II–6.05, II–6.28, II–11.01
Finance Act.. I–7.40
s 47... I–7.32
Sch 11... I–7.32
Sch 12... I–7.36
Finance (No 2) Act
s 7... I–7.07
Infrastructure Act I–5.38, I–6.45–8, I–9.35, II.3.11
s 41..................I–5.28, I–5.33, I–5.34, I–5.35, I–6.16, I–6.19
s 42... I–5.28

ss 43–48.. I–9.120
s 50.. I–9.27, I–9.28
s 56.. I–9.120
Small Business, Enterprise and Employment Act...... II–10.88
2016   Energy Act I–5.33, I–5.36, I–6.15, I–6.16, I–6.19,
           I–6.21, I–6.75, I–12.06, I–12.47, I–12.50, I–12.63, II–3.10,
                                          II–3.11, II–9.03
s 1... I–6.18
s 19.. II–15.10
ss 19–26... II–15.10
s 22.. I–6.21, I–6.37
s 23(4) II–15.10
s 24... I–6.21
s 26.. II–15.09
s 34................................... I–6.21, I–6.43, I–6.44
    (4) I–5.35
s 42.. I–6.46, I–6.48
    (3)(a) I–5.35
    (3)(b) I–5.35
    (4) I–5.37
s 43.. I–5.37
ss 44–46.. I–5.37
s 47.. I–5.37
s 48.. I–5.37
s 49.. I–5.37
ss 50–52... II–15.09
s 58.. II–15.09
s 70.. I–6.46
s 71.. I–6.46
Sch 1 paras 63–72 .... I–6.18
Finance Act
s 46.. I–7.07
s 58.. I–7.08
s 140(1) I–7.09
Petroleum Act I–12.75
Scotland Act
s 36.. I–12.82
s 47
    (1) I–9.124
    (2) I–9.125
    (3) I–9.125
s 49(1) I–9.125

TABLE OF STATUTORY INSTRUMENTS

1935   Petroleum (Production) Regulations (SR&O
      1935/426) I–10.05
1964   Continental Shelf (Designation of Areas) Order
      (SI 1967/697)... I–8.07
      Petroleum (Production) (Continental Shelf and
      Territorial Sea) Regulations (SI 1964/708)................ I–10.05
        Sch 2, Cl 18... I–10.05
1972   Mineral Workings (Offshore Installations) Act 1971
      (Commencement) Order (SI 1972/644)..................... I–10.16
      Offshore Installations (Logbooks and Registration
      of Death) Regulations (1972/1542) I–10.16
      Offshore Installations (Managers) Regulations
      (SI 1972/703)... I–10.16
      Offshore Installations (Registration) Regulations
      (SI 1972/702)... I–10.16
1973   Offshore Installations (Inspectors and Casualties)
      Regulations (SI 1973/1842) I–10.16
1974   Offshore Installations (Construction and Survey)
      Regulations (SI 1974/289) I–10.16
      Offshore Installations (Public Inquiries) Regulations
      (SI 1974/338)... I–10.16
1976   Employment Protection (Offshore Employment)
      Order (SI 1976/766) II.14.31
      Offshore Installations (Emergency Procedures)
      Regulations (SI 1976/1542) I–10.16
      Offshore Installations (Operational Safety, Health
      and Welfare) Regulations (SI 1976/1019) I–10.16
      Petroleum (Production) Regulations (SI 1976/276)..... I–4.36
      Submarine Pipe-lines (Diving Operations)
      Regulations (SI 1976/923) I.10.16
1977 Health and Safety at Work Act 1974 (Application
      Outside Great Britain) Order (SI 1977/1232) I–10.22
      Offshore Installations (Life-Saving Appliances)
      Regulations (SI 1977/486) I–10.16
      Safety Representatives and Safety Committees
      Regulations (SI 1977/500) I–10.26
      Submarine Pipe-lines (Inspectors) Regulations (SI
      1977/835) I.10.16

1978 Offshore Installations (Fire-Fighting Equipment)
Regulations (SI 1978/611) I–10.16
1980 Offshore Installations (Well Control) Regulations
(SI 1980/1759).. I–10.16
1981 Employment Protection (Offshore Employment)
(Amendment) Order (SI 1981/208)........................... II.14.31
1982 Petroleum (Production) Regulations (SI 1982/1000)
Sch 5 .... I–7.29
1984 Control of Industrial Major Accident Hazard
Regulations (SI 1984/1902) I–10.38
1985 Deposits in the Sea (Exemption) Order
(SI 1985/1699)... I–11.49, I–11.72
reg 14... I–11.37
reg 15... I–11.37
reg 15A I–11.37
1986 Insolvency Rules (SI 1986/1925)
r 2.87 ....II.10.103
r 15.34(4)..II.10.120
Insolvency (Scotland) Rules (SI 1986/1915)
r 1.16A(2) II.10.120
r 2.39B II.10.100
r 4.67 ....II.10.100
1988 Petroleum (Production) (Seaward Areas) Regulations
(SI 1988/1213)..I–4.19, I–A.7
reg 3(1) I–4.03, I–4.41
reg 7(5) I–4.23
Sch 1 .... I–4.03, I–4.41
Sch 4 .... I–A.20, I–A.29
Model Cl 16...I–A.20
Model Cl 17...I–A.29
1989 Offshore Installations (Pipe-line Valve) Regulations
(SI 1989/680)... I–10.45
Offshore Installations (Safety Representatives and
Safety Committees) Regulations
(SI 1989/971).................................. I–10.39, I–10.50
1992 Health and Safety (Display Screen Equipment)
Regulations (SI 1992/2792) I–10.46
Management of Health and Safety at Work
Regulations (SI 1992/2051) I–10.46
Manual Handling Operations Regulations
(SI 1992/2793).. I–10.46
Offshore Installations (Safety Case) Regulations
(SI 1992/2885).................................... I–9.16, I.10.41
reg 4
(1).. I–10.41

(2).. I–10.41
reg 5 .... I–10.41
reg 6 .... I–10.41
reg 7 .... I–10.41
reg 8 .... I–10.41
reg 9 .... I–10.41
reg 10.. I–10.41, I–10.42
reg 13 .... I–10.41
reg 14 .... I–10.41
Sch 3 .... I–10.41
Sch 4 .... I–10.41
Sch 5 .... I–10.41
Personal Protective Equipment of Work Regulations
(SI 1992/2966).. I–10.46
Personal Protective Equipment Regulations
(SI 1992/3139).. I–10.46
Provision and Use of Work Equipment Regulations
(SI 1992/2932).. I–10.46
1993 Act of Sederunt (Sheriff Court Ordinary Cause
Rules) (SI 1993/1956)
Ch 40 .... II–15.75
r 9.13 .... II–15.43
r 40.12(3)(m) II–15.75
1994 Act of Sederunt (Rules of the Court of Session)
(SI 1994/1443)
Ch 47 .... II–15.75
Ch 77 .... II–15.76
Ch 78 .... II–15.76
r 47.11(1)(e).. II–15.75
Conservation (Natural Habitats) Regulations
(SI 1994/2716).. I–4.25
1995 Borehole Sites and Operations Regulations
(SI 1995/2038)...................................... I–9.17, I–9.19
reg 7(2) I–9.17
Hydrocarbons Licensing Directive Regulations
(SI 1995/1434).. I–4.16
reg 3
(1)...................................... I–4.36, I–4.37
(1)(a) I–4.38
(1)(b) I–4.38
(1)(d) I–4.38, I–4.49
(2).. I–4.37
(4).. I–4.37
reg 4
(1).. I–4.16

(2)... I–4.16
reg 5 .... I–4.37
Offshore Installations (Management and
Administration) Regulations (SI 1995/738) I–10.42
reg 5 .... I–10.42
reg 6 .... I–10.42
(b).. I–10.42
reg 7 .... I–10.42
reg 8 .... I–10.42
reg 9 .... I–10.42
reg 10 .... I–10.42
reg 11 .... I–10.42
reg 12 .... I–10.42
reg 13 .... I–10.42
reg 14 .... I–10.42
reg 15 .... I–10.42
reg 16 .... I–10.42
reg 17 .... I–10.42
reg 18 .... I–10.42
reg 19 .... I–10.42
reg 20 .... I–10.42
reg 21 .... I–10.42
Offshore Installations (Prevention of Fire and
Explosion, and Emergency Response) Regulations
(SI 1995/743)... I–10.43, I–10.71
reg 4
(1).. I–10.43
(2).. I–10.43
reg 5 .... I–10.43
reg 6 .... I–10.43
reg 7 .... I–10.43
reg 8 .... I–10.43
reg 9 .... I–10.43
reg 10 .... I–10.43
reg 11 .... I–10.43
reg 12 .... I–10.43
reg 13 .... I–10.43
reg 14 .... I–10.43
reg 15 .... I–10.43
reg 16 .... I–10.43
reg 17 .... I–10.43
reg 18 .... I–10.43
reg 19 .... I–10.43
reg 20 .... I–10.43
reg 21 .... I–10.43

Petroleum (Production) (Landward Areas)
Regulations (SI 1999/1436)
    Sch 3 .... I–4.79
Petroleum (Production) (Seaward Areas)
(Amendment) Regulations (SI 1995/1435)
    reg 6 .... I–4.23
1996 Merchant Shipping (Prevention of Oil Pollution)
Regulations (SI 1996/??) I–11.87
Offshore Installations and Wells (Design and
Construction) Regulations (SI 1996/913) I–9.19,
                                                       I–9.21, I–10.44
    Pt II I–10.44
    Pt III I–10.44
    Pt IV I–10.44
    reg 2(1) I–10.44
    reg 4
        (1) I–10.44
        (2) I–10.44
    reg 5 .... I–10.44
    reg 6 .... I–10.44
    reg 7 .... I–10.44
    reg 8 .... I–10.44
    reg 9 .... I–10.44
    reg 10 .... I–10.44
    reg 13
        (1) I–10.44
        (2) I–10.44
    reg 14 .... I–10.44
    reg 15 .... I–10.44
    reg 16 .... I–10.44
    reg 17 .... I–10.44
    reg 18 .... I–10.44
    reg 19 .... I–10.44
    reg 20 .... I–10.44
    reg 21 .... I–10.44
    reg 22 .... I–10.44
    Sch 1 .... I–10.44
Pipelines Safety Regulations (SI 1996/825) I–9.16,
                                                       I–10.45, I.12.58
1997 Copyright and Rights in Databases Regulations
(SI 1997/3032) II.12.12
Diving at Work Regulations (SI 1997/2776) I–10.45
1998 Civil Procedure Rules (SI 1998/3132) II.15.43
    Ch 78 .... II.15.77
    Pt 36 .... II.15.43

Pt 58 ....II.15.75
r 1.4(2)(e) II.15.43
r 26.4(1) II.15.43
r 44.4.(3)(a)(i) II.15.43
r 44.5(a) II.15.43
Lifting Operations and Lifting Equipment
Regulations (SI 1998/2307) I–10.45
Management of Health and Safety at Work
Regulations (SI 1996/2306) I–10.46
Merchant Shipping (Oil Pollution Preparedness,
Response and Co-operation Convention)
Regulations (SI 1998/1056) I–11.81, I–11.82–5, I–11.87
    reg 2 ....I–11.82
    reg 3
        (2) I–11.84
        (a) I–11.83
    reg 4
        (1)(c) I–11.83
        (5)(b) I–11.83
        (7) I–11.83
        (a)(iii) I–11.83
        (a)(iii)(bb) I–11.83
    reg 5
        (1) I–11.85
        (2) I–11.85
    reg 7(1) I–11.83
    reg 8 ....I–11.84
Provision and Use of Work Equipment Regulations
(SI 1998/2306) I–10.46
Working Time Regulations (SI 1998/1833) II–14.19,
                    II–14.23, II–14.44–82
    reg 2 ....II.14.46, II.14.49, II.14.54, II.14.56, II.14.78
        (1)(b) II–14.25
    reg 4 ....II.14.47
        (1) II.14.72
    reg 5 ....II.14.65
        (2) II.14.65
        (3) II.14.65
    reg 6 ....II.14.56–9
        (1) II.14.58
        (2) II.14.58
        (3) II.14.56
        (7) II.14.58
    reg 7 ....II.14.59
    reg 8 ....II.14.60

reg 9 .... II.14.61
reg 10 .... II.14.50
    (1) II–14.50
reg 11 .... II.14.51
    (1) II–14.51
    (2) II–14.51
reg 12 .... II.14.52
    (1) II.14.52
reg 13 .... II.14.62–4
reg 13A II.14.62
reg 18 .... II.14.45
reg 21 .... II–14.51, II.14.58, II.14.66
    (a) II–14.50, II.14.52, II.14.58
reg 22 .... II.14.53–5
reg 23 .... II.14.67
    (a) II.14.56
reg 24 .... II–14.50, II.14.52, II.14.53, II.14.58, II.14.66,
                                              II.14.67, II.14.74
reg 25B II.14.47, II.14.48
reg 28 .... II.14.69
reg 29 .... II.14.70
reg 30 .... II.14.71
reg 36 .... II.14.46
Sch 3 .... II.14.70
Working Time (Northern Ireland) Regulations
(SI 1998/386) II.14.56
1999   Control of Major Accident Hazard (COMAH)
Regulations (SI 1999/743) I.10.38, I–10.84
Management of Health and Safety at Work
Regulations (SI 1999/3242) I–10.46
    reg 3 .... II.14.57
Merchant Shipping (Marine Equipment) Regulations
(SI 1999/1957) I–11.67, I–11.69
National Minimum Wage (Offshore Employment)
Order (SI 1999/1128) II.14.31
Offshore Petroleum Production and Pipelines
(Assessment of Environmental Effects) Regulations
(SI 1999/360) I–11.16, I.12.67
    reg 4 .... I–11.16
    reg 5 .... I–11.16
        (4)(b)(ii) I–11.19
    reg 6 .... I–11.45
    reg 9 .... I–11.17
    Sch 2
        (a) I–11.17

(a)(i) I–11.17
(a)(ii) I–11.17
(a)(iii) I–11.17
(b) I–11.17
(c)(i) I–11.17
(c)(ii) I–11.17
(d) I–11.17
(e) I–11.17
Petroleum (Current Model Clauses) Order (SI 1999/160)
    Schs 1–14 .... I–4.17, I–5.01
2000   Employment Relations (Offshore Employment)
    Order (SI 2000/1828) II.14.31
2001   Offshore Combustion Installations (Prevention and
    Control of Pollution) Regulations
    (SI 2001/1091) I–11.58–61
    reg 2 .... I–11.58, I–11.59, I–11.60
    reg 3 .... I–11.59
    reg 4 .... I–11.61
      (1) I–11.60
      (2)(g)(i) I–11.60
      (2)(g)(ii) I–11.60
      (2)(g)(iii) I–11.60
    reg 5 .... I–11.59
      (3) I–11.59
    reg 7
      (3) I–11.60
      (4) I–11.60
    reg 9 .... I–11.61
    reg 10 .... I–11.61
    regs 13–16 .... I–11.61
    reg 18 .... I–11.61
      (4) I–11.61
Offshore Petroleum Activities (Conservation of
Habitats) Regulations (SI 2001/1754) I–4.25, I–11.08–15
    reg 2 .... I–4.25
      (1)(a) I–11.08
      (1)(b) I–11.08
      (1)(c) I–11.08
      (1)(d) I–11.08
      (1)(e) I–11.08
      (1)(f) I–11.08
      (2) I–11.08
    reg 4
      (1)(a) I–11.13
      (1)(b) I–11.13

(1)(c)... I–11.13
reg 5.. I–11.08, I–11.14
(1)... I–4.25, I–11.08, I–11.09
(2)... I–4.25, I–11.09
(3)... I–4.25, I–11.09
(4).. I–11.09
reg 6
(1)(a) I–4.25, I–11.09
(1)(b) I–4.25, I–11.09
(2)(a) I–4.25
(2)(b) I–4.25
reg 7.. I–11.12
(1)(a) I–11.10
(1)(c) I–11.10
(2)(a) I–11.10
reg 10.. I–11.11, I–11.12
reg 11.. I–11.11, I–11.12
reg 11(a).. I–11.11
reg 12.. I–11.12
reg 14.. I–11.11
reg 16.. I–11.12
reg 17.. I–11.12
reg 18.. I–11.12
reg 19
(2).. I–11.12
(3).. I–11.12
(4).. I–11.12
2002   Merchant Shipping (Hours of Work) Regulations
(SI 2002/2125)..II.14.45
National Emission Ceilings Regulations
(SI 2002/3118).. I–11.62
reg 2.. I–11.64
(2)(a) I–11.64
(2)(b) I–11.64
reg 3.. I–11.64
reg 4(1) I–11.64
Offshore Chemical Regulations (SI 2002/1355)....... I–11.38,
I–11.44, I–11.45–8, I–11.87
reg 2.......................... I–11.38, I–11.45, I–11.46, I–11.47
reg 3
(1).. I–11.45
reg 4
(1).. I–11.45
(1)(a) I–11.45
(2).. I–11.45

reg 5
(1).. I–11.46
(2).. I–11.46
reg 6
(1)(a–d).. I–11.46
reg 7
(1).. I–11.45
(2).. I–11.45
(3).. I–11.45
reg 10
(1).. I–11.46
(2).. I–11.46
(3).. I–11.46
(5).. I–11.46
reg 11.. I–11.47
(2).. I–11.47
(4).. I–11.47
reg 12.. I–11.47
(2).. I–11.47
(7).. I–11.47
reg 13
(1).. I–11.48
reg 14
(1).. I–11.46
(2).. I–11.46
reg 15
(1)(a) I–11.48
(1)(b) I–11.48
(2).. I–11.48
reg 16.. I–11.48
reg 18.. I–11.48
(4).. I–11.48
(6).. I–11.49
Offshore Installations (Emergency Pollution
Control) Regulations (SI 2002/1861).........I–11.38, I–11.42,
I–11.81, I–11.90–3
reg 2.. I–11.90
reg 3
(1)(a) I–11.91
(1)(b) I–11.91
(1)(c).. I–11.91
(2).. I–11.91
(3).. I–11.91
(3)(a) I–11.91
(3)(b) I–11.91

(3)(c)...I–11.91
(4)...I–11.91
(4)(b) I–11.91
(4)(c)...I–11.91
reg 5
(2)...I–11.93
(4)...I–11.93
Offshore Safety (Miscellaneous Amendments)
Regulations (SI 2002/2175) I–10.42
Renewables Obligation Order (SI 2002/914).............. I–3.63
art 3 ....I–3.63
art 6 ....I–3.63
Sch 1 ....I–3.63

2003 Financial Collateral Arrangements (No 2)
Regulations (SI 2003/3226) II–10.90, II-10.91
reg 8...II–10.90
Working Time (Amendment) Regulations
(SI 2003/1684)...II.14.45

2004 Environmental Information Regulations (SI 2004/3391)
reg 12(5)(e) I–11.28
Petroleum Licensing (Exploration and Production)
(Seaward and Landward Areas) Regulations
(SI 2004/352)...............................I–4.52, I–11.22–9
reg 9(1)(e) I–11.28, I–11.29
Sch 1 .... I–11.22, I–11.23
Model Cl 1(2)...................................... I–11.23
Model Cl 2 .... I–11.22, I–11.24
Model Cl 3 ....I–11.22
Model Cl 7(1)...................................... I–11.24
Model Cl 7(2)...................................... I–11.24
Model Cl 7(4)...................................... I–11.24
Model Cl 7(5)...................................... I–11.24
Model Cl 7(6)...................................... I–11.24
Model Cl 7(7)...................................... I–11.24
Model Cl 9(1).............................. I–11.24, I–11.26
Model Cl 9(2)...................................... I–11.25
Model Cl 9(3)...................................... I–11.26
Model Cl 10 .... I–11.26
Model Cl 11(1)..................................... I–11.28
Model Cl 12 ....I–A.20
Model Cl 12(1)(a) I–11.28
Model Cl 12(2)..................................... I–11.28
Model Cl 13 ....I–A.29
Model Cl 14 .... I–11.28
Model Cl 14(b)..................................... I–11.28

Model Cl 14(c) I–11.28
Model Cl 14(d).. I–11.28
Model Cl 15(a).. I–11.28
Model Cl 16(a).. I–11.29
Model Cl 16(b).. I–11.29
Model Cl 17 .... I–11.29
Model Cl 20(2)(a) I–11.29
Model Cl 20(2)(c).. I–11.29
Model Cl 20(2)(e).. I–11.29
Model Cl 20(2)(f) I–11.29
Model Cl 20A I–11.29
Model Cl 23(1).. I–11.27
Model Cl 23(2).. I–11.27
Model Cl 23(a).. I–11.27
Model Cl 23(c) I–11.27
Schs 1–7 .... I–4.17
Schs 2–4 .... I–11.22
Sch 4 .... I–4.43, I–4.54, I–4.57, I–4.58, I–A.20, I–A.29
Sch 6 .... I–4.79, I–11.22
Sch 7 .... I–11.22
Renewables Obligation Order (SI 2004/924).............. I–3.63
2005 Offshore Installations (Safety Case) Regulations
(SI 2005/3317)...........................I.1.16, I–10.03, I–10.48–54,
                                    I–10.64, I.12.58
reg 2...I-10.51, I.10.52
reg 5.. I–10.51
reg 6.. I–10.49
reg 9(1) I–10.49
reg 10... I–10.49
reg 11... I–10.49
reg 12... I–10.50
reg 13... I–10.49
reg 14... I–10.49
reg 24... I–10.49
Sch 2, para 3 .... I–10.50
Offshore Petroleum Activities (Oil Prevention and
Control) Regulations (SI 2005/2055)....................I–11.35–6,
                                    I–11.81, I–11.86–9
reg 2... I–11.35, I–11.86
reg 3
(1)... I–11.35, I–11.86
(2)(a) I–11.87
(2)(b) I–11.87
(2)(c)... I–11.87
(4)... I–11.87

reg 4
  (1).. I–11.87, I–11.88
  (2)... I–11.88
  (3)... I–11.88
  (4)... I–11.88
  (a)... I–11.88
reg 5
  (1)(a) I–11.87
  (1)(b) I–11.87
  (1)(c).. I–11.87
  (2)... I–11.87
reg 7
  (1)... I–11.88
  (2)... I–11.88
reg 8... I–11.88
reg 9
  (1)(a) I–11.87
  (1)(b) I–11.88
reg 12.. I–11.89
  (1)(a) I–11.35, I–11.89
  (1)(b) I–11.35, I–11.89
reg 14
  (1)... I–11.89
  (2)... I–11.89
reg 16.. I–11.89
  (2)(a) I–11.89
  (2)(b) I–11.89
  (3)(b) I–11.89
Renewables Obligation Order (SI 2005/926)............. I–3.63

2006   Petroleum Licensing (Exploration and Production)
(Seaward and Landward Areas) (Amendment)
Regulations (SI 2006/784) I–4.17, I–4.79
Renewables Obligation Order (SI 2006/1004)............ I–3.63
Transfer of Undertakings (Protection of
Employment) Regulations
(SI 2006/246)...................................II.10.117, II.14.02
Working Time (Amendment) (No 2) Regulations
(SI 2006/2389)...II.14.45

2007   Large Combustion Plants (National Emissions
Reduction Plan) Regulations (SI 2007/2325) I–3.37
  reg 8.. I–3.37
Marine Works (Environmental Impact Assessment)
Regulations (SI 2007/1518)
  reg 2(1) I–11.72
Offshore Petroleum Production and Pipelines

(Assessment of Environmental Effects) (Amendment)
Regulations (SI 2007/933) I–11.16, I.12.67
Renewables Obligation Order 2006 (Amendment)
Order (SI 2007/1078) I–3.63
2008 Merchant Shipping (Prevention of Air Pollution
from Ships) Regulations (SI 2008/2924) I–11.51–7
    Pt III... I–11.55
    reg 2
        (1)...................... I–11.52, I–11.53, I–11.55, I–11.56
        (5)(b) I–11.54
    reg 3
        (5)... I–11.57
        (13)(c).. I–11.56
        (13)(c)(1) I–11.56
        (13)(e).. I–11.55
    regs 5–15... I–11.54
    reg 9(1) I–11.54
    reg 16(1) I–11.57
    reg 18... I–11.57
    reg 20
        (1)... I–11.55
        (3)... I–11.55
        (4)... I–11.55
    reg 21... I–11.55
    reg 22... I–11.55
        (3)(a) I–11.55
        (3)(b) I–11.55
    reg 23... I–11.56
    reg 24
        (4)... I–11.56
        (5)... I–11.56
    reg 25... I–11.56
        (1)(c).. I–11.56
    reg 28(1) I–11.57
Merchant Shipping (Prevention of Pollution by
Sewage and Garbage from Ships) Regulations
(SI 2008/3257)...I–11.65–72
    reg 2
        (1).....................I–11.65, I–11.66, I–11.67, I–11.68,
                                 I–11.69, I–11.70
        (4)... I–11.67, I–11.69
        (6)(b) I–11.68
    reg 7... I–11.67, I–11.68
    reg 8... I–11.68
    reg 9(1) I–11.68, I–11.69

reg 21
  (1).. I–11.69
  (1)(a) I–11.67
  (1)(b) I–11.67
  (2).. I–11.67, I–11.69
reg 23
  (1).. I–11.67
  (3).. I–11.67
reg 24 .... I–11.67, I–11.69
  (a).. I–11.67
  (b).. I–11.67
  (c).. I–11.67
reg 25 .... I–11.67, I–11.69
  (1)(a) I–11.67
  (1)(b) I–11.67
  (2).. I–11.67
  (3).. I–11.67
reg 26(1) I–11.70
reg 29... I–11.70
  (2).. I–11.70
reg 32 .... I–11.70
reg 36 .... I–11.71
reg 38 .... I–11.71
Petroleum Licensing (Production) (Seaward Areas)
Regulations (SI 2008/225) I–4.12, I–4.38, I–5.05,
      I–11.30–3, I–11.62–3, I–12.77, I–A.20, II–2.05
reg 23
  (3)(a) I–11.62
  (4).. I–11.63
  (7).. I–11.62
  (7)(a) I–11.62
  (7)(b) I–11.62
reg 40(1) II.10.75
Sch I–4.44, I–5.06
  Model Cls.. I–4.14, II–3.10
  Model Cl 1(2)... I–11.30
  Model Cl 2 .... I–4.12, I–4.41, II–3.10
  Model Cl 3 .... I–4.52
  Model Cl 3(2).. I–4.53
  Model Cl 4 .... I–4.13
  Model Cl 4(2)(b) I–4.50
  Model Cl 6(3).. I–4.54
  Model Cl 7 .... I–4.13, I–4.56
  Model Cl 8(1).. I–4.57
  Model Cl 8(3)(a) I–4.57

Model Cl 8(3)(b) I–4.57
Model Cl 8(3)(c)... I–4.57
Model Cl 12 .... I–4.45
Model Cl 12(2)... I–4.51
Model Cl 14 .... I–4.13, I–4.51, I–4.60
Model Cl 15 .... I–4.60
Model Cl 16 .... I–4.13, I–4.83, I–A.7, I–A.28, I–A.29
Model Cl 16(2)......................... I–4.51, I–A.20, I–A.28
Model Cl 16(3)... I–4.51
Model Cl 16(4)... I–4.51
Model Cl 16(4)(a) I–4.51
Model Cl 16(4)(b) I–4.51
Model Cl 16(6)... I–4.51
Model Cl 16(7)... I–4.51
Model Cl 17.... I–4.13, I–4.83, I–11.30, I–A.7, I–A.28,
        I–A.29, I–A.30, I–A.31, I–A.47, I–A.50, II–3.12
Model Cl 17(1)... I–A.28
Model Cl 17(1)(a) I–11.30
Model Cl 17(1)(b) I–11.30
Model Cl 17(2)............. I–A.28, I–A.31, I–A.32, I–A.47
Model Cl 17(2)(c)... I–11.30
Model Cl 17(3).................................... I–A.31, I–A.47
Model Cl 17(4)... I–A.28
Model Cl 17(4)(c)(ii) I–11.30
Model Cl 17(5)... I–A.28
Model Cl 17(6)... I–A.28
Model Cl 17(9)... I–11.30
Model Cl 18 ...... I–4.83, I–5.06, I–A.7, I–A.28, I–A.29,
                                          I–A.31, II–3.12
Model Cl 18(6)... I–11.30
Model Cl 19 .... I–4.60, I–11.31
Model Cl 19(12)(a) I–11.31
Model Cl 19(12)(b) I–11.31
Model Cl 20 .... I–4.60, II–3.12
Model Cl 21 .... I–4.13, I–4.60, I–11.32
Model Cl 21(4)... I–11.32
Model Cl 23 .... I–4.13, I–4.60, I–5.06, I–11.32
Model Cl 23(3)(a) I–11.32
Model Cl 23(7)(a) I–11.32
Model Cl 23(9)... I–11.32
Model Cl 24 .... I–4.13, I–4.60, II–2.24
Model Cl 24(1)... II–9.31
Model Cl 24(2)... II–9.36
Model Cl 27 .... II–3.12
Model Cl 27(1)... II–3.13

Model Cl 27(2)..II–3.13
Model Cl 27(4).....................................II–3.13, II–3.16
Model Cl 27(5)..II–3.13
Model Cl 28 ....II–3.12, II–3.50
Model Cl 29 ....I–4.60
Model Cl 30 ....I–4.60
Model Cl 31 ....I–4.60
Model Cl 40(1)..II–9.30
Model Cl 41 ....I–4.13
Model Cl 41(2)(b) I–4.51
Model Cl 41(3)................................... II–9.34–5
Model Cl 41(4)...................................II–9.34
Model Cl 42 ....I–4.13
Model Cl 43 ....II–3.13
Model Cl 43(1)...I–4.13
Model Cl 44 ....I–4.60
Model Cl 45 .... I–4.13, I–4.60
Model Cl 176 ....I–4.57

2009   Offshore Exploration (Petroleum and Gas Storage
and Unloading) (Model Clauses) Regulations
(SI 2009/2814)
       Sch I–4.39–40
       Model Cl 2 .... I–4.39
       Model Cl 3 .... I–4.39
       Model Cl 4 .... I–4.39
       Model Cl 9 .... I–4.40
       Model Cl 11 .... I–4.40
       Model Cl 13 .... I–4.40
       Model Cl 22 .... I–4.40
       Model Cl 23 .... I–4.40
       Petroleum Licensing (Amendment) Regulations (SI
       2009/3283)......................... I–4.12, I–4.44, I–5.05
       Renewables Obligation Order (SI 2009/785)............. I–3.63
       Sch 1 .... I–3.63
       Renewables Obligation (Northern Ireland) Order
       (SI 2009/154)................................... I–3.63
       Renewables Obligation (Scotland) Order
       (SI 2009/140)...................................... I–3.63
2010   Environmental Permitting (England and Wales)
       Regulations (SI 2010/675) I–9.26
       Equality Act 2010 (Offshore Work) Order
       (SI 2010/1835)..................................II.14.31
       Merchant Shipping (Prevention of Air Pollution
       from Ships) (Amendment) Regulations
       (SI 2010/895)...................................... I–11.51

Merchant Shipping (Prevention of Pollution by
Sewage and Garbage from Ships) (Amendment)
Regulations (SI 2010/897) I–11.65
2011 Offshore Petroleum Activities (Oil Prevention and
Control) (Amendment) Regulations (SI 2011/983) ... I–11.81
2012 Oil Stock Order (SI 2012/2862) I–3.34, I–3.35, I–3.67
        reg 3.. I–3.34
Pollution Prevention Control (Scotland) regulations
(SI ??)... I–3.40
2013
Environmental Permitting (England and Wales)
(Amendment) Regulations (SI 2013/390).................... I–3.40
Pollution Prevention and Control (Industrial
Emissions) Regulations (Northern Ireland)
(SI 2013/160)... I–3.40
2014 Merchant Shipping (Maritime Labour Convention)
(Hours of Work) (Amendment) Regulations
(SI 2014/308)... II–14.45
Petroleum (Exploration and Production) (Landward
Areas) Regulations (SI 2014/1686) I–9.125, II–3.10
        s 4I–9.09
        Sch 1 .... I–9.125
        Sch 2 .... I–4.79, I–9.11
        Sch 3 .... I–4.79
        Model Cl 2.. II–3.10
2015 Control of Major Accident Hazard (COMAH)
Regulations (SI 2015/483) I–10.63
Infrastructure Act (Commencement No 1)
Regulations (SI 2015/481)
        reg 3(b) I–5.35
Offshore Installations (Offshore Safety Directive)
(Safety Case) Regulations
(SI 2015/398).....................................I–10.64–76, I–11.97
        reg 2
            (1)... I–10.67
            (7)... I–10.67
            (8)... I–10.67
        reg 4... I–10.64
        reg 6... I–10.75
        reg 7
            (2)(a) I–10.65
            (2)(b) I–10.65
        reg 8
            (1)... I–10.66
            (2)... I–10.66

(3)...I–10.66
reg 9
  (1)...I–10.67
  (1)(a) I–10.67
  (1)(b) I–10.67
  (2)...I–10.67
  (3)...I–10.67
  (4)...I–10.67
reg 10(1) I–10.67
regs 10–13...I–10.68
reg 11(2)(b) I–10.68
reg 21
  (3)...I–10.68
  (4)...I–10.68
  (6)...I–10.68
reg 26...I–10.75
reg 28
  (3)...I–10.69
  (4)...I–10.69
reg 29
  (1)...I–10.70
  (2)...I–10.70
reg 30
  (1)(a) I–10.71
  (1)(b) I–10.71
  (3)...I–10.72
  (14)..I–10.71
reg 31
  (1)(a) I–10.72
  (1)(b) I–10.72
reg 32(1) I–10.73
reg 33...I–10.74
reg 34(1) I–10.65
Sch 1 ....I–10.65
Sch 2 ....I–10.66
  para 2..I–10.65
Sch 3 ....I–10.66
Sch 4
  Pt 1..I–10.67
  Pt 2..I–10.68
Sch 11 ....I–10.73
Sch 13, paras 33–40.....................................I–10.64
Offshore Petroleum Licensing (Offshore Safety
Directive) Regulations (SI 2015/385)
  reg 1...I–10.64

reg 7... I–10.76
reg 8(2) I–10.76
Renewables Obligation Order
(SI 2015/1947)... I–3.63, I–3.73

2016   Energy Act 2016 (Commencement No 2 and
Transitional Provisions) Regulations (SI 2016/920)
reg 2(b) I–5.35
Insolvency (England and Wales) Rules (SI 2016/1024)
r 3.51 ....II.10.100
r 14.1 ....II.10.103
r 15.34 ....II.10.120
Oil and Gas Authority (Fees) Regulations
(SI 2016/904)... I–4.34
Onshore Hydraulic Fracturing (Protected Areas)
Regulations (SI 2016/??) I–9.28
reg 2.. I–9.27
reg 3.. I–9.27
Petroleum (Exploration and Production) (Landward
Areas) (Amendment) Regulations (SI 2016/1029)
reg 2.. I–4.79
Petroleum (Transfer of Functions) Regulations (SI
2016/898) I–5.33
reg 2.. I–4.10

TABLE OF EUROPEAN LEGISLATION

Treaties

1957   Treaty Establishing the European Community......... II–11.09

2009   Treaty of the Functioning of the European Union
(TFEU)... II–11.09

      Art 4(3).. II–11.62

      Art 101 .... I–6.64, II–11.09, II–11.10–17, II–11.61,
II–11.64

      Art 101(1).................. II–11.14, II–11.31, II–11.41, II–11.43,
II–11.45, II–11.47–9

      Art 101(3).................. II–11.17, II–11.30, II–11.32, II–11.35,
II–11.61

      Art 102 .... I–6.64, II–11.09, II–11.18–20, II–11.51,
II–11.53, II–11.58

Directives

1968   68/414/EEC imposing an obligation to maintain
minimum stocks of crude oil and/or petroleum
products... I–3.26, I–3.31

      Art 1 .... I–3.26

      Art 6 .... I–3.26

      Art 7 .... I–3.26

1972   72/425/EEC of 19 December amending Directive
68/414/EEC I–3.27

1973   73/238/EEC of 24 July on measures to mitigate the
effects of difficulties in the supply of crude oil and
petroleum products... I–3.28

      Art 1 .... I–3.28

      Art 3 .... I–3.28

      Art 5 .... I–3.28

1979   79/409/EEC Birds Directive I–11.09

      Art 1 .... I–11.11

      Art 4(1)... I–11.08

      Art 4(2)... I–11.08

1982   82/501/EEC Seveso Directive I–10.38

1985   85/337/EEC Environmental Impact
Assessment... I–11.16, I–12.67

      Art 1 .... I–4.26

      Art 2(1)... I–4.26

Art 4(2).. I–4.26
Arts 5–7.. I–4.26
Annex 1 .... I–11.73
Annex 2 .... I–11.72
Annex 2(2)(f).. I–4.26
Annex 2(2)(g) I–4.26
1989 89/391/EEC Framework Safety at Work
Directive I–10.46
Art 2... II–14.44
Art 16(1).. I–10.46
1989 89/655/EEC on minimum safety and health
requirements for the use of work equipment by
workers at work I–10.46
1989 89/656 Personal Protective Equipment at Work
Directive I–10.46
1989 89/686/EC Personal Protective Equipment
Directive I–10.46
1990 90/269/EEC Manual Handling Directive I–10.46
1990 90/270/EC Display Screen Equipment Directive I–10.46
1992 92/43/EEC Habitats Directive........I–4.24–5, I–4.27, I–11.09
Art 4... I–11.08
Art 4(2)... I–11.08
Art 6(3)... I–11.14
Art 6(4)... I–11.14
Annex IV(a) I–11.11
1992 92/85/EEC Pregnant Workers Directive I–10.46
1992 92/91/EEC Extractive Industries
Directive I–10.46, I–10.59
1993 93/104/EC Working Time Directive II–14.44, II–14.73,
II–14.80
Art 1(3)... II–14.44
1994 94/22/EC Hydrocarbons Licensing Directive I–4.11,
I–4.14, I–4.36
Art 3 .... I–4.19
Art 3(2)(a) I–4.20
Art 3(2)(b) I–4.22
1994 94/33/EC on the protection of young workers.......... I–10.46
1995 95/63/EC I–10.46
1996 96/82/EC Seveso II Directive................................... I–10.38
1997 97/11/EC I–4.26, I–11.16, I–12.67
1998 98/93/EC of 14 December amending Directive
68/414/EEC I–3.31
Preamble, para 2... I–3.31
Preamble, para 9... I–3.31
Preamble, para 11... I–3.31

   Art 1(1)... I–3.31
   Art 1(4)... I–3.31
   Art 1(4)(3) I–3.31
   Art 1(6)... I–3.31
   Art 1(7)... I–3.31
1999 1999/63/EC Seafarers Directive II–14.45
   Art 1 .... II–14.45
   Art 2 .... II–14.45
2000 2000/34/EC Working Time Directive II–14.44, II–14.45
   Art 20(2)... II–14.47
2001 2001/42/EC Strategic Environmental Assessment
   Directive I–4.24, I–4.27
   Art 2(b).. I–4.27
   Art 3(1).. I–4.27
   Art 3(2)(a) I–4.27
   Art 3(2)(b) I–4.27
   Art 3(3).. I–4.27
   Art 5 .... I–4.27
   Art 6 .... I–4.27
   Art 7 .... I–4.27
   Art 8 .... I–4.27
2001 2001/77/EC on the promotion of electricity
   produced from renewable energy sources in the
   internal electricity market I–3.30, I–3.63, I–3.73
   Art 3 .... I–3.30
2001 2001/80/EC on the limitation of emissions of certain
   pollutants into the air from large combustion
   plants.. I–3.37
   Art 2(10)... I–3.37
2003 2003/35/EC I–4.26
2003 2003/105/EC I–10.38
2004 2004/35/EEC Environmental Liability Directive I–10.60,
                I–11.105, I–11.106
2006 2006/67/EC imposing an obligation on Member
   States to maintain stocks of crude oil and/or
   petroleum products..................................... I–3.31, I–3.32
2008 2008/52/EC on certain aspects of mediation in civil
   and commercial matters....................................... II–15.46
2008 2008/98/EC Waste Directive I–11.105
2009 2009/24 on the legal protection of computer programs
   Recital 3 .... II–12.23
2009 2009/28 Renewable Energy Directive I–3.30, I–3.73
   Art 3(1).. I–3.30
   Art 3 (2) I–3.30
   Art 17.. I–3.63

Art 19 .... I–3.63
Annex 1 .... I–3.30
2009   2009/31/EC on the geological storage of carbon
       dioxide I–4.26
2009   2009/119/EC I–3.31, I–3.32
       Art 3(1) I–3.31
       Art 5 .... I–3.31, I–3.35
       Art 7 .... I–3.35
       Art (9)(2) I–3.35
       Art 13 .... I–3.31
       Art 19 .... I–3.31
2010   2010/75/EU Industrial Emissions Directive I–3.40
       Art 32 .... I–3.40
       Art 33 .... I–3.40
2013   2013/20/EU Platform Directive I–11.04, I–11.95, I–11.97
       Art 1 .... I–11.04
       Art 3 .... I–11.95
       Art 6 .... I–11.95
       Art 6(6) I–11.95
       Art 7 .... I–11.96
       Art 8(3) I–11.95
       Art 11 .... I–11.95
       Art 28 .... I–11.96
2013   2013/30/EU Offshore Safety Directive I.1.16, I–10.60–76,
                                                              I–11.105
2014   2014/89/EU Framework Directive on maritime
       spatial planning I–4.30
       Art 8(2) I–4.30
2016   2016/943 Trade Secrets Directive II–12.31
       Art 2 .... II–12.31

Regulations

2000   1346/2000 on Insolvency Proceedings II–10.96
2003   1/2003 Modernisation Regulation II–11.17, II–11.53
       Art 1 .... II–11.21
       Art 23(2)(a) II–11.01
2006   1907/2006 Registration, Evaluation, Authorisation
       and Restriction of Chemicals (REACH)
       Regulations I–9.26
2010   330/2010 on block exemption for vertical
       agreements II–11.17, II–11.56
       Art 3 .... II–11.56
       Art 4(a) II–11.58
2010   1217/2010 on block exemption for research and
       development agreements II–11.17

2010 1218/2010 on block exemption for specialisation
      agreements...II–11.17, II–11.48
      Art 3 II–11.48
      Art 4(a).. II–11.48
2014 316/2014 Technology Transfer Block Exemption II–12.27
      Art 1(1)(i) II–12.27
      Art 1(1)(i)(i)... II–12.27
      Art 1(1)(i)(ii)... II–12.27
      Art 1(1)(i)(iii).. II–12.27
2015 848/2015 on Insolvency Proceedings II–10.96

Decisions
1977 Council Decision 77/706/EEC of 7 November on
      the setting of a Community target for a reduction in
      the consumption of primary sources of energy in the
      event of difficulties in the supply of crude oil and
      petroleum products.. I–3.29
      Preamble... I–3.29
      Art 1 .... I–3.29
2011 Commission Decision 2011/13/EU I–3.63
2012 Commission Decision M6477 (BP/Chevron/Eni/
      Sonangol/Total/JV).. II–11.27
2013 Commission Decision M6801 (Rosneft/TNK) II–11.27
2013 Commission Decision M6910 (Gazprom/
      Wintershall/Target Companies)............................... II–11.27
2014 Commission Decision M7316 (DNO/Marathon)..... II–11.27
2014 Commission Decision M7318 (Rosneft/Morgan
      Stanley Global Oil Merchanting Unit) II–11.27
2015 Commission Decision M7631 (Royal Dutch Shell/
      BG Group)... II–11.27

TABLE OF INTERNATIONAL INSTRUMENTS

1948 Inter-Governmental Maritime Consultative
Organization Convention
Art 1 .... I–12.10
Art 29 .... I–12.10
1956 CMR (Convention on the Contract for International
Carriage of Goods by Road) II–2.29, II–2.30
1958 Continental Shelf Convention I–8.04–5, I–8.07, I–8.11,
I–8.16, I–8.34, I–10.04, I–12.06, I–12.09
Art 1 .... I–8.05, I–8.16, I–8.31, II–12.15
Art 2 .... I–8.04
(1) I–4.09, I–12.07
Art 5
(1) I–12.07
(2) I–12.07
(5) I–12.07
Art 6 .... I–8.07
(1) I–8.08
(2) I–8.08
Art 60(3) I–12.08
Art 80 .... I–12.08
New York Convention (Convention on the
Recognition and Enforcement of Foreign Arbitral
Awards) II.15.81
Art V(1) II.15.68
Art V(2)(b) II.15.81
Territorial Sea Convention I–8.31
Art 10 .... I–8.31, I–8.32
Art 11 .... I–8.31
1960 OECD Convention I–3.16
1965 Denmark-Norway Continental Shelf Delimitation
Agreement I–8.08
UK-Netherlands Continental Shelf Delimitation
Agreement I–8.08, II–3.52–7
Art 1 .... II–3.52
Art 2 .... II–3.52, II–3.55
UK-Norway Continental Shelf Delimitation
Agreement I–8.08, II–3.58–9
Art 4 .... II–3.58, II–3.60

1966 ICSID (International Convention on the Settlement
of Investment Disputes) II.15.48
UK-Denmark Continental Shelf Delimitation
Agreement I–8.08
1968 Sweden/Norway Continental Shelf Delimitation
Agreement I–8.08
1969 International Convention on Civil Liability for Oil
Pollution Damage I–11.104
1970 Patent Co-operation Treaty....................................... II–12.15
1971 Denmark-Germany-Netherlands Continental Shelf
Delimitation Agreement... I–8.12
International Convention on the Establishment of
an International Fund for Compensation for Oil
Pollution Damage I–11.104
UK-Germany Continental Shelf Delimitation
Agreement I–8.12
1972 London Dumping Convention I–12.06, I–12.11–12,
I–12.22, I–12.37
Annex I... I–12.11, I–12.12
Annex II.. I–12.11
Annex III I–12.12
  (a)(ii) I–12.12
Annex IV(1).. I–12.12
Oslo Convention for the Protection of Marine
Pollution by Dumping from Ships and Aircraft I–12.13,
I–12.17, I–12.22, I–12.24, I–12.36
1973 European Patent Convention II–12.15
MARPOL 73/73 (International Convention for the
Prevention of Pollution from Ships)
Annex IV I–11.65, I–11.71
Annex V I–11.65, I–11.71
Annex VI I–11.51
1974 IEP Agreement I–3.16–21, I–3.32
Art 1... I–3.17
Art 2
  (1) I–3.17
  (2) I–3.17
Art 5.. I–3.18
Art 7
  (3) I–3.19
  (4) I–3.18
Art 8.. I–3.19
Art 13.. I–3.18
Art 17(1).. I–3.19
Ch III.. I–3.19

Paris Convention for the Prevention of Marine
Pollution from Land-based Sources I–12.13
1976   Frigg Field Agreement...................... II–3.59–61, II–3.63
Art 2... II–3.60
1979   Murchison Field Agreement......................... II–3.59, II–3.63
Statfjord Field Agreement II–3.59, II–3.62, II–3.63
1980   Rome Convention on the Law Applicable to
Contractual Obligations II-14.38
1982   UNCLOS (UN Convention on the Law of the Sea) ... I–8.14,
I–8.16, I–8.18, I–12.06, I–12.81
Preamble... I–12.08
Art 57.. I–4.01
Art 60(3)... I–12.10, II–2.93
Art 74.. I–8.23
Art 76.......................... I–8.16, I–8.17, I–8.24
(3) I–8.17
(8) I–8.23
(10) I–8.23
Art 77(3)... I–8.23
Art 83.. I–8.23
Art 84.. I–8.17
Art 121(3).. I–8.32, I–8.33
Art 123.. II–3.51
Art 298(1)... I–8.35
Art 311(1)... I–8.15
Pt V   II–3.15
Pt XIV I–8.29
Pt XV... I–8.35
Annex II Art 9 .... I–8.24
1985   UNCITRAL Model Law on International
Commercial Arbitration............................... II–15.66
1988   UK/Ireland Continental Shelf Delimitation
Agreement I–8.21
1992   Markham Agreement................................. II–3.53–7
Art 5.. II–3.54
Art 10.. II–3.55
Art 16.. II–3.54
Art 23.. II–3.55
OSPAR Convention (Convention for the Protection
of the Marine Environment in the North-East
Atlantic)..............I–5.23, I–11.36, I–11.38, I–11.44, I–11.45,
I–11.46, I–11.47, I–12.06, I–12.13–18, I–12.19,
I–12.36, II–2.93
Art 1
(a)... I–12.13

(f) I–12.13
(g) I–12.13
Art 2
   (1)(a) I–12.14
   (2)(a) I–12.15
   (2)(b) I–12.15
   (3)(b) I–12.15
Annex III
   Art 5(1) I–12.16
   Art 5(3) I–12.17
   Art 6 .... I–12.18
   Art 8 .... I–12.18
   Art 10 .... I–12.18
App 1 .... I–12.15

1994 Energy Charter Protocol on Energy Efficiency and
      Related Environmental Aspects I–3.22
      Energy Charter Treaty I–3.22–5, I–3.68, I–4.13
      Annex 1 .... I–3.22
      Art 2 .... I–3.22
      Art 19 .... I–3.22
        (2) I–3.22
      Art 26 .... II.15.48

1995 WTO GATS (General Agreement on Trade in Services)
      Art XXIII II.15.48

1996 Protocol to London Convention I–12.06, I–12.37
      Art 3
        (1) I–12.37
        (2) I–12.37
      Art 4(1.2) I–12.37
      Annex I I–12.37
      Annex II I–12.37

2001 International Convention on Civil Liability for
      Bunker Oil Pollution Damage I–11.104

2002 UNCITRAL Model Law on International
      Commercial Conciliation II–15.66

2005 UK-Norway Framework Treaty II–3.63–5
      Art 1
        (14) II–3.64
        (15) II–3.64
      Art 3 .... II–3.64
        (1)(1) II–3.64
        (2)(1) II–3.64
        (2)(2) II–3.64
        (3) II–3.64
        (4) II–3.64

(7) II–3.64
(9)(1) II–3.64
(12) II–3.64
Art 5 .... II–3.64
Annex D II–3.64
(2) II–3.64
(4) II–3.64
(5) II–3.64
2010 Vienna Convention (Convention on Contracts for
the International Sale of Goods) II–8.02
Arts 61–70 .... II–8.02

INTRODUCTION AND CONTEXT

CHAPTER I-1

OIL AND GAS LAW ON THE UNITED KINGDOM CONTINENTAL SHELF: CURRENT PRACTICE AND EMERGING TRENDS IN PUBLIC AND REGULATORY LAW

Greg Gordon, John Paterson, Emre Üşenmez and James Cowie

INTRODUCTION

When the second edition of this book was published some seven **I-1.01** years ago, the editors commenced the introduction by saying, "If there is one word that best describes the United Kingdom Continental Shelf (UKCS) at the beginning of the second decade of the 21st century, it is *mature*."[1] One would hardly expect that to have become less true over the last seven years, and indeed, in parts of the North Sea area, the expression "ultra-maturity" has come into usage as production rates have continued to decline and some major assets have either edged closer to, or entered, the cessation of production phase. However, in the last seven years, things have moved on in other parts of the province, too. The area to the west of Shetland – formerly described as a frontier area – now needs to be described differently. Almost the entire area either is or has been under licence; infrastructure is developing, as is a knowledge base of the geological characteristics of the area, and significant new finds are being made.[2] It is approaching maturity, not in the way that this word tends to

---

[1] G Gordon, J Paterson and E Üşenmez, *Oil and Gas Law: Current Issues and Emerging Trends* (2nd edn, 2011), para 1.1.

[2] See eg Hurricane Energy's discovery of a very significant find in the Greater Lancaster Area: A Cramb, "Oil exploration firm in 'largest undeveloped discovery' on the UK Continental Shelf", *The Telegraph*, 27 March 2017, available at www.telegraph.co.uk/news/2017/03/27/oil-exploration-firm-largest-undeveloped-discovery-uk-continental (accessed 12 May 2017).

be used upon the UKCS – as a euphemism for nearing the end of production – but in the way that Norwegian oil and gas lawyers tend to use it, as meaning that the area has become fully established. In the meantime, the true offshore frontier has moved out further into the Atlantic margin. There is a new frontier too, rather closer to home, as the shale gas industry seeks to establish itself in the onshore area. Thus, if there is one word that best describes the UK and the United Kingdom Continental Shelf (UKCS) at the beginning of the second decade of the 21st century, it is "diverse", a description that implies complexity and may seem to call for greater flexibility and sophistication of approach.

I-1.02   It is not just the lifecycle of the various assets and areas of the UKCS that have moved on in the last seven years. In that time we have lived through one of the oil industry's periodic cycles of boom and bust, with oil price topping out at over $120 per barrel in 2012 before plunging to a low of less than $30 per barrel for a time in 2015. The boom masked, to some extent, the challenges of maturity and helped to extend the production life of some assets which, although producing with less than optimal efficiency, were still profitable enough, in a time of high price, to be viable. After extended (and, in many cases, painful) cost-cutting, the industry may have ridden out the worst of the short-term storm, and should be reasonably well placed to grow again in the event of even a modest increase in price. However, the intense focus on cost-cutting and efficiency has had and will have a major impact upon law, regulation and commercial arrangements.[3]

I-1.03   It is not, however, just – or even mainly – through the passage of time and the vicissitudes of the oil market that the UKCS has undergone change. Two major reviews of the UKCS's mode of governance have been undertaken: the Wood Review and the Fiscal Review. Of these, the Wood Review is the one which has garnered more attention, a trend that will be continued in this book as, because of its pervasive nature,[4] and radical departure from the UKCS's previous mode of governance,[5] it calls for lengthier discussion and

---

[3] See eg the discussion at paras I-5.07 to I-5.08, I-7.06 to I-7.14 and II-5.

[4] As well as having its own dedicated chapter, it needs substantial discussion in almost half of the other chapters within the book, and is at least adverted to in virtually all the others.

[5] It has led to the creation of a new, better-resourced and largely industry-funded resource management regulator, the Oil and Gas Authority (OGA), which has been given the mission of being a more assertive and hands-on regulator, actively involving itself in matters such as development strategy, challenging the industry to work more collaboratively and innovatively, and generally trying to ensure that the maximum possible aggregate economic value of oil is produced from the UKCS. One of the principal tools at the new regulator's disposal will be its ability to enforce a new legal obligation, to comply with the principal objective of maximising economic recovery upon the UKCS, a complex

analysis. However, the Fiscal Review led to a major reboot of the UKCS tax system, in which the Government simplified the system and, fundamentally, accepted that it would have to accept a significantly lower share of direct tax take if the industry were to survive. In practical terms, the Fiscal Review may prove to have been equally significant to Wood. Other significant changes, not least of all to the regulation of offshore health and safety, have also come about in the period since the last edition; however, they are somewhat more self-contained and are introduced in the individual chapter summaries below.

There has also been significant constitutional change since the last edition of this work. In Scotland's independence referendum of 2014, the people voted, by a clear but by no means overwhelming majority, to remain part of the United Kingdom. Following that vote, in implementation of a pledge made by the leaders of the main political parties, further powers were devolved to the Scottish Parliament and Government; among them, the power to act as the licensing authority for oil and gas developments in the Scottish onshore area. This development is discussed in Chapter I-1.09. A 2016 referendum provided a narrow majority for the UK to leave the European Union. At the time of writing, the form and modalities of Brexit are not known and the precise impact it will have upon the legal regulation of the UK oil industry is not clear. As a result, apart from some passing references, Brexit has been substantially ignored in this edition. However it is reasonable to suppose it will feature prominently in the event that the book runs to a fourth edition. I-1.04

THE STRUCTURE OF THE BOOK

The "whirlwind of change", to use an expression borrowed from one of our authors, Uisdean Vass, that we have endeavoured to sketch above has had an inevitable effect upon this book. In particular, the need to discuss the Wood Review and its implementation has added around 30,000 words, not just in the new chapter that is dedicated to the topic, but in the extensive revisions that have been necessary in other existing chapters.[6] The editors had, in any event, decided to take the opportunity offered by a third edition to expand the book's scope, by adding a significant number of new chapters to the work. As a result, it has been necessary to divide the book into two volumes. Broadly speaking, the dividing line is that between public law (the I-1.05

---

regulatory device that, as we shall see, requires significant unpicking and calls for in-depth discussion and analysis.

[6] See in particular (in this volume) Chapters I-4, I-6 and I-12 and (in the second volume) Chapters II-5, II-11 and II-12.

subject of this volume) and private and commercial law (discussed in Volume II), but as we shall see below, there are some exceptions to this.

CHAPTER-BY-CHAPTER OUTLINE AND INDICATION OF CROSS-CUTTING THEMES

I-1.06 This volume is divided into three parts. The introductory and contextual section of which this chapter forms part[7] is followed by a group of five chapters unified by the theme of resource management. These chapters relate to licensing, the Wood Review and maximising economic recovery as a legal concept, the UKCS fiscal regime, third-party access to infrastructure, and maritime boundaries in the North Sea and Atlantic Margin. The third and final part first deals primarily with regulatory law.[8] It commences with a discussion of the UK's nascent onshore shale gas industry before going on to consider the topics of Health and Safety at Work Offshore, Offshore Environmental Law and Regulation, and Decommissioning, to which two chapters are devoted: one on International and Regulatory Issues, and the other on Decommissioning Security.[9]

I-1.07   This volume also contains, as an appendix, a reprint of the chapter on Mature Province Initiatives that was contained in the second edition of this work. We decided to include because although what we called the Mature Province Initiatives (ie Fallow Fields and Stewardship) had been rendered redundant by the measures taken to implement the Wood Review, these initiatives were nevertheless an important stage in the evolution of the UK's system of resource management.

I-1.08   After this introduction, the book commences with Chapter I-2, Alex Kemp and Linda Stephen's "Rejuvenating Activity in the North Sea Oil and Gas Industry: The Role of Tax Incentives in Context". This chapter is concerned not with law, but with key issues in the

---

[7] The other two chapters in this section are concerned with economic issues and energy security.

[8] Some areas raise blended issues of resource management and regulation: see eg shale gas, further discussed below. Environmental Law and Regulation on the UKCS is primarily directed towards regulation but also includes a brief afterword on compensation arrangements for those affected by pollution emanating from an offshore installation. Strictly, that would seem not to be a matter of regulatory law as such, but a question of the rights of the injured party resulting from the pollution incident; on balance, however, this seemed the most logical place for that discussion. Neither is categorisation an exact science. Competition law lies at the intersection of regulatory and commercial law. We have included it within the second volume, but a strong case could have been made for including it among the regulatory chapters.

[9] A strong case could have been made for including this within the commercial chapters of Volume 2 but in the end we thought it sensible to keep the two decommissioning chapters together.

economic context. From a recent high of $115 a barrel in June 2014, the price of Brent Crude plummeted to around $28 in January 2016 and now (as at May 2017) sits at $50 per barrel. What has been described as the "new normal", "lower for longer"[10] oil price has had a significant economic impact. In 2016 alone, 16 UK oil and gas companies became insolvent.[11] It is estimated that 120,000 UK jobs have been lost (across the supply chain and connected industries) in the sector since their peak in 2014.[12] The cause of the crash is both a supply-side and a demand-side story: the explosion of US shale gas onto the market, and reluctance amongst OPEC countries to compromise market share has meant there is no shortage in supply, while slowdowns in important consuming economies such as China and Europe have restricted the market for this oil. Globally, the low price has had consequences for nations such as Iran, Nigeria, Venezuela and Russia, who all require prices above $100 to balance budgets. Nevertheless, some optimism can be taken from Oil & Gas UK's latest reports showing production has increased by 16 per cent since 2014, and improvements in efficiency have seen operating costs slashed by 48 per cent on the 2014 figures.[13] It is also hoped that recent changes to the fiscal regime and the new obligation to Maximise Economic Recovery (MER) will secure the future of what is a very important industry for the UK: supporting 330,000 UK jobs, with estimates of up to 20 billion barrels of oil and gas still to be recovered from the basin.[14] It is against this background that Kemp and Stephen present the results of their economic analysis of the UKCS. They argue that recent tax changes alone may not be sufficient to secure the long-term future of the UKCS as an oil and gas province, but that further cost savings and technological innovation could provide the platform necessary for production to continue to 2050 and beyond.

Chapter I-3, by Emre Üşenmez, James Cowie and Greg Gordon,   I-1.09
is on energy security in the UK, a topic that is in constant evolution. On Friday 21 April 2017, the UK went a full day without using coal to generate electricity. Coal, which has powered the UK since the

[10] BBC News, "BP boss: Oil price will rise", available at www.bbc.co.uk/news/business-35363066 (accessed 8 May 2017).

[11] Moore Stephens, "UK based oil and gas sector insolvencies hit a new high", available at www.moorestephens.co.uk/news-views/january-2017/uk-based-oil-and-gas-sector-insolvencies (accessed 8 May 2017).

[12] Oil & Gas UK, "Oil & Gas UK figures show impact of oil price downturn on jobs", available at http://oilandgasuk.co.uk/oil-gas-uk-figures-show-impact-of-oil-price-downturn-on-jobs (accessed 8 May 2017).

[13] Oil & Gas UK, "Business Outlook 2017", available at http://oilandgasuk.co.uk/businessoutlook.cfm (accessed 8 May 2017).

[14] *Ibid.*

Industrial Revolution, now accounts for a historic low of 9 per cent of the UK's electricity generation. With the closure of the last UK coal mine in 2015 and the move towards cleaner sources of energy, including gas and renewables, coal's role in the UK energy mix is far less significant than it once was. It is against this backdrop, together with recent developments in shale gas, liquefied natural gas (LNG) and the changing regulatory landscape, that the authors discuss energy security in the UK. Production, exportation and importation of oil and gas in the UK are examined, before the international and European dimensions of the UK's energy security are analysed. In the next sections, the authors discuss law and policy in the context of the UK as both a producing and a consuming country. With an eye to the future, the chapter concludes by looking at ways in which the UK can maintain security of supply and the proper functioning of its energy networks.

I-1.10   Moving now into the group of chapters that relate to the theme of resource management, the first chapter in this section is "Petroleum Licensing" by Greg Gordon. Here, Gordon discusses the essential features of the UK's licensing system, including its hybrid nature (and the complications that flow from that) and its essentially discretionary character. He then goes on to consider the basic technical details upon which the licence allocation system is built, including the grid system and licensing rounds, as well as the extent to which this is increasingly impacted by environmental law, and the extent to which marine spatial planning could make a future contribution in this area. The second half of Gordon's chapter consists of a commentary on the UK oil and gas licences with particular reference to Production Licences. Practice in this area is in the process of changing markedly as a result of the Oil and Gas Authority's (OGA) adoption of a new licence (the Innovate Licence). However, previous practice continues to be highly relevant as the great majority of the licences currently extant upon the UKCS were granted under previous regimes. Gordon discusses the essential features of both the legacy and the present licensing regime, offering a preliminary appraisal of the desirability or otherwise of the changes (and in particular the increased flexibility) that the new licence introduces into the system. Noting that this is an area where the change of regulatory ideation can be seen to be having an early impact, he goes on to discuss the interaction between the new MER UK obligation and the licence obligations, a discussion that sets the scene for the following chapter.

I-1.11   Chapter I-5, co-authored by Greg Gordon, John Paterson and Uisdean Vass, continues the discussion of the Wood Review and the MER UK obligation that emerged from it. The chapter does not seek to offer a comprehensive overview of all aspects of the OGA's operations, or even of all aspects of the MER UK obligation; as has been noted above, MER UK is too all-encompassing to be confined to one chapter, but needs to be discussed pervasively throughout the work. Instead, the chapter locates the Wood Review within its historical context, as the latest in a line of attempts to reboot the system of governance upon the UKCS, offers a detailed account of the key findings of the Wood Review and discusses the somewhat unusual way in which the MER UK obligation has been implemented, as well as offering some comments on the complexities that arise as a result of MER UK as a regulatory philosophy and the particular implementation strategy that has been adopted.

Vass returns, this time as sole author, in Chapter I-6, on the perennially thorny problem of third-party access to infrastructure. The Wood Review, in 2014, reported more than 20 instances in the preceding three years where the inability to agree terms for access to essential infrastructure had led to suboptimal developments, significant delays and, in some instances, stranded assets. The issue of access to infrastructure is of huge importance to the future of the UKCS as a mature province. The majority of new discoveries in the basin are not large enough to justify purpose-built infrastructure networks, and instead must be tied back to existing pipelines and processing facilities. Ensuring there is an effective regime to facilitate these agreements is vital to achieving the UK's resource management objectives. Vass begins his chapter with an analysis of the commercial and technical factors that can inhibit access negotiations. Where negotiations fail, access-seekers can have recourse to the legislative regime under which the new regulator, the OGA, can impose terms of access. The third-party access provisions under the Energy Act 2011 are examined together with the regulator's guidance and the industry's voluntary code (ICOP) which is intended to guide parties' negotiations. In what is a very important update to the second edition of this book, the author provides a detailed set of comments on the impact of the new MER obligation on access talks in the UKCS. Amidst what is essentially a story of change, Vass notes the continuation of two all-too-familiar themes: the problem of complexity and the extreme difficulty, in a negotiation-led system, that the regulator faces in the identification and consistent application of a coherent set of principles.

In Chapter I-7, Claire Ralph, a new author for this edition, considers the fiscal regime applicable to the upstream sector in the UK.[15] She focuses in particular upon the recent changes that have

I-1.12

I-1.13

---

[15] The other new authors for this edition are Constantinos Yiallourides, Tina Hunter, Stephen Latta and Valerie Allan. The editors are delighted to welcome all the new contributors to the fold.

occurred following the price crash, the Wood Review and the Fiscal Review, and upon those elements of taxation that are specific to the oil and gas industry. She discusses the changes to the ring fence corporation tax, supplementary charge and petroleum revenue tax and considers their technical characteristics, also commenting on how particular policy decisions that can now be considered as errors (such as the historic abolition, as opposed to a rate cut to zero, of petroleum revenue tax for fields receiving field development consent after 1993) have cast a long shadow on subsequent practice, introducing an unfortunate legacy of complexity. She subsequently explores the potential tax issues at both the exploration and decommissioning stages and the issues surrounding the trading of mature upstream assets. She concludes by underlining the importance of the stability of and predictability within the fiscal regime going forward, while maintaining optimism in the simplified regime that has emerged following these recent changes.

I-1.14   Chapter I-8 has been written by another new author, Constantinos Yiallourides. It addresses the topic of the continental shelf boundaries within the North Sea and Atlantic margin areas. For coastal states, the importance of the sea can hardly be overestimated. The UK enjoys a particularly rich marine heritage. Since the 1960s, the UKCS has also played host to the offshore oil and gas industry. However, to what extent can the UK claim sovereignty and jurisdiction over the North Sea and the petroleum lying beneath the sea bed, given that it shares North Sea coastlines with, among others, Norway, Denmark, France and Germany? This is a fundamental question. If the UK does not hold the sovereign rights to the petroleum exploited, it cannot legally license its exploration and production. In this chapter, Yiallourides charts the development of international agreements among coastal states to govern territorial claims to the world's waters. What follows is an examination of the importance of the United Nations' Conventions and the International Court of Justice's (ICJ) decisions on the delimitation of maritime boundaries in the North Sea. The author goes on to present an application of international law on the disputed Hatton-Rockall plateau and islet of Rockall on the other side of the British Isles, in the North Atlantic. As it remains a live issue, with interest in oil and gas operations driving the discussions on boundaries in the region, this case study serves to illustrate the importance of this area of law.

I-1.15   Moving now into the strand of chapters primarily concerned with regulatory law, Chapter I-9, a new chapter commissioned for the third edition and co-authored by Tina Hunter, Steven Latta and Greg Gordon, addresses the politically charged and legally complex topic of shale gas. Oil and gas operations in the UK have, in the recent past, been primarily confined to the offshore. However, with conventional offshore production falling and the UK becoming progressively more dependent upon gas imports, there exist both financial and energy security drivers to rapidly investigate the extent to which a domestic shale gas industry may be developed. However, shale gas development is controversial. It poses difficult questions related to both regulation and public acceptance, and does so from multiple standpoints: in individual cases, local concerns might be overcome by, for example, the provision of a generous community bounty; however, such a measure will not in any way address the concerns of climate change activists, for example. Against this background, the authors discuss the current Great Britain-wide licensing[16] and regulatory regimes, as well as the planning regimes present in Scotland and England and Wales. The chapter also addresses the particular political challenges to shale gas development that arise in the Scottish context, noting the difficult choice that may lie ahead for the Scottish Government in this regard.

Chapter I-10, by John Paterson, traces the evolution of the I-1.16 regulatory approach to health and safety at work offshore. The evolution is divided into four phases. The first phase covers the period when health and safety at work was dealt with under the licence and focuses upon, first, the Continental Shelf Act 1964 and then the findings of the inquiry into the Sea Gem accident. The second phase covers the period during which the detailed prescriptive regime recommended by the Sea Gem Inquiry was gradually developed and implemented. It accordingly considers the Mineral Workings (Offshore Installations) Act 1971; the tension in due course between this Act and the Health and Safety at Work, etc. Act 1974; and the findings of the Burgoyne Committee on Offshore Safety. The third phase covers the period during which the permissioning approach now in place was developed. Thus, it considers the Piper Alpha disaster and the findings of the subsequent public inquiry, the Offshore Safety Act 1992 and the subsequent safety case and goal-setting regulations; and the revised Offshore Installations (Safety Case) Regulations 2005. The fourth phase runs from the Macondo disaster in the US to the present, including the Offshore Safety Directive and the impact it has had upon the architecture and orientation of the UK's regulatory approach. This treatment of the issue of health and safety at work offshore serves a number of purposes. First, by contrasting the current permissioning approach with the foregoing approaches it highlights the particularity of the

---

[16] As noted above, this chapter considers hybrid issues and could have been located within either of the book's two main sections.

means by which health and safety at work offshore is now regulated. This is of interest not only to those based in the UK, but also to the increasing number of people from other jurisdictions who are involved in the development of health and safety regulation in the offshore industry and who are looking to the UK for inspiration. Second, it reveals the extent to which the problems that the current approach seeks to deal with were evident from a relatively early stage but were never adequately dealt with in the recommendations of inquiries or in legislators' discussions. This serves also to bolster the new approach in the face of criticism from those who regret the passing of the prescriptive regime. Finally, it allows an appraisal of the ability of the permissioning approach to respond to the emergent challenges posed by the UKCS. Selected developments in criminal law are then considered before the concluding remarks to this chapter ask the question of whether the troubling findings that have periodically emerged in relation to the implementation of the safety case approach in the context variously of an investigation by the regulator, a Select Committee inquiry and the criminal courts will finally be addressed by the latest developments under the Offshore Safety Directive or whether, if anything, the challenges are only becoming more intense such that a concerted effort to deal with what appears to be a continuing weakness is now required.

I-1.17   Chapter I-11, co-authored by Luke Havemann and Tina Hunter, addresses the issue of environmental regulation. Noting the interconnection and overlap between the regulatory steps taken to prevent major safety and major environmental incidents, the chapter goes on to note the existence, too, of a separate regime governing more day-to-day, operational harms which, while not as dramatic as the major event, can, over the passage of time, build up to cause adverse environmental consequences. The chapter provides an overview of the key legislative and regulatory instruments concerned with preventing or mitigating each of these two categories of harm. It ends with an afterword that briefly addresses the topic of civil liability for environmental harm. It is hoped that this section can be further developed into a full chapter if the book should go into a fourth edition.

I-1.18   Chapter I-12 addresses the increasingly relevant and significant issue of decommissioning, and is authored by John Paterson. The first of two chapters on the topic, this one focuses in particular upon the international and public law dimension of the subject. It begins by tracing the evolution of the international legal regime relevant to decommissioning, both at the global and regional levels, prior to the Brent Spar case. There is thus consideration of the Convention on the Continental Shelf 1958, the London Dumping Convention 1972, the United Nations Convention on the Law of the Sea 1982 and the International Maritime Organization's Guidelines and Standards 1989, and the OSPAR Convention 1992. Thereafter the extraordinary events of the Brent Spar case and its implications are examined. This involves consideration of the initial regulatory approach, the Greenpeace protest, the Stakeholder Dialogue initiated by Shell and the impact of the case on OSPAR and the UK Government. This is followed by a review of international and domestic legal developments post-Brent Spar with particular attention paid to the 1996 Protocol to the London Convention, OSPAR Decisions 98/3 and the Petroleum Act 1998 as amended by the Energy Acts 2008 and 2016. Noting that the Department has not utilised its regulatory powers under the 1998 Act but rather has preferred to operate on the basis of guidance, the chapter then examines its updated Guidance Notes for Industry in some detail, specifically the treatment of Section 29 notices and the decommissioning programme process. The interaction of the Department's responsibilities in relation to decommissioning with the Oil and Gas Authority's responsibilities in relation to the MER Strategy are then considered. A continuing area of uncertainty related to residual liabilities is briefly discussed before concluding remarks are made.

The book concludes with a further chapter on decommissioning I-1.19 that focuses, this time, upon the issue of decommissioning security. This topic was previously dealt with as a sub-section of the general chapter on decommissioning. Such is the topic's relevance in practice, however, that for this edition we thought it would merit a chapter of its own, and this has been written by Judith Aldersey-Williams, one of the architects of the UK's current arrangements. The chapter explains the difficulties that the spectre of future decommissioning liabilities present, when viewed either from the Government's standpoint or from the perspective of commercial deal-making, noting, in particular, the complicating issue of the extent to which companies making provision for such liabilities are entitled to presume that historic or present tax reliefs will still be available at the point in the future that decommissioning will take place. Aldersey-Williams offers a comprehensive commentary on the Decommissioning Security Agreement and the Decommissioning Relief Deed, the latter of which is an innovative device intended to provide some much-needed certainty to the taxation element of the question.

CHAPTER I-2

REJUVENATING ACTIVITY IN THE NORTH SEA OIL AND GAS INDUSTRY: THE ROLE OF TAX INCENTIVES IN CONTEXT

Alexander Kemp and Linda Stephen

INTRODUCTION

I-2.01   Activity in the UK Continental Shelf (UKCS) is currently exhibiting the signs of advancing maturity. Thus, production peaked at 4.55 mmboe/d in 1999 and declined briskly thereafter to 1.49 mmboe/d in 2014 despite the rising oil price over the period. However, production has recently increased to some 1.7 mmboe/d. The exploration effort has also fallen dramatically this century, even before the collapse in the oil price. Thus, 14 exploration wells were drilled in 2014, 13 in 2015 and 14 in 2016. By comparison 75 were drilled in 1986, 74 in 1987 and 93 in 1988. These were all years following a major price collapse. The peak years for exploration effort were 1990, when 157 wells were drilled, and 1991 when 103 were drilled, though it should be recognised that there were special circumstances in these years, namely the execution of work programme promises made by BP relating to its takeover of Britoil.

I-2.02   Another manifestation of the maturity of the province is the decline in the average size of discovery and development. Currently the average size of discovery is around 20 mmboe. The most likely size is less than this given the lognormal distribution. In the first half of the 1970s the average exceeded 320 mmboe.

I-2.03   The number of significant new discoveries has also fallen considerably in recent years. Using the definition established by the Department of Energy and Climate Change (DECC) and the Oil and Gas Authority (OGA),[1] there were four in 2013, one in 2014 and

---

[1] The description "significant" generally refers to the flow rates that were achieved (or would have been achieved) in well tests (15 mmcfgd or 1000 BOPD). It does not indicate

four in 2015. By comparison, there were 14 in 1986 and 20 in 1987. Both were years of low oil prices. There were 79 in 1989. Another indicator of maturity is the number of field development approvals. There were only eight in 2014, five in 2015 and two in 2016, far below the average for the long period 1970–2015.[2]

Another feature consistent with maturity is the steep rise in unit costs. Field investment costs averaged $20.40 per boe in 2014 and $17.30 per boe in 2015. Unit operating costs averaged $29.30 in 2014 and $20.95 in 2015.[3] These figures relate to a very wide range of operations. Thus, on very old fields serviced by large old platforms, where production is now extremely low, operating costs per barrel can be very high indeed. The general dramatic degree of cost inflation in the industry across the world, on top of the ageing platform structures, plus declining production, produced some extremely high unit operating costs prior to recent cost reductions.

I-2.04

INTERPRETATION OF RECENT EXPERIENCE

The above observations appear straightforward, but further analysis is needed to enhance understanding of the recent behaviour of the sector. The subject of production decline rates in the oil and gas sector has probably received insufficient attention. There is the frequently accepted view that field decline rates are exponential in character. Kemp and Kasim found that the logistic curve produced the best fit across fields for the UKCS in the period up to the early years of this century, with incremental investments moderating the rates of decrease in the more mature years of field life.[4] Since that study was undertaken, many further new field developments have occurred. Reflecting the maturity of the province, they are generally much smaller in terms of reserves. But, in addition, their decline rates are noticeably faster than those exhibited by the earlier generation of much larger fields.

I-2.05

Figs I-2.1 and I-2.2 show the behaviour of oil and gas depletion by field vintage of first production. It is clear that the decline rates

I-2.06

---

the commercial potential of the discovery. See www.gov.uk/government/uploads/system/uploads/attachment_data/file/549380/Significant_Discoveries_August_2016.pdf (accessed 2 May 2017).

[2] Field Development Plans Consents and Field Development Plan Addenda Consents by the OGA in 2016 and 2017 to date, available at www.ogauthority.co.uk/data-centre/data-downloads-and-publications/field-data (accessed 2 May 2017).

[3] See Oil & Gas UK (2016), *Activity Survey 2016*, p 58. Available online at http://oilandgasuk.co.uk/wp-content/uploads/2016/02/Oil-Gas-UK-Activity-Survey-2016.pdf (accessed 4 August 2017).

[4] A G Kemp and A S Kasim, "Are Production Decline Rates Really Exponential? Evidence from the UKCS", 26(1) (2005) *The Energy Journal* 27.

in fields of more recent vintage are significantly faster than those of earlier ones. This has contributed to the brisk rate of overall decline in the UKCS this century. It is likely that the rate of decline in the newer generation of generally smaller fields is inherently faster than that in the larger older ones. But aggregate production is also a function of other factors. These include the numbers of new fields coming on stream and the production efficiency achieved across all fields. Production efficiency is the ratio of actual production to that at the maximum efficient rate. The Department for Business, Industry and Industrial Strategy (BEIS) (formerly the DECC) has calculated that the ratio has fallen from 80 per cent in 2004 to 60 per cent in 2012.[5] This has made a significant contribution to the fast decline rate. A main cause has been the substantial unplanned shutdowns relating to technical problems on the producing facilities. The increased interdependence of fields, using infrastructure such as processing hubs and pipelines, has sometimes produced major knock-on effects. When an important processing hub platform has to shut down, the fields that feed into it will also have to stop production.

I-2.07   Virtually all production projections made at the beginning of this century have turned out to be substantially over-optimistic.

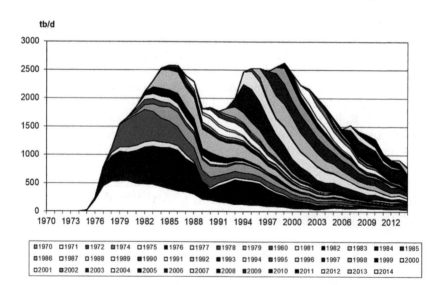

Figure I-2.1 Historic UKCS Oil Production by Production Start Date

---

[5] Department for Energy and Climate Change (DECC) (now BEIS), *Production Efficiency Survey 2013*.

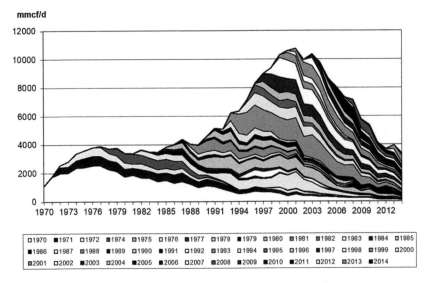

Figure I-2.2 Historic UKCS Gas Production by Production Start Date

For example, the Joint Industry and UK Government Task Force set up in 1999 to assess and make recommendations on the future prospects of the sector delivered a production target for 2010 of 3 mmboe/d.[6] The outcome was around 2.4 mmboe/d.

The Task Force of 1999, the Wood Review of 2013–14[7] and the follow-up work by the industry and the OGA have all diagnosed the issue very effectively. Progress has been made. As a notable example, production efficiency has increased markedly and is currently estimated by the OGA to be around 71 per cent.[8] This has contributed significantly not only to the recent reversal of production decline but also to a sustained increase in production.   I-2.08

The industry group, which examined the subject of production efficiency in depth, has expressed confidence that the improvement can continue over the next several years.[9] Their estimates are shown in Fig. I-2.3.   I-2.09

---

[6] The Oil and Gas Industry Task Force Report, *A Template For Change* (September 1999), available at http://webarchive.nationalarchives.gov.uk/20101227132010/www.pilottask-force.co.uk/docs/aboutpilot/atemplateforchange.pdf (accessed 2 May 2017).

[7] Wood Review (2014), UKCS Maximising Recovery Review, available at www.woodreview.co.uk/documents/UKCS%20Maximising%20Recover%20Review%20FINAL%2072pp%20locked.pdf (accessed 2 May 2017).

[8] See www.ogauthority.co.uk/news-publications/publications/2016/ukcs-production-efficiency (accessed 2 May 2017).

[9] See http://oilandgasuk.co.uk/cost-efficiency.cfm (accessed 2 May 2017).

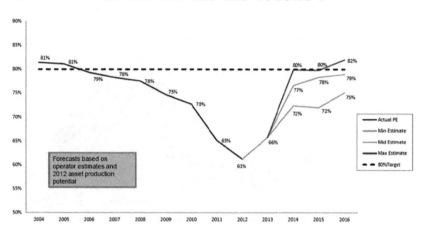

Figure I-2.3 Production Efficiency – UK Continental Shelf (Actual and Forecast)

I-2.10   However, the increases in production and its efficiency may reflect other issues. The Health and Safety Executive (HSE) has noted that the increased production in the recent past has coincided with a significant increase in the backlog of safety critical maintenance. This may open the prospect of future production problems. The postponement of maintenance work may foster short-term production gains, but it also increases the risks of disruption in the future. This issue remains to be fully understood.

ECONOMIC MODELLING PROCEDURE

I-2.11   The present authors have built a large financial simulation model, incorporating the Monte Carlo technique for risk assessment, to analyse the prospects for exploration, development and production. To examine possible aggregate activity, the modelling has been conducted with a large field database. This incorporates key individual field data on: historic production, investment costs (drilling and facilities), operating costs (tariffs separately) and decommissioning costs, plus estimates of future values under the same headings relating to sanctioned fields, unsanctioned probable and possible fields, and incremental projects. There are over 370 sanctioned fields, some 170 incremental projects and approximately 40 fields in the probable or possible categories. There is an additional database, categorised as technical reserves, containing some 250 fields where there are no current development plans. Some were previously in the probable or possible categories.

I-2.12   Future exploration activity and its fruits were also modelled. Historic exploration success rates over recent years were calculated, as were appraisal successes. This was done separately for each main region of the UKCS, namely Southern North Sea (SNS), Central North Sea/Moray Firth (CNS/MF), Northern North Sea (NNS), West of Shetland (WoS) and Irish Sea (IS). The success rates, sizes of discoveries, types of resource (oil, gas or condensate), and exploration and development costs for the discoveries all vary according to geographic region.[10]

Using the above information, the Monte Carlo technique was I-2.13 employed to project discoveries in each of the five regions to 2045. It was assumed that the distribution of field sizes was lognormal following historic evidence. The standard deviation (SD) was set at 50 per cent of the mean value, which was assumed to decline in accordance with historic evidence. The Monte Carlo technique was also employed to calculate field development costs for new discoveries. For each region the average development cost per boe sanctioned in recent years, but prior to the cost reductions, was calculated. The SD was set at 20 per cent of the mean value in the Monte Carlo simulations.

Investment hurdles reflecting the capital rationing experienced I-2.14 in recent years were employed to determine whether a new field or incremental project were developed or not. Two cases were modelled. The first is where the ratio of post-tax NPV@10 per cent/pre-tax to I@10 per cent is greater than 0.3. The second is where the ratio exceeds 0.5. This latter may be described as a situation of very serious capital rationing. It should be noted that the use of NPV/I > 0.3 as the hurdle often excludes projects where the NPV@10 per cent is clearly positive.

To facilitate understanding of the long-term prospects the I-2.15 modelling was initially undertaken for conditions before both the oil price collapse and cost reductions.

RESULTS FOR LONG-TERM PROSPECTS BEFORE THE OIL PRICE COLLAPSE

Using the case of an investment screening oil price of $70 per barrel I-2.16 and 45 pence for gas in real terms, the production prospects with investment hurdles of NPV/I > 0.3 and > 0.5 are shown in Figs I-2.4 and I-2.5 on the assumption that the production efficiency problem discussed above is partially resolved. It is seen that there is a short-term upturn followed by a long-term decline at a fairly

---

[10] For fuller details, see A G Kemp and L Stephen (2015), *Prospective Returns to Exploration in the UKCS with Cost Reductions and Tax Incentives*, North Sea Study Occasional Paper No. 134, 81, available at www.abdn.ac.uk/research/acreef/working-papers (accessed 2 May 2017).

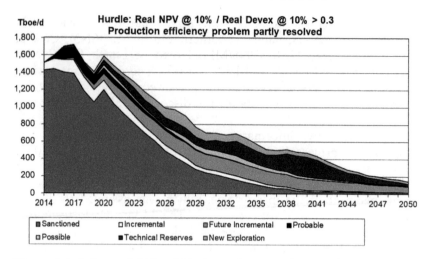

Figure I-2.4 Potential Total Hydrocarbon Production $70/bbl and 45p/therm

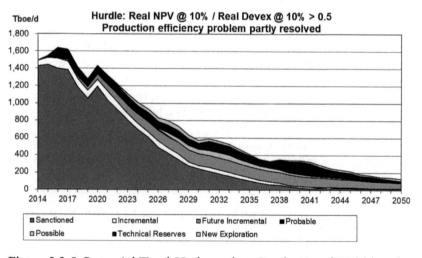

Figure I-2.5 Potential Total Hydrocarbon Production $70/bbl and 45p/therm (Production Efficiency Problem Partly Resolved)

brisk pace. Over the period 2014–2050 cumulative production is 11 bnboe with the NPV/I hurdle of 0.3 and 9.5 bnboe when it is > 0.5. The field expenditures with the lower hurdle rate are shown in Fig. I-2.6. A key feature is the sharp fall in field investment over the next few years. Over the period to 2050 cumulative field investment is £81.4 billion, cumulative operating costs £135 billion

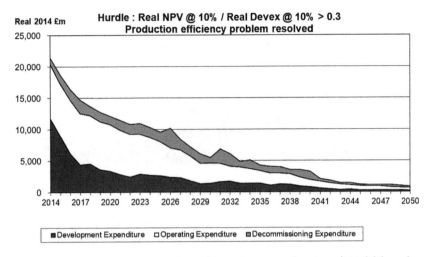

Figure I-2.6 Potential Total Hydrocarbon Production $70/bbl and 45p/therm (Production Efficiency Problem Resolved)

and cumulative decommissioning costs £41.8 billion, all at 2014 prices.[11]

Further modelling was undertaken to assess the effects of cost reductions on long-term activity levels. Two key effects were identified. The first was the impact on the costs of new projects and ongoing activities which in any case would have continued without the cost reductions. This effect applies both to new investment projects and to ongoing operations in existing producing fields. From the viewpoint of the supply chain this is a negative effect. The second effect relates to extra activity induced by the cost reductions. This refers to new field developments in particular. The induced effects relate to the extra investment, operating and decommissioning costs, and production. These are the positive effects with respect to activity.   **I-2.17**

A case of 15 per cent reduction in *all* costs was modelled. Key results are shown in Figs I-2.7, I-2.8 and I-2.9 respectively for the changes in production, development expenditures and operating expenditures under the $70, 45 pence investment screening price scenario. Over the period to 2050 the induced extra cumulative production is 2.9 bnboe. This is a major enhancement over the   **I-2.18**

---

[11] Another simulation was undertaken with a real oil price of $90 and gas price of 58 pence. With an investment hurdle of NPV/I > 0.3 over the period to 2050 cumulative production was found to be in the range of 14–15 bnboe depending on the extent of the improvement in production efficiency. When the investment hurdle was NPV/I > 0.5 cumulative production was in the range 11.6–12.6 bnboe. Field investment was very much higher with a cumulative total of £122 billion.

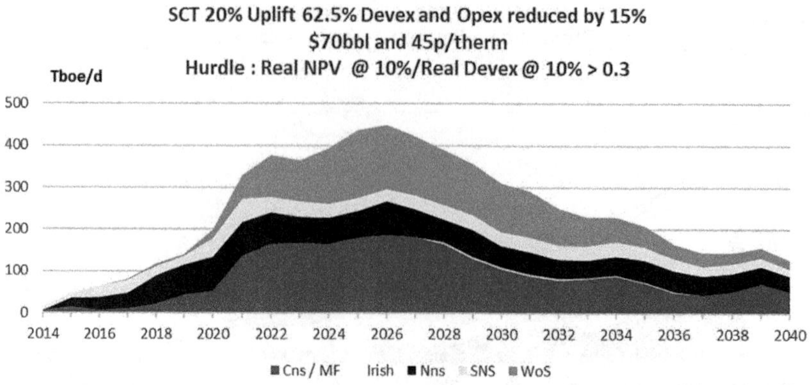

SCT 20% Uplift 62.5% Devex and Opex reduced by 15%
$70bbl and 45p/therm
Hurdle : Real NPV @ 10%/Real Devex @ 10% > 0.3

Tboe/d

■ Cns / MF Irish ■ Nns SNS ■ WoS

Figure I-2.7 Change in Potential Hydrocarbon Production

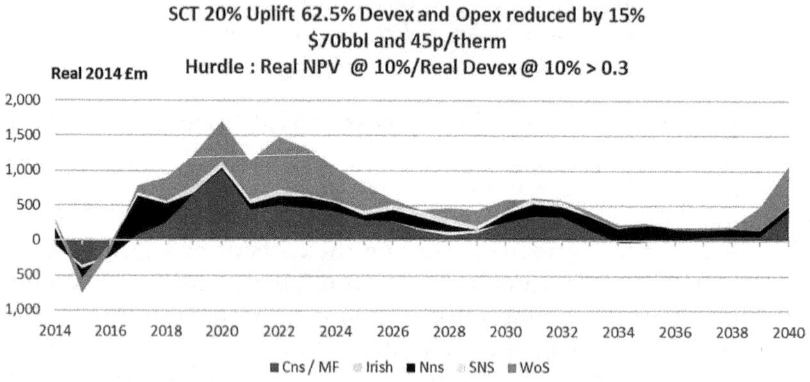

SCT 20% Uplift 62.5% Devex and Opex reduced by 15%
$70bbl and 45p/therm
Hurdle : Real NPV @ 10%/Real Devex @ 10% > 0.3

Real 2014 £m

■ Cns / MF Irish ■ Nns SNS ■ WoS

Figure I-2.8 Change in Potential Development Expenditure

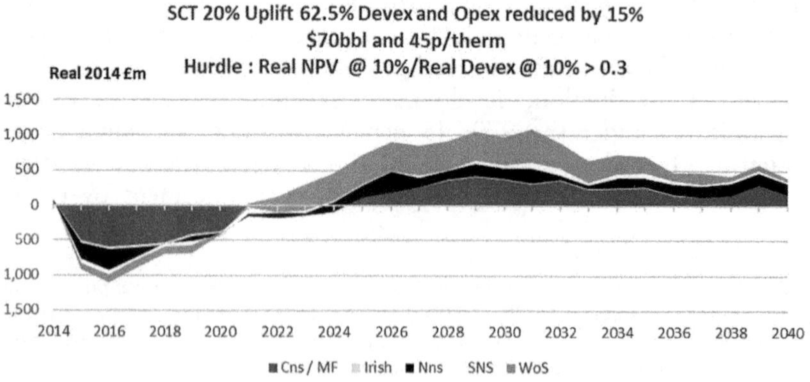

SCT 20% Uplift 62.5% Devex and Opex reduced by 15%
$70bbl and 45p/therm
Hurdle : Real NPV @ 10%/Real Devex @ 10% > 0.3

Real 2014 £m

■ Cns / MF Irish ■ Nns SNS ■ WoS

Figure I-2.9 Change in Potential Operating Expenditure

11–12 bnboe in the absence of the cost reductions. The extra cumulative field investment to 2050 is £22 billion at 2014 prices. It is noticeable from Fig. I-2.8 that a major part of the increase comes in the relatively near future. Over the period to 2050 there is a net increase in field operating expenditures of £23.4 billion at 2014 prices. It can be seen in Fig. I-2.9 that there is a major decrease over the next few years. The operating cost reductions apply to all the existing producing fields. However, over the longer term the positive effects of the expenditures on new fields outweigh the reductions on the existing ones. Over the whole period there is also a net increase in expenditure on decommissioning of £2.8 billion, reflecting the gains from the induced extra field developments.

I-2.19   It is useful to compare these results with the latest estimates of the remaining potential produced by the OGA. These are shown in Table I-2.1.

I-2.20   No dates or oil and gas prices are attached to the recovery of the resources, but the long-run estimates of the present authors are generally consistent with the remaining potential as seen by the OGA. Currently oil and gas prices are well below the levels employed in the modelling (though not out of line with long-run estimates produced by other bodies such as the International Energy Agency and the US Department of Energy). Also, much more ambitious cost reductions are planned by the industry which would increase both the near-term negative effects on the supply chain and the size of the positive longer-term induced effects. The effects of lower prices are discussed in detail in paras I-2.23 to I-2.48 below.

MATURITY AND SIZE DISTRIBUTION OF UNDEVELOPED DISCOVERIES

I-2.21   In the results of the modelling shown in the previous section, many existing discoveries were either uneconomic pre-tax or uncommercial after tax. A feature of a mature petroleum province is the

Table I-2.1 UK Oil and Gas Reserves and Resources (bnboe)

	Low	Central	High
Reserves	3.9	6.3	8.2
Contingent resources	0.6	1.4	2.6
PAR	1.5	3.6	7.2
Undiscovered resources (risked)	1.9	6.0	9.2

Source: OGA, July 2016

decrease in the most likely sizes of discovery. There are diminishing returns to the exploration effort. The distribution of sizes of current undeveloped discoveries in the authors' database is shown in Fig. I-2.10. Altogether there are 7.375 bnboe in 287 fields. The average is 25.7 mmboe, but the distribution is highly skewed. Thus there are 63 fields where the potentially recoverable resources are in reservoirs of less than 5 mmboe, and there are 71 fields where the resources are in reservoirs in the 5–10 mmboe range. There are 37 fields where the reserves are in reservoirs in the 10–15 mmboe range, and there are 42 fields where the reserves are in reservoirs in the 15–20 mmboe range. Thus 1.175 bnboe are in fields where the reserves are less than 15 mmboe and 1.9 mmboe are in fields where the reserves are less than 20 mmboe. A key current challenge is how to facilitate the development of typical fields in the context of current oil and gas prices and costs. Several issues rise here including further cost reductions, tax incentives, technological progress and more effective collaboration such as with respect to access to infrastructure. There has been much debate regarding tax incentives and these are discussed in paras I-2.22 to I-2.42 below.

TAX INCENTIVES AND NEW FIELD DEVELOPMENTS

I-2.22   New field developments are subject to Ring Fence Corporation Tax (CT). For some years, the CT rate was 30 per cent. Allowances for exploration, appraisal and development are all on a 100 per

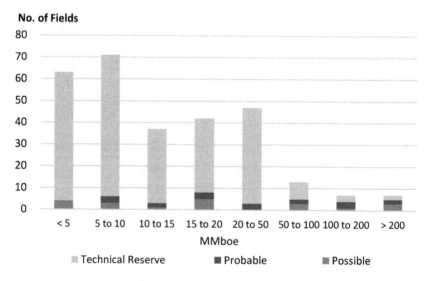

Figure I-2.10 Undeveloped Discoveries

cent first-year basis. Supplementary Charge (SC) also applies to new field developments. The rate has varied upwards and downwards since its introduction in 2002. In 2015 it was reduced from 32 per cent to 20 per cent and in 2016 it was reduced further to 10 per cent. Allowances for exploration, appraisal and development are all on a 100 per cent first-year basis. In addition, there is an Investment Allowance (IA) equal to 62.5 per cent of field investment. Loan interest is not deductible. The ring fence applies to all activities in the UKCS. Thus a licensee can set the capital allowances relating to a new field development against income received from other fields and, given that allowances are on a 100 per cent first-year basis, he receives speedy tax relief. This depicts the situation of an investor with tax shelter, here termed an "ongoing investor". In current circumstances where production losses are not uncommon many investors do not have this tax shelter. In that event the investor in a new field can benefit from the Ring Fence Expenditure Supplement (RFES) which means that he can carry forward his allowances for investment and operating costs at 10 per cent compound interest for up to ten years starting from the initial claim period. In this paper this is termed a "project investor". In the modelling the positions of both investor perspectives are analysed.

I-2.23   The modelling was undertaken on a set of 18 representative oil and gas fields. They are representative in terms of (1) size and (2) costs after substantial cost reductions reflecting the position at the summer of 2016. They are based on approved developments in recent years in the four main regions of the UKCS. The detailed modelling for a cross-section of these fields is discussed here.[12]

I-2.24   To highlight the complex issues involved in tax design and effects on investment several tax schemes were modelled. These are (1) the scheme of 2015 with CT at 30 per cent and SC at 20 per cent; (2) the scheme of 2016 with CT at 30 per cent and SC at 10 per cent; (3) CT at 20 per cent and SC at 20 per cent; (4) CT at 30 per cent and SC at 0 per cent; and (5) CT at 20 per cent and SC at 0 per cent. The IA for SC at 62.5 per cent is incorporated in all of the schemes.

I-2.25   To understand the effect of the tax system on investment incentives it is necessary to distinguish the effects of (1) the tax on income and (2) the relief for the investment expenditure. The two effects are shown in Table I-2.2 for the 2015 and 2016 systems.

---

[12] Full details of all the 18 fields are in A G Kemp and L Stephen (2016), *Field Development Tax Incentives for the UK Continental Shelf (UKCS)*, North Sea Study Occasional Paper No. 136, 66, available at www.abdn.ac.uk/research/acreef/working-papers (accessed 2 May 2017).

Table I-2.2 Rates of Tax on Income and Rates of Relief for Investment in the UKCS

	Tax on income	Relief for investment
2015 terms	0.3 + 0.2 = 0.5	0.3 + 0.2 + 0.625(0.2) = 0.625
2016 terms	0.3 + 0.1 = 0.4	0.3 + 0.1 + 0.625(0.1) = 0.4625

I-2.26   To reflect the current problem of serious capital rationing the results discussed here highlight the post-tax NPV@10 per cent/ pre-tax I@10 per cent ratios. These are widely employed in the industry. Historically a threshold of NPV/I > 0.3 was considered to be widely acceptable, but in current circumstances OGUK suggests that a threshold of 0.5 may be appropriate.[13]

I-2.27   The economic modelling found that at oil prices of $30 and $40 the representative fields were generally uneconomic before tax. Thus the modelling presented here concentrates on price scenarios of (1) $50 per barrel and 40 pence per therm, both in real terms, and (2) $60 and 45 pence. Not all the projects are viable before tax even at the $60 price. Those selected for detailed analysis here do not include the most uneconomic ones.

Very small pool (oil) in the CNS

I-2.28   The pre-tax returns on a representative oil field of c. 10 mmbbls in the CNS under the various tax schemes at the $50 price case are shown in Fig. I-2.11 in terms of NPV/I ratios. There are several noteworthy features. The project is unlikely to be acceptable to investors under all the tax arrangements. Under the 2015 system the returns to the ongoing investor are much higher than those for the project investor. The former obtains early tax relief against income from other fields. He is able to utilise all his IA for the SC. On the other hand, the project investor carries forward his allowances with interest against the income from the new field. But he has insufficient field income against which to offset all his allowances. He cannot fully utilise the IA for SC.

I-2.29   It is noteworthy that the returns to the ongoing investor are lower under the 2016 system compared to the 2015 one, despite the fact that the headline tax rate is reduced from 50 per cent to 40 per cent. This is because the reduction in the value of the investment relief from 62.5 per cent to 46.25 per cent is worth more than the reduction in the tax rate on income from the field from 50 per cent

---

[13] See Oil & Gas UK (2016), *Activity Survey 2016*, p 58. Available online at http:// oilandgasuk.co.uk/wp-content/uploads/2016/02/Oil-Gas-UK-Activity-Survey-2016.pdf (accessed 4 August 2017).

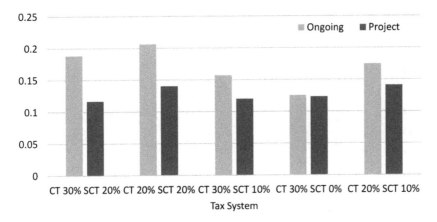

Figure I-2.11 CNS Oil 10 Mboe; Real Post-tax NPV @ 10% / Real Devex @ 10%; Oil Price $50/bbl Gas Price 40p/therm

to 40 per cent. On the other hand, the returns to the project investor under the 2016 system are close to those under the 2015 terms. They remain below those for the ongoing investor because the latter still has the advantage of early relief for his initial investment and fuller utilisation of the IA for the SC.

It is noteworthy from Fig. I-2.11 that the post-tax returns are higher with a scheme of CT at 20 per cent and SC at 20 per cent compared to both the 2016 and the 2015 ones. The effective rates of relief for investment are higher with CT at 20 per cent and SC at 20 per cent. A given reduction in the rate of CT is more potent than the same reduction in the rate of SC because there is less loss of investment relief. I-2.30

The returns to investors on the 10 mmbbls field are shown in Fig. I-2.12 under the $60 price scenario. To set the context the pre-tax NPV/I ratio is 0.5. Under the 2015 tax system the NPV/I ratio for the ongoing investor is nearly 0.35 and for the project investor 0.3. The difference in returns is much less compared to the $50 price case. The larger revenues permit the project investor to more fully utilise his allowances including the IA for SC. Under the 2016 tax terms there is little difference in the NPV/I ratio for the ongoing investor compared to the 2015 terms. The larger revenues permit more benefits to be received from the reduced tax rate on income. With the $60 price the returns to the project investor under the 2016 tax terms are closer to those for the ongoing investor and higher than the return under the 2015 terms. The larger income permits a fuller utilisation of the allowances and some benefit from the reduced tax on income. The investment project is clearly acceptable if the hurdle is NPV/I > 0.3. It is also noteworthy that a higher ratio is still I-2.31

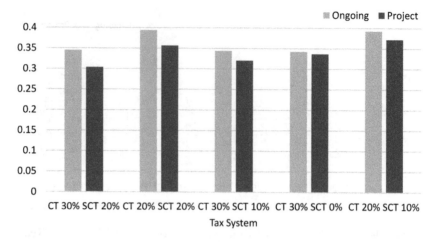

Figure I-2.12 CNS Oil 10 Mboe; Real Post-tax NPV @ 10% / Real Devex @ 10%; Oil Price $60/bbl Gas Price 45p/therm

achieved with a tax scheme of CT at 20 per cent and SC at 20 per cent because of the extra relief for the investment. At the $60 price it is also noteworthy that the highest return for the project investor is with CT at 20 per cent and SC at 10 per cent. The lower rate of tax on income is worth more in this case.

Typical small field (oil) in the CNS

I-2.32   In Fig. I-2.13 the post-tax returns to investment in a representative oil field of 20 mmbbls in the CNS are shown under the $50 price. The context is that the pre-tax NPV/I ratio is 0.35. Under the 2015 tax scheme the project is unlikely to be commercially viable to an investor in a full tax-paying position and less likely to be acceptable to a project investor. The difference in returns between the two investors is less on this field compared to the 10 mmbbls one because the larger revenues permit more effective utilisation of allowances including the IA by the project investor. The 2016 tax terms reduce the returns to the full tax-paying investor because the reduction in the value of the relief for investment still exceeds the benefit of the lower tax rate on income. The position of the project investor is slightly improved compared to the 2015 tax terms but the project remains sub-marginal. It is also seen from Fig. I-2.13 that reducing the CT rate enhances returns but the project remains very marginal.

I-2.33   In Fig. I-2.14 the post-tax returns are shown under the $60 price scenario. In context the pre-tax NPV/I ratio is 0.75. It is seen that under the 2015 tax terms the ratios are 0.48 for the ongoing investor and 0.44 for the project investor. The difference is relatively small

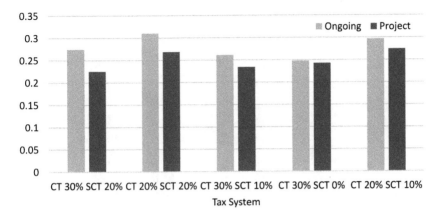

Figure I-2.13 CNS Oil 20 Mboe; Real Post-tax NPV @ 10% / Real Devex @ 10%; Oil Price $50/bbl Gas Price 40p/therm

because the larger revenues permit the project investor to recover his costs and utilise the IA. With the 2016 tax terms it is seen that the NPV/I ratio is 0.5 for the ongoing investor and 0.48 for the project investor. The larger revenues mean that there are greater benefits from the reduction in tax rates.

Significant new field (oil) in the CNS

In Fig. I-2.15 the post-tax returns to investment in a field of 100 mmbbls in the CNS are shown under the $50 oil price. Before tax I-2.34

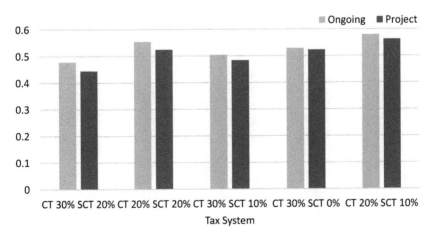

Figure I-2.14 CNS Oil 20 Mboe; Real Post-tax NPV @ 10% / Real Devex @ 10%; Oil Price $60/bbl Gas Price 45p/therm

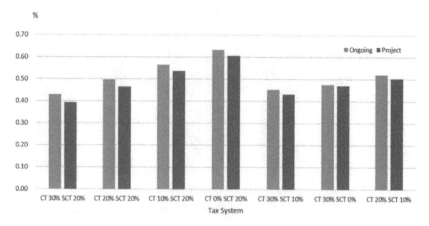

Figure I-2.15 CNS Oil 100 Mboe; Real Post-tax NPV @ 10% / Real Devex @ 10%; Oil Price $50/bbl Gas Price 40p/therm

the NPV/I ratio is 0.68. Under the 2015 tax terms the ratio is 0.425 for the ongoing investor and 0.39 for the project investor. Under the 2016 tax terms the ratios become 0.44 and 0.425 respectively. The project could well be commercially viable. The substantial size of the field means that the project investor recovers his costs and the benefit of the IA and still benefits to a worthwhile extent from the reduced tax rate on the income. There is thus only a minor difference between the returns to the investors in different tax positions.

I-2.35   The post-tax returns to the 100 mmbbls field are shown in Fig. I-2.16 under the $60 price. The pre-tax NPV/I ratio is 1.15. The

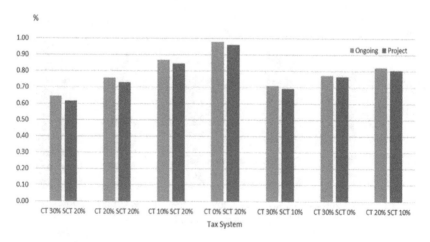

Figure I-2.16 CNS Oil 100 Mboe; Real Post-tax NPV @ 10% / Real Devex @ 10%; Oil Price $60/bbl Gas Price 45p/therm

project is clearly acceptable to both types of investors under both the 2015 and the 2016 tax terms. The 2016 terms enhance the returns because the large revenues permit substantial benefits to be received from the lower tax rate. Sadly, there are currently very few remaining undeveloped discoveries in the above category.

Typical field (gas) in the CNS

A gas project in the CNS with reserves of around 20 mmboe was I-2.36 modelled at the 40 pence price. It was found that the project was hopelessly uneconomic before tax and the results are not shown here. The results with a price of 45 pence are shown in Fig. I-2.17. The returns under both the 2015 and the 2016 tax systems are inadequate in a capital-constrained environment. There are advantages to the investor in a tax-paying position as he receives fuller and earlier relief for his costs.

Typical field (oil) in WoS

In Fig. I-2.18 the post-tax results are shown for an oil field of I-2.37 100 mmbbls developed in WoS under the $50 price. The project is clearly not commercially viable in a capital- constrained world under any of the tax arrangements. Indeed, it is uneconomic before tax. Under the 2015 tax scheme there is a very large difference in the NPV/I ratios between the ongoing and project investors. The former obtains the early benefit of relief for his allowances against other income. The high costs in relation to the field income inhibit

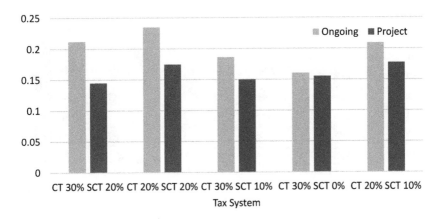

Figure I-2.17 CNS Gas 20 Mboe; Real Post-tax NPV @ 10% / Real Devex @ 10%; Oil Price $60/bbl Gas Price 45p/therm

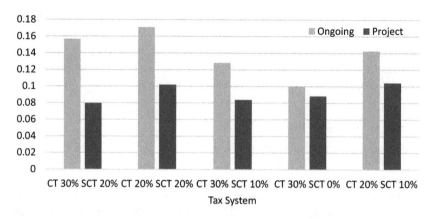

Figure I-2.18 WoS Oil 100 Mboe; Real Post-tax NPV @ 10% / Real Devex @ 10%; Oil Price $50/bbl Gas Price 40p/therm 000

the use of all the allowances by the project investor. Under the 2016 tax terms the ongoing investor is worse off because of the reduction in the value of his allowances. His expected return still exceeds that of the project investor in this generally sub-economic situation.

I-2.38   The returns from the same field under the $60 price are shown in Fig. I-2.19. With this price the pre-tax NPV/I ratio is 0.49. Under the 2015 tax terms the post-tax ratio is 0.32 for the ongoing investor and 0.27 for the project investor. The ongoing investor benefits from early utilisation of allowances against other income. Under the 2016

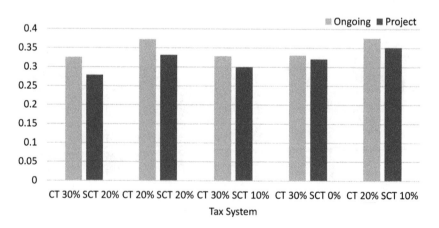

Figure I-2.19 WoS Oil 100 Mboe; Real Post-tax NPV @ 10% / Real Devex @ 10%; Oil Price $60/bbl Gas Price 45p/therm

tax terms the return to the ongoing investor is unchanged while the project investor's NPV/I ratio increases to 0.3. The investment project is now quite marginal after tax.

Typical field (oil) in the NNS

A representative oil field of 50 mmbbls in the NNS was also I-2.39 examined. Under the $50 price this project was found to be hopelessly uneconomic before tax and the results are not displayed here. The results under the $60 price are shown in Fig. I-2.20. It is seen that the returns in a capital-constrained situation are below those likely to be needed. Under the 2015 tax system the NPV/I ratio for the ongoing investor is just below 0.2 while for the project investor it is 0.124. Under the 2016 tax terms the ratio for the ongoing investor is reduced to 0.17 while the project investor's ratio increases very slightly. The investment project remains unlikely to pass the hurdle in a capital constrained world.

Typical field (gas) in the SNS

The returns to a representative gas field of 20 mmboe in the SNS I-2.40 were also modelled. At the 40 pence price the project was hopelessly uneconomic and the results are not shown here. They are shown for the 45 pence price in Fig. I-2.21. The pre-tax NPV/I ratio is 0.27. Under the 2015 tax system the ratio for the ongoing investor is 0.24 and for the project investor 0.176. Under the 2016 tax terms the ratio for the ongoing investor is reduced to 0.22 while for the project investor it increases very slightly.

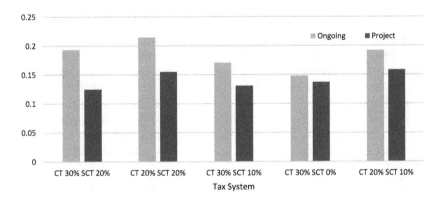

Figure I-2.20 NNS Oil 50 Mboe; Real Post-tax NPV @ 10% / Real Devex @ 10%; Oil Price $60/bbl Gas Price 45p/therm

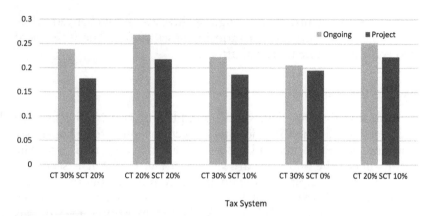

Figure I-2.21 SNS Gas 50 Mboe; Real Post-tax NPV @ 10% / Real Devex @ 10%; Oil Price $60/bbl Gas Price 45p/therm

MODELLING EXPLORATION ECONOMICS IN THE UKCS

I-2.41   A Monte Carlo financial simulation model has been constructed to estimate the distribution of expected monetary values (EMVs) from a specified exploration effort. In the modelling the investor undertakes exploration with a success rate determined by recent experience. When a discovery is made it is appraised. There is again a success rate determined by recent experience. Appraisal success means that there is a potential commercial development. The consequences of developing the discovery are assessed with the use of the Monte Carlo technique. Key stochastic variables are the size of the discovery, the development costs, and oil and gas prices.

I-2.42   The time taken from initial exploration to first production has a significant effect on the full-cycle returns when expressed in present-value terms. The returns also depend on the extent and costs of the exploration and appraisal efforts required. In this study two scenarios were modelled reflecting the experience and performance of the industry over the past few years. For ready convenience these are termed the "fast" and "slow" cycle cases. The phasing under the fast cycle case is from first exploration in year T_0 to first production in T_5. Under the slow case the time from first exploration to first production is T_0 to T_7. In the results below the fast cycle case is shown.

I-2.43   The prospective returns obviously depend on the costs at the various stages of the cycle. It is assumed that the industry succeeds in its present cost-reduction initiatives. After examining the experience in 2015 estimates of exploration and appraisal (E and A) costs were derived at levels considerably below those of 2014. The study examines the SNS, CNS, NNS and WoS separately. For the SNS, E and A costs per well were estimated at 50 per cent of the average for the UKCS. For the WoS region the costs were estimated at 1.25 times the average for the UKCS. The values employed in the study are shown in Table I-2.3 for each of the four regions.

Development costs also vary markedly across the four regions I-2.44 studied. Separate estimates were made for each region, again taking into account the reductions felt to be plausible from recent reported experiences. For modelling purposes development costs per barrel or boe were calculated. The average size of significant discovery was calculated for the period 2005–14, details of which are shown in Table I-2.3. The absolute costs for WoS are higher than elsewhere but the larger volumes pull down the relative unit costs. Development costs were phased over two to five years depending on the size of discovery. Annual operating costs were modelled as a percentage of accumulated development costs with the percentage increasing as the size of field decreased, reflecting economies of scale.

The modelling employs the Monte Carlo technique to reflect the I-2.45 uncertainties facing the explorationist and field developer. The mean values were made part of distributions of the stochastic variables which determine the returns facing the explorationist. The details of the input distributions obviously vary across each of the four regions, but have some common features. Thus the distribution of field sizes is taken to be lognormal with a standard deviation expressed as 50 per cent of the mean. The distribution of development costs per boe is taken to be normal with a common standard deviation of 20 per cent as a percentage of the mean. The mean oil price was set at \$55 per barrel in real terms with the assumption that it follows a mean-reverting behaviour through time. The standard deviation was set at 20 per cent of the mean. (Minimum and maximum values from the modelling were \$11 per barrel and \$99 per barrel respectively in real terms.) The mean gas price was set at 40 pence per therm in real terms with a standard deviation of 10 per cent of the mean. Mean-reverting behaviour is assumed. (The minimum value from the modelling was 24 pence and the maximum 56 pence, both in real terms.)

Other modelling assumptions relate to exploration and appraisal I-2.46 success rates. Significant discoveries are defined as all those published by the DECC plus others known to the authors covering the period 2008–2014 inclusive. Appraisal success covers all fields for which development has been started, firmly planned or contemplated. This definition excludes discoveries for which no field development plan is currently contemplated. All financial values in Table I-2.3 are in real terms.

Table I-2.3 Assumptions for Monte Carlo Modelling by Region After Cost Reductions

		Central North Sea	Southern North Sea	Northern North Sea	West of Shetland
Exploration success		34.2%	35.3%	40%	50%
Chance of oil		82%	0%	88%	75%
Chance of gas		18%	100%	12%	25%
Appraisal success		47.4%	30%	50%	55.6%
Reserves	Average	39.1 mmboe	16.4 mmboe	16.5 mmboe	112.6 mmboe
	Minimum significant size	8.5 mmboe	3.55 mmboe	3.6 mmboe	24.4 mmboe
	Maximum significant size	110 mmboe	50 mmboe	50 mmboe	320 mmboe
Well costs for E&A		£24.68m	£14.1m	£24.68m	£30.85m
Average devex per boe		$23.67	$11.392	$17.152	$15.82
Minimum devex per boe		$9.47	$4.56	$6.86	$6.33
Maximum devex per boe		$37.88	$18.23	$27.44	$25.32

Results of exploration modelling

The distribution of post-tax EMVs@10 per cent for the CNS is shown   **I-2.47**
in Fig. I-2.22. The mean expected value is –£4.6 million. There is a 68
per cent chance that the EMV will be in the range –£20.6 million to
+£10.1 million, and a 95 per cent chance that it will be in the range
–£43.2 million to +£30.7 million. There is a 58 per cent chance that
the EMV is negative. The upside potential is very limited. There is a
30 per cent chance that the EMV will exceed +£3.86 million.

In Fig. I-2.23 the distribution of post-tax EMVs is shown for   **I-2.48**
the explorationist in the NNS. The mean expected value is +£3.99

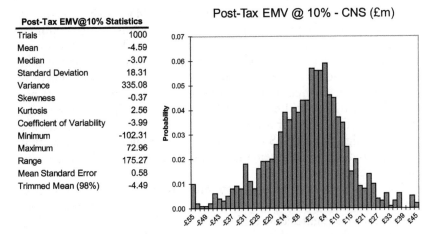

Post-Tax EMV@10% Statistics

Trials	1000
Mean	-4.59
Median	-3.07
Standard Deviation	18.31
Variance	335.08
Skewness	-0.37
Kurtosis	2.56
Coefficient of Variability	-3.99
Minimum	-102.31
Maximum	72.96
Range	175.27
Mean Standard Error	0.58
Trimmed Mean (98%)	-4.49

Figure I-2.22 CNS Project Fast Limited IA Initial Price $55 p/b
40p/therm Reduced Costs

Post-Tax EMV@10% Statistics

Trials	1000
Mean	3.99
Median	3.63
Standard Deviation	8.95
Variance	80.10
Skewness	0.96
Kurtosis	3.84
Coefficient of Variability	2.24
Minimum	-20.92
Maximum	68.09
Range	89.01
Mean Standard Error	0.28
Trimmed Mean (98%)	3.84

Figure I-2.23 NNS Project Fast Limited IA Initial Price $55 p/b
and 40p/therm Reduced Costs

million. There is a 68 per cent chance that the EMV will be in the range –£4.4 million to +£11.7 million, and a 95 per cent chance that it will be in the range –£10.9 million to +£24.6 million. There is a 33 per cent chance that the EMV will be negative and a 30 per cent chance that it will exceed +£7.64 million.

I-2.49   In Fig. I-2.24 the distribution of post-tax EMVs is shown for the explorationist in the SNS. The mean expected value is +£3.42 million. There is a 68 per cent chance that the EMV will be in the range +£1.1 million to +£9.7 million, and a 95 per cent chance that it will be in the range –£0.5 million to +£9.7 million. While there is only a 4 per cent chance that the EMV is negative there is a 30 per cent chance that it will exceed +£4.3 million.

I-2.50   In Fig. I-2.25 the distribution of post-tax EMVs is shown for the explorationist in WoS. The mean expected value is +£71.7 million. There is a 68 per cent chance that the EMV will be in the range +£10.1 million to +£134.6 million, and a 95 per cent chance that it will be in the range –£52.2 million to +£253.5 million. There is an 11 per cent chance that the EMV will be negative and a 30 per cent chance that it will exceed +£96.9 million. It should be stressed that the absolute exploration, appraisal and development costs are relatively high in the WoS region. In the other regions it is clear that the expected returns are generally unexciting in a capital constrained environment.

I-2.51   The UK Government has responded to the fall in exploration activity by funding the provision of seismic data in frontier areas where the potential is regarded as under-explored. The freely available data can be regarded as a public good provided to all explorers. The extra

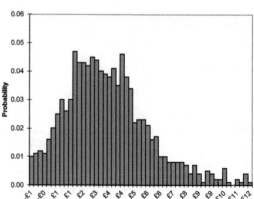

Post-Tax EMV@10% Statistics	
Trials	1000
Mean	3.42
Median	3.10
Standard Deviation	2.51
Variance	6.32
Skewness	1.10
Kurtosis	2.39
Coefficient of Variability	0.73
Minimum	-2.16
Maximum	17.48
Range	19.64
Mean Standard Error	0.08
Trimmed Mean (98%)	3.37

Figure I-2.24 SNS Project Fast Limited IA Initial Price $55 p/b and 40p/therm Reduced Costs

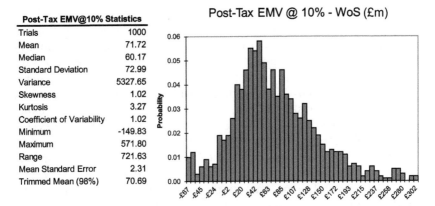

Post-Tax EMV@10% Statistics	
Trials	1000
Mean	71.72
Median	60.17
Standard Deviation	72.99
Variance	5327.65
Skewness	1.02
Kurtosis	3.27
Coefficient of Variability	1.02
Minimum	-149.83
Maximum	571.80
Range	721.63
Mean Standard Error	2.31
Trimmed Mean (98%)	70.69

Figure I-2.25 WoS Project Fast Limited IA Initial Price $55 p/b and 40p/therm Reduced Costs

information might in due course increase the exploration success rate and subsequently produce national benefits in terms of enhanced development and production activity. Given the combination of the extremely low current exploration effort and the substantial estimates of yet-to-find resources the public investment is defensible.

CONCLUSIONS: REINFORCING THE MER STRATEGY

From the analysis of the economics of new field investments and exploration in current circumstances in the UKCS it is clear that at $50 and $60 prices there are many marginal investment situations which will not attract investment capital. The tax rate reductions introduced in 2015 and 2016 have two effects. First, they enhance cash flows on existing operations. In a situation where the industry as a whole is cash-flow negative this is undoubtedly appropriate. But with respect to new field investments the effects are more complex. The effect on incentives and returns to investors depends on the combined effects of the reduction in the tax rate on income and the reduction in the rate of relief for the investment costs. It was found that, on small fields where the pre-tax returns were quite modest, the reduced rate of relief could be more important than the reduction in the tax rate on income. On larger fields and on small fields with higher oil prices the reduced tax on income is more important than the reduced rate of relief. Reductions in the rate of CT rather than SC were found to be more potent in incentivising new investments. In current circumstances there is a case for reducing the CT rate, which at 30 per cent is now far above the non-North Sea rate.

I-2.52

I-2.53   Tax incentives alone cannot ensure the revitalisation of the UKCS. The painful cost reductions currently being implemented are a regrettable necessity. To facilitate the development of the many uneconomic fields, including small pools, technological advances are necessary. Expenditure on R and D in the fossil fuels segment of the energy sector has been relatively low for a considerable number of years. The long-term trend is shown in Fig. I-2.26. There is a need to enhance this if the recovery factor is to be significantly improved. The new Oil and Gas Technology Centre seeks to be a major catalyst in this area.

I-2.54   The Wood Review has emphasised the need for more collaboration among licensees and contractors to enhance economic recovery. This relates to several areas such as third-party access to infrastructure and sharing of information, such as that relating to decommissioning. This recommendation is within the context of traditional competition among licensees and among contractors. The UK also has competition laws to which all companies have to adhere. To facilitate collaboration without falling foul of competition laws can be a challenge. To reduce uncertainty in this area it is suggested that the OGA and Competition and Markets Authority issue joint guidance notes which would clarify which collaborative agreements are consistent with competition laws and which are inconsistent.

I-2.55   It is also suggested that the emphasis on objectives could be geared to maximisation of total value added from the whole sector including the supply chain. This would include exports from the supply chain. The UK/Scottish supply chain has become increasingly active in overseas markets over the last two decades. For the Scottish supply chain this is indicated in Fig. I-2.27.

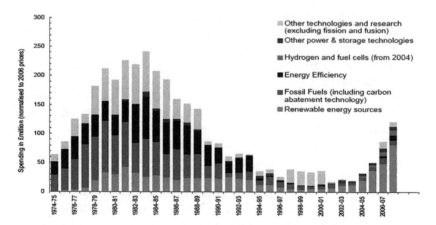

Figure I-2.26 R and D in the UK Energy Sector
Source: M Wicks, *Energy Security: A National Challenge in a Changing World* (2009), DECC

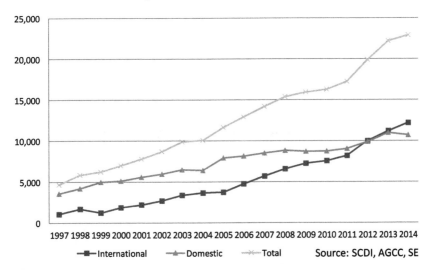

Figure I-2.27 Scottish Oil and Gas Supply Chain. International and UK Market Sales 1997–2014, £m (MoD) (including overseas sales of Scottish subsidiaries)

It can be seen that sales to export markets have grown markedly **I-2.56** to some 53 per cent of the total. As the UKCS declines but the world market continues to grow export activity should also continue to grow. But a continued healthy home market will be necessary to encourage companies to maintain bases in the UK.

CHAPTER I-3

THE UK'S ENERGY SECURITY

Emre Üşenmez, James Cowie and Greg Gordon

I-3.01   On 18 November 2015, eight months before the Department of Energy and Climate Change (DECC) was subsumed into the Department for Business, Energy and Industrial Strategy (BEIS),[1] the last Secretary of State for Energy and Climate Change, Amber Rudd, outlined a new direction for the UK's energy policy.[2] In her speech she stated that energy security had "to be the number one priority" because it was "fundamental to the health of the [UK] economy and the lives of [the British] people", and that it was the key to the "foundation of [the UK's] future economic success".[3]

I-3.02   A year later, as the first Secretary of the State for the newly created BEIS, Greg Clark echoed the position of Amber Rudd,[4] and emphasised his focus on cleaner energy. In his speech at the annual Energy UK Conference, he said that the goal of the Government was to "create the right framework" by harnessing "new technology to deliver more secure, cleaner energy at a lower cost".[5]

I-3.03   The importance of energy security for the UK was also highlighted in the most recent National Security Strategy (NSS).[6] The NSS Report

---

[1] On 14 July 2016, BEIS was created by the merger of the DECC and the Department for Business, Innovation and Skills.

[2] A Rudd, "Amber Rudd's Speech on a New Direction for UK Energy Policy' (DECC, 18 November 2015). Available for download at www.gov.uk/government/speeches/amber-rudds-speech-on-a-new-direction-for-uk-energy-policy (accessed 24 August 2017).

[3] *Ibid.*

[4] Albeit less strongly than Amber Rudd.

[5] Greg Clark's speech at the annual Energy UK Conference (10 November 2016), available at www.gov.uk/government/speeches/greg-clark-speech-at-energy-uk (accessed 5 May 2017).

[6] Prime Minister's Office, *National Security Strategy and Strategic Defence and Security Review 2015: A Secure and Prosperous United Kingdom*, November 2015. Available

emphasised how global energy security could be "affected by regional disputes, instability, terrorism and cyber threats, and more positively by technological developments",[7] and how the Government would be tackling "energy security challenges robustly"[8] by "managing the risks posed by regional instability, climate change, natural events and rising global demand".[9]

From this, it is evident that the issues surrounding environmental concerns and energy security are intertwined, and as priority areas for policymakers, they are being discussed at the highest levels of the UK Government. But what exactly is meant by "energy security"? **I-3.04**

The definitions of energy security vary depending on which aspect[10] or perspective[11] is being discussed. Towards the broader end of the spectrum of definitions, it is defined as "a condition in which a nation and all, or most, of its citizens and businesses have access to sufficient energy resources at reasonable prices for the foreseeable future free from serious risk of major disruption of service".[12] This is an ideal definition from the perspective of a consumer state. However, viewed from the perspective of the UK as a producer country, the definition is incomplete. The description in the Wicks Report, on the other hand, is just broad enough to include both perspectives. The report regards energy security as a tripartite concept: a balance of "physical", "price" and "geopolitical" security which energy policies "must aim at achieving".[13] Under these definitions, an energy policy must, respectively: avoid "involuntary interruptions of supply"; provide "energy at reasonable prices to consumers"; and ensure "the UK retains independence in its foreign policy through avoiding dependence on particular nations".[14] These views were echoed in the most recent UK Energy Security Strategy: **I-3.05**

---

for download at www.gov.uk/government/publications/national-security-strategy-and-strategic-defence-and-security-review-2015 (accessed 24 August 2017).

[7] *Ibid*, para 3.39.

[8] *Ibid*, para 4.141.

[9] *Ibid*, para 4.140.

[10] For example, trade and investment issues, political issues, consumption/transmission/production efficiency issues, environmental issues, etc.

[11] That is, perspectives of an energy producer nation or a consumer nation.

[12] B Barton, C Redgwell, A Rønne and D N Zillman, "Introduction", in B Barton, C Redgwell, A Rønne and D N Zillman (eds), *Energy Security: Managing Risk in a Dynamic Legal and Regulatory Environment* (2004), p 5.

[13] M Wicks MP, *Energy Security: A national challenge in a changing world* (5 August 2009), available for download at http://130.88.20.21/uknuclear/pdfs/Energy_Security_Wicks_Review_August_2009.pdf (accessed 24 August 2017) (hereinafter "Wicks, *Energy Security*"), p 8.

[14] *Ibid*, p 8.

"[t]here is no perfect definition of energy security. When discussing energy security the Government is primarily concerned about ensuring that consumers have access to the energy services they need (physical security) at prices that avoid excessive volatility (price security). Energy security must be delivered alongside achievement of our legally binding targets on carbon emissions and renewable energy".[15]

I-3.06   Towards the more specific end of the spectrum, the approach to defining energy security becomes more systematic. The Energy Charter Secretariat interprets it "as a triad consisting of security of supplies, security of infrastructure and security of demand, including thereafter, the issues of access to resources, infrastructure and markets".[16] The Center for Strategic & International Studies (CSIS)[17] has a similarly systematic approach to the concept. In a recent study, it identified "eleven factors closely associated with energy security: diversity of energy supplies; diversity of suppliers; import levels; security of trade flows; geopolitics and economics; reliability; risk of nuclear proliferation; market/price volatility; affordability; energy intensity (energy used per unit of gross domestic product); and feasibility".[18]

I-3.07   Although all of the above-mentioned definitions are valid, this chapter aligns itself with the interpretation of the Energy Charter Secretariat as it is comprehensive enough to take all the perspectives into consideration and broad enough to include all the major aspects of energy security. The rest of the chapter is divided into five sections. The first will set out the UK energy context. The second and third sections will look at the European Union (EU) and the international dimensions of the UK's energy security. These will be followed by sections discussing the energy security issue from the viewpoints of the UK both as a producing country and as a consumer nation. Finally, the chapter will end with a brief discussion on the current and proposed steps towards alleviating the UK's energy insecurity.

---

[15] DECC (now BEIS), *Energy Security Strategy* (November 2012). Available for download at https://www.gov.uk/government/publications/energy-security-strategy (accessed 24 August 2017).

[16] A Konoplyanik, "Energy security: the role of business, governments, international organizations and international legal framework", (6) 2007 *IELTR* 85.

[17] A bipartisan, non-profit organisation headquartered in Washington, DC, CSIS conducts research and analysis and provides policy solutions to decision makers in government, international institutions, the private sector and civil society.

[18] B Childs Staley, S Ladislaw, K Zyla and J Goodward, *Evaluating the Energy Security Implications of a Carbon-Constrained U.S. Economy* (January 2009), CSIS Issue Brief: Energy Security and Climate Change, at p 1. For full definitions of each of the factors see Annex II of the Issue Brief.

ENERGY IN THE UK

Prior to the 1970s, the vast majority of the UK's domestic primary   I-3.08
energy production was in the form of coal. In 1958, for example,
the domestic coal industry was producing much the same calorific
value of energy as the domestic oil industry did at its peak in 1998,
and supplied 81 per cent of the UK's energy consumption.[19] The first
important petroleum discovery on the UKCS was in 1965, but it was
not until the end of the 1970s that production reached significant
levels. In 1965, total crude oil production was at around 83,000
tonnes of oil equivalent (toe), which increased to about 410,000
toe in 1974, and then to a sizeable 77.8 million toe in 1979.[20] This
increase was the result of the surge in exploration activity following
the fourth licensing round[21] in 1971, and the OPEC oil embargo
two years later. With these events, and the quadrupling of oil prices
in 1973 as a result of the embargo, "[s]uddenly, the possibility of
economic development in water depths of 400–500 ft was realised,
and this initiated what was to become one of the great technological
achievements in oil exploration and production".[22]

However, it was not until 1981, at the production level of 89.5   I-3.09
million toe, that the UK became a net exporter of crude oil.[23] That
same year, the UK also became a net exporter of overall energy[24]
with an export surplus of 12.6 million toe of fuel.[25] The UK
remained a net exporter of energy until the end of 2003, except
for "the period between 1989 and 1992 when North Sea oil
production dipped in the aftermath of the Piper Alpha disaster".[26]

---

[19] Authors' calculation, using figures from DECC, Digest of United Kingdom Energy
Statistics: 60th Anniversary Edition (2009), available at www.gov.uk/government/uploads/
system/uploads/attachment_data/file/65896/1_20090729135638_e___dukes60.pdf, 5–6
(Tables 1 and 2). Domestically produced oil, from the small number of onshore fields then
in production, supplied a tenth of 1 per cent of the UK's energy supply.

[20] DECC Statistics, *Crude oil and petroleum products: production, imports and exports
1890 to 2008*. One tonne of oil equivalent is approximately 7.4 barrels of oil equivalent.

[21] For discussion on the licensing rounds please see Chapter I-4.

[22] J M Bowen, *25 Years of UK North Sea Exploration*, Geological Society, London,
Memoirs (1991), 14, pp 1–7, available at http://mem.lyellcollection.org/cgi/reprint/14/1/1.
pdf (accessed 5 May 2017).

[23] DECC Statistics 2009, DUKES Table 3.1.1, "Crude oil and petroleum products:
production, imports and exports 1970 to 2008" (hereinafter "DUKES Table 3.1.1").

[24] In addition to the crude oil, it includes coal, natural gas and primary electricity. See
DECC Statistics, DUKES Table 1.1.2, "Availability of consumption of primary fuels and
equivalents (energy supplied basis) 1970 to 2008".

[25] DECC Statistics 2009, DUKES Table 1.1.3, "Comparison of net imports of fuel with
total consumption of primary fuels and equivalents, 1970 to 2008" (hereinafter "DUKES
Table 1.1.3").

[26] Wicks, *Energy Security*, para 1.9. For the discussion on the Piper Alpha disaster see
Chapter I-10.

Both crude oil production and net exports of energy peaked in 1999 at approximately 137 million toe[27] and 51.5 million toe[28] respectively. They have been in decline since; and from 2004 onwards the UK became, once again, a net importer of energy (see Fig. I-3.1).

I-3.10   To say the UK was a net exporter of energy does not mean that it was energy independent. Since the 1920s, at no point did the UK reach energy self-sufficiency.[29] "It relied on imports to meet some of its coal needs and imported uranium to fuel its nuclear power stations, while the oil it consumed was traded and priced in an international market."[30] Even when it was exporting crude oil, the UK was simultaneously importing it (see Fig. I-3.1). This was due to the "demand by different refiners for different types of oil to North Sea crude, and to take advantage of pipeline infrastructure shared with some Norwegian fields".[31]

I-3.11   Nevertheless, since 2004, the share of the net fuel imports in the UK's energy consumption levels has been increasing rapidly. The

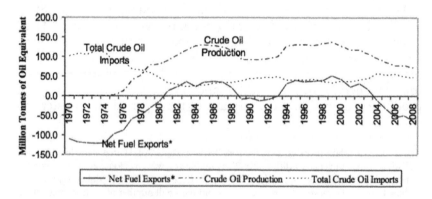

Figure I-3.1 UK Net Energy Export and Crude Oil Production Levels 1970–2008

Source: DECC Statistics

*Negative figures represent net energy imports

---

[27] DECC Statistics 2009, DUKES Table 3.1.1.

[28] DECC Statistics 2009, DUKES Table 1.1.3.

[29] Presumably this is true for the period before 1920 also, but the data is available only from 1920 onwards. See DECC Statistics, *Crude oil and petroleum products: imports by product 1920–2008*. It is important to note that "energy self-sufficiency" does not mean energy security, nor is it one of the goals of energy security.

[30] Wicks, *Energy Security*, para 1.9.

[31] *Ibid*, para 3.6. For the discussion on the shared infrastructure with Norwegian fields, see Chapter I-6.

UK's import dependency[32] rose from 4.5 per cent in that year to more than 46 per cent in 2014.[33] It subsequently dropped to 38 per cent in 2015.[34] These figures, though, do not necessarily signal energy insecurity. A country can be completely reliant on imports and still not suffer from a lack of energy security. This is dependent on several things: diverse sources of energy supply; high energy efficiency; and/ or, in the absence of diverse suppliers, a highly reliable, dependable and stable source. Japan and Korea, with near 100 per cent import dependency, are good examples. Japan has "one of the lowest energy-intensity economies in the world", and they both have a diverse supplier base, and diverse energy-generating sources.[35]

The picture in the UK, though, is different from that in Japan or Korea. Since the mid-1980s, a significant proportion of the imported crude oil and natural gas has been supplied from one source: Norway. In 2015, Norway's share in total imported crude was approximately 50 per cent (having been 65 per cent for many years), while some 39 per cent of imported crude was from the OPEC countries.[36] That same year, Norway supplied approximately 60 per cent of the UK's natural gas imports. Despite being a single major source for crude oil and natural gas imports, Norway is not an energy security concern. It is a reliable, dependable and stable supplier; and, in addition to the geographical proximity, both countries enjoy an active bilateral relationship. Recently, the Secretary of State for the DECC (now BEIS) and his counterpart in Norway released a Joint Ministerial Statement, agreeing to "intensify their co-operation" in carbon capture and storage (CCS),[37] in energy issues including improvements

I-3.12

---

[32] Calculated as the percentage ratio of net imports of fuel over gross inland consumption of primary fuels (inclusive of marine bunkers).

[33] BEIS Statistics 2016, DUKES Table 1B, "Net import dependency 2013 to 2015".

[34] *Ibid.*

[35] For a very brief look into Japan's and Korea's energy security questions see Wicks, *Energy Security*, Box 10: Japan, p 65; and Box 5: Republic of Korea, p 35.

[36] BEIS Statistics 2016, DUKES Chart 3.2, "Sources of UK crude oil imports 2000–2015".

[37] In simple terms CCS, as a climate change mitigation process, entails capturing carbon dioxide emissions from industry and other sources and permanently storing them in sub-seabed geological formations. In 2005, UK's Minister for Energy and Norway's Minister of Petroleum and Energy jointly established the North Sea Basin Task Force with the aim to "develop broad common principles that could form the basis for regulating the storage of CO2 in the North Sea and to provide a consistent basis for managing this activity". In recent years, significant spending on CCS schemes in the UK has looked likely. From 2012 to 2015, the UK Government ran a competition, worth £1 billion of public funding, for the design, construction and operation of the UK's first commercial-scale CCS projects. Two preferred bidders were announced in 2013: the White Rose Project and the Peterhead Project. See www.gov.uk/guidance/uk-carbon-capture-and-storage-government-funding-and-support (accessed 5 May 2017). However, in 2015, the funding and competition was pulled on the grounds of public spending concerns, but not without £100 million being spent on the competition to 2015; see www.bbc.co.uk/news/

in the ultimate levels of oil recovery and further development of the "relationship between Norway as a gas supplier and the United Kingdom as an important gas consumer" and in renewable energy.[38]

I-3.13　　However, it should be noted that although Norway as a primary source does not in itself pose a threat to the UK's energy security, the concentration in the infrastructure conveying natural gas from Norway into the UK may. There are three entry points of Norwegian gas: the Vesterlad pipeline, the Tampen Link pipeline[39] – both of which deliver gas to St Fergus in Scotland – and the Langeled pipeline received at Easington in England.[40] Any prolonged disruptions to these three pipelines, or to the two receiving terminals, could potentially cause a considerable natural gas shortage.[41] In 2005, imports of liquefied natural gas (LNG) began. These ramped up considerably in 2009 following the completion of new terminal facilities (Dragon and South Haven) in Milford Haven. Since then, imports of LNG have accounted for between a quarter and a half of the UK's gas imports, and have at times rivalled and in one year (2011) exceeded imports from Norway.[42] Qatar presently accounts for the great majority of LNG imports. LNG imports are expected to increase as indigenous UK and Norwegian supplies decline. With four UK import terminals and a diversity of international supply, energy security could be positively impacted, although industrial action or a major accident at either of the Milford Haven terminals could still cause significant difficulties, at least in the short term.

---

uk-scotland-scotland-business-38687835 (accessed 5 May 2017). The UK Government's position is that the door has not closed on CCS technology.

[38] Norwegian Ministry of Petroleum and Energy, and DECC, *One North Sea: Joint Ministerial Statement on Climate Change and Energy Security*, Bergen, Norway, 28 May 2009. Available at https://www.gov.uk/government/news/one-north-sea-joint-ministerial-statement-on-climate-change-and-energy-security (accessed 29 August 2017).

[39] Tampen Link ties into the UK's FLAGS system.

[40] The entire gas transport system within, and from, the Norwegian sector of the Continental Shelf has been under the operatorship of Gassco since 2002. Established in mid-2001, Gassco is a wholly state-owned company. See www.gassco.no (accessed 5 May 2017).

[41] This is not beyond possibility either. Disruption could be caused by technical failure or industrial action. In terms of the former, see http://news.sky.com/story/uk-gas-hit-as-norway-pipeline-supply-cut-10447303 (accessed 5 May 2017). In terms of the latter, as the home to two brand-new LNG terminals, and as a supplier of 25 per cent of the UK's refined petroleum products it was widely believed that a strike at Milford Haven could disrupt the oil and gas supply to the UK. See J Guthrie, "Port strike risks oil and gas disruption", *Financial Times*, 16 February 2010. A similar action in either St Fergus or Easington could also have serious disruptive effects.

[42] For supporting statistics, see DECC, DUKES 2016, available at www.gov.uk/government/uploads/system/uploads/attachment_data/file/553601/DUKES_2016_INTERNET_FINAL.pdfp.171 (accessed 7 May 2017).

The UK has also been improving its energy efficiency. Its energy   I-3.14
intensity is approximately 43 per cent that of the 1970 level. In
1970, the UK was consuming 389.5 toe per £1 million of GDP.
That number was successfully reduced to 169.1 toe by 2008.[43] Final
consumption in 2015 was 145.7 toe.[44]

Crude oil exports increased in 2015 to reach nearly 30 million   I-3.15
tonnes, which is less than half the peak of 87 million tonnes exported
in 2000. The UK is heavily reliant on EU-based customers. Within the
EU, in 2015, approximately 50 per cent of the total crude oil exports
were destined for the Netherlands and Germany. The remaining
non-EU crude oil exports were predominantly exported to the USA.
However, since 2008, non-EU exports of crude oil have found other
markets beyond the US, which now represents a bigger market than
the US.[45] Similarly, in 2015, the UK's natural gas exports were almost
exclusively to the EU.[46] Again, the EU being the sole customer in
itself does not constitute an energy security concern. EU countries,
especially the Netherlands, Germany and the Republic of Ireland,
have been, and still are, reliable, dependable and stable customers.

THE INTERNATIONAL DIMENSION OF THE UK'S ENERGY SECURITY

International Energy Agency

The impact of international law on the UK's energy security was first   I-3.16
felt in the aftermath of the oil embargo imposed by the Arab members
of OPEC on states supporting Israel in the Fourth Arab-Israeli War
in 1973. The energy crisis provoked by this embargo prompted
collective action on the part of consumer countries. The industrialised nations in Europe, North America and the Far East were
already joined together under the umbrella of the Organisation for
Economic Co-operation and Development (OECD). Established in
December 1960,[47] the OECD provided a platform for the industrialised nations[48] to compare and co-ordinate policies predominantly

---

[43] DECC Statistics 2009, DUKES Table 1.1.4, "Primary energy consumption, gross domestic product and the energy ratio 1970 to 2008".
[44] BEIS Statistics 2016, DUKES Chart 1.5, "Final Consumption 2015".
[45] BEIS Statistics 2016, DUKES Chart 3.3, "Destination of UK crude oil exports 2000 to 2015".
[46] BEIS Statistics 2016, DUKES Chart 4.4, "Exports of Natural Gas 2000 to 2015".
[47] *Convention on the Organisation for Economic Co-operation and Development*, Paris, 14 December 1960.
[48] The initial signatories on 14 December 1960 were Austria, Belgium, Canada, Denmark, France, Germany, Greece, Iceland, Ireland, Italy, Luxembourg, the Netherlands, Norway, Portugal, Spain, Sweden, Switzerland, Turkey, the United Kingdom and the United States.

on economic and development issues.[49] It was, however, deemed to be "wholly inadequate for the management of the risks of more troublesome hardships in the future",[50] including oil supply disruptions. On the other hand, it proved to have an adequate structure for establishing a permanent international body within the Organisation that could manage risks sufficiently. On 18 November 1974, therefore, the member countries signed an international treaty, the Agreement on an International Energy Program (IEP Agreement), establishing the International Energy Agency (IEA).[51] The Agency was established under the OECD because it was "regarded as being fully in harmony with the OECD Convention and traditions":

> "There was an expectation that the Agency would also contribute a dynamic element to the Organisation and that the other bodies of the OECD would gain from close association with the work of the Agency, which in turn would benefit from its close association with the OECD. The establishment of the IEA would reduce neither the general co-operative efforts within the OECD nor the work of its various bodies on energy policies and related questions."[52]

I-3.17   The IEP Agreement obliged the members to hold emergency reserves equivalent to, first, 60 days of oil consumption levels,[53] then, from 1975 onwards, 90 days of consumption levels.[54] Despite this stock-holding requirement, however, during the 1973–1974 crisis the participating countries' stock levels were "standing only at about

---

Since then the following ten countries have joined the organisation: Australia, the Czech Republic, Finland, Hungary, Japan, Korea, Mexico, New Zealand, Poland, and Slovakia.

[49] OECD, *About OECD: Our Mission*, available at www.oecd.org/pages/0,34 17,en_36734052_36734103_1_1_1_1_1,00.html (accessed 5 May 2017).

[50] R Scott, *The History of the International Energy Agency, Volume 1: Origins and Structure* (1994) (hereinafter "Scott, *The History of the IEA, Vol. 1*"), pp 29–30, available at www.iea.org/media/about/1ieahistory.pdf (accessed 5 May 2017). For OECD's response to the 1973–1974 oil crisis see pp 33–38 of the same volume.

[51] *Ibid*, p 20. "Although the Program could have been adopted with binding effect in the Council Decision, the treaty form was thought to provide advantages flowing from parliamentary commitment, and from the treaty's formality, visibility and fully independent legal standing." *Ibid*, p 55. For the establishment of the IEA, see Agreement on an International Energy Program (as amended 9 May 2014), Art 1, read together with Chapter IX. Available online at www.iea.org/about/faqs/organisationandstructure (accessed 29 August 2017). Initially, only the 16 members of the OECD signed the IEP Agreement. These were Austria, Belgium, Canada, Denmark, Germany, Ireland, Italy, Japan, Luxembourg, the Netherlands, Spain, Sweden, Switzerland, Turkey, the UK and the US. Currently, all the OECD member states with the exception of Iceland and Mexico are members of the IEA.

[52] Scott, *The History of the IEA, Vol. 1*, p 52.

[53] Agreement on an International Energy Program (as amended 25 September 2008), Art 2(1).

[54] *Ibid*, Art 2(2).

a seventy day supply".[55] This crisis clearly demonstrated the inadequate preparation for a serious supply disruption. "Their failure to co-ordinate the use of their stocks during the course of the crisis was a further failure of organisation which weakened their response."[56] This organisational failure was probably why the IEP Agreement also established further response measures based on the co-operation of the participating countries.[57]

Under these additional response measures, the emergency "demand restraint" provisions[58] obliged the participating countries to have individual contingency plans to reduce consumption. When, as a group, the participating countries could "sustain" a reduction of certain levels "in the daily rate of its oil supplies", the IEP Agreement provided that they should each implement these contingency measures to reduce their "final consumption". If that sustainable reduction level in oil supplies reached 7 per cent of the "average daily rate of [the group's] final consumption", then each country had to reduce its own final consumption by 7 per cent.[59] However, if a 12 per cent reduction level could be sustainable, each participating country had to reduce its own final consumption by 10 per cent only. These limited levels of final consumption were referred to as "permissible consumption".[60]

Together with the emergency demand restraint measures, the IEP Agreement also provided allocation measures for available oil.[61] While these allocation measures used the same trigger levels as the demand restraint measures, ie 7 per cent and 12 per cent, the Agreement additionally provided that if the reduction in any of the participating country's final consumption levels were more than 7 per cent, then the available oil had to be allocated to that country.[62] A participating country consequently had an "allocation obligation" if, during an emergency, its total available oil reserves[63] were more than its supply right.[64] Similarly, a participating country had an

I-3.18

I-3.19

[55] Scott, *The History of the IEA, Vol. 1*, p 37.

[56] *Ibid*, p 37.

[57] Participating countries defined as "States to which [the IEP Agreement] applies provisionally and States for which the Agreement has entered into and remains in force". Agreement on an International Energy Program (as amended 9 May 2014), Art 1.

[58] *Ibid*, Art 5.

[59] *Ibid*, Art 13.

[60] *Ibid*, Art 7(4).

[61] *Ibid*, Chapter III.

[62] *Ibid*, Art 17(1), read together with Art 8.

[63] Defined as the sum of normal domestic production and actual net imports available.

[64] Agreement on an International Energy Program (as amended 9 May 2014), Art 7(3). Supply right is defined as the remaining level of permissible consumption after the emergency reserve drawdown obligation is subtracted. The emergency reserve drawdown is calculated by dividing the emergency reserve commitment (90-day consumption level)

"allocation right", if its total available oil reserves were less than its supply right.[65] In simple terms, therefore, the group's aggregate available oil supply was allocated among participating countries based on a specific calculation mechanism aiming for an equitable distribution.

I-3.20   These emergency response measures were tested during the 1979–1981 oil crisis that began with the political turmoil in Iran.[66] Even though this crisis "caused severe economic damage to the IEA countries",[67] "at no point had the supply loss approached 7 per cent for the IEA group as a whole";[68] thus, it failed to trigger the emergency response mechanism. The crisis demonstrated clearly that supply disruptions below the 7 per cent benchmark could also have serious consequences. The IEA, therefore, introduced in 1984[69] an additional set of measures called the Co-ordinated Emergency Response Measures (CERM) to complement the existing mechanism. While the allocation regime "was designed to ensure activation by virtue of an *administrative* rather than a prior *political* decision" the CERM required a "political decision".[70] The CERM were employed only once, during the 1990–1991 Gulf Crisis. As a response to invasion of Kuwait by Iraq, the United Nations embargoed all the oil exports from those two countries, resulting in 4.3 million barrels less oil available in the global market per day.[71] With the additional expected disruption to the oil supply due to a military campaign and the uncertainty surrounding the reaction of Arab oil producers to such a campaign, the IEA issued a formal notice on 17 January 1991 for the member states "to activate the Co-ordinated Energy Emergency Response Contingency Plan to make available to the market 2.5 million additional barrels of oil per day within 15 days' time".[72]

---

by the total emergency reserve level of the group and then multiplying it by the group supply shortfall.

[65] Agreement on an International Energy Program (as amended 9 May 2014), Art 7(2).

[66] The events surrounding the Iranian revolution and Iran-Iraq War are beyond the scope of this chapter. For further information on IEA's response to the oil crisis see R Scott, *The History of the International Energy Agency, Volume 2: Major Policies and Actions* (1994) (hereinafter "Scott, *The History of the IEA, Vol. 2*"), pp 114–132, available at www.iea.org/media/about/2ieahistory.pdf (accessed 5 May 2017).

[67] Scott, *The History of the IEA, Vol. 2*, p 114.

[68] *Ibid*, p 119. Although the 7 per cent limit was reached temporarily for some of the countries, it was never the case for the group as a whole.

[69] IEA, *Decision on Stocks and Supply Disruptions*, 11 July 1984, IEA/GB(84)27, Item 2(a)(ii), Annex 1 and Appendices.

[70] Scott, *The History of the IEA, Vol. 2*, p 128.

[71] *Ibid*, pp 133–134. Note that approximately "two-thirds of that loss directly affected IEA countries".

[72] *Ibid*, p 134. "Two million barrels were to come from participants' oil stocks, 400,000

Initially the CERM were being implemented for disruptions below I-3.21
the 7 per cent threshold. From 1995 onwards, however, the IEA
Governing Board decided that the CERM could be implemented
even for the supply disruptions above that benchmark.[73] Today, the
CERM are considered together with other emergency measures.[74]

Energy Charter Treaty

The emergency response measures were not the only method I-3.22
developed by the industrialised consumer nations to cope with
energy supply problems. Had it been possible during the oil crises of
1973–1974 and 1979–1981, diversification of the suppliers would
also have reduced the impact of supply disruptions. Unfortunately,
the opportunity for the OECD to diversify did not come until much
later – with the collapse of the Soviet Union in 1990. In June of that
year, at the meeting of the European Council in Dublin, the Prime
Minister of the Netherlands suggested that "cooperation in the
energy sector with the eastern European and former Soviet Union
countries" would be mutually beneficial: these countries would
benefit from faster economic growth and recovery while the EU
would benefit from the strengthening of its energy supply security.[75]
After a series of negotiations, in December 1994, the Energy Charter
Treaty (ECT) and the Energy Charter Protocol on Energy Efficiency
and Related Environmental Aspects were signed.[76] The Treaty was
"designed to promote East-West industrial co-operation by providing
legal safeguards in areas such as investment, transit and trade".[77]
The purpose of the Treaty was to establish a "legal framework in
order to promote a long-term cooperation in the energy field, based
on complementarities and mutual benefits, in accordance with the

---

barrels from demand restraint measures designed to reduce oil consumption, and 100,000
barrels from fuel switching out of oil and the use of spare capacity."

[73] See www.iea.org/topics/energysecurity/subtopics/energy_security_emergency_response0
(accessed 5 May 2017).

[74] Including drawdown of oil stocks, demand restraint, fuel-switching, surge oil production
and the allocation of available supplies.

[75] Europa, *Summaries of EU legislation: European Energy Charter*, under Background,
30 January 2007, available at http://europa.eu/legislation_summaries/energy/external_
dimension_enlargement/l27028_en.htm (accessed 5 May 2017).

[76] Energy Charter Treaty (Annex 1 to the Final Act of the European Energy Charter
Conference), Currently, there are 53 members to the treaty. There are 51 countries that
have either signed or acceded the Treaty along with the European Community and
Euratom.

[77] The Energy Charter Treaty and Related Documents: A Legal Framework for
International Energy Cooperation, *Final Act of the European Energy Charter Conference*,
II- Background.

objectives and principles of the Charter".[78] The ECT was unique in that it provided a multilateral legal framework for investment, trade and other related issues for a single sector – energy. Although "many of the Treaty's rights and obligations [were] of a "hard law' nature, enforceable in legally binding arbitration or through GATT-type dispute resolution process",[79] there were also some soft law provisions that were not legally binding but rather left to the individual member's political will.[80]

I-3.23　　As the largest hydrocarbon producer in the region, the attitude of Russia to the ECT has been crucial to its success or failure. While the country applied the Treaty provisionally, longstanding disagreements with the EU over a number of issues including transit meant that it failed to ratify it. Finally, on 20 August 2009, Russia announced that it was not going to become a party to the Treaty. From 19 October onwards, Russia ceased to apply the Treaty provisionally.[81] However its legacy with the ECT continues with the Yukos arbitration. A full discussion of the arbitration is outwith the scope of this chapter. Suffice to say, there has been a protracted dispute as to the provisional application of the dispute resolution provisions in the ECT. The facts concern the 2003 expropriation of assets belonging to the Yukos oil company. Investors initiated arbitration proceedings, worth $50 billion, against Russia under the ECT. The investors initially succeeded in the arbitration, receiving $50 billion, the largest arbitral award in history.[82] However, the award has now been set aside by the domestic Dutch courts (the Netherlands having been the seat of arbitration) on the basis that as Russia had only ever applied the Treaty provisionally, it was only bound by provisions compatible with Russian law; and as the dispute concerned relations of a public-law nature, Russian law did not permit international arbitration.[83] Enforcement proceedings based on the original award are still live in numerous jurisdictions, but standing the decision of the Dutch court,

[78] Energy Charter Treaty (Annex 1 to the Final Act of the European Energy Charter Conference), Art 2.

[79] C Bamberger, the Legal Counsel of the IEA, *The Energy Charter Treaty: a Description of its Provisions* (1995), p 8.

[80] For example, Energy Charter Treaty, Art 19: Environmental Aspects. Art 19(2): "At the request of one or more Contracting Parties, disputes concerning the application or interpretation of provisions of this Article shall, to the extent that arrangements for the consideration of such disputes do not exist in other appropriate international fora, be reviewed by the Charter Conference aiming at a solution."

[81] Energy Charter Secretariat, *Russia*, available at www.energycharter.org/who-we-are/members-observers/countries/russian-federation (accessed 5 May 2017).

[82] See eg C Gibson, "*Yukos Universal Limited (Isle of Man)* v *The Russian Federation: A Classic Case of Indirect Expropriation*", 30 (2015) ICSID Review, at 303.

[83] An unofficial English translation of the case is available at https://uitspraken.rechtspraak.nl/inziendocument?id=ECLI:NL:RBDHA:2016:4230 (accessed 8 May 2017).

it must be doubtful if the investors will receive any compensation. More generally, the decision calls into serious question the practical usefulness of arbitrating against Russia relative to a breach of the ECT during the period of provisional application.

Given that very little Russian gas is consumed in the UK, this I-3.24 did not pose a direct energy security problem for the UK. However, the UK's "interconnection with the continent and the geopolitical impacts of [potential] energy supply disruptions" meant that "this [was nevertheless] a UK concern too".[84] The impact of Russia's decision on the viability of the ECT in its current form as the most ambitious international legal instrument concerned with energy security is very much an open question.

Norway, the UK's biggest oil and natural gas supplier, has not I-3.25 ratified the Treaty, though it applies it provisionally. However, because of the bilateral agreements and the traditionally strong ties enjoyed between the two countries it is, again, of little concern for the UK.

THE EU DIMENSION OF THE UK'S ENERGY SECURITY

While the international community at large had not taken collective I-3.26 action in respect of energy security until 1973, the European Economic Community (EEC), as it then was, had become concerned about this issue some years earlier following the Arab oil embargo against states supporting Israel at the time of the Third Arab-Israeli War in 1967.[85] Directive 68/414/EEC[86] underlined the significance of imported crude oil and petroleum products for Europe and how "any difficulty, even temporary, having the effect of reducing supplies of such products imported from third States could cause serious disturbances in the economic activity of the Community".[87] It therefore obliged member states to hold stocks of "at least 65 days' average daily internal consumption".[88] This obligation was

---

[84] Wicks, *Energy Security*, para 1.9. In the event that Russia completely cuts off gas supplies to Europe, the natural gas supplies to UK may be indirectly disrupted as the demand in the EU for the gas surges.

[85] The events surrounding the Six Day War and the oil embargo are beyond the scope of this work. For an excellent and concise account, see D Yergin, *The Prize: The Epic Quest For Oil, Money & Power* (1991) (hereinafter "Yergin, *Prize*"), Chapter 27.

[86] Council Directive 68/414/EEC of 20 December 1968 imposing an obligation on member states of the EEC to maintain minimum stocks of crude oil and/or petroleum products, OJ L 308, 23 December 1968, 14–16.

[87] *Ibid.*

[88] *Ibid*, Art 1. For an excellent and concise account of stock holding obligations under the EU and the International Energy Agency see S S Haghighi, "Obligation to Hold Stocks of Crude Oil and/or Petroleum Products: The European Community and the International

reduced by a maximum of 15 per cent for the states that had indigenous production.[89] One Government could also individually agree – save for the Commission commenting on the drafts of those bilateral agreements prior to an agreement being reached – to hold stocks within its territory for another, as long as the former would not restrict access to the owner of the stocks.[90] Unless there was a "particular urgency" or a requirement to "meet minor local needs", the only time a member state could draw from the stocks was when "difficulties" arose, and only after a consultation with the member states.[91] However, neither "urgency", "local needs" nor "difficulties" were defined.

I-3.27   The stock-holding requirement was increased to 90 days with a Council Directive in 1972, "owing to changes in the pattern of oil supplies in Western Europe during recent years".[92]

I-3.28   The UK acceded to the EEC on 1 January 1973. On 24 July 1973, the Council issued a new Directive that introduced another set of mitigation strategies against supply disruptions.[93] In addition to the stock-holding requirements, it provided for restricting energy consumption. It obliged member states to set up an authority within their jurisdictions that would not only act on these reserves in the event of a disruption, but also "impose specific or broad restrictions on consumption ... give priority to supplies of petroleum products to certain groups of users" and "regulate prices in order to prevent abnormal price rises".[94] Additionally, the Directive provided for a group of delegates that would be working under the Commission to be consulted "in order to ensure coordination" of these measures.[95] These were to be in force, by way of law, regulation or administrative action, by 30 June 1974.[96] This deadline proved to be too long as the then ongoing crisis in the Middle East developed in October 1973 into the second Arab oil embargo.[97]

Energy Agency Compared", OGEL 5(4) (hereinafter "Haghighi, 'Obligation to Hold Stocks'"), available at www.ogel.org/article.asp?key=2656 (accessed 5 May 2017).

[89] Directive 68/414/EEC, Art 1.

[90] *Ibid*, Art 6.

[91] *Ibid*, Art 7.

[92] Council Directive 72/425/EEC of 19 December 1972 amending the Council Directive of 20 December 1968 imposing an obligation on member states of the EEC to maintain minimum stocks of crude oil and/or petroleum products, OJ L 291, 28 December 1972, at 154.

[93] Council Directive 73/238/EEC of 24 July 1973 on measures to mitigate the effects of difficulties in the supply of crude oil and petroleum products, OJ L 228, 16 August 1973, at 1–2.

[94] *Ibid*, Art 1.

[95] *Ibid*, Art 3.

[96] *Ibid*, Art 5.

[97] It is likely the drafters of the Directive were well aware of the developments in the

A Council Decision in 1977 added a target-based approach to the consumption reduction requirement.[98] Under this approach, the Commission, after consulting the group of delegates, could impose a reduction of up to 10 per cent in the entire energy consumption of the Community for up to two months.[99] This approach was deemed to be necessary "in order to safeguard the unity of the [common] market and to ensure that all users of energy within the Community [bore] a fair share of the difficulties arising from [a supply disruption]".[100]

I-3.29

By the late 1990s, the EU was actively pursuing the goal of sourcing at least some of its energy from renewables not only to respond to climate change concerns but also as a means of reducing import dependency, thus contributing to the security of supply. In its White Paper[101] in 1997, the EU identified the need for a "clear and comprehensive strategy [for renewables] accompanied by legislative measures",[102] and set an "indicative objective" that by 2010 12 per cent of the EU's electricity consumption had to be from renewables.[103] These recommendations were duly accepted, and were endorsed in Directive 2001/77/EC.[104] Directive 2001/77/EC obliged member states to establish ten-year "indicative targets" which were to be set every five years. These targets had to be "consistent with the global indicative target of 12% of gross national energy consumption by 2010 and in particular with the 22.1% indicative share of electricity produced from renewable energy sources in total Community electricity consumption by 2010".[105] The Renewable Energy Directive (2009/28/EC), which replaced Directive 2001/77/EC, imposes legally binding targets on member states to increase

I-3.30

Middle East at the time. For an excellent and concise account of the events leading up to the oil embargo of 1973 and the aftermath, see Yergin, *Prize*, Chapters 29 and 30. However, this is beyond the scope of this chapter.

[98] 77/706/EEC: Council Decision of 7 November 1977 on the setting of a Community target for a reduction in the consumption of primary sources of energy in the event of difficulties in the supply of crude oil and petroleum products, OJ L 292, 16 November 1977, at 9–10.

[99] *Ibid*, Art 1.

[100] *Ibid*, Preamble.

[101] Communication from the European Commission, *Energy for the Future: Renewable Sources of Energy*, White Paper for a Community Strategy and Action Plan, COM(97)599 final, 26 November 1997.

[102] *Ibid*, at 6.

[103] *Ibid*, s 1.3.1.

[104] Directive 2001/77/EC of the European Parliament and of the Council of 27 September 2001 on the promotion of electricity produced from renewable energy sources in the internal electricity market, OJ L 283, 27 October 2001, at 33–40.

[105] *Ibid*, Art 3. Note that the directive also obliges member states to provide an "analysis of success in meeting the national indicative targets" every two years.

consumption of energy from renewable sources.[106] Collectively, the targets amount to a 20 per cent share of EU energy consumption from renewable sources by the year 2020. The Directive forms one part of the EU's 2020 targets, the other objectives being a reduction of EU greenhouse gas (GHG) emissions by 20 per cent on a 1990 baseline and a 20 per cent improvement on energy efficiency. The targets for individual member states are set out in Annex I of the Directive and reflect national circumstances (the ability of each state to exploit renewable energy) and progress made under the preceding 2001 Directive. The Directive also sets out indicative trajectories which are not binding, however member states are legally obliged to implement measures calculated to follow their trajectory.[107] The UK has a binding target under the Directive to supply 15 per cent of final energy demand from renewable sources by 2020. Its plan to achieve this is to source 30 per cent of electricity, 12 per cent of heat and 10 per cent of transport fuel consumption from renewable sources by 2020.[108] It remains extremely doubtful whether it will reach its targets. While the UK is firmly on track to achieve its electricity sub-target, meeting the heat and transport targets looks very unlikely: "On its current course, the UK will fail to achieve its 2020 renewable energy targets. This would be an unacceptable outcome given the UK's reputation for climate-change leadership."[109]

I-3.31   In 1998, the initial European oil stockholding legislation, Directive 68/414/EEC, was amended for the second time.[110] While reiterating the minimum 90-day consumption level requirement for the stock-holding obligation, the new Directive increased the percentage of reduction in this obligation for the indigenous producers to 25 per cent.[111] It also defined "difficulty" to include significant price hikes.[112] It introduced a more structured approach to the statistical calculation of the stocks[113] and to stockholding and maintenance.[114] Directive 98/93/EC also provided that the member states would

---

[106] Directive (2009/28), Art 3(1).

[107] *Ibid*, Art 3(2).

[108] DECC, *National Renewable Energy Action Plan*, July 2010.

[109] House of Commons Energy and Climate Change Committee, *2020 Renewable Heat and Transport Targets*, Second Report of Session 2016–17, 9, available at www.publications.parliament.uk/pa/cm201617/cmselect/cmenergy/173/173.pdf (accessed 7 May 2017).

[110] Council Directive 98/93/EC of 14 December 1998 amending Directive 68/414/EEC imposing an obligation on member states of the EEC to maintain minimum stocks of crude oil and/or petroleum products, OJ L 358, 31 December 1998, at 100–104.

[111] *Ibid*, Art 1(1).

[112] *Ibid*, Preamble, para 2.

[113] *Ibid*, Arts 1(6) and (7).

[114] *Ibid*, Art 1(4).

have to "ensure that fair and non-discriminatory conditions apply in [the] stockholding arrangements"; and when member states hold stocks together they would be "jointly responsible for the obligations deriving from this Directive".[115] Consequently, in this last regard, where two or more member states decided to hold the stocks jointly, the Directive provided for the option "to have recourse to a joint stockholding body or entity".[116] This possibility of recourse was available "in order to organise the maintenance of stocks"[117] which would ensure both that the stocks would be available and that the consumers could access them.[118] It was envisaged that, within these organisational arrangements, the balance of stocks would be "maintained by refiners and other market operators"[119] and, therefore, "partnership between the government and the industry [would be] essential to operate an efficient and reliable stock-holding mechanism."[120] The final version of this Directive was codified in 2006 as Council Directive 2006/67/EC. This has since been replaced with Council Directive 2009/119/EC (2009 Directive). The 2009 Directive provides that:

(a) EU countries must maintain emergency stocks of crude oil and/or petroleum products equal to at least 90 days of net imports or 61 days of consumption, whichever is higher.[121]

(b) Stocks must be readily available so that in the event of a crisis they can be allocated quickly.[122]

(c) States must send the European Commission a statistical summary of their stocks at the end of each month.[123]

(d) In the event of a supply crisis, the Commission will be responsible for organising a consultation between EU countries.[124]

Currently, the UK is obliged to hold 61 days of average daily inland consumption. However, as the level of production decreases over time, and the UK becomes increasingly reliant on net imports, the target will shift to a requirement to hold emergency stocks of crude oil amounting to 90 days of net imports. As a member of the EU and a founding member of the IEA, the UK has obligations under both the 2009 Directive and the IEP Agreement. The

I-3.32

---

[115] *Ibid*, Art 1(4)(3).
[116] *Ibid*, Art 1(4)(3).
[117] *Ibid*, Preamble, para 11.
[118] *Ibid*, Preamble, para 9.
[119] *Ibid*, Preamble, para 11.
[120] *Ibid*, Preamble, para 11.
[121] Directive 2009/119/EC, Art 3(1).
[122] *Ibid*, Art 5.
[123] *Ibid*, Art 13.
[124] *Ibid*, Art 19.

second edition of this book noted some of the differences between Council Directive 2006/67/EC and the IEP Agreement relating to the type and calculation of stocks under the two documents.[125] However, with the coming into force of the 2009 Directive, and its objective to harmonise the obligations imposed on member states who are also participants to the IEP, problems of reconciliation are less apparent. The 2009 Directive adopts the same method-ology as that which the IEA uses for calculating the minimum stockholding obligation (and compliance thereof).[126] However, the 2009 Directive adopts a narrower definition of "stocks", restricting compliant oil to "emergency stocks" held for meeting the Directive obligation.

I-3.33   According to the latest BEIS DUKES Statistics, analysing 2015 data, the UK's obligated stocks are, on average, at around 60 days of average daily inland consumption (12.23 mtoe). This is marginally less than the 61-day requirement under the 2009 Directive (12.24 mtoe). Commercial oil stocks in the UK (stock held under normal commercial arrangements) are recorded at 3.5 mtoe. However, following the implementation of the 2009 Directive, these commercial oil stocks no longer count towards the UK's Compulsory Stockholding Obligation (CSO).[127]

I-3.34   The UK[128] implemented its CSO through the Energy Act 1976.[129] Under this Act, the Secretary of State may oblige any company that "produces, supplies or uses" crude oil or petroleum products to hold oil stocks,[130] and may "prescribe" the type and the method of calculation of the stocks.[131] Compliance with the 2009 Directive is secured through the Oil Stocking Order 2012[132] which provides the authority for the Secretary of State to implement its powers under the 1976 Act.[133]

I-3.35   Under Article 7 of the 2009 Directive, member states may set up a Central Stockholding Entity (CSE) to hold oil on the State's behalf. However, the UK has no CSE. Instead, under Article 5 of the 2009 Directive, and under the terms of the Oil Stocking Order

---

[125] For an analysis of similarities and differences between the two EU Directives (pre-2009) and IEP documents see Haghighi, "Obligation to Hold Stocks", 11–25.

[126] See IEA, *Energy Security Supply 2014 Part 1* and specifically *Part 4 Annex D*.

[127] BEIS Statistics 2016, DUKES Chart 3.3, "UK Oil Stocks, 2001–2015".

[128] For discussion, see G Gordon, A McHarg and J Paterson, "Energy Law in the United Kingdom", in M Roggenkamp, C Redgwell, A Rønne and I Del Guayo (eds), *Energy Law in Europe* (2016), pp 1053–1136.

[129] Energy Act 1976 (Chapter 76).

[130] *Ibid*, s 6.

[131] *Ibid*, s 6(6).

[132] SI2012/2862.

[133] *Ibid*, Art 3.

2012, obligated companies[134] are required to hold one-third[135] of their stocks, pursuant to the lists in the 2009 Directive.[136] While this is the model that currently operates in the UK, change looks likely. In 2013, the Government consulted on proposals to reform the UK's CSO. Specifically, views were invited on the establishment of an industry-owned and operated CSE. Respondents submitted that a CSE could bring several benefits:

> "A CSE could manage the CSO more strategically, allowing companies to more effectively plan meeting their obligations. It would help improve transparency of CSO costs. Also, a CSE could help incentivise the development of storage capacity in the UK, through enhanced co-operation and better access to finance. A CSE would also be able to better manage the impact on the CSO of any future changes to the market, helping to ensure that obligations continue to be met effectively and efficiently. Almost all respondents argued that for such benefits to be realised membership of the CSE would have to be mandatory for all obligated companies."[137]

In its response, the Government stated that it agreed that an industry- I-3.36
owned and operated CSE should be established in the UK. However, before the Government could agree to legislate for this, obligated parties were required to work together to prepare a roadmap for the CSE. If the UK Government was convinced it presented a robust approach, then "it [would] seek to take forward the necessary legislation as soon as parliamentary time allows".[138]

THE EMISSIONS DIMENSION OF UK ENERGY SECURITY

While EU law in relation to the holding of oil stocks has had a I-3.37
significant positive impact on the energy security of member states, it is arguably the case that a more important negative effect has been caused by EU law directed at climate change. On 23 October 2001 the European Parliament and the Council of the European Union adopted the Directive on the limitation of emissions of

---

[134] "Obligated companies" are those companies that supply at least 50,000 tonnes of oil to the UK market.

[135] Refiners are required to hold 67.5 days of stock whereas non-refiners (including importers and significant traders) are required to hold 58 days of stock (primarily due to different operational models). See www.gov.uk/government/news/uk-s-oil-stocking-system-to-be-reviewed (accessed 5 May 2017).

[136] 2009 Directive, Art 9(2).

[137] DECC (now BEIS), Future Management of the Compulsory Stocking Obligation in the UK: Government Response to Consultation, at 1.8, available at www.gov.uk/government/uploads/system/uploads/attachment_data/file/301957/Governement_Response_to_CSO_consultation_FINAL_08042014.pdf (accessed 6 May 2017).

[138] *Ibid*, at 1.17.

certain pollutants into the air from large combustion plants (Large Combustion Plants Directive – LCPD).[139] The aim was to tackle the acidification of air by reducing the emissions of acidifying pollutants, particularly sulphur dioxide (SO_2) and nitrogen oxides (NO_x).[140] In the UK, this Directive was implemented in late 2007 as the Large Combustion Plants Regulations.[141] It introduced annual emissions allowances and the procedures for the transfer of these allowances between the participating large combustion plants (LCPs).[142] Ninety-four existing[143] and operational LCPs opted for the National Emissions Reduction Plan.[144]

I-3.38   The concerns over energy security relevant to the implementation of the LCPD and the associated plant closures were raised in the 2009 Economic Report of Oil & Gas UK:

"As far as security of supply is concerned, the biggest concern is power generation. There are so many coal (plus the few remaining oil) fired and nuclear power stations to be closed in the next 15 years, on account of both age and more stringent emissions' limits $(NO_x$ and $SO_x)$ [due to the LCPD], that it has been estimated that replacement of this generating capacity with a mixture of coal and gas fired nuclear power stations will require some £30 billion of investment by mid-2020s. Taking account of the target for renewable energy supplies in 2020, however, will raise this already very large sum to of the order of £100 billion."[145]

I-3.39   In 2003, the Director of the Institution of Civil Engineers revealed that the energy deficiency created by these closures will not be mitigated by renewable energy; instead "the outstanding balance will have to be replaced by gas-fired power stations, importing 90 per cent of their fuel".[146] In 2009, the Wicks Report reiterated this, stating that as "the cheapest and quickest technology", gas-fired

---

[139] Directive 2001/80/EC of the European Parliament and of the Council of 23 October 2001 on the limitation of emissions of certain pollutants into the air from large combustion plants, OJ L 309, 27 November 2001.

[140] *Ibid*, p 1.

[141] The Large Combustion Plants (National Emission Reduction Plan) Regulations 2007, SI 2007/2325.

[142] *Ibid*, Art 8.

[143] Defined as "any combustion plant for which the original construction licence or, in the absence of such a procedure, the original operating licence was granted before 1 July 1987", Directive 2001/80/EC, Art 2(10).

[144] The Environment Agency, *National Emissions Reduction Plan (NERP)*, 02 October 2009, available at www.environment-agency.gov.uk/business/topics/pollution/32230.aspx (accessed 5 May 2017). NERP has now been replaced with the Transitional National Plan.

[145] Oil & Gas UK, *2009 Economic Report*, July 2009, Chapter 4, p 15.

[146] Institution of Civil Engineers, Press Release, "Short-sighted energy planning threatens bleak future", 1 July 2003, available at www.ice.org.uk/knowledge/newsdetail_ice.asp?PressID=238&NewsType=Press&FacultyID=3 (accessed 7 May 2017).

power stations would replace these closing coal- and nuclear-powered plants, which would mean the UK "would potentially be locking in import-dependence at an uncomfortable level".[147] Over the period 2012–2017, the number of coal- fired stations has halved and coal's share of UK electricity generation has fallen by 75 per cent.

The LCPD was replaced with the Industrial Emissions Directive **I-3.40** (2010/75/EU)[148] (IED) from 1 January 2016 and lowered emission limits. From 1 January 2016, three compliance options were open to plants:

- Meet the emissions limits from 1 January 2016;
- Participate in the Transitional National Plan (TNP), whereby non-compliant plants are restricted to 1,500 hours of operating per year, and must achieve compliance with the emissions limits by July 2020;[149] or
- Utilise the Limited Lifetime Derogation (LLD), under which plants are limited to 17,500 hours of operation per year until December 2023, after which they must close.[150]

At the time of writing, there are eight coal-fired power stations left **I-3.41** in the United Kingdom, representing around 9 per cent of the UK's total generating capacity.[151] Only one of these stations (Ratcliffe) has installed the equipment necessary to comply with the IED's emissions limits. The station at Eggborough has made use of the LLD and will close by 2023. The remaining six stations are participating in the TNP and therefore have until 2020 to comply with the IED's emission limits. It is against this backdrop that the UK Government has consulted on the future of coal power in the UK.[152] Views were invited on proposals to put the closure of unabated coal into effect, including extension of the emission requirements already prescribed to newly constructed plants to existing stations or, alternatively, "constraining the operation of coal fired power stations in the

---

[147] Wicks, *Energy Security*, para 6.32.

[148] Implemented in the UK via Environmental Permitting (England and Wales) (Amendment) Regulations 2013, Pollution Prevention Control (Scotland) Regulations 2012 and the Pollution Prevention and Control (Industrial Emissions) Regulations (Northern Ireland) 2013.

[149] IED, Art 32.

[150] *Ibid*, Art 33.

[151] BEIS, March Energy Trends 2017, available at www.gov.uk/government/uploads/system/uploads/attachment_data/file/612492/Energy_Trends_March_2017.pdf (accessed 7 May 2017).

[152] BEIS, "Coal Generation in Britain: The pathway to a low carbon future", available at www.gov.uk/government/uploads/system/uploads/attachment_data/file/577080/With_SIG_Unabated_coal_closure_consultation_FINAL__v6.1_.pdf (accessed 5 May 2017).

years leading up to 2025 in order to smooth the rate of unabated coal power station closures".[153] The Government's stated priority is to replace unabated coal-fired power stations with less emission-intensive forms of generation (particularly gas and renewables) whilst maintaining energy security.[154]

I-3.42   Suffice to say, the future of coal power is uncertain beyond 2020. If energy security is to be maintained in a post-coal UK, gas-powered plants would appear to be best placed to fill the coal void. Attention will now turn towards ways in which the UK can promote energy security.

THE UK AS AN ENERGY PRODUCER

I-3.43   To reduce its gas import dependency, the UK needs to maximise its indigenous production. To this end, the Government provides incentives necessary to attract enough investment.

I-3.44   One of the main incentive mechanisms the Government employs is the fiscal arrangements.[155] The first of these is in the form of capital allowances for onshore oil and gas activities[156] aiming to incentivise shale gas production.[157] For those onshore sites[158] that received their development consent after 4 December 2013,[159] 75 per cent of the capital expenditures[160] incurred are removed from calculations of

---

[153] *Ibid*, at p 6.

[154] *Ibid*, and see the Impact Assessment accompanying the Consultation, available at www.gov.uk/government/uploads/system/uploads/attachment_data/file/567057/WITH_ SIG_Unabated_coal_closure_consultation_Impact_Assessment_FINAL.pdf (accessed 7 May 2017).

[155] See Chapter I-7.

[156] For a discussion on the tax treatment of onshore oil and gas activities see T Hunter, E Üşenmez and J Paterson, "Future Trends in Shale Law & Policy in the UK", in T Hunter (ed.), *Handbook on Shale Gas Law & Policy* (2016).

[157] HM Treasury, "Harnessing the potential of the UK's natural resources: a fiscal regime for shale gas", 19 July 2013; and HM Treasury, "A fiscal regime for shale gas: summary of responses", 10 December 2013; see also Chapter I-9 for a detailed discussion on the UK's shale commitment.

[158] As defined in Corporation Tax Act 2010, s 356BC as inserted by Finance Act 2014, s 70 and Sch 15.

[159] Corporation Tax Act 2010, s 356C as inserted by Finance Act 2014, s 70 and Sch 15.

[160] HMRC states that these capital expenditures are likely to include "seismic acquisition and interpretation, planning and permit costs (successful and unsuccessful applications), planning and permitting judicial review costs, drilling wells, site setup costs, hydrocarbon treatment facility, pipelines, compression facilities. Leasing costs will qualify depending on the nature of the lease (for example, whether it is a long funding lease). However, for the purposes of the allowance, leasing costs incurred in bringing into existence a capital asset (for example, rig leasing costs in order to drill a well) are likely to qualify. As a general rule decommissioning costs will not qualify as they are not incurred for the purposes of onshore oil-related activities in accordance with S356BA. However costs of midlife

Supplementary Charge[161] liability.[162] This allowance is available until production from the site either reaches or is expected to reach 7 million tonnes.[163]

The second set of fiscal incentives lies in the Treasury's response **I-3.45** to the Wood Review.[164] From December 2014, following its own review of the tax regime applicable to oil and gas business in the UK, the Treasury implemented a series of tax reforms that not only simplified the fiscal mechanisms but also lightened the tax burden on investors.[165] The marginal tax rate is now 40 per cent following these reforms, down from 81 per cent for the fields liable for Petroleum Revenue Tax (PRT) and 62 per cent for all other fields.

The UK is also making concerted efforts to increase its domestic **I-3.46** supply through regulatory design and other Government initiatives. The most obvious and striking example of this lies in the Government's implementation of the Wood Review. As we will see in greater detail in Chapter I-5, a new arms-length regulator, the Oil and Gas Authority (OGA), has been created. Better resourced than its forebears, it has a greater range of powers and has been specifically charged with becoming a more hands-on regulator, involving itself in commercial discussion-making, and encouraging (and perhaps compelling) the development of infrastructure hubs and regional development strategies. At the root of all this activity lies the concept of maximising economic recovery in the UK – a long-standing policy objective that has now, as we shall see, been elevated to a legal obligation imposed upon the regulator itself as well as all relevant players in the UKCS. This has all been done with the express aim of getting the most producible hydrocarbons out of the UKCS. Wood is the latest and most striking of the UK's attempts to secure that aim. Previously, the Government has focused, for instance, on ways to encourage exploration and production activities in those areas lying to the West of the Shetlands and Hebrides.

---

plugging and abandonment (for example, of a well sidetrack) may qualify it is in the course of say further extraction activities." HMRC, "Corporation tax ring fence: onshore allowance – generation of the onshore allowance", *Oil Taxation Manual* – OT21515.

[161] Supplementary Charge is one of the three fiscal pillars that apply to upstream oil and gas activities in the UK. For a detailed discussion on the current regime see Chapter I-7.

[162] Corporation Tax Act 2010, s 356C as inserted by Finance Act 2014, s 70 and Sch 15.

[163] Corporation Tax Act 2010, s 356CA as inserted by Finance Act 2014, s 70 and Sch 15; "For the purposes of this section 1,100 cubic metres of gas at a temperature of 15 degrees Celsius and pressure of one atmosphere is to be counted as equivalent to one tonne". Corporation Tax Act 2010, s 356CA(3) as inserted by Finance Act 2014, s 70 and Sch 15.

[164] Non-fiscal aspects of the Wood Review's recommendations are discussed immediately below. For a fuller account of the Wood Review and its implementation, see Chapter I-5.

[165] See Chapter I-7 for a discussion on the current regime following the fiscal reforms. For a detailed analysis and an in-depth discussion of the reforms see E Üşenmez, *Redesigning Petroleum Taxation: Aligning Government and Investors in the UK* (forthcoming).

Potentially 3 to 4 billion boe, and "some 10 to 15% of remaining UK gas reserves" are estimated to be located there.[166] In recognition of this potential, in 2004 a new set of licences, Frontier Licences, were created specifically for these areas with a set of incentives including a 90 per cent reduction in area rental charges.[167] However, the lack of infrastructure[168] has been the main deterrent for large investment, and almost all of "the discoveries made to date [have not been] of a scale to justify the necessary infrastructure on their own".[169] In order to assess and suggest solutions to this problem the Government established the West of Shetland Task Force in 2006 as a joint Government-industry initiative.[170] The Task Force has been aiming to "find a technical and economic solution which will allow for infrastructure (pipelines) to be put in place that could allow the development and exploration of this area".[171] This was a very encouraging development and it was argued that the Government should continue to strengthen its dialogue with the industry and further support the work of the Task Force. Development of this area is deemed vital in realising the full potential of the UKCS. The Government has been called to "stand ready to play a catalytic role in bringing together the interests necessary for its further development, including an adequate gas transportation capacity".[172]

I-3.47   These calls were duly heard. HM Treasury's extension of the field allowance to the gas developments in this region was found to be "most encouraging".[173] Then, in 2010, Total E&P UK Limited announced the go-ahead for the development of its West of Shetland fields, Laggan and Tormore.[174]

---

[166] Energy and Climate Change Committee, "UK offshore oil and gas", First Report of Session 2008–2009, Volume II: Oral and Written Evidence, 17 June 2009, Ev 69, para 25.

[167] Initially the model clauses for the Frontier Licences were in Petroleum Licensing (Exploration and Production) (Seaward and Landward Areas) Regulations 2004, Sch 2. However, for the licences issued after 6 April 2008, these model clauses are no longer applicable. Instead the new model clauses are to be found in the Petroleum Licensing (Production) (Seaward Areas) Regulations 2008 (SI 2008/225). For further discussion on Frontier Licences and licences in general, see Chapter I-4.

[168] The area is around 400km from the nearest gas terminal.

[169] Energy and Climate Change Committee, "UK offshore oil and gas", First Report of Session 2008–2009, Volume I: Report, together with formal minutes, 17 June 2009, para 83.

[170] Ibid, para 84.

[171] Department of Energy and Climate Change, *Oil and Gas: West of Shetlands Task Force*, available at www.og.decc.gov.uk/UKpromote/wos_task.htm (accessed 7 May 2017).

[172] Wicks, *Energy Security*, Recommendations: point 2, p 112.

[173] D Odling, "Storage is not the whole answer", *Wireline*, 12 (March 2010), 5.

[174] Total News Release, "United Kingdom: Total launches the development of the Laggan and Tormore gas fields", 17 March 2010, available at www.total.com/en/media/news/press-releases/royaume-uni-total-lance-le-developpement-des-champs-gaziers-de-laggan-et-tormore (accessed 7 May 2017).

On 19 March 2010, when the DECC gave its consent for the I-3.48
development of this project, Lord Hunt, the Energy Minister,
expressed the importance of this infrastructure "for the wider development of the West of Shetland area" and for maintaining "secure
energy supplies", while Lord Mandelson pointed out the impact of
the recent fiscal incentives: "The recent initiative by the Treasury in
extending Field Allowance to such fields [ie Laggan and Tormore]
has been particularly important."[175] The industry representative, Oil
& Gas UK, shared the sentiment:

> "[This project] will stimulate exploration as it will enhance the
> viability of future discoveries in this frontier area. This can only lead
> to greater recovery of the UK's oil and gas resource. This move further
> underlines the importance of domestic oil and gas in [the UK's] energy
> supply. It's a positive example of what can be achieved by industry
> and Government working together to deliver the maximum recovery
> of the nation's oil and gas reserves."[176]

As part of the £3.5 billion project, Total constructed a new 140km I-3.49
gas pipeline to the Shetland Islands and a gas processing plant at
Sullom Voe on Shetland. Production from the fields began in 2016,
and will produce 90,000 boe/d.[177]

One other area in which the oil and gas companies, particularly I-3.50
the smaller independents, are having difficulty is access to infrastructure.[178] For a field that is economical enough to produce but
not large enough to justify its own infrastructure, access to existing
infrastructure is fundamental. The Oil and Gas Independents'
Association (OGIA) argues that the infrastructure owners can
have "disproportionate" demands "by creating delay or offering
inappropriate tariffs and liabilities in relation to the risks they
take"; and that the existing remedial arrangements are failing
to make "any significant differences and many bad behaviours
and practices still remain" in the North Sea.[179] As we shall see in

---

[175] DECC, Press Release, "Go-ahead given for gas development West of Shetland", 19
March 2010, available at http://webarchive.nationalarchives.gov.uk/20100430155414
(accessed 7 May 2017).

[176] Oil & Gas UK Press Release, "Government Approval for Gas Development West of
Shetland will Help Secure UK Energy Supply, says Oil & Gas UK", 19 March 2010,
available at http://oilandgasuk.co.uk/government-approval-for-gas-development-west-of-
shetland-will-help-secure-uk-energy-supply-says-oil-gas-uk-2 (accessed 7 May 2017).

[177] See www.total.uk/en/home/media/list-news/total-starts-production-laggan-tormore-west-
shetland (accessed 7 May 2017).

[178] Discussion on access to infrastructure is beyond the scope of this chapter. See Chapter I-6.

[179] Energy and Climate Change Committee, "UK offshore oil and gas", First Report of
Session 2008–2009, Volume II: Oral and Written Evidence, 17 June 2009, Ev 102, para
1.3.2. These difficulties faced by the small independents and potential remedies in place
are discussed in Chapter I-6.

Chapter I-6, despite ongoing regulatory reform, some difficulties remain.

I-3.51   In the 2012 "Energy Security Strategy", the UK Government stressed that it was "firmly committed to its efforts to ensure that the conditions are right for investment in new nuclear power in the UK".[180] These policy goals were reiterated in subsequent publications including "Long Term Nuclear Energy Strategy"[181] and "The UK's Nuclear Future".[182] In the latter, an ambitious vision of achieving 40–50 per cent of electricity generation from nuclear by 2050 was set out.[183] Most recently, Andrea Leadsom, then Minister for the DECC, stressed the UK Government's belief in the importance of nuclear for energy security. In her speech at the 8th Nuclear New Build Forum, she set out the Government's intention to support the nuclear supply chain, develop the nuclear skills base and promote investment in nuclear research and development.[184] In this regard, the approval in 2016 of the Hinkley Point C nuclear power station in Somerset should signal greater investment in new nuclear projects, which, despite its critics, will help to diversify the supply of energy in the UK.[185]

I-3.52   Another area with significant potential is the development of onshore oil and gas. Although the great majority of the UK's oil and gas production has historically come from offshore conventional fields, the Government and commercial interests in the UK have looked upon the US's shale gas revolution with considerable interest. Geological studies have indicated that several onshore areas of the UK may have potential for shale gas development,[186] and some test drillings have been undertaken in an attempt to better understand the extent of that potential. Government has seen in shale a potential means of improving the UK's domestic supply and energy security

---

[180] DECC (now BEIS), Energy Security Strategy (November 2012), p 23.

[181] UK Government, "Long Term Nuclear Energy Strategy", available at www.gov.uk/government/uploads/system/uploads/attachment_data/file/168047/bis-13-630-long-term-nuclear-energy-strategy.pdf (accessed 8 May 2017).

[182] UK Government, "The UK's Nuclear Future", available at www.gov.uk/government/uploads/system/uploads/attachment_data/file/168048/bis-13-627-nuclear-industrial-strategy-the-uks-nuclear-future.pdf (accessed 8 May 2017).

[183] *Ibid*, p 10.

[184] A Leadsom, "Realising the vision for a new fleet of nuclear power stations", DECC, 20 April 2016.

[185] See www.bbc.co.uk/news/business-37369786 (accessed 8 May 2017).

[186] See eg the British Geological Survey's Bowland Shale Gas Study, available at www.gov.uk/government/publications/bowland-shale-gas-study; the the British Geological Survey's Midland Valley of Scotland Shale Gas Study, available at www.bgs.ac.uk/research/energy/shaleGas/midlandValley.html (both accessed 7 May 2017).

situation.[187] In a Treasury Consultation in 2013, the benefits were put thus:

> "Shale gas ... could greatly increase our energy security – making us less dependent on imported gas supplies and reducing our exposure to geopolitical risks and lengthening supply chains. [188] UK net gas imports are set to rise from 45 per cent of demand in 2011 to 76 per cent by 2030, according to the latest central projections from the Department of Energy and Climate Change. The cost of these imports is expected to increase from £5.9 billion to around £16 billion (in 2012 prices) over the same timeframe. It is therefore important that the government provides the right conditions to allow companies to explore fully the potential of domestic gas resources."[189]

As we shall see further in Chapter I-9, the UK Government remains   I-3.53
committed to shale development, providing (often contentious) support through the planning system and by streamlining regulation. These developments have, however, remained mired in controversy, and both the Northern Irish and Scottish Governments currently maintain moratoria which mean that it is only in England and Wales that the UK Government's pro-shale policy can have effect. This controversy and the lack of concrete data on the prospectivity of the shale gas means that is hard to predict what impact this area will have on the UK's energy security.

THE UK AS AN ENERGY CONSUMER

It is evident that the only short-term viable option for replacing the   I-3.54
energy loss that will come about from the planned plant closures is by building replacement gas-fired plants. The UK, however, is import-dependent in terms of gas, supplied predominantly by Norway. It therefore first and foremost needs to maintain its good bilateral

---

[187] It has also seen it as a potential means of addressing the perennial political problem of high energy price: see eg (then Prime Minister) D Cameron, "We Cannot Afford to Miss Out on Shale Gas", *The Telegraph*, 11 August 2013, available at www.telegraph. co.uk/news/politics/10236664/We-cannot-afford-to-miss-out-on-shale-gas.html. However, at least in the short to medium term, there is no realistic prospect of shale gas materially reducing price. "David Cameron was wrong to raise public's hopes on fracking, says energy expert", *The Telegraph*, 7 November 2013, available at http://www.telegraph. co.uk/news/earth/energy/fracking/10433041/David-Cameron-was-wrong-to-raise-publics-hopes-on-fracking-says-energy-expert.html (both accessed 7 May 2017).
[188] Other perceived benefits adverted to included job-creation and revenue-generation for the state.
[189] HM Treasury, "Harnessing the potential of the UK's natural resources: a fiscal regime for shale gas" (2013), para 2.1, available at www.gov.uk/government/consulta-tions/harnessing-the-potential-of-the-uks-natural-resources-a-fiscal-regime-for-shale-gas/harnessing-the-potential-of-the-uks-natural-resources-a-fiscal-regime-for-shale-gas (accessed 7 May 2017).

relationship with Norway to preserve its "attractive customer" status. In addition to the mature infrastructure between the two countries, the UK's liberal and transparent market and the "ease with which gas could be exported from the UK to continental Europe" are the main reasons why Norway finds the UK attractive.[190] There have accordingly been calls for the UK to continue to preserve the liberal structure of its gas market, and ensure that its regulatory system is "stable to enable partners to make strategic commitments to the UK market".[191]

I-3.55   Despite the relative security of Norway as a natural gas supplier, the limited number of receiving terminals increases the potential for disruptions. The increase in the UK's LNG infrastructure is therefore a welcome step in diversifying its natural gas suppliers.[192] "For geographic reasons, the UK draws most of its gas from the Atlantic Basin and from the Middle East."[193] It imports, or has imported at one time, LNG from Algeria, Egypt, Qatar and Trinidad & Tobago.[194] Strengthening the ties with these countries would therefore be beneficial. The UK is already enjoying a strong relationship with Qatar, the largest supplier of LNG in the world.[195] Qatargas is rapidly expanding its fleet and its production to meet the rising global LNG demand. "Yet, with their geographical position enabling Qatar to supply both Asia and Europe with gas, the UK will face increasing competition for Qatari supplies". There exists for the UK an opportunity to build on its good relationship and strengthen its political and commercial ties in order to get the "Qataris to commit to supply to the UK".[196]

I-3.56   On the other hand, because of an increased competition for Qatari LNG supply in the horizon, there would also be an opportunity to tap into other markets as well. As well as liquefied shale gas from the USA (which has become an exporter of shale LNG), there are new and significant developments in West Africa, particularly in Sierra Leone, Liberia, Côte d'Ivoire, São Tomé and Príncipe, and Ghana, but also in more established producers like Nigeria, Angola

[190] Wicks, *Energy Security*, para 5.40.
[191] *Ibid*, para 5.40.
[192] "LNG's share of total gas imports have risen from 25 per cent in 2009 to 47 per cent in 2011, but fell to 20 per cent in 2013. Despite this, LNG remains an important component of the UK's energy mix. In 2015 LNG imports increased 23 per cent on 2014, making up 31 per cent of all gas imported. In 2015, Qatar accounted for 93 per cent of LNG imports", BEIS Statistics 2016, DUKES, p 100.
[193] Wicks, *Energy Security*, para 5.46.
[194] The UK imported from Egypt and Qatar in 2006 and 2007 only. DECC Statistics 2008, DUKES Table G.6, "Physical imports and exports of gas, 1997 to 2008".
[195] Wicks, *Energy Security*, paras 5.43–5.44.
[196] *Ibid*, paras 5.43–5.44.

and Gabon.[197] By entering into a bilateral or regional dialogue with these West African states the parties would assess the ways in which the commercial supply of natural gas can be mutually beneficial. In this regard, the UK-Nigeria Energy Working Group is a welcome development.[198]

A third method that may possibly be employed in preparing for a supply shortage is gas storage. The Energy Act 2008 provides for a licensing regime for gas and LNG storage.[199] By simplifying and clarifying the regulatory framework for gas storage it seeks to attract the private sector to invest in commercial gas storage ventures.[200] The UK currently has an existing total gas storage capacity of 4.7 billion cubic metres in six sites.[201] Since the last edition of this book, a number of projects have been cancelled. However, an additional 8.7 billion cubic metres of gas storage capacity is potentially available in projects that either have received planning permission and are waiting for final investment decisions, or have applied for planning permission and are waiting for the outcome.[202]   I-3.57

These figures do not explain the whole picture, though. The majority – 3.3 billion cubic metres – of the existing storage capacity is in the Rough facility located in the Southern North Sea.[203] On 16 February 2006, there was an accidental fire on the Rough platform which resulted in the operator Centrica declaring a force majeure that lasted for about nine months, until 20 November 2006. Had there been a disruption in the natural gas supplies to the UK during this period, it would have left the country in a difficult position. Indeed, in March 2015, Centrica was forced to reduce storage capacity for six months in order to undertake investigative work on a technical issue. Again, in June 2016, further issues were discovered and new injections were suspended.[204] At the time of writing, it has been announced that Rough will not accept any new injections until May 2018, while essential maintenance is undertaken of the ageing   I-3.58

---

[197] "What the latest Ghanaian discovery and Sierra Leone wildcat demonstrate is that hydrocarbon resources in the region are far richer than many ... had assumed." J Cresswell, "Hydrocarbons hunt heating up: Eyes turn to little-explored part of the West Africa sector", *Energy Voice*, 5 October 2009, available at www.energyvoice.com/oilandgas/19209/hydrocarbons-hunt-heating-up (accessed 7 May 20917).

[198] Wicks, *Energy Security*, para 5.46.

[199] The Energy Act 2008 (2008 c 32), Chapter 1: Gas Importation and Storage Zones.

[200] The Energy Act 2008, Chapter 2: Importation and Storage of Combustible Gas, read together with Explanatory Notes, Energy Act 2008 (2008 c 32), para 17.

[201] National Grid, "Gas Transportation: Ten Year Statement 2016", Table 5.4A – Existing UK storage.

[202] *Ibid*, Table 5.4B.

[203] *Ibid*, Table 5.4A – Existing UK storage.

[204] *Ibid*, p 51.

wells.[205] The additional projects that are currently under development are, therefore, a welcome step in alleviating this vulnerability.

I-3.59   However, even after factoring in the future storage projects under construction, the total capacity is still significantly lower than comparable markets in Europe. It has been suggested that one way of increasing the storage capacity "would be for the Government to contract for Strategic Storage", in a similar method to the emergency oil stock system of the IEA.[206] However, the feasibility of this proposal is debatable. The cost of storing gas is about five to seven times that of "a comparable energy content of oil",[207] making it hardly an attractive proposition.

I-3.60   As a fourth method the UK simultaneously needs to focus on increasing energy efficiency in order to lower energy demand without seriously affecting consumption patterns. This was one of the priorities in the Government's plan[208] to switch to a low-carbon economy. The UK Low Carbon Transition Plan (the "Transition Plan") itself was the product of the Climate Change Act which was enacted about eight months earlier, in November 2008.[209] The Climate Change Act was a defining moment in that it set statutory targets for carbon reduction.[210] It obliged the Secretary of State not only to ensure that the net UK greenhouse gas emissions "for the year 2050 [were] at least 80% lower than the 1990" levels[211] but also to set emission targets, or "carbon budgets", for the UK, for five-year "budgetary periods" from the period 2008–2012 and onwards,[212] with a restriction that the "carbon budget" for the 2018–2022 period must be "at least 26% lower than the 1990 baseline".[213] It also established a Committee on Climate Change to provide advice to the Secretary of State.[214]

---

[205] See www.ft.com/content/ca377e3c-1f9f-11e7-a454-ab04428977f9 (accessed 7 May 2017).

[206] Wicks, *Energy Security*, para 6.55.

[207] *Ibid*, para 6.56.

[208] DECC, "The UK Low Carbon Transition Plan".

[209] Climate Change Act 2008 (2008 c 27), ss 12–14.

[210] As opposed to the indicative targets of policies.

[211] Climate Change Act 2008, s 1.

[212] *Ibid*, s 4.

[213] *Ibid*, s 5.

[214] Part 2, s 33(3) of the Climate Change Act 2008 obliged the Climate Change Committee to publish an advisory report before 31 December 2008. This report was published as Committee on Climate Change, *Building a low-carbon economy – The UK's contribution to tackling climate change: The First Report of the Committee on Climate Change*, December 2008. The Climate Change Act, s 36(1) also obliges the Committee to report on progress. This second report was published as Committee on Climate Change, *Meeting Carbon Budgets – the need for a step change: Progress report to Parliament Committee on Climate Change*, 12 October 2009. Both of these documents, however, are beyond the

By setting out the "UK's first ever comprehensive low carbon I-3.61 transition plan to 2020"[215] the DECC's Transition Plan introduced some very encouraging measures. The Plan identified that 13 per cent of all the UK's greenhouse gas emissions came from heating homes and household water,[216] 12 per cent from the energy used in workplaces,[217] 20 per cent from the transport sector[218] and another 7 per cent from "farming and changes in land use".[219] Through the energy efficiency measures proposed,[220] by 2020 these emissions could fall by 29 per cent,[221] 13 per cent,[222] 14 per cent[223] and 13 per cent[224] respectively from the 2008 levels. It provided that if all its proposals were to be implemented, the UK's gas demand could be reduced by 29 per cent by 2020.[225]

The obligation of an 80 per cent emission reduction in 1990 I-3.62 levels by 2050 also has a direct impact on the power sector. Coal- and gas-powered plants currently supply about three- quarters of the electricity consumed in the UK.[226] The Transition Plan envisages a 22 per cent cut in 2008 emission levels by 2020.[227] To achieve this, the Government intends to generate about 40 per cent of the electricity from low-carbon technologies.[228] The fifth method, perhaps more as a medium- to long-term plan, therefore needs to focus on the consumption of less fossil-fuel energy and more low-carbon sources.

The UK has been active in policies for diversifying its energy mix I-3.63

---

scope of this chapter. For a detailed discussion on both the Climate Change Act and the work of Committee on Climate Change see Chapter I-11.

[215] DECC, *The UK Low Carbon Transition Plan*, Executive Summary, p 4.

[216] *Ibid*, p 80.

[217] *Ibid*, p 112.

[218] *Ibid*, p 134.

[219] *Ibid*, p 152.

[220] Energy efficiency measures combined with other proposed measures. For all the measures proposed for the households see DECC, *The UK Low Carbon Transition Plan*, Chapter 4; the measures for workplaces are provided in *Ibid*, Chapter 5; the measures for the transport sector are provided in *Ibid*, Chapter 6; the measures for the agriculture sector and land use in general are provided in *Ibid*, Chapter 7.

[221] *Ibid*, p 82. There was also a consultation recently on heat- and energy-saving measures for the long term.

[222] *Ibid*, p 116.

[223] *Ibid*, p 136.

[224] *Ibid*, p 155.

[225] *Ibid*, p 103.

[226] *Ibid*, p 54.

[227] *Ibid*, p 55.

[228] These technologies include "renewables, nuclear and fossil fuel fired generation fitted with carbon capture and storage technology". There is also a need for a "bigger, smarter electricity grid that is able to manage a more complex system of electricity supply and demand". *Ibid*, p 54. This chapter, however, will only focus on renewables.

for some time. In 2002, the Renewables Obligation Order[229] implemented Directive 2001/77/EC and set varying percentage targets for the amount of electricity to be supplied from renewables.[230] The 2002 Order has been modified several times.[231] The most recent of these modifications, the Renewables Obligation Order 2015,[232] which consolidates and reenacts the Renewables Obligation Order 2009,[233] set the UK's renewables obligation level at 15.4 per cent from 2015–2016.[234]

CONCLUSION

I-3.64   As North Sea production declines, the UK finds itself increasingly dependent on energy imports. Since 2004, the UK has been a net importer of energy. Today, it imports more than a quarter of its fuel consumption, predominantly from Norway. This in itself, however, does not constitute energy insecurity. Norway is a stable and reliable supplier and there is an active bilateral relationship between the two countries further strengthened by the recent Joint Ministerial Statement. However, a strike or other disruption at one of the limited number of entry points for Norwegian gas would be a serious concern for the UK's short-term energy security.

I-3.65   Similarly, the UK does not export its oil and gas to a diverse range of consumers. Its biggest oil and gas customer is the EU. Again, the EU being the main importer does not constitute an energy security concern because the EU members are reliable and stable customers.

I-3.66   Because the UK is not in isolation, a discussion on its energy security must take into account the measures also adopted by the EU

---

[229] The Renewables Obligation Order 2002 (SI 2002/914).

[230] For the target levels and the time periods, see the Renewables Obligation Order 2002, Sch 1, read together with Arts 3 and 6.

[231] It was modified by the Renewables Obligation Order 2004 (SI 2004/924), then revoked and re-enacted with modifications by the Renewables Obligation Order 2005 (SI 2005/926) which itself was revoked and re-enacted with modifications by the Renewables Obligation Order of 2006 (SI 2006/1004), which was amended by the Renewables Obligation Order 2006 (Amendment) Order 2007 (SI 2007/1078), and which was finally revoked and re-enacted with modifications by the Renewables Obligation Order 2009 (SI 2009/785).

[232] The Renewables Obligation Order 2015 (SI 2015/1947) which also implements Arts 17 and 19 of Directive 2009/28/EC and Commission Decision 2011/13/EU.

[233] It implements the changes provided in the Energy Act 2008, Part 2: Electricity from Renewable Sources. Please note that this Order only applies to England and Wales. There are complementary Orders for Scotland and Northern Ireland, The Renewables Obligation (Scotland) Order 2009 (SI 2009/140) and the Renewables Obligation Order (Northern Ireland) 2009 (SI 2009/154), respectively. The UK Renewables Obligation is the combination of the three.

[234] The Renewables Obligation Order 2009, Sch 1.

and the OECD. In response to the disruptions to the supplies from the Middle East in the late 1960s and the 1970s, the EEC obliged its members first to hold stocks to be used in emergencies, and then to restrict their energy consumption. From the late 1990s onwards, the EU started to pursue sourcing some of its energy from renewables to respond to climate change concerns,[235] further reducing its import dependency and to contribute to security of supply.

Due to the same problems in the Middle East, the OECD **I-3.67** members signed a treaty, the Agreement on an International Energy Program, in late 1974, establishing the IEA. In a similar way to the EU measures, the IEA obliged the participating countries first to hold emergency stocks and then also to restrict their consumption. The IEA additionally introduced a stock allocation mechanism and a further set of complementary measures called the CERM. Today, the CERM are being used together with the other emergency measures including drawdown of oil stocks, demand restraint, fuel-switching, surge oil production and the allocation of available supplies. These IEA obligations, together with the EU measures, were implemented in the UK in the Energy Act 1976 and the Oil Stocking Order 2012.

The effects of these supply disruptions could have been less severe **I-3.68** had the EU imported its energy from a diverse range of producers. However, the opportunity to diversify the suppliers did not come until the collapse of the Soviet Union. In 1994, the Energy Charter Treaty was signed to bring the energy co-operation between the hydrocarbon-rich former Soviet Union countries and the EU within a legal framework. Russia's recent actions, however, raise serious questions about the future of the ECT.

The UK's energy security is further affected by the issues **I-3.69** surrounding climate change and other environmental concerns. In order to tackle the acidification of the air, for example, the EU has enacted the LCPD which subsequently has been replaced by the IED. As a result, six out of the eight remaining[236] coal-fired power stations are now participating in the Transitional National Plan, adjusting the plants to comply with the lower emissions limit by 2020.[237] The other two plants either have the necessary equipment to lower the emissions to levels that are compliant with the Directive or are participating in the LLD scheme and therefore will close down by the end of 2023.[238] The only feasible short-term solution to fill the

---

[235] With a binding target of 20 per cent of renewables-based energy generation by 2020. See para I-3.30 above.
[236] Coal-fired power generation was halved in the last five years. See para I-3.39.
[237] See para I-3.40.
[238] See para I-3.41.

energy gap that will be created from these limited productions and closures is the construction of new gas-fired plants. This, in turn, means an increase in the demand for gas which will have to be either imported or sourced domestically. This is one of the many reasons why the UKCS production needs to be maximised. The necessity for fiscal or other policy and regulatory incentives to keep both the UKCS and onshore production relatively attractive for the industry appears clear.

I-3.70   In the meantime, maintaining the good bilateral relationships with Norway and in the medium-term building stronger relationships with the LNG producers would also be beneficial. In this, direct engagement, either through bilateral or regional deals, with North American and West African producers would provide the opportunity to tap into these markets as well. The UK has the necessary technical expertise to offer when seeking ways in which the commercial supply of natural gas from these regions could be mutually beneficial.

I-3.71   Although significantly more expensive, the UK can assess ways to increase and diversify the current gas storage capacity. The Energy Act 2008 provides for a licensing regime for gas and LNG storage. This is a welcome development that needs to be built upon.

I-3.72   Simultaneously, the UK needs to increase its energy efficiency to lower the energy demand levels. Both the Climate Change Act and the UK Low Carbon Transition Plan are also significant and welcome developments in laying the framework for increasing the energy efficiency of the UK.

I-3.73   A final but very important development is the Renewables Obligation Order. As a consequence of EU Directives 2001/77 and 2009/28, the most recent Order obliges the UK to generate about 15 per cent of its electricity from renewable sources by 2015. According to the Transition Plan, the Government expects this number to increase to 40 per cent by 2020. The measures to decrease the use of hydrocarbons from the UK's energy mix in the medium to long term is a very crucial step in tackling the UK's long-term import dependency.

I-3.74   It is evident that energy security issues will be a priority on British policymakers' agenda for some time. The UK is trying to balance its energy security policies on hydrocarbon production and consumption, with its climate and environmental concerns. As the National Security Strategy suggests, the UK will

"promote investment in renewable, shale and other innovative technologies to increase domestic production…[and] work with the EU to shape the single energy market, helping reduce the EU's energy dependence on Russia. [The UK] will lead efforts to evolve international energy governance, reducing market distortions and better integrating major non-OECD consumers into decision making. "[239]

There is no doubt that both internationally and domestically, policies I-3.75 and laws have been adopted successively over the past few decades that have made a significant contribution to the multi-dimensional challenges of the UK's energy security. The unanswered question is whether the energy security architecture now in place will be sufficient for what looks to be an even more challenging future.

---

[239] UK Cabinet Office, "The National Security Strategy of the United Kingdom", para 4.142, available to download at www.gov.uk/government/publications/the-national-security-strategy-of-the-united-kingdom-security-in-an-interdependent-world (accessed 7 May 2017).

RESOURCE MANAGEMENT

CHAPTER I-4

PETROLEUM LICENSING

Greg Gordon[1]

INTRODUCTION

The comparative study of petroleum laws discloses that, almost I-4.01
universally,[2] a nation state will claim some manner of right to the
oil and gas deposits situated within its borders or located beneath
the Continental Shelf to the outer limit of its Exclusive Economic
Zone.[3] Such is the economic and strategic importance of oil and gas
(and indeed of energy-yielding minerals in general)[4] that this propo-

---

[1] The content on Marine Spatial Planning contained at paras I-4.28 to I-4.31 was
authored by Anne-Michelle Slater, Senior Lecturer, University of Aberdeen.

[2] A major exception is the USA: see eg B Taverne, *Petroleum, Industry and Governments*
(1999) (hereinafter "Taverne"), at para 5.1.3.1. In the USA, no over-arching state claim is
made to oil and gas *in situ*. Licensing systems exist, but apply only to offshore areas and
onshore territory within public ownership. See Taverne, at paras 6.1.2 to 6.1.3.1.

[3] As determined formerly in accordance with the Convention on the Continental Shelf
done at Geneva on 29 April 158 (hereinafter "Convention on the Continental Shelf"), Art
1 and now by the United Nations Convention on the Law of the Sea done at Montego
Bay, Jamaica, 10 December 1982 (hereinafter "UNCLOS"), Art 57.

[4] Coal deposits are also commonly reserved to the state or at least to a public corporation: in the UK, coal deposits in strata were vested in the Coal Corporation by the Coal
Act 1938, s 3. By virtue of the Coal Industry Act 1994, s 1(1)(a) the Coal Authority is
the most recent statutory successor to the Corporation. The Atomic Energy Act 1946, ss
6 and 7 (as amended) empowers the state to search for and acquire rights to work radioactive minerals: while there are no active uranium workings in the UK, geological surveys
have revealed apparently good-quality, commercially viable deposits in the northern part
of the Scottish mainland and in Orkney: T D Colman and D C Cooper, *Exploration
for Metalliferous and Related Minerals in Britain, a Guide* (2nd edn, 2000), Appendix
2, p 69. Proposals for an opencast mine by Stromness in Orkney led to significant
public protest. The development was not proceeded with, but is said to have acted as
a spur to the development of the Orcadian interest in renewables. If verified, the claim
would provide a rare instance of a community faced with an undesirable development

sition holds true even in legal systems, such as those of Scotland and of England and Wales, which ordinarily provide that subterranean minerals are owned by the proprietor of the overlying land.[5] States vary in the extent to which they become directly involved in the task of exploiting these resources. Frequently, the state will be actively involved, commonly through the vehicle of a national oil company.[6] The UK utilised such an approach in the early days of offshore petroleum exploration and development, but this approach was discontinued as a result of a lack of political will. Alternatively, the state will effectively delegate this function to the private-sector oil and gas industry, and this is the approach that the UK has taken from the early 1980s to the present day. Throughout this period, the UK has been known for its "light-touch" approach to resource management issues.[7] However, with the implementation of the Wood Review and the introduction of a new, better-resourced and more interventionist regulator in the form of the Oil and Gas Authority (OGA), this position is presently undergoing evolution, or perhaps a revolution. This issue will be further discussed throughout this chapter and in Chapter I-5.

I-4.02   Where the private sector is involved, the state needs to choose the vehicle through which its industry partners are to be involved. Some states favour a model based upon the grant of licences or concessions. Others enter into contractual arrangements such as production-sharing agreements or service contracts,[8] or hybrid

---

responding not with a simple "no", but a "no, but …". See Orkney Sustainable Energy, *Nuclear or Renewables?*, available at www.orkneywind.co.uk/orkney-wind-forecast.html (accessed 2 May 2017).

[5] For the position in Scotland, see W Gordon and S Wortley, *Scottish Land Law* (3rd edn, 2009) (hereinafter "Gordon and Wortley, *Scottish Land Law*"), para 5.02; for England, see Lord Mackay of Clashfern (General Editor), *Halsbury's Laws of England* (4th edn, 2003 Reissue) (hereinafter "*Halsbury*"), vol 31, para 363 *et seq*.

[6] This approach is commonly utilised by Middle Eastern oil-producing states and in Latin America and West Africa but is not unknown in the West: see M Bunter, *The Promotion and Licensing of Petroleum Prospective Acreage* (2002) (hereinafter "Bunter, *Promotion and Licensing*"), at 19–20. Norway's National Oil Company, Statoil, was founded in 1972 and although now part-privatised, it continues to play an instrumental role in the development of the oil and gas industry in that province. For a detailed account, see M Thurber and B Tangen Istad, *Norway's Evolving Champion: Statoil and the Politics of State Enterprise*, Stanford University Program on Energy and Sustainable Development Working Paper 92, available at http://iis-db.stanford.edu/pubs/22919/WP_92,_Thurber_and_Istad,_Statoil,_21May2010.pdf (accessed 29 August 2017).

[7] Sir I Wood, *UKCS Maximising Recovery Review: Final Report* (hereinafter "the Wood Review"), p 9, available at www.gov.uk/government/uploads/system/uploads/attachment_data/file/471452/UKCS_Maximising_Recovery_Review_FINAL_72pp_locked.pdf (accessed 7 May 2017).

[8] A detailed treatment of such agreements is beyond the scope of this book. For an introductory account, see Taverne, Chapter 7.

models which involve elements of licence and contract. Whichever model is chosen, the state will expect to receive a return in the form of a share of produced hydrocarbon[9] and/or financial benefits, such as a cash premium paid in exchange for the grant of the licence,[10] rental payments in respect of the licensed area, cash royalties and/ or revenue from taxation.[11] The private-sector oil industry player(s) involved will also expect to be rewarded for the work they have carried out and the risk they have undertaken.

The UK, in common with the other Western democracies **I-4.03** possessing oil and gas reserves, has adopted a licensing model.[12] From rather primitive beginnings, the UK's licensing system has evolved into one of considerable complexity,[13] the level of which is about to further increase, as, from the 29th offshore round, licences are to be granted subject to a greater degree of individual tailoring and negotiation.[14] The main current regime[15] recognises a distinction

---

[9] This is a classic feature of production-sharing agreements but is not unknown in licensing regimes. For a time the UK Government inserted provisions in its petroleum licences which permitted the taking of royalty in kind: see eg Petroleum and Submarine Pipelines Act 1975, Sch 2, Pt II, Model Cl 11, "Delivery of petroleum instead of royalty".

[10] Cash premium bidding – which is to say, the letting of acreage to the highest bidder – is a feature of a number of licensing regimes, notably the USA: T Daintith, *Discretion in the Administration of Offshore Oil and Gas* (2006) (hereinafter "Daintith, *Discretion*"), at paras 9109 and 9110. It has also been used, albeit infrequently, in the UK: see para I-4.32.

[11] Various permutations of rental, royalties and sundry different forms of taxation have all been used at different times in the UKCS: see paras I-4.32 to I-4.34. For a detailed discussion of the current tax regime, see Chapter I-7. For a comparative discussion on royalties and related financial payments see Daintith, *Discretion*, Chapter 9.

[12] The licensing approach has also been adopted by the USA, Canada and Australia. Each of these jurisdictions is discussed at length throughout Daintith, *Discretion*. Licensing systems have also been adopted by eg Norway (see Taverne, Chapter 6.4) and the Netherlands (see Taverne, Chapter 6.5). There is no reason in principle why they could not adopt a different model, such as a production sharing agreement. Historically, however, production sharing agreements have tended to be favoured in jurisdictions coming to terms with the legacy of the post-colonial era, one of the drivers for the development of the PSA being a dissatisfaction at the apparent loss of control (or even sovereignty) associated with a concessionary or licensing model. That particular driver is not present in the same way in Western countries, even those having a colonial past, although the colonial legacy can manifest itself in a different manner in those countries, for instance in the context of exploration and production activities in areas where an aboriginal land right has been claimed.

[13] For a concise account of the evolution of the UK oil and gas licensing regime, see T Daintith, G Willoughby and A Hill, *United Kingdom Oil and Gas Law* (3rd edn, looseleaf, 2000–date) (hereinafter "Daintith, Willoughby and Hill"), Chapter 1; and/or G Gordon, "British Hydrocarbon Policies and Legislation", in E Pereira and H Bjorneybe, *North Sea and Beyond* (2018). For a full account of the development of the UK licensing regime prior to 1993, see also A Kemp, *Official History of North Sea Oil and Gas*, vol 1: The Growing Dominance of the State, and vol 2: Moderating the State's Role (2011).

[14] See paras I-4.41–42 and I-4.75–78.

[15] Separate licensing regimes are applied to onshore Northern Ireland (but not its terri-

between *seaward* and *landward* licences.[16] These categories are then subdivided: landward into the Petroleum Exploration and Development Licence and the supplementary seismic survey licence, and seaward into Exploration and Production Licences. This latter subcategory has, in the recent past, been further subdivided into a range of variations: the traditional Production Licence, the Frontier Licence (of which there are now two variants named after the length of their initial terms – the six-year and nine-year Frontier Licences) and the Promote Licence. From the 29th licensing round onwards, however, all newly granted Seaward Production Licences are to be designated Innovate Licences (although those granted under the 29th licensing round will be issued using the present model clauses; see further the discussion at para I-4.75). The Innovate Licence is intended to allow for more flexibility in key licence terms and conditions[17] than was generally present in the previous system – although as we shall see under the heading of "bespoke licences" below it was always possible for the state to issue a licence tailored to meet particular circumstances, if good reason existed. We will discuss these various licences in turn after briefly introducing the legal concept of licensing and discussing the legal basis for the UK oil and gas licensing regime.

LICENSING AS A LEGAL CONCEPT

I-4.04   A licence is a permission that authorises an activity the conduct of which would otherwise be unlawful. Licences take a number of forms. Many licences are best classified as "administratively granted

---

torial waters) and the Isle of Man and its territorial waters. Northern Ireland's licensing system derives from powers conveyed by the Petroleum (Production) Act (Northern Ireland) 1964 and is administered by the Devolved Executive's Department of Enterprise, Trade and Investment. This chapter's focus is on the principal regime and although this chapter continues the general practice of referring to "United Kingdom" oil and gas licensing regime, "British" would be strictly more correct, at least when discussing onshore developments.

[16] The terms describe the areas' relative position to the low-water line, which is set by the Petroleum (Production) (Seaward Areas) Regulations 1988 (SI 1988/1213), Reg 3(1), read together with Sch 1. An area lying to the seaward side of the line is a seaward area; an area lying towards the land, a landward area. Although commonly used throughout the industry, and indeed by the Government in its Guidance notes and informal documentation such as press releases, the terms "onshore" and "offshore" are not formally used in the licensing regime and are technically inaccurate, as significant areas of inland waters lie landward of the line, and are therefore subject to what is colloquially called the onshore regime.

[17] There is greater flexibility than there was previously on, for instance, the size of area licensed, the length of the initial term and the need to relinquish certain areas at certain key stages.

exemptions from legislative prohibitions".[18] Only the state, or a person to whom the state has delegated authority, can grant such licences. Most of the licences that a private individual encounters in day-to-day life,[19] as well as many commercial licences,[20] fall into this category. The efficacy of such licensing systems depends upon the existence of a punitive sanction in the event that a licence is either not obtained or, if obtained, is not complied with. Second, a legal entity – whether the state, a business organisation or a private individual – holding property rights may grant a licence permitting the licensee to make some kind of use of the licensor's property. The power to grant such licences stems from the legal status of ownership, not from the authority of the state. Licences to occupy land or premises are a feature of the land law of some jurisdictions.[21] Licences which stem from the licensee's status as a property owner are also commonly encountered in the context of intellectual property rights. It is not necessary, even when granted by an emanation of the state, for such licences to be fenced with a punitive sanction as the breach or non-observance of the licence's terms may result in its being terminated,[22] and a range of civil remedies ranging from damages to injunctive relief and actions of ejection will be available against any party who makes unauthorised use of the licensor's property. Landward petroleum licences are proprietary licences in the sense that they emanate from the Crown's ownership of hydrocarbon deposits *in strata*. The legal character of the Seaward Production Licences is somewhat less clear: see the discussion at para I-4.12 below.

Licences may also be *exclusive* or *non-exclusive*. It has been said   I-4.05
that the holder of an exclusive licence has the comfort of knowing that "so long as the licence is valid, no person other than the licensee itself is authorised to exercise the rights conferred".[23] However, as a matter of property law this is not strictly correct. As a licence confers no real right,[24] an exclusive licence confers on the licensee only a personal right to preclude the granter from issuing further licences in respect of the same geographic area. In so far as a petroleum

---

[18] Daintith, Willoughby and Hill, at para 1-302.

[19] For instance, driving licences (Road Traffic Act 1988, s 87) and television licences (Communications Act 2003, s 363).

[20] For instance, a licence is required in order to carry on a consumer credit business: Consumer Credit Act 1974, s 21.

[21] For the English position, see *Halsbury*, vol 27(1), paras 6–16; for the position in Scots law, see Gordon and Wortley, *Scottish Land Law*, at para 18-17.

[22] It is however possible to do this: consider, for example, the various offences contained in Chapter 6 of the Copyright, Designs and Patents Act 1988.

[23] Taverne, at para 5.2.1.

[24] For a discussion of what is meant by a "real right", see para II-15.07.

Production Licence is contractual,[25] the OGA would be in breach of contract if it were to grant a second licence; however, that licence would be valid, albeit voidable. By contrast, a non-exclusive licence holder is merely one of a number of persons who may have received concurrent permissions from the licensor. As we shall see, Production Licences (whether landward or seaward) are exclusive in nature. Exploration Licences are non-exclusive.

I-4.06   It is also worth noting here that while licences grant rights to licensees, they do not operate to take away the pre-existing rights of third parties. Given the legal status of the Continental Shelf, this is of little consequence outside the territorial sea. However, within territorial waters, rights of private property may very well exist. Thus, in addition to a statutory licence, grants of rights may also be needed from parties such as landowners or salmon fishers in coastal waters.[26]

THE LEGAL BASIS FOR THE UK OIL AND LICENSING REGIME

Domestic law

The law's evolution

I-4.07   Early UK oil and gas licences issued in accordance with the provisions of the Petroleum (Production) Act 1918 were clearly of the administrative exemption type. The 1918 Act avoided the then-contentious question of who owned oil and gas deposits *in situ* by providing that only those holding a licence from His Majesty could "search or bore for or get" petroleum. Section 1(1) of the 1918 Act provided that anyone undertaking such works without authority would forfeit any petroleum so obtained and in addition pay a penalty of three times its value.

I-4.08   The Petroleum (Production) Act 1934 fundamentally altered the nature of the British petroleum licence.[27] Section 1(1) of the 1934 Act expressly vested "property in petroleum, existing in its natural condition in strata in Great Britain"[28] in the Crown, and stated that the Crown held "the exclusive right of searching and boring for and getting such petroleum". The 1934 Act did not provide for penalties

---

[25] See further the discussion at paras I-4.12 to I-4.13.

[26] See para II-15.01.

[27] The Act was introduced as the Government claimed that petroleum exploration activities had been unduly hampered by the deficiencies of the earlier legislation: Daintith, Willoughby and Hill, para 1-104. See also *Star Energy Weald Basin Ltd v Bocardo SA* [2010] UKSC 35, [2010] 3 WLR 654, at para 90 per Lord Brown.

[28] That is, the land mass comprising England, Scotland and Wales, and the territorial seas appertaining thereto.

in the event of breach but, as the owner of petroleum *in situ*, the Crown would be in a position to take action to prevent any attempt at expropriation, or seek compensation therefor.[29]

The realisation in the late 1950s that geological formations **I-4.09** associated with the Groningen gas field in the Netherlands might extend into the Continental Shelf underlying the North Sea[30] prompted the UK, in common with other North Sea coastal states, to promulgate a legal framework authorising the exploration for, and production of, petroleum.[31] The UK did this by exporting the existing but largely untested landward licensing regime into the seaward United Kingdom Continental Shelf (UKCS). Section 1(1) of the Continental Shelf Act 1964 vested in the Crown "any rights exercisable by the United Kingdom outside territorial waters with respect to the sea-bed and subsoil and their natural resources", except coal.[32] Section 1(3) of the 1964 Act applied most of the key licensing provisions[33] of the 1934 Act to the UKCS. However, Sections 1 of the 1934 Act (vesting of property in petroleum *in strata*) was not so applied as international law conveyed upon the state only a "sovereign right"[34] to exploit natural resources, not a right of full ownership.[35] The British approach is consistent with that taken by certain other coastal states.[36] However, the opacity of the nature of the right enjoyed by the state has led to debate over the legal character not just of the state's rights, but also of the rights

---

[29] See the discussion at para I-4.04.

[30] Taverne: see the note appended to para 6.5.1 at the foot of p 238.

[31] For example, the Netherlands (on which see Taverne at para 6.5.2.1) and Norway (Taverne: see the note appended to para 6.4.1).

[32] Coal was reserved to the National Coal Board: s 1(2). The Coal Authority is its modern successor in title: see para I-4.1.

[33] That is, ss 2 (licences), 3 (compulsory acquisition of the right to enter land), 4 (power to supply natural gas), 5 (receipts and expenditure) and 6 (power to make regulations).

[34] Convention on the Continental Shelf, Art 2.1.

[35] P Cameron, *Property rights and sovereign rights: the case of North Sea oil* (1983) (hereinafter "Cameron, *Property rights and sovereign rights*"), at pp 46–50; S Jayakumar, "The Continental Shelf Regime under the UN Convention on the Law of the Sea: Reflections after Thirty Years", in M Nordquist et al. (eds) *Regulation of Continental Shelf Development* (2013).

[36] For example Australia: see the Seas and Submerged Lands Act 1973, s 10A. Norway, by contrast, having initially claimed sovereign rights (see the discussion at I-8.02) now makes a straightforward proprietary claim: "The Norwegian State has the proprietary right to subsea petroleum deposits and the exclusive right to resource management." Act 29 November 1996 No 72 relating to petroleum activities (as amended), section 1-1. Danish law makes a similar proprietary claim: see Consolidated Act No 889 of 4 July 2007, *Consolidated Act on the Use of the Danish Subsoil*, ss 1 and 2. At first blush this may appear more straightforward than the UK approach, but this approach poses its own questions, most notably: if international law does provide only for a lesser right, on what basis can the state assert a right of full ownership?

enjoyed by the holder of a Seaward Production Licence.[37] Taverne asserts that, within the UKCS as much as within Great Britain, full ownership of petroleum *in situ* is vested in the Crown.[38] However, given what has already been said above about the 1964 Act, the present author would respectfully suggest that this must be incorrect. At the opposite end of the spectrum of opinion, Marriage argues that oil and gas *in strata* within the UKCS is *res nullius*, which is to say, wholly ownerless.[39] Daintith and Hill adopt an intermediate position, namely that oil and gas deposits within the UKCS cannot properly be considered *res nullius* "as a determinate person, the Crown, has ... the right to reduce them to possession and exclude others from so doing". Thus they argue that, although the Crown's right is not one of full ownership, it must still be essentially proprietorial in nature[40] (and presumably, therefore, admitting of proprietorial remedies in the event of unauthorised activity). From the standpoint of principle, this seems questionable. For example, an owner of land has the right, for so long as they are situated within his territory, to reduce wild birds and fish to possession and exclude others from so doing, but the mere fact that he is entitled to take this course of action conveys no ownership or cognate proprietary right in those things. However, from the standpoint of policy, Daintith, Willoughby and Hill's view has much to commend it. They note that to state that the right is otherwise involves the assertion that the whole of the Government's offshore licensing scheme was "fundamentally misconceived".[41] If that were to be found to be so, the practical consequences would indeed be enormous. However, alternative positions would seem to be available to a court called upon to adjudicate on this matter. It may be possible to conclude, for example, that while the licence is not technically proprietorial, the fact that sovereign rights are recognised as exclusive must mean that they are supported by some

---

[37] The practical significance of these issues is highlighted by the series of questions posed by Daintith, Willoughby and Hill, at para 1-344.

[38] Taverne, at para 6.3.1.

[39] P Marriage, "North Sea Petroleum Financing in the United Kingdom" 5 (1977) *Int Bus Lawyer* 207, at 209.

[40] Daintith, Willoughby and Hill, at paras 1-345 to 1-346.

[41] *Ibid.* Given the dreadful state of eg the UK's offshore health and safety regime (see para I-10.18) and fiscal regime (para I-5.5) in the early 1970s, it could be observed that if the licensing were indeed to have been fundamentally misconceived, it would not have been the only part of the UK's offshore regulatory and resource management system to have been in poor order. Nor would this be surprising. The health and safety system only improved as a result of repeated major accidents which illustrated the deficiencies of the system. The mere fact that no company has thus far considered it to be in its interests to test the nature of the licensing system does not mean that there is no question to be asked.

framework of vindicatory rights, even if the court is required to discover them.

The current law

The licensing provisions of the 1934 Act were, together with a   I-4.10 number of other statutory provisions bearing on oil and gas law,[42] consolidated and re-enacted in the Petroleum Act 1998. The 1998 Act leaves unchanged the essential features of the 1934 Act's licensing regime. Issues of the devolution to the Scottish Parliament and Government aside,[43] all oil and gas licences to "search and bore for and get" petroleum within Great Britain, its territorial sea and the UKCS are now granted by the Crown under powers conferred by Section 3(1) of the 1998 Act. Prior to the implementation of the Wood Review, these licences were granted by the Department of Energy and Climate Change (DECC) and its predecessors, but this function has now been transferred to the OGA.[44]

European Union law

As a member state of the EU, the UK is bound by EU law.[45] The   I-4.11 Hydrocarbons Licensing Directive[46] has had a considerable impact upon UK oil and gas licensing law. The main purpose of the Directive is to prevent the licensing authority from distorting competition by discriminating against persons from other members of the EU.[47] The Directive and its implementing Regulations will be further considered below.[48]

---

[42] But not the Continental Shelf Act 1964, which is not petroleum-specific but instead bears more generally on the state's right to regulate the use of and exploit the Continental Shelf.

[43] On the devolution issue, see further the discussion at I-9.111 to I-9.126.

[44] The Petroleum (Transfer of Functions) Regulations 2016 (SI2016/898), Reg 2.

[45] In the absence, at the time of writing, of a clear conclusion as to the form Brexit will take, this chapter does not address the question of how UK licensing law may change following Brexit.

[46] Directive 94/22/EC of the European Parliament and of the Council of 30 May 1994 on the conditions for granting and using authorisations for the prospection, exploration and production of hydrocarbons [1994] OJ L164/3 (hereinafter "the Hydrocarbons Licensing Directive").

[47] See the recitals to the Hydrocarbons Licensing Directive: "Whereas steps must be taken to ensure the non-discriminatory access to and pursuit of activities relating to the prospection, exploration and production of hydrocarbons under conditions which encourage greater competition ... [and] it is necessary to set up common rules for ensuring that the procedures for granting authorizations for the prospection, exploration and production of hydrocarbons must be open to all entities possessing the necessary capabilities".

[48] See paras I-4.14, I-4.16, I-4.20, I-4.22 and I-4.36–37.

KEY FEATURES OF THE UK OIL AND GAS LICENSING REGIME

A regulatory and contractual hybrid

I-4.12   A casual examination of a UK oil and gas licence might lead one to the conclusion that the licence is simply a commercial contract. The licence's opening narration states that it is "made between the OGA … on the one part and the companies listed … on the other part".[49] It is executed by both parties.[50] It contains an arbitration clause,[51] a common feature of commercial agreements. Model Cl 2 (Grant of Licence)[52] is careful to narrate that payments have been made, and will continue to be made, in exchange for the licence's grant, thus satisfying the requirements of the doctrine of consideration which forms part of English contract law.[53]

I-4.13   Daintith and Hill are therefore fully justified in describing the licence as "contractual in form".[54] However, the same authors are also surely correct to recognise that the licence also fulfils another function. A close examination of the licence makes it clear that it is as much a regulatory instrument as a document recording the terms of a commercial deal. Much of the licence is given over to imposing a set of controls and obligations upon the licensee,[55] the breach or non-observance of which may – among other lesser sanctions[56] –

---

[49] Copy sample licences provided to the author.

[50] *Ibid.*

[51] See Model Cl 43.

[52] The focus of the discussion in this work is the current set of Seaward Production Licence model clauses to be found in the Schedule annexed to the Petroleum Licensing (Production) (Seaward Areas) Regulations 2008 (SI 2008/225), as amended by the Petroleum Licensing (Amendment) Regulations 2009 (SI 2009/3283) (hereinafter "the 2008 Regulations as amended"). These governed the 25th to 29th offshore licensing rounds, and, at the time of writing, continue to the be the extant model clauses, although, as we shall see, the OGA has indicated that new ones will be issued in time for the 30th offshore round. Unless the context requires a contrary interpretation, any reference within this chapter to a numbered model clause is to the model clause as contained in the Schedule to SI2008/225. However, earlier sets of model clauses continue to govern a large number of licences: see the discussion at para I-4.17.

[53] The doctrine is unknown in Scots law: see eg W Gloag, *The Law of Contract* (2nd edn, 1929), at p 48.

[54] Daintith, Willoughby and Hill, at para 1-323.

[55] For a short discussion of these controls, see paras I-4.59 to I-4.60. Two of the most significant controls, those relating to the licensee's obligation to submit work programmes (Model Cl 16) and production and development programmes (Model Cl 17), are discussed at length elsewhere in this chapter and in the Appendix.

[56] Prior to the implementation of the Wood Review via the Energy Act 2016, the sanctions or remedies for breach were limited: see further G Gordon, "Production Licensing on the UK Continental Shelf: Ministerial Powers and Controls", 4 (2015) *LSU Journal of Energy Law and Resources*, 74, at 79–81. The Minister could revoke, but this was a nuclear

result in the licence being revoked.[57] By contrast, very few positive
obligations are imposed upon the OGA. Moreover, by no means all
disputes under the licence may be referred to arbitration: a large
and important set of areas of potential disagreement is excluded
from the scope of the arbitration clause.[58] In these circumstances
it has been argued that in the event of a dispute between state
and licensee, the licensee will, in addition to any recourse it may
have in private law, be entitled to pursue a judicial review.[59] The
possibility also exists that the actions of the OGA may in certain
circumstances be susceptible to challenge under the Human Rights
Act 1998.[60]

An essentially discretionary system

The licensing provisions of the Petroleum Act 1998 do not set out I-4.14
a code which stipulates in fine detail how the state is to administer
petroleum exploration or production activities.[61] Instead, the Act

---

option, and nothing else was expressly provided for. The contractual aspect of the licence
suggested that an action could conceivably lie in the hands of the Minister in contract in
the event of breach causing loss, but this idea remained untested. However, as we shall see
in Chapter I-5, when faced with a breach of a licence term, the OGA is now empowered
to impose a range of graduated sanctions, including financial penalties.

[57] The Minister's powers of revocation are set forth in Model Cl 41. For a discussion of
the enforceability of the revocation provision, see Daintith, Willoughby and Hill, paras
1-342 to 1-343. A new power of partial revocation was introduced in 2008 and may be
found in Model Cl 42. This power was in addition inserted into all existing licences with
retroactive effect; see further the discussion at para I-4.17.

[58] Model Cl 43(1) states that the arbitration provisions shall not apply to any matter
or thing expressly said to be "determined, decided, directed, approved or consented to
by the Minister". Some examples of ministerial determinations, decisions, directions,
approvals or consents can be seen in Model Cll 4 (potential for discretionary decision
as to term), 5 (potential for a ministerial direction that a Frontier Licence which would
otherwise lapse should continue), 7 (power to determine that first or second licence
term shall be extended), 14 (power to direct that measuring devices be tested and, if
found faulty, to determine how long the fault will be deemed to have subsisted), 17
(power to make directions relative to production and development plans), 21 (power
to approve a programme of Completion Work), 23 (ministerial consent to potentially
harmful modes of working), 24 (power to approve the appointment of operator) and
45 (power to determine that debris potentially dangerous to the fishing industry should
be removed).

[59] S Dow, "Energy", in *The Laws of Scotland: Stair Memorial Encyclopaedia* (Reissue,
2000), at para 19; T Daintith, "Contractual Discretion and Administrative Discretion: A
Unified Analysis", 68 (2005) *MLR* 554, at 592: "Courts in ... the United Kingdom will
not however accept even [unfettered] discretions ... as simply unreviewable".

[60] Daintith, *Discretion*, para 7304.

[61] In this respect, the position in the UK can be distinguished from that in a number of
other jurisdictions, where the regime is more prescriptive and less discretionary in nature:
eg, Australian petroleum law, discussed in some detail in Daintith, *Discretion*. See also
Bunter, *Promotion and Licensing*, at 19–25.

provides the OGA with a number of broad enabling powers: the OGA may grant licences to such persons as it thinks fit,[62] on such terms and conditions as it thinks fit[63] and for such consideration as it, with the consent of the Treasury, may determine.[64] Thus the system has been described as one of discretionary allocation, in that no single criterion is determinative of the OGA's choice of whether and to whom a licence is to be awarded.[65] But it is not just in the allocation of licences that a high degree of discretion is evident. Instead, discretion is a theme which recurs throughout the whole administration of the licensing system.[66] As we shall see throughout the next section of this chapter, however, the OGA's powers, while still extensive, have over time come to be limited in some significant respects by both the Hydrocarbons Licensing Directive and domestic delegated legislation. Additionally, the Government[67] has always issued a significant quantity of guidance on the practical operation of its procedures. At the time of writing,[68] we are in something of an interregnum, with the OGA now the licensing authority but still operating within the framework of the 2008 Model Clauses and Guidance largely inherited from BEIS. By the time of the 30th round, the OGA will have had more of an opportunity to put its own stamp on the system. A comparison between the DECC-managed 28th, transitional 29th and forthcoming OGA-managed 30th rounds will say much about how resource management has changed on the UKCS post-Wood.

---

[62] Petroleum Act 1998, s 3(1).

[63] *Ibid*, s 3(3).

[64] *Ibid*, s 3(3).

[65] Daintith Willoughby and Hill, para 1-317.

[66] A fact demonstrated throughout Daintith, *Discretion*.

[67] Over the 50 years of offshore licence grants, the relevant government department has changed frequently, often upon a change of government. Broadly speaking, the changes have come when Labour Governments have been more inclined to view power or energy as a thing apart, requiring its own ministry, or when Conservative Governments have been more inclined to view the sector as a sub-species of trade, although examples can be found of each party tolerating, at least for a time, arrangements put in place by the other. Initially, the relevant department was the Ministry of Power; thereafter it was the (short-lived) Ministry of Technology, the Department of Trade and Industry, the Department of Energy, the Department of Trade and Industry again and the Department of Business, Enterprise and Regulatory Reform. From 2008, the relevant department was the Department of Energy and Climate Change (DECC, still the author of much of the guidance in use to govern licensing policy and administration). Since 2016, it has been the Department of Business, Energy and Industrial Strategy (BEIS; pronounced, I am told, "to rhyme with Amadeus").

[68] April 2017.

General remarks on the licence's terms and conditions

The discussion under this heading is restricted to selected points I-4.15
common to all UK petroleum licences. For a more detailed commentary
on some of the more significant traditional Production Licence terms
and conditions, see paras I-4.43 to I-4.60. For a discussion of the
terms and conditions of Frontier and Promote Licences, see paras
I-4.61 to I-4.68 and I-4.69 to I-4.73 respectively. For a discussion of
the new Innovate Licence, see paras I-4.41 to I-4.42.

Since the entry into force of the Hydrocarbons Licensing Directive I-4.16
Regulations,[69] which adopted the Hydrocarbons Licensing Directive
into domestic law, there has been a restriction upon the types of
terms that the OGA is entitled to include in a licence. The only terms
and conditions permissible are those justified exclusively for the
purpose of:

(a) ensuring the proper performance of the activities permitted by
the licence;
(b) providing for the payment of consideration for the grant of
the licence; and
(c) for certain operational and other purposes set out in Reg
4(2).[70]

Some of the most significant and commercially sensitive terms I-4.17
found in a licence have little or no statutory basis but are either
simply intimated in advance and imposed upon the licensee in
accordance with the licensing authority's usual practice[71] or are
the subject of specific agreement between the parties.[72] These terms
are supplemented by the model clauses which Section 4(1)(e) of
the Petroleum Act 1998 provides shall be prescribed by regulations and shall be incorporated into petroleum licences, subject to
the OGA's discretion to modify them in particular circumstances.[73]
The Act empowers the OGA to prescribe different model clauses
for different types of licence.[74] The OGA's predecessors as licensing
authority duly exercised this power: regulations provide one set of
model clauses for Exploration Licences[75] and another for Seaward

---

[69] Hydrocarbons Licensing Directive Regulations 1995 (SI 1995/1434) (hereinafter
"Hydrocarbons Licensing Directive Regulations 1995").

[70] *Ibid*, Reg 4(1).

[71] For example, area rentals, discussed further at paras I-4.45 to I-4.46.

[72] For example, the exploration activities the licensee intends to carry out in its initial
work programme: see para I-4.47.

[73] For a discussion on the occasions when the Minister has chosen to modify the model
clauses, see para I-4.74.

[74] Petroleum Act 1998, s 4(2).

[75] Discussed further at para I-4.39.

Production Licences.[76] Additionally, it is important to note that the licence does not incorporate the model clauses "as they are or may come to be"; it incorporates the relevant set of model clauses in force at the time when the licence is granted.[77] Thus, a large number of different sets of model clauses, each to a greater or lesser extent different in their terms, will be in force at any given time. Only the most recent licences will incorporate the most up-to-date iteration: licences of long standing will be governed by an elderly set of model clauses.[78] If these model clauses (or indeed any other terms within the licence) are subsequently thought to be undesirable, then they can be changed only by the agreement of the licensee or by the enactment of retroactive amending legislation.[79] This provides the licensee with a degree of protection against unilateral change and, as we shall see in Chapter I-5, has made the implementation of the Wood Review a more complex process than it would have been had the state reserved a right to retrospectively amend licences, such as exists in, for example, Denmark and Norway. It should be remembered, however, that the licence does not comprise the totality of the licensee's relationship with the state. A whole host

---

[76] Previously, more extensive use was made of this power. Immediately prior to the current set of model clauses the practice was to provide a suite of different sets of model clauses, one for each of the different types of Production Licences available: see Schs 1–7 of the Petroleum Licensing (Exploration and Production) (Seaward and Landward Areas) Regulations 2004 (SI 2004/352), as amended by the Petroleum Licensing (Exploration and Production) (Seaward and Landward Areas) (Amendment) Regulations 2006 (SI 2006/784).

[77] Daintith, Willoughby and Hill, para 1-305.

[78] The 25th round of licences was governed by the Petroleum Licensing (Exploration and Production) (Seaward and Landward Areas) Regulations 2004 (SI 2004/352), as amended by the Petroleum Licensing (Exploration and Production) (Seaward and Landward Areas) (Amendment) Regulations 2006 (SI 2006/784). The 24th round was governed by the 2004 Regulations without amendment. The principal sets of preceding offshore model clauses may be found in the Petroleum (Current Model Clauses) Order 1999 (SI 1999/160), Schs 1–14.

[79] The Petroleum and Submarine Pipelines Act 1975, s 18, was a particularly contentious example of retroactive legislation in the oil and gas context: see further the discussion on the Fallow Areas Initiative and Stewardship in the Appendix. More recently, s 77(2) of the Energy Act 2008 incorporated into all extant oil and gas licences with retrospective effect the changes contained in Sch 3 to that Act. However, these changes, which introduced the requirement to provide contact details to the Minister, a ministerial power of partial revocation of a licence, provisions intended to ensure that the Minister retains the right to revoke the licence even if the parties thereto have changed without the appropriate ministerial consent and enhanced ministerial powers relative to the plugging and abandonment of wells, had been the subject of advance consultation with the industry and this particular use of retroactivity did not prove to be contentious. A further recent example of retroactive change lies in the transfer from the UK Government to the Scottish Government of the rights and obligations subsisting in respect of existing landward Production Licences within the Scottish onshore area. See further paras I-9.111 to I-9.126.

of further issues are governed not by the licence but by primary or secondary legislation, which may from time to time be amended so as to become more or less burdensome. In previous editions of this work, the foregoing statement was directed primarily towards fiscal matters and regulatory concerns such as Health and Safety at Work. However, in the post-Wood era, it must also be taken to refer to the MER UK obligation, which has been implemented by means of a combination of primary legislation and binding strategy documentation, and which, it would seem, now forms a new species of economic regulatory law – one which very much strays into what has traditionally been seen as the territory of the licence.

SELECTED ISSUES IN ALLOCATION AND ADMINISTRATION OF LICENCES

Grid system

The petroleum law of some states contains provisions prescribing the size of areas to be offered under licence.[80] By contrast, UK petroleum law views this matter as a purely administrative one within the discretion of the OGA: there is nothing compelling it to adopt a particular grid or block pattern.[81] In practice, standard UK blocks measure approximately 10km by 25km.[82] However, it would be a mistake to imagine that licensed interests in the UKCS are currently held in uniform 250km² blocks. Relinquishments of parts of licensed areas[83] and the fact that consideration is now given to requests for part blocks[84] mean that the block pattern has become increasingly disjointed. This can be seen graphically by examining a map relative to one of the UKCS longer-standing petroleum-producing areas such as Quadrant 22.[85]

I-4.18

---

[80] See eg Australia, where the Offshore Petroleum Act 2006, s 16 provides for a grid system consisting of blocks of 5 minutes latitude by 5 minutes longitude.

[81] Daintith, *Discretion*, para 2105.

[82] PILOT Progressing Partnership Work Group considered this a relatively large block area and considered this fact had at least contributed to certain problems with the UK licensing regime discussed further in Chapter I-4: PILOT PPWG, *The Work of the Progressing Partnership Work Group* (2002), at para 3.2.1.1. While Australian blocks comprise an area of around a third of that size (Daintith, *Discretion*, para 2101), in reality the UK block size lies towards the lower end of the range that one tends to find: see Bunter, *Promotion and Licensing*, at 167.

[83] Discussed further at paras I-4.54 to I-4.55.

[84] See OGA, *Applications for Production Licences: General Guidance* (hereinafter "OGA, *Applications Guidance*"), available at www.ogauthority.co.uk/media/1434/29r_guidance_general.pdf (accessed 9 May 2017), at para 29.

[85] See DECC, Overview map of quadrant 22, available for download at www.og.decc.gov.uk/information/bb_updates/maps/Q22.pdf (accessed 7 May 2017).

Public announcement of available acreage

I-4.19 It has always been the practice of the relevant Minister to advertise publicly the availability of acreage for exploration and development;[86] however, it was not until the entry into force of the Hydrocarbons Licensing Directive that the Minister was formally bound to do so.[87]

Licensing rounds

I-4.20 Petroleum Production Licences are generally issued within licensing rounds. Each round is presaged by the publication, at least 90 days before the closing date for applications, of a notice specifying which areas are to be made available and inviting bids from interested parties. Since the advent of the Hydrocarbons Licensing Directive, as part of the process of ensuring that enterprises located within other member states of the EU receive an equal opportunity to apply for acreage, this notification must be placed in the *Official Journal* of the EU.[88]

I-4.21 At the time of writing, there have been 28 completed seaward licensing rounds since the UKCS was opened for oil and gas exploration. The 29th round has closed but the licences awarded have yet to be announced.[89] Early licensing rounds were held at irregular intervals but, in 2004, the Department then having responsibility for licensing matters (the DECC) stated that it was committed to a regular timetable of one seaward and one landward licensing round per year. This timetable was, however, only ever an aspirational target and was not rigidly adhered to.

Out-of-rounds applications

I-4.22 The OGA may also grant licences outside the regular rounds-based process when presented with compelling reasons to do so. The onus of establishing the existence of compelling reasons lies on the company approaching the Department seeking an out-of-round grant.[90] Even

---

[86] See eg Department of Trade and Industry, *Notice Inviting Applications* under the Petroleum (Production) (Seaward Areas) Regulations 1988, *London Gazette*, 29 June 1990, at 11243 to 11245, available for download at www.gazettes-online.co.uk (accessed 7 May 2017).

[87] Hydrocarbons Licensing Directive, Art 3.

[88] *Ibid*, Art 3(2)(a). See eg the announcement of the 26th Offshore Licensing Round in [2010] OJ C 12/32, available for download via http://eur-lex.europa.eu/JOIndex.do?ihmlang=en (accessed 7 May 2017). Previously, the announcement was made in the *London Gazette*.

[89] OGA, *Licensing Rounds*, available at www.ogauthority.co.uk/licensing-consents/licensing-rounds (accessed 7 May 2017).

[90] OGA, *Applications Guidance*, at p 10: "Out of Rounds Applications".

if the licensing authority is persuaded that there are compelling reasons for making an out-of-rounds award, there is no prospect of a private deal being done. In all but the case of neighbouring or "contiguous" blocks the Hydrocarbons Licensing Directive requires the OGA to advertise the area in the *Official Journal* on 90 days' notice.[91] This is to prevent the use of out-of-rounds applications as a vehicle for a company having acreage let to it on a non-competitive footing. Thus, out-of-rounds applications can come to resemble "a mini licensing round".[92]

In the case of contiguous blocks only, regulations[93] provide for the possibility of a less formal allocation system: where the OGA is satisfied that it is justified for "geological or production reasons" it may choose not to advertise in the *Official Journal*, but only to write to the licensees of the contiguous blocks, inviting applications within a timescale of the OGA's choosing. Thus, even in this case there is the possibility of some competition, albeit only within a limited category of persons.   I-4.23

Acreage selection: environmental issues

The decision as to which areas to invite applications for a particular area in the UKCS is a matter for the OGA's discretion, and in the Ministerial era practice varied greatly.[94] However, the OGA's discretion is now subject to the limitations imposed by two significant EU environmental protection measures, namely Council Directive 92/43/EEC of 21 May 1992 on the conservation of natural habitats and of wild fauna and flora (the "Habitats Directive") and Directive 2001/42/EC of the European Parliament and of the Council of 27 June 2001 on the assessment of the effects of certain plans and programmes on the environment (the "Strategic Environmental Assessment Directive").   I-4.24

Habitats Directive
Put broadly, the effect of the Habitats Directive is to prevent the licensing of acreage where the activities that would be carried under the licence might have a significant impact upon a site of a type which the Directive protects.[95] The Government initially   I-4.25

---

[91] Hydrocarbons Licensing Directive, Art 3(2)(b).
[92] Daintith, Willoughby and Hill, para 1-316.
[93] Petroleum (Production) (Seaward Areas) Regulations 1988 (SI 1988/1213), Reg 7(5), as amended by the Petroleum (Production) (Seaward Areas) (Amendment) Regulations 1995 (SI 1995/1435), Reg 6.
[94] Daintith, *Discretion*, at paras 2403 and 2407.
[95] *Ibid*, at para 2403.

implemented the Habitats Directive by Regulations which applied it not to the UKCS but only to the outer limits of the territorial sea.[96] However, this interpretation was successfully challenged in a judicial review mounted by the environmental pressure group Greenpeace, in which it was held that the Regulations had not adequately implemented the Habitats Directive and that this would now be applied throughout the UKCS under the doctrine of the direct effect of EU Directives.[97] New implementing regulations were promulgated following the judicial review.[98] The relevance of the Regulations for present purposes[99] is that they forbid the OGA to grant any petroleum licence "where [it] considers that anything that might be done or any activity which might be carried on pursuant to such a licence ... is likely to have a significant effect on a relevant site"[100] without first making an appropriate assessment of the conservation implications for the site.[101] In making its assessment, the licensing authority is obliged to consult with the appropriate nature consultation body[102] and may, if it thinks it appropriate, in addition consult with the general public.[103] Subject to the exception described immediately below, the OGA may grant a licence only if it is satisfied that "nothing that might be done and no activity that might be carried out pursuant thereto" would have an adverse effect on the integrity of a relevant site.[104] The OGA is entitled to

---

[96] Conservation (Natural Habitats, etc) Regulations 1994 (SI 1994/2716).

[97] R v Secretary of State for Trade and Industry, ex parte Greenpeace Ltd (No 2) [2000] Env L R 221. The case was brought by Greenpeace as part of its attempt to prevent the development of the West of Shetland Atlantic margin: see S Tromans, "European Environmental Law Goes Offshore" (2000) IELTR 75. Greenpeace have not been wholly successful in securing this objective: subsequent licensing developments relating to this area are discussed at paras I-4.61 to I.4.67.

[98] Offshore Petroleum Activities (Conservation of Habitats) Regulations 2001 (SI 2001/1754) (as amended).

[99] The Directive and its associated Regulations are not just relevant at the time when licences are granted, but also apply to inter alia consents granted pursuant to a UKCS licence: Reg 5(1) read with Reg 2. This aspect of the matter will be further considered at paras I-4.59 to I-4.60 and in Chapter I-11.

[100] Defined by Reg 2 so as to include inter alia special areas of conservations and various sites listed under the Habitats or Wild Birds Directives or which, following consultation with the appropriate nature conservation body (discussed further below), is likely to be included in any forthcoming listing of such sites.

[101] Reg 5(1).

[102] In England and within English territorial waters, the Environment Agency; in Scotland and within Scottish territorial waters, the Scottish Environmental Protection Agency; for the remainder of the UKCS, the Joint Nature Conservancy Council (JNCC). For further information about the JNCC's activities in this regard, see www.jncc.gov.uk/page-1374 (accessed 7 May 2017).

[103] Reg 5(2).

[104] Reg 5(3).

authorise the activity only if, in its opinion, there is "no satisfactory alternative"[105] and if it has certified that the project should be carried out for "imperative reasons of overriding public interest".[106] Where the reason of overriding public interest appertains to human health, public safety or to the "beneficial consequences of primary importance for the environment", the OGA is entitled to issue the certificate itself.[107] However, as an important check against the abuse of OGA discretion, in all other cases it must obtain a consenting opinion from the European Commission.[108]

Environmental Impact Assessment Directive

An EU environmental impact regime which requires relevant author- I-4.26 ities to consider the environmental effects of "certain public and private projects"[109] has existed since 1985. That regime continues to be in force[110] and stipulates that certain projects "likely to have significant effects on the environment by virtue *inter alia*, of their nature, size or location" require to be the subject of an environmental impact assessment (EIA) before the relevant authority can consent to their being undertaken.[111] The extraction of petroleum is included among the activities listed in the annex to the Directive comprising projects which may require an assessment at the discretion of the member state.[112] The EIA regime does not forbid development in the event of an unfavourable EIA being received, or even seek to define what an unfavourable EIA is; instead, it seeks to ensure that the authorities in question are provided with relevant environment information when exercising their decision-making function.[113]

---

[105] Reg 6(1)(a).

[106] Reg 6(1)(b).

[107] Reg 6(2)(a). Social and economic reasons are both expressly recognised as potentially valid.

[108] Reg 6(2)(b).

[109] Council Directive 85/337/EEC of 27 June 1985 on the assessment of the effects of certain public and private projects on the environment (hereinafter "Environmental Impact Assessment Directive"), Art 1.

[110] The original Directive of 1985 has been amended by Directives 97/11/EC, 2003/35/EC and 2009/31/EC; all references are to the Regulations as so amended. The effect of the 2009 Regulations is to bring carbon capture, transportation and storage schemes under the ambit of the regime. For further information see European Commission, *Environmental Impact Assessment*, available at http://ec.europa.eu/environment/eia/eia-legalcontext.htm (accessed 7 May 2017).

[111] Art 2(1).

[112] Art 4(2), read with Annex 2, para 2(f) and (g).

[113] This aspect of the Directive is demonstrated by the provisions of Arts 5–7. See also C Reid, *Nature Conservation Law* (2nd edn, 2002) (hereinafter "Reid, *Nature Conservation Law*"), at para 8.3.13: an EIA "regulates the process by which decisions are reached, not substantive outcomes".

Strategic Environmental Assessment Directive

I-4.27 One of the perceived weaknesses of the EIA system is that it applies
on a project-by-project basis. The decisions reached in this way
may lack the wider view of the policy made possible by taking
decisions at a more strategic level; thus there is the danger of a lack
of appropriate focus on environmental issues and of an incremental
erosion of environmental protection.[114] The Strategic Environmental
Assessment Directive,[115] which required to be implemented by all
member states by the end of July 2004, attempts to address this
deficiency by providing that a strategic environmental assessment
(SEA)[116] shall be carried out for draft plans and programmes which
are likely to have significant environmental effects.[117] Certain plans
or programmes (including those pertaining to the extraction of
petroleum[118] and those determined to require an assessment under
the Habitats Directive)[119] are deemed, subject to a *de minimis*
provision,[120] to be likely to have significant environmental effects.
The SEA process, like the EIA, is essentially concerned with gathering
information as an aid to informed decision-making.[121] The authority
is not bound by a SEA but must take it into account.[122] However, this
does not mean that the process is without effect: in the 24th offshore
licensing round, 21 blocks which would otherwise have been made
available were withheld for environmental reasons, at least "for the
present".[123]

Marine spatial planning

I-4.28 Although not, at present, of direct relevance to oil and gas licensing,
such matters being excluded from its ambit, it is appropriate to
give some consideration to the emergent issue of marine spatial
planning (MSP). MSP is a planning process for the sea which aims

---

[114] Reid, *Nature Conservation Law*, para 8.3.14.

[115] Directive 2001/42/EC.

[116] Defined in Art 2(b). The term "strategic environmental assessment" does not appear in
the Directive itself but has been adopted by commentators in order to differentiate these
assessments from those required under the EIA regime.

[117] Art 3(1).

[118] Specified by Art 3(2)(a) as being "those prepared for agriculture, forestry, fisheries,
energy, industry, transport, waste management, water management, telecommunications,
tourism, town and country planning or land use and which set the framework for future
development consent of projects listed in Annexes I and II to [the Environmental Impact
Directive]".

[119] Art 3(2)(b).

[120] For which see Art 3(3).

[121] See eg Arts 5, 6 and 7.

[122] Art 8.

[123] For details and consultations, see www.gov.uk/guidance/offshore-energy-strategic-
environmental-assessment-sea-an-overview-of-the-sea-process (accessed 8 May 2017).

to consider the "big picture" of ocean management in a process of assessing and balancing existing and future marine activities, often through the ecosystem approach.[124] Whilst this concept is increasingly regarded as important at both a global and a local level, it is highly complex,[125] and in the past has lacked any internationally recognised definition, meaning different things to different people.[126] Worldwide common practices and processes, however, are now emerging, building on existing law, new statutes and the development of policy.

Although the UK can be regarded as well advanced in the implementation of MSP, the concept (as in most other jurisdictions) has evolved around existing offshore oil and gas developments rather than those developments being assimilated into it. The licensing of oil and gas developments in the UK, acts pertaining to the construction or maintenance of petroleum pipelines, or anything done for the purposes of establishing or maintaining an offshore installation are all exempted from the MSP process.[127] With increasing competition for areas of the sea as a result of the activities of a more diverse range of sea-users, however, both the legal and the policy landscape around MSP will increasingly impinge on offshore oil and gas development; and some have argued that bringing oil and gas developments within the ambit of the process will provide benefits, not just for other sea-users, but also for the industry itself.[128] The principal benefit identified is the avoidance of duplication of processes and the resultant avoidance of effort and cost – an issue which has particular resonance in the post-Wood environment. Future editions of this work may therefore require to address in greater detail the nexus between oil and gas activities and MSP. Nor should it be thought that the existence of other regulatory provisions makes MSP irrelevant to the oil and gas industry. Offshore renewable energy installations already require marine licences under the terms of the Marine and Coastal Access Act 2009; and, in Scotland, the Marine (Scotland) Act 2010 can trigger processes in relation to the Habitats

I-4.29

---

[124] P M Gilliland and D Laffoley, "Key elements and steps in the process of developing ecosystem – based marine spatial planning", 32 (2008) *Marine Policy* 787.

[125] S Jay et al., "Coastal and Marine Spatial Planning: International Progress in Marine Spatial Planning" 27 (2013) *Ocean Year Bulletin* 171.

[126] N Soininen and D Hassan, "Marine spatial planning as an instrument of sustainable ocean governance", in D Hassan, T Kuokkanen and N Soininen (eds), *Transboundary Marine Spatial Planning in International Law* (2015).

[127] Marine and Coastal Access Act 2009, s 77.

[128] See eg RSPB, "Report: Potential benefits of marine spatial planning to economic activity in the UK" (2004), available at www.rspb.org.uk/Images/MSPUK_tcm9-132923. pdf (accessed 7 May 2017), para 6.4.

Directive and EIA requirements that (as discussed above) are already relevant to oil and gas licensing.[129]

I-4.30   The marine planning process can be regarded as a vehicle through which other consents and approvals are pulled together and considered in a holistic manner. Decisions on marine licensing have to be taken in accordance with the relevant marine plans. Marine plans are beginning to emerge in UK waters with the passing of the Marine and Coastal Access Act 2009 and subsequent legislation by the devolved administrations. Further impetus for the development of the process was provided by Directive 2014/89/EU, establishing a European framework for maritime spatial planning.[130] Oil and gas have been included as possible activities for inclusion in plans and as part of a marine planning process, but without prejudice to member states' competences.[131] Relevant policy already exists, with the Marine Policy Statement appearing in 2011 as a requirement of the Marine and Coastal Access Act 2009.[132] The Statement sets out high-level policy objectives for the UK waters as a whole and can be regarded as providing a framework within which further, more detailed marine spatial plans can be created.[133] The Marine Management Organisation (MMO) is developing marine plans for English waters.[134] Marine Scotland has created Scotland's National Marine Plan, published in 2015 and covering both inshore and offshore areas.[135] Regional plans are now being created for Scotland's 11 Marine regions.[136]

I-4.31   Oil and gas licensing and regulation have developed incrementally, responding to demands of the industry, and environmental and

---

[129] See I-4.25 to I-4.26.

[130] EU Directive 2014/89/EU of the European Parliament and of the Council of 23 July 2014 establishing a framework for maritime spatial planning.

[131] Article 8 (2) installations and infrastructures for the exploration and extraction of oil, of gas and other energy resources, of minerals and aggregates and for the production of energy from renewable sources. Brexit may, of course, make the specific European dimension of this question redundant, but much will depend on the timing and nature of Brexit. It should not be assumed that the Directive is a dead letter in the UK.

[132] Marine and Coastal Access Act 2009, s 46.

[133] HM Government, 2011. *UK Marine Policy Statement*, HM Government, Northern Ireland Executive Scottish Government Welsh Assembly Government 2011, 3.3.7 to 3.3.15. Available for download at www.gov.uk/government/publications/uk-marine-policy-statement (accessed 29 August 2017).

[134] See https://marinedevelopments.blog.gov.uk/2017/04/12/production-of-marine-plans-an-update (accessed 7 May 2017).

[135] Scottish Government, *Scotland's National Marine Plan: A Single Framework for Managing Our Seas* (2015). Available for download at www.gov.scot/Publications/2015/03/6517 (accessed 29 August 2017).

[136] See www.gov.scot/Topics/marine/seamanagement/regional/Boundaries/SMRmap (accessed 8 May 2017).

other controls. It is nevertheless a major marine use and activity and, with decommissioning commencing in the North Sea, it continues to evolve and develop in ways that impact on other users and the wider marine environment. Decommissioning consultation has already created a public participation element that has not otherwise been embedded within the oil and gas regulatory processes to date.[137] It is submitted that the further maturing of MSP plans and processes, backed by international commitments, should result in offshore oil and gas activities being increasingly embedded within MSP. If it is not, as far as UK waters are concerned, the UK regime will lack the holistic approach that is of the essence of good management of and effective planning for its marine area.

Consideration

Section 3(3) of the Petroleum Act 1998 provides that licences shall be granted in exchange for "such consideration (whether by way of royalty or otherwise) as the OGA with the consent of the Treasury may determine". The discretion granted to the OGA by this piece of legislation has not been fettered, either by statutory instrument or otherwise. The width with which this provision has been drafted is perhaps best demonstrated by noting that it has permitted the licensing authority occasionally to utilise cash premium bidding, not usually a feature of UK oil and gas law, without any requirement for primary, or even secondary, legislation.[138]

I-4.32

In the early years of oil and gas production, royalty payments, consisting of a state claim to 12.5 per cent of the market value of produced oil and gas,[139] constituted a major element of the UK's revenue from oil and gas production.[140] However, amid concerns that royalty had an inhibiting effect upon exploration and development activity, a succession of provisions were enacted to exclude from the requirement to pay royalty the licensees of fields which received their development consent after 1 April 1982[141] before a declaration was issued "irrevocably revoking" royalty payments with effect from 1 January 2003.[142] Ministerial decree by press release is an unorthodox

I-4.33

---

[137] See Chapter II-11.

[138] The UK experimented in the 4th, 8th and 9th licensing rounds with a two-tier system whereby cash premium bidding – effectively an auction – was utilised in relation to a relatively small number of fields alongside the usual discretionary arrangements. A variant upon this system was in addition utilised in the 7th licensing round. For a fuller account, see Daintith, Willoughby and Hill, at paras 1-317 to 1-322.

[139] Daintith, *Discretion*, para 9213.

[140] See the statistics at HMRC, *Government revenues from UK oil and gas production*, available at www.hmrc.gov.uk/stats/corporate_tax/table11_11.pdf (accessed 7 May 2017).

[141] Daintith, *Discretion*, para 9107.

[142] Daintith, Willoughby and Hill, para 1-336/1.

way of effecting a change to a legal instrument but the Government has declined to take measures to effect a more formal amendment of the relevant licences.[143]

I-4.34   The UK Government's revenue from petroleum now consists solely of licence fees (comprising application fees and area rental) and various elements of taxation.[144] Application fees are set by delegated legislation.[145] Area rental fees are not set by regulation but are determined, historically by the Minister in accordance with promulgated guidance,[146] and now by the OGA, advertised in advance and included as an annexation to the licence. Although area rental fees are sufficiently high to have been believed to act as a barrier both to certain types of development[147] and to development by certain types of company,[148] the state's revenue from this source has historically been dwarfed by the sums raised through taxation. However, in view of falling rates of production and the sharp and prolonged drop in the oil price between 2014 and 2017, tax revenue from oil has plummeted. Indeed, when decommissioning tax relief is taken into account, tax flow is, at the present time, around zero.[149]

Assessing competing applications

I-4.35   The manner in which the licensing authority assesses applications has altered markedly over time. Two main factors can be identified for this change: the requirements of European law and the industry's own desire for greater transparency. These will be discussed in turn.

EU law

I-4.36   In the event of competition between two or more interested parties, the OGA uses a variety of published criteria in order to determine to whom a petroleum licence should be granted. Many of the criteria that have been used by the OGA over the years are uncontroversial from the standpoint of EU anti-discrimination law and continue to

---

[143] Daintith, *Discretion*, para 9213.

[144] For further information on the tax regime within the UKCS, see Chapter I-7.

[145] The principal application fees were updated by The Oil and Gas Authority (Fees) Regulations 2016 (SI2016/904). The current application fee is £2,100 for Seaward Production Licences and £1,400 for (Landward) Petroleum Exploration and Development Licences.

[146] OGA, *Applications Guidance* per Annexe 1. A schedule (in current practice Sch 2) containing the rental charges is incorporated into the licence by Model Cl 12.

[147] This is one of a number of factors which led to the creation of Frontier Licences, discussed further at paras I-4.61 to I-4.68.

[148] This is one of a number of factors which led to the creation of the Promote Licence, on which see paras I-4.69 to I-4.73.

[149] Office for Budget Responsibility, *Oil and Gas Revenues*, available at http://budget-responsibility.org.uk/forecasts-in-depth/tax-by-tax-spend-by-spend/oil-and-gas-revenues (accessed 7 May 2017).

be used today.[150] However, some of the factors have proven to be more problematic. In 1985, the UK Government was warned that the OGA's willingness to take into account the prospective licensee's contribution to the UK economy, readiness to involve UK organisations to participate in research and development activities, and readiness to offer a full and fair opportunity to UK firms to compete for orders for goods and services were potentially discriminatory.[151] The initial effect of this warning was not particularly salutary: the Government stopped publishing detailed information on how it would determine applications.[152] However, the adoption of the Hydrocarbons Licensing Directive changed this position. The implementing regulations provide that "every application" for a licence shall be determined on the basis of the following criteria:

(a) the technical and financial capability of the applicant;
(b) the way in which the applicant proposes to carry out the activities that would be permitted by the licence;
(c) in a case where tenders are invited, the price the applicant is prepared to pay in order to obtain the licence; and
(d) where the applicant holds, or has held, a licence of any description under the Petroleum (Production) Act 1934, any lack of efficiency and responsibility displayed by the applicant in operations under that licence.[153, 154]

The licensing authority is specifically empowered to refuse an application for a licence, even where there is no competition.[155] In the event that two or more applications for a licence have equal merit when assessed according to the principal criteria, "other

I-4.37

---

[150] For example, the technical and financial capability of the licensee has been used since the early days of the regime: see para 5(a) of the Notice pursuant to the Petroleum (Production) Regulations 1976 of 8 August 1978: 1978 *London Gazette*, issue 47612, at 9508, available by searching www.london-gazette.co.uk

[151] Daintith, Willoughby and Hill, para 1-317. See eg the Notice of 8 August 1978: 1978 *London Gazette*, issue 47612, at 9508, paras 5(e) and (h).

[152] Ibid, para 1-317. See eg the Notice of 25 July 1986: 1986 *London Gazette*, issue 50609, at 9837, para 7.

[153] A nice point, which does not yet seem to have been tested, is whether the "and" in Reg 3(1)(d) requires to be interpreted adjunctively or disjunctively; in other words, whether a company can lose standing only as a result of exhibiting both "a lack of responsibility and efficiency" or whether a lack of either "responsibility" or "efficiency" will be sufficient. This may be material. It is not difficult to conceive of a set of circumstances which could allow one to hold a licensee inefficient, but not irresponsible.

[154] Hydrocarbons Licensing Directive Regulations 1995, Reg 3(1). In practice the third criterion is usually omitted from notices as the UK does not generally operate a cash auction system. See eg the announcement of the 21st Offshore Licensing Round in [2003] OJ C27/03, para 9.

[155] Hydrocarbons Licensing Directive Regulations 1995, Reg 3(1).

relevant criteria" may be applied.[156] Neither the principal criteria nor any such "tie-break" criteria may be applied in a discriminatory manner.[157] All criteria to be used in determining applications must be included by the OGA in the notice inviting applications.[158] The OGA retains the power to refuse to grant a licence on the basis of national security where the applicant is effectively controlled by a foreign state other than a member state of the EU (or by nationals of such a state).[159]

Enhanced transparency

I-4.38   The Oil Industry Taskforce[160] published its report *A Template For Change* in 1999. Among other innovations directed towards reducing the cost of oil and gas activity in the UKCS and creating "a climate for the UKCS to retain its position as a pre-eminent active centre of oil and gas exploration, development and production",[161] the report identified the need to improve licence administration.[162] This recommendation led to further work by the Taskforce which in turn led to it recommending the introduction of a marking scheme for applications. This recommendation was duly adopted by the Government. Recent practice has been for the marking scheme against which applications will be measured is now published in advance of each licensing round, and for the licensing authority to base its assessment upon the technical understanding demonstrated by the applicant at interview, the generation of valid prospectivity derived from evaluation of available data, the quality of the work that it has already done and the proposed work programme.[163] Thereafter, the marks awarded to the successful applications in blocks in respect of which there was competition have been published.[164] No marks have

---

[156] *Ibid*, Reg 3(2).

[157] *Ibid*, Reg 3(4).

[158] Hydrocarbons Licensing Directive Regulations 1995, Reg 5. In practice the Government has not always formally publicised "tie-break" criteria: see eg the Notice announcing the 21st Offshore Licensing Round at [2003] OJ C27/03 which does no more than recast the Reg 3(1) criteria in slightly different language.

[159] Hydrocarbons Licensing Directive Regulations 1995, Reg 3(4).

[160] The Oil Industry Taskforce was a group consisting of Government Ministers, and civil servants, members of other oil industry groups and initiatives, and senior oil executives. See Oil Industry Taskforce, *A Template for Change* (1999) (hereinafter "Oil Industry Taskforce, *Template*"), Appendices 2 and 3. PILOT is the group's successor organisation.

[161] Oil Industry Taskforce, *Template*, Foreword by Rt Hon Stephen Byers, Secretary of State for Trade and Industry.

[162] *Ibid*, at 12.

[163] OGA, *Applications Guidance*, at para 41. The OGA's *Technical Guidance* and *Financial Guidance* is available at www.ogauthority.co.uk/licensing-consents/licensing-rounds (accessed 8 May 2017).

[164] The winning marks for the 29th Seaward Round process may be seen in OGA,

been awarded in respect of the Reg 3(1)(a) criteria of technical and financial capability. These are threshold criteria which are either met or not: any deficiencies in these respects cannot be made up by exceptional strengths in other areas of the application. In practice, the licensing authority's published guidance on the marks scheme focuses wholly upon the details of what activities the licensee proposes to carry out: in other words, the criterion set out in Reg 3(1)(b). The Guidance said nothing about the extent to which the Department would attach weight to the Reg 3(1)(d) criterion, i.e. any lack of efficiency and responsibility historically displayed by the applicant.

COMMENTARY ON RECENT FORMS OF UK OIL AND GAS LICENCES

Seaward licences

Exploration Licence

An Exploration Licence is a licence to search for petroleum "in any **I-4.39** seaward area and in those parts of any landward area below the low water line"[165] – in other words, within the territorial sea and to the outer limit of the UK's Exclusive Economic Zone. Areas in respect of which a Production Licence is extant are excluded from the territorial extent of the Exploration Licence unless the production licensee agrees to the proposed exploration activities.[166] Exploration Licences are non-exclusive in nature[167] and now run for a period of three years, renewable for a further three years on the provision of three months' notice.[168] The rights granted under such a licence are tightly circumscribed.[169] In practice such licences are used to enable the carrying out of seismic surveys and other methods of geological prospecting. An Exploration Licence does not convey a right to "get" petroleum, to drill wells for the production of petroleum, or indeed to drill any well with a depth greater than 350m below the seabed.[170] Thus, despite the potentially confusing similarity in terminology, Exploration Licences carry insufficient rights to authorise

---

Winning Marks By Block List, available at www.gov.uk/government/uploads/system/uploads/attachment_data/file/370434/28R_MarksByBlock (accessed 7 May 2017).

[165] The Offshore Exploration (Petroleum, and Gas Storage and Unloading) (Model Clauses) Regulations 2009 (SI2009/2814), Sch, Model Cl 2.

[166] Model Cl 2.

[167] The licence and liberty is granted "in common with all other persons to whom the like right may have been granted or may hereinafter be granted". See Model Cl 2.

[168] Model Cl 4.

[169] Model Cl 3.

[170] *Ibid.*

meaningful exploration drilling.[171] A Production Licence is required for that activity.

I-4.40   Because of the limited nature of the rights conferred by an Exploration Licence, many of the provisions found in other sets of model clauses are unnecessary and therefore omitted. That said, the model clauses still run to 24 paragraphs and, among other provisions, require the licensee to avoid harmful methods of working, keep records and samples, file regular returns with the OGA and provide advance notice of certain activities to the Ministry of Defence and representatives of the local fishing industry.[172]

Production licence

I-4.41   The Production Licence is the classic licence to "search or bore for or get"[173] petroleum situated on the seaward side of the low water line.[174] It is an exclusive licence.[175] Initially, only one type of Production Licence was offered, ostensibly on rigid terms and conditions, but subject to the possibility of being offered on bespoke terms if the circumstances could be shown to justify it. However, as time has gone by, the state has responded to the challenges associated with development in frontier areas and the need to attract new entrants to the sector given the UK's status as a maturing oil and gas province by providing greater flexibility in the system. Initially, this was done by providing the petroleum industry with a suite of variant licence types from which to choose, each having its own dedicated set of model clauses.[176] More recently, all the model clauses were collapsed into one more flexible set of clauses, although, by guidance and practice, the variant licence types were informally maintained. The recently completed 29th round can be seen as a transitional round where the new Innovate Licence is being offered, supported by the same single set of model clauses in combination with the upfront announcement of the options available to potential licensees and the practices to

---

[171] Oil and gas is not generally encountered on the UKCS in meaningful quantities at such shallow depths. Most UKCS finds have been encountered at depths in excess of 1,000m.

[172] Model Cll 9, 11 and 13, 12, 22 and 23 respectively. For a fuller discussion of the extent of the environmental protections provided by this set of model clauses, see paras I-11.23 to I-11.29.

[173] The Petroleum Licensing (Production) (Seaward Areas) Regulations 2008 (SI2008/225) Sch Model Cl 2.

[174] That is, the line determined by the Petroleum (Production) (Seaward Areas) Regulations 1988 (SI 1988/1213), Reg 3(1) read together with Sch 1.

[175] Model Cl 2 states that the licensee has "exclusive licence and liberty during the continuance of this licence and subject to the provisions hereof to search and bore for, and get, Petroleum in the sea bed and subsoil" of the licensed area.

[176] That is, the traditional licence and the Frontier and Promote Licences.

be adopted by the OGA.[177] By the time of the 30th licence round, it is anticipated that a new set of model clauses will have emerged, specifically designed to support the Innovate Licence.

I-4.42 In these circumstances, it might be thought that there is little point in discussing the licence types that preceded the advent of the Innovate Licence. However, this is not so. As has already been noted, the terms of UK petroleum licences are fixed at the point of the licence's grant; instances of express retroactive amendment apart, the issuing of a new set of model clauses has no bearing upon the terms of existing licences. Thus the overwhelming majority of Seaward Production Licences extant upon the UKCS will, for many years, be one or another of the historic types.

I-4.43 **Traditional (or standard) Production Licence.** The great majority of the Production Licences presently at large are traditional Production Licences.[178] Although the traditional Production Licence has been complemented by the introduction of alternative licensing options, it continues to be a highly significant licensing vehicle. In the last seaward licensing round to maintain the distinction between different types of Production Licence (the 28th), of the 172 licences awarded, the great majority – 119 – were issued on traditional terms.[179]

I-4.44 The terms on which traditional Production Licences are offered have evolved over time. Some of the developments shall be noted in passing, but this discussion will focus on the 2008 set of model clauses.[180] However, as has already been noted, the terms of UK

---

[177] See the announcement of the round at [2016] OJ C 244/05; DECC, Applications for Production Licences: General Guidance (2014), available at www.gov.uk/government/uploads/system/uploads/attachment_data/file/540557/29R_Guidance_General.pdf (accessed 24 August 2017), at pp 3–4.

[178] When the alternative licensing models further described at paras I-4.61 to I-4.68 were first introduced, "standard" was the term used in statutory instrument to describe the pre-existing licence format to which the new forms provided an alternative: see the Petroleum Licensing (Exploration and Production) (Seaward and Landward Areas) Regulations 2004, Sch 4. However governmental guidance and press releases (see eg DTI, *North Sea oil and gas applications, 35-year high continues*, dated 19 June 2006, commonly referred to such Production Licences as "traditional" ones. Despite its lack of statutory basis the term "traditional" appears to have taken root and so that term will be used throughout this chapter.

[179] These figures are arrived at using data available from two databases, "Potential Awards in the 28th Round – By Block" and "Potential Awards in the 28th Round – By Block – Second Tranche", both available at the 28th Offshore Licensing Round tab in the OGA website www.ogauthority.co.uk/licensing-consents/licensing-rounds (accessed 7 May 2017).

[180] That is, the model clauses incorporated into licences granted under the 26th Seaward Licensing Round: those to be found in Schedule to the Petroleum Licensing (Production) (Seaward Areas) Regulations 2008, as amended by the Petroleum Licensing (Amendment) Regulations 2009 (SI 2009/3283).

petroleum licences are fixed at the point when the licence is granted; thus many traditional Production Licences continue on terms and conditions which are, to some extent, different from those which will be described below.

I-4.45   *Area rental payments.* Area rental payments[181] for traditional licences – as with all the legacy licensing variants – were not determined either by primary or secondary legislation but are instead published in guidance, disclosed by the licensing authority in the notice inviting applications, and then included in the licence itself (under present practice, by Model Cl 12 read together with Sch 2 to the licence). The area rental payments for the most recent traditional Production Licences commenced at £150 per km², staying at that level throughout the licence's initial term before doubling in the first year of its second term and escalating by annual increments of £900 until reaching £7,500 per area factor, subject always to the possibility that it will increase through the accompanying indexation provisions.[182]

I-4.46   The strategy of starting rental acreages at a relatively low figure was logical in that it provided the licensee with some comfort in the early years of the licence, where only exploration and development work will be taking place and no oil, and therefore no revenue stream, will yet be flowing. It also provides the licensee with some measure of incentive, first, to explore and develop quickly, and, second, to surrender increasingly expensive acreage which is not being utilised. However, the standard consideration regime was found to be inadequate in itself to achieve any of those purposes. This is one of a number of factors which led both to the introduction of variant types of licence variants in 2003 and 2004[183] and to the Fallow Area and Stewardship initiatives.[184]

I-4.47   *Work programme.* The term "work programme" (as used both in the traditional licence and in current licensing practice, under the Innovate Licence) refers not to the totality of the work that will be carried out during the life of the licence, but only to exploration activities.[185] Moreover, it is used in two distinct senses. The "initial work programme" narrates the exploration work that the licensee

---

[181] Area rentals are an aspect of consideration, which was discussed in more general terms at para I-4.32.

[182] *Ibid*, Annexe 1, para 3.

[183] Discussed at paras I-4.61 to I-4.68 and I-4.69 to I-4.73.

[184] Discussed in the Appendix.

[185] The work to be undertaken in later phases of the licences will be described not in a work programme but in a development and production programme. These are discussed in greater detail in the Appendix.

has agreed to undertake in exchange for receiving the grant of licence, short particulars of which are published on the licensing authority's website at the time the licence is allocated.[186] In addition, the licensee might be called upon by the OGA to prepare a further "appropriate work programme". These categories will be discussed in turn.

Initial work programmes are effectively agreed between the licensee and the licensing authority. When applying for a licence, the licensee states the exploration work it intends to carry out; by agreeing to offer a licence to the applicant, the authority (which, for these legacy licences, would have been the DECC or one of its predecessors) agrees that this work programme is acceptable. I-4.48

Such work programmes are now somewhat more detailed than was formerly the case. Many elderly licences contain surprisingly vague work programmes (expressed, in accordance with the then current terminology, as "working obligations"), obliging the licensee to carry out, without further definition or qualification, "seismic survey work" and/or "drill at least one exploration well".[187] More modern work programmes are expressed with a greater degree of particularisation. For example, they are likely to specify a minimum quantity and type of seismic data which must be acquired and/or to state the depth to which the licensee is expected to drill.[188] Moreover, work programmes have come to differentiate between differing degrees of commitment.[189] A "firm" well drilling commitment is an unequivocal undertaking to drill, appropriate only where the licensee is certain that they intend to execute these works.[190] A "contingent" well commitment is in effect a firm commitment which the OGA may waive in the event that a specified study or evaluation establishes that drilling is likely to be a futile exercise.[191] "Drill or drop" is a looser commitment by which the licensee undertakes either to proceed with drilling or to surrender the licence. The principal benefit of drill or drop is that, contrary to the position with the other degrees of commitment, the licensee does not lose good standing with the OGA in the event that it decides not to carry out the works.[192] By contrast, I-4.49

---

[186] See eg DECC, *Potential Awards in the 28th Round*, initially published by DECC but now available at www.ogauthority.co.uk/media/1513/28r_potential_awards_by_block. xlsx (accessed 7 May 2017).

[187] Sample licences provided to the author.

[188] Sample licences provided to the author. The level of detail contained within the licences is greater than in the abbreviated work obligations posted on the OGA website.

[189] Although this discussion takes place in the context of the legacy traditional licence, it is true also for other licence variations and the new Innovate Licence.

[190] OGA, *Technical Information*, para 9, read with Annexe 1, Definitions, at para 4.

[191] *Ibid*, para 9, read with Annexe 1, Definitions, at para 5.

[192] *Ibid*, para 9, read with Annexe 1, Definitions, at para 6.

a failure to carry out without good reason either a firm commitment or a contingent commitment where the condition precedent has been satisfied is a factor which can lead the OGA to conclude that there has been a lack of efficiency and responsibility on the part of the licensee.[193]

I-4.50   As was noted in the discussion at para I-4.38, the materiality of proposed work programmes weighs heavily in the marking scheme used to choose between competing applications. This provided a powerful incentive for the licensee to offer to carry out an extensive work programme, at least in respect of acreage where the applicant knows or anticipates it will face competition. However, when the licence is granted, the work programme will be incorporated into it,[194] and the continuance of the licence from its initial term into a second term is made contingent upon the work programme's successful completion, subject only to the authority's discretion to waive or modify this requirement.[195] This, and the potential loss of good standing discussed above, should have served to discourage companies from illegitimately securing acreage by promising much exploration activity and then delivering little. However, historically there was at least some concern within the industry that certain prospective licensees might be securing acreage by offering firm work commitments in circumstances where only a lesser degree of commitment was justified, then were failing to carry their commitments through and attempting to justify their failings by making reference to supposedly unforeseen difficulties which were, in fact, perfectly predictable.[196] It may be that the better resourced and more hands-on OGA will be better placed than its predecessors to combat any sharp practice in this regard.

I-4.51   In addition to this initial agreed work programme the model clauses state that the OGA is entitled, at any time before the end of the licence,[197] to demand the preparation and submission by the licensee of "an appropriate work programme".[198] A work programme is "appropriate" for these purposes if it is one which would be prepared by a licensee seeking to exploit its licence rights

---

[193] Hydrocarbons Licensing Directive Regulations 1995, Reg 3(1)(d), discussed further at para I-4.36.

[194] Under current practice, the work programme will be annexed to the licence as Sch 3.

[195] Model Cl 4(2)(b).

[196] Daintith, *Discretion*, para 3227.

[197] Model Cl 16(2). Where the licensee has carried out a programme during a part of the term of this licence, the Minister may serve notice in pursuance of Model Cl 16(2) in respect of another part of that term: Model Cl 16(7).

[198] Model Cl 16(2). This power was one of a number of provisions controversially introduced with retrospective effect by Petroleum and Submarine Pipelines Act 1975: s 18, read with Sch 2, Pt II, Model Cl 14.

to the best commercial advantage, and who had the competence and resources necessary so to do, and if it is one which could reasonably be expected to be completed by such a licensee before the end of the life of the licence.[199] This is an exacting standard, and one which takes no account of the licensee's other commitments. If no programme is served the OGA may revoke the licence or part thereof.[200] If a programme is served, the OGA may reject it if it is not satisfied that it is "appropriate"; if the parties are in dispute on this point the matter may be referred to arbitration.[201] The OGA may revoke the licence in whole or in part in the event of repeated failures to submit an appropriate programme[202] or a failure to undertake the works set out in an approved programme.[203] This is an important and powerful provision which effectively means that – quite apart from the extensive powers that the OGA enjoys as a result of the implementation of the Wood Review's recommendations on maximising economic recovery[204] – it is entitled to demand that the licensee carry out exploration activities well into the planning and production stages of the licence. This model clause is discussed further in the context of the Fallow Areas initiative in the Appendix.

Term and relinquishments. In some states, separate licences are   I-4.52 granted to regulate the distinct phases of the life of an oil and gas development. Under such regimes, a licensee will hold two or more licences in succession as the discovery progresses from one phase to the next.[205] The traditional Production Licence, by contrast, covers the various phases of work on the discovery. Most of the traditional licences extant upon the UKCS are divided into three distinct terms: the initial term, which is focused on exploration; the second term, which may be viewed as a phase of development and preparation that acts as a bridge between the exploration and production phases and the production period.[206] This format, which was devised by the PILOT Progressing Partnership Work Group (PILOT PPWG) following a review of UK licensing law and practice,[207] was introduced[208] in 2002 for the 20th licensing round and was in use

---

[199] Model Cl 12(2).
[200] Model Cl 41(2)(b), read together with Model Cl 16(2) and (6).
[201] Model Cl 16(4)(a).
[202] Model Cl 41(2)(b), read together with Model Cl 16(3), (4) and (6).
[203] Model Cl 41(2)(b), read together with Model Cl 16(3), (5) and (6).
[204] On which see further Chapter I-5.
[205] See eg Canada and Australia: see Daintith, *Discretion*, at para 4001. Until relatively recently this approach was also taken by the UK in relation to landward licences: see para I-4.79.
[206] Model Cl 3.
[207] PILOT PPWG, *The Work of the Progressing Partnership Work Group* (2002), para 3.2.
[208] Rather irregularly: revised model clauses were not prepared in advance of the 20th

continuously until the advent of the Innovate Licence.[209] Before
the three-term structure was settled on, a variety of configurations
of terms had been used,[210] none of which was found to be wholly
satisfactory and some of which were criticised for failing properly
to secure the state's interests due to their lack of sophistication and
excessive aggregate length.[211]

I-4.53   Prior to the introduction of the Innovate Licence, the general
philosophy of the modern traditional Production Licence has been
that, by the time that any given term comes to an end, certain tasks
or activities should have been carried out. If they have, and if the
licensee notifies the OGA that it wishes the licence to continue and
is willing to surrender a certain amount of acreage, then the licence
will carry on. Otherwise, the licence becomes susceptible to being
brought to an end as a result of the licensee's failure to comply with
its obligations.[212]

I-4.54   Broadly speaking, the initial term is intended to permit the
licensee to carry out the exploration work set out in the licence's
work programme. Under the previous set of model clauses, the
licensee was permitted a period of four years to carry out this
work and was under a strict obligation, at the end of that period,
to surrender an area comprising at least half of the initial licensed
area,[213] subject only to the proviso that the licensee would be
entitled to retain at least 30 sections.[214] In the 2008 iteration of the
model clauses, however, the position changed markedly. The initial
term was no longer fixed by statutory instrument at four years,[215]

---

or 21st rounds; instead the notices inviting applications indicated that licences would
be issued in accordance with the then-extant model clauses subject to the deletion of
certain clauses and the incorporation of a number of bespoke provisions: see, for the 20th
Offshore Round, United Kingdom Government notice concerning Directive 94/22/EC of
the European Parliament and Council of 30 May 1994 on the conditions for granting and
using authorisations for the prospection, exploration and production of hydrocarbons
[2002] OJ C12/03, at paras 2 and 13–16; and for the 21st round, and the like notice
of 5 February 2003, [2003] OJ C27/03. In due course the changes were formally introduced into UK petroleum law by the Petroleum Licensing (Exploration and Production)
(Seaward and Landward Areas) Regulations 2004.

[209] As we shall see, the Innovate Licence has some flexibility around term structure,
envisaging that where exploration is not necessary, the licence can progress directly to the
second term. However the three-term structure is generally retained, although there are
significant alterations to the structure of the first term. See paras I-4.75 to I-4.78.

[210] See Daintith, *Discretion*, para 3502.

[211] PILOT PPWG, *Work*, paras 3.1.2, and 3.2.1.1 and 3.2.1.2.

[212] Model Cl 3(2).

[213] 2004 Regulations as amended, Sch 4, Model Cl 4(4).

[214] *Ibid*, Sch 4, Model Cl 4(5).

[215] Strictly, these matters were never entirely "fixed" in that the Minister always enjoyed
the power to modify or restrict the application of the model clauses in a particular case:
Petroleum Act 1998, s 4(1)(e). Thus the system provided for a set of general norms subject

but became one of the many matters which fell to be determined by the OGA and included in Sch 5 to the licence.[216] The extent of the licensee's surrender obligation is made similarly unclear by the 2008 model clauses: the obligation is to surrender no less than the "Mandatory Surrender Area", but what is meant by that expression was a matter for the OGA to determine in Sch 5. The purpose of a surrender requirement is two-fold. First, it acts as an incentive to thorough and timely exploration.[217] The licensee will wish to be as certain as it can be that it is surrendering the least promising part of its territory, an assessment which can be safely made only after detailed exploration. Second, requiring a surrender within this relatively short period allows the area to be offered for licence again within a reasonable period of time. Rapid recycling of discarded acreage is considered to serve the state's interest in securing the swift and thorough exploitation of the UKCS's hydrocarbon reserves. An examination of the DECC's[218] own summary of its understanding of its licensing powers (now withdrawn) suggested that the 2008 amendments were not intended radically to change the position which existed under the 2004 regime: the expectation was that the initial term would, for traditional Production Licences, continue to be four years, and the mandatory surrender area usually 50 per cent. If regard is had to the contents of the traditional Production Licences issued at this time, it can be seen that in practice they generally continue with the 4-4-18 structure that had been stipulated in the previous set of model clauses.[219] However, some examples exist of licences having non-standard structures; for instance, the term structure for Production Licence P2301 is 3-4-18.[220] Here, we

---

to a ministerial over-ride which could be applied in special circumstances. That set of baseline norms has now disappeared from the UK's petroleum law.

[216] Model Cl 6(3) imposes the obligation to surrender no less than the Mandatory Surrender Area. Model Cl 1 provides that "Mandatory Surrender Area" means whatever it is said to mean by Sch 5 to the licence. The flexibility inherent in the licence structure is one of the features that allowed the "soft launch" of the Innovate Licence, prior to the development of dedicated model clauses.

[217] PILOT PPWG, Work, para 3.2.1.1.

[218] The DECC was the licensing authority at the material time.

[219] See DECC, Licence Data Portal, linked to from the OGA's website and available at https://itportal.decc.gov.uk/information/licence_reports/offshorebylicence.html (accessed 10 May 2017).

[220] Scanned copy licences are available for download from DECC, Licence Portal Recent Licences, linked to from the OGA's website and available at https://itportal.decc.gov.uk/web_files/recent_licences/oglicences.htm (accessed 10 May 2017). The author is not privy to why the initial term in this licence was three years but one could imagine a situation where the licensee of eg a previously licensed area approached the licensing authority with a request for a shorter term on the basis that, given its state of knowledge, it did not need a full initial term and wished to proceed expeditiously onto second-term activities.

may see some of the seeds of the flexible approach adopted in the
Innovate Licence.

I-4.55   In the second edition of this book, I indicated my dissatisfaction
at the fact that a matter so important as what the licence term was
to be had "been rendered so opaque and open textured", and noted
the risk that this might lead to the Department being over-burdened
by requests for special treatment.[221] In retrospect, this may look
very much like a typical lawyer's approach, seeking certainty above
all; certainly I do not seem to have been sensitive to the extent to
which the system was to travel either in the direction of flexibility
or pro-active resource management. To some extent, however, the
comment was justified at the time. We are in a new era now; given
how understaffed the DECC was in 2008, a flood of requests for
individualised terms might well have created difficulties. Moreover,
and as I also noted, the stated purpose of which was "to make the
position simpler, clearer and more transparent",[222] and this the
change manifestly failed to do. The pursuit of flexibility as a thought-
through policy aim – a sign of developed and rational governance,
and a selling point making the UKCS more attractive – is one thing;
dressing up opacity as the converse is quite another.

I-4.56   The 2008 model clauses also provided the licensing authority for
the first time with a discretionary power to extend the initial term
for a period and subject to such terms and conditions as he shall see
fit,[223] another example of an increased emphasis upon flexibility.

I-4.57   The second term was granted in order to provide the licensee with
time to develop a programme for the production of petroleum from
the discoveries made during the initial term. If the licensee wished
the licence to continue beyond the second term into the production
period it must, no later than three months before the end of the
second term, provide the licensing authority with written notice of
this intention.[224] At the end of the second term, the licence will be
permitted only to continue in respect of the "producing part".[225] In
the ordinary course of events, the producing part will be an area
in respect of which the licensee has submitted, and the OGA has
accepted, a detailed development and production programme that
contains the licensee's proposals for how it is to get and transport
the hydrocarbon it has discovered.[226] In previous practice, the second

---

[221] G Gordon, J Paterson and E üşenmez, *Oil and Gas Law: Current Practice and
Emerging Trends* (2nd edn, 2011), para 4.49.
[222] *Ibid.*
[223] Model Cl 7.
[224] Model Cl 8(1).
[225] *Ibid.*
[226] Model Cl 8(3)(b). Alternatively, the Minister may grant a consent (Model Cl 8(3)(a))

term, like the first, was of a fixed duration of four years.[227] The changes introduced by the 2008 model clauses again had the effect of making the duration of the second term a matter for Sch 5 to the licence.[228] In a continuation of previous practice,[229] the 2008 model clauses provided the licensing authority with a discretionary power to extend the second term.[230]

The purpose of the production period (third term) is to permit **I-4.58** the licensee to carry out the function of getting and conveying away the hydrocarbons which were discovered, and in respect of which development and production programmes were created, in the earlier phases of the licence. In previous practice, the third term was of an initial duration of 18 years,[231] a period which represented a significant reduction on the production phases permitted by earlier licences, and in practice that continued to be the period granted in licences awarded under the 2008 set of model clauses. In the event that the field is still producing as it approaches the end of the third term, the licensees may seek an extension to the production period. Such an extension must be sought by applying to the OGA in writing no less than three months before the end of the production period. The OGA has discretion as to whether to grant such an extension and, if so, on which terms and conditions.[232]

Provisions permitting intervention and operational control. It is **I-4.59** sometimes colloquially said that a certain oil company "owns" a given oil field. In some respects this view is understandable. Oil and gas companies are commonly associated with a given field for decades, and need to invest massively in order to develop their licensed interests, at no little financial risk. In other respects, however, such thinking is misleading. The state may have chosen to make use of the capital and expertise of the international oil and gas industry; however, the fact remains that it is the Crown that ultimately holds the right to exploit oil and gas deposits *in strata*.[233] As we have seen, the Crown has a range of legitimate interests in relation to oil and gas deposits.[234] It needs to ensure that these interests are properly protected; so too other interests not so less intimately associated

---

or himself serve a production and development plan upon the licensee (Model Cl 8(3)(c), read with Model Cl 176).

[227] 2004 Regulations, Sch 4, Model Cl 1: definition of "Second Term".

[228] See further the discussion at para I-4.54.

[229] See eg 2004 Regulations, Sch 4, Model Cl 5(3)(d).

[230] Model Cl 7.

[231] 2004 Regulations, Sch 4, Model Cl 6.

[232] Model Cl 9.

[233] See paras I-4.08 to I-4.11.

[234] See Chapters I-1 to I-3.

with oil and gas policy but which may be impacted by oil and gas operations, such as the rights of other parties who make use of the sea. Thus, the state needs to exercise some degree of control over operational matters.

I-4.60   The licence contains a range of provisions whereby the OGA requires the provision of information and/or seeks to exercise a degree of control over operations. The OGA's power under Model Cl 16 to demand a secondary work programme has already been discussed.[235] The licensee's obligation to submit a production and development plan has been noted in passing and will be further referred to in the Appendix. The OGA is also entitled to decline to approve the appointment of an operator, or to revoke a previously given approval, on the basis that the operator is incompetent to exercise that function.[236] Many other examples of OGA involvement or control may be given. For instance, the licensee is required:

- to measure the petroleum it extracts from the licensed area (Model Cl 14), keep "full and correct" accounts (Model Cl 15) and retain sundry records (Model Cl 29) and samples (Model Cl 31);
- to furnish quarterly and annual returns detailing (*inter alia*) all geological work and the results thereof (Model Cl 30);
- to liaise with the Ministry of Defence (Model Cl 44) and local fishing organisations (Model Cl 45) before undertaking certain works;
- to obtain consent before drilling or abandoning wells, to comply with conditions imposed in any given consent, and, when abandoning wells, to plug them to the licensing authority's specification and in a good and workmanlike manner (Model Cl 19) and to observe further specific requirements in relation to development wells (Model Cl 21);[237]
- to avoid harmful methods of working by maintaining appliances, apparatus and wells in good repair and condition and by executing all operations in a proper and workmanlike manner in accordance with methods customarily used in good oilfield practice (Model Cl 23);[238]
- to refrain from drilling wells in close proximity to boundaries

---

[235] See paras I-4.48 to I-4.51.

[236] Model Cl 24.

[237] In assessing whether to grant such consent, the Minister must bear in mind the appropriate environmental requirements: see paras I-4.24 to I-4.27.

[238] "Good oilfield practice" is not defined by the Regulations. It is described in OGA, *Guidance notes on procedures regulating offshore oil and gas developments*, available at www.og.OGA.gov.uk/regulation/guidance/reg_offshore/reg_offshore_guide.doc (accessed 10 May 2017), at p 8, as relating "largely to technical matters within the disciplines of

(Model Cl 20), to unitise when required (Model Cl 27) and to follow directions concerning cross-border developments (Model Cl 24).[239]

Frontier Licence. Geological exploration in the UKCS has histori-   I-4.61
cally tended in the northern (principally oil-producing) part of the province to be focused around the North Sea Basin lying between Great Britain and Norway, and in the region lying between Great Britain and Denmark and the Netherlands in the southern (princi-pally gas-producing) part of the province. Certain other areas of considerable potential – most notably, the area to the West of Shetland and (even more so) the Atlantic margin – have lagged behind in terms of exploration.[240] There are several reasons for this. First, investors have naturally sought to shorten the odds of discovering oil and gas by exploring in areas where discoveries have already been made. Second, a well-developed system of infra-structure has gradually been built up in the longer established parts of the province. This is lacking in the frontier areas, although it can now be seen to be beginning to emerge in the West of Shetland area, at least, as the area moves from being a true frontier area and starts to approach maturity. The capital cost of building new infrastructure and/or the increased unit cost of moving oil not by pipeline but by tanker acts as a barrier to development, and means that small discoveries that might have been viable in the established regions are likely to be unprofitable in frontier areas.[241] Third, while no offshore development is ever free from difficulty, developments in the estab-lished basins are generally not as technically challenging as those in the frontier areas, where licensees require to deal with deep water,[242] difficult geology[243] and a harsh physical environment.[244]

---

geology and reservoir, petroleum and facilities engineering and to the impact of the devel-opment on the environment".

[239] For a further discussion of these topics, see Chapter II-3.

[240] House of Commons Energy and Climate Change Committee, *UK Offshore Oil and Gas: First Report of Session 2008–09*, vol II: *Oral and Written Evidence* (17 June 2009) (hereinafter "Commons, *First Report*, vol 2") Ev 16, Q99 and Ev 23, Q133.

[241] Commons, *First Report*, vol 2, Ev 16, Q99. Note that Total E&P UK, in partnership with DONG Energy, is developing the Laggan-Tormore area and constructing a new gas pipeline system and a gas plant on Shetland, in Sullom Voe. See House of Commons Energy and Climate Change Committee, *UK Deepwater Drilling – Implications of the Gulf of Mexico Oil Spill: Second Report of Session 2010–11*, vol 1: *Report, together with formal minutes, oral and written evidence* (6 January 2011) (HC 450-I) (hereinafter "Commons, *Second Report*, vol 1"), para 13.

[242] Early developments in the North Sea were commonly in depths of water of 50–100m. Modern developments in the Atlantic Margin commonly have to deal with water depths between 350 and 400m.

[243] Commons, *First Report*, vol 2, Ev 7, Q54.

[244] *Ibid*, Ev 58, para 2.1.1.

I-4.62   Given these risks and difficulties, frontier exploration has generally been the preserve of major oil and gas companies, although in more recent years, several Frontier Licences having an initial work programme comprising solely seismic surveying work have been granted to smaller companies.[245]

I-4.63   As long ago as 1991, when a group of "frontier area licences"[246] was awarded relative to certain parts of the Atlantic Margin, it was recognised that, in order to encourage the opening up of new parts of the province, licences might have to be granted on special terms. After a period of quiescence, this recognition led in 2004 to the creation of the Frontier Licence (now known as the "six-year Frontier Licence") that continued to be utilised until the 28th round. In addition, a new variant (the "nine-year Frontier Licence") was created in the 26th licensing round to encourage further exploration activity in the Atlantic margin area to the west of Scotland. Although the Frontier Licence has been withdrawn as a specific licensing instrument, the new Innovate Licence has been designed with sufficient flexibility to allow individual licences to take cognisance of the difficulties inherent in conducting exploration and production activities in the frontier area.

I-4.64   In most respects the Frontier Licence was very similar to the traditional Production Licence. The principal differences between the licences lay in their respective terms, area rental costs and surrender provisions. When first introduced in 2004, the Frontier Licence had four terms, as opposed to the traditional licence's three: a two-year initial term, a four-year second term, a six-year third term and a production period which, like that of the traditional Production Licence, lasted for 18 years. The structure of the licence changed for the 26th licensing round, as with the other licensing variants, meaning that the model clauses no longer set out the duration of each of the terms; this instead becomes a matter for Sch 5 to the licence.[247] Moreover, the licence, from the 26th round onwards, consisted of three terms, not four; what were previously the initial term and the second term have been collapsed together into one. Notwithstanding these changes, the broad thrust of the licence remained the same with the DECC indicating in now-withdrawn guidance that the licence would comprise a six-year initial term, a six-year second term and

---

[245] P Carter, "The Regulator's Dilemma: how to regulate yet promote investment in the same asset base – the UK experience" [2007] IELTR 62 (hereinafter "Carter, 'The Regulator's Dilemma'").

[246] These were issued on different terms to the current frontier area licences and are not to be confused with them. These were a form of bespoke licence and are discussed further at para I-4.74.

[247] For a discussion of the issues raised by this policy, see paras I-4.54 to I-4.58.

an 18-year third term. As before, the acreage rental of the licence was greatly reduced for the first two years of its existence, and the expectation was that the licensee would let a large initial area, invest heavily in exploration and relinquish a large proportion of the licensed area.

A variant on the Frontier Licence was introduced for the 26th I-4.65 licensing round, the nine-year Frontier Licence. This was similar in most material respects to the six-year Frontier Licence but provided an even longer (nine-year) initial term to take cognisance of the still greater difficulties associated with oil and gas exploration west of Scotland.[248] DECC guidance (now withdrawn) indicated that the licence would be made available only in relation to work obligations presented on a "drill or drop" basis, with the decision on whether to proceed to drill being taken within the first six years of the initial term.

Limitations were imposed upon the number of blocks which may I-4.66 be held by a licensee on frontier terms,[249] and it appears that the licensing authority required large areas of the licence to be relinquished at a relatively early stage. In the case of both the six-year and nine-year Frontier Licence, seven-eighths of the area initially let will have to be relinquished by the end of the initial term. At the end of the second term, the surrender provisions come to mirror those which apply to a traditional Production Licence at the end of its second term.

The Frontier Licence was essentially an attempt, long before I-4.67 the Wood Review, at maximising economic recovery on the UKCS by facilitating the opening up of prospective but underdeveloped areas. It exhibited a flexibility in approach which has only increased with the introduction of the new Innovate Licence, and involved an attempt by the Government to strike a reasonable balance between the interests of the oil company (exclusivity over a large area at a modest cost; enough time to explore and develop prior to production) and those of the state (undertaking of exploratory activity; enough pressure to ensure the licensee felt an incentive to undertake works expeditiously).

Assessing the success of the Frontier Licence. By the author's calcu- I-4.68 lations, 34 Frontier Licences have been awarded since the licence form's introduction in 2004 to its abandonment after the 28th round. This is a very low figure when compared with either the

---

[248] OGA, *Licensing: Licence Types.*
[249] Licensees are not permitted to hold more than ten contiguous blocks per Frontier Licence, with an aggregate total of 40 blocks per applicant per round: OGA, *Applications for Production Licences General Guidance*, at para 29.

promote or the traditional Production Licences, where over the same period there have been 341 and 668 grants respectively. However, it would be dangerous to judge the success of this licensing variant purely on the strength of the number of licences granted. First, it has to be noted that the initial area let under a Frontier Licence will be much greater than that let on either traditional or promote terms. Moreover, it is generally accepted that, although substantial finds are by no means impossible in the well-explored areas of the UKCS,[250] there is a greater scope for making very substantial discoveries in frontier areas than in the previously explored areas of the province.[251] Much of the area West of Shetland either now is or has been held on a Frontier Licence. This would suggest that this variation has been a success.

I-4.69   **Promote Licence.** The introduction of the Promote Licence followed a period of consultation, formally[252] begun in September 2002, between the Department and a broad cross-section of industry interests and stakeholders in the UKCS.[253] The purpose of the consultation was to assess the level of industry support for a licence tailored to meet the needs of the smaller or "niche" enterprise possessing considerable geo-technical ability but lacking the financial and/or technical and environmental capabilities which had hitherto been necessary to hold a Production Licence. The consultation document proposed that certain relaxations should be offered to such companies in order to attract them to the UKCS.[254] The Department received widespread (but not universal)[255] support for the proposal, and Promote Licences were first awarded during 2003 in the 21st seaward licensing round.

I-4.70   The Promote Licence comprised only a very slight variation to the traditional Production Licence. The initial term was divided into two parts by a break-point occurring on the second anniversary of the licence's grant: the work programme[256] is divided into Parts I and II; the licence survives beyond the break-point only if the licensee

---

[250] See eg Buzzard in the Moray Firth; Catcher in the Central North Sea.

[251] See eg Carter, "The Regulator's Dilemma", at 63 where frontier areas are described as "high risk/high potential".

[252] As the industry organisation PILOT had been involved in the formulation of the proposal in one sense the consultation process could be said to have started earlier.

[253] See DTI, *Possible Introduction of "Promote" to Encourage Exploration Activity on the UKCS: Consultation Document* 2002 (hereinafter "DTI, *Possible Introduction of Promote*").

[254] DTI, *Possible Introduction of Promote*, paras 1–4. See also DTI, *Possible Introduction of "Promote" Licence to Encourage Exploration Activity on the UKCS: Open Consultation October/November 2002: Summary of Responses* (hereinafter "DTI, *Promote: Summary of Responses*"), paras 1 and 2.

[255] DTI, *Promote: Summary of Responses*, at para 3.

[256] See the further discussion of work programme at paras I-4.47 to I-4.48.

has taken all action described in Part I of the work plan and has undertaken to complete before the expiry of the initial term the work described in Part II of the work programme; and the requirement found in the traditional Production Licence that, for a licence to continue into its second term, its "work programme" must be completed before the end of the first term, is in practice replaced by a reference to "Part II of the work programme". If a Promote Licence survived into a second term, it effectively converted into a traditional Production Licence.

These superficially modest departures from the traditional licence I-4.71 were supplemented by ministerial guidance and practice. In particular, for the first two years of a Promote Licence, the annual rental was reduced by 90 per cent compared with a traditional Production Licence. This discount was offered in order to make the Promote Licence affordable to its target market, on the basis that the benefits that will accrue by expanding the pool of players in the UKCS will outweigh the potential loss of acreage rental.[257]

In addition, the DECC confirmed that an application containing I-4.72 a limited workscope restricted to activities such as data acquisition and evaluation will initially be acceptable during Part I of the work programme, but that it expected that, by the time Part II commences at the end of Year 2, there must be "a firm drilling (or agreed equivalent equally substantive activity) commitment". The Department has also confirmed that while it requires full financial, technical and environmental capacity to be achieved within the first two years, it is not necessary – or, indeed, appropriate[258] – for the promote licensee to possess these capabilities at the time of the grant. In practice, the initial promote licensee sought to attain the requisite capabilities by entering into a joint venture with one or more established oil company players, or by divesting itself of the asset entirely by selling it to a more established oil company after completing initial appraisal work.[259]

The relaxations offered by this licensing variant provided the I-4.73 promote licensee with a set of concessions not available to the more established holder of a traditional licence. There was therefore unequal treatment between the two classes of licensee. However,

---

[257] As was noted earlier, the contribution of acreage rental to the state's take from oil and gas production in the UKCS is relatively small: see paras I-4.32 to I-4.34.

[258] If the licensee does possess such capabilities, the Department expected it to seek a traditional licence, rather than a promote one. Where Promote Licence applications are in competition with frontier or traditional licence applications, the acreage was awarded to the traditional/frontier applicant in all but the most exceptional circumstances, because of the confidence that the DECC could draw from the proof of financial and operating capacity that a traditional/frontier applicant must provide.

[259] DECC, *Possible Introduction of Promote*, para 10.

there seems to have been little disquiet about this among the more established companies.[260] The super-majors who operate on the global scale will be disinterested in the relatively small developments which tend to attract the attention of the promote licensee and, as we have already seen, other established companies may perceive the Promote Licence initiative as an opportunity to permit them, in due course, to acquire, with at least a somewhat reduced risk, some of the fruits of the promote licensee's labours. Overall, as with the Frontier Licence, the introduction of the Promote Licence was an early example of a flexible approach to resource management which has continued with the Innovate Licence and which, as we have seen, resulted in a very significant number of licence awards.

I-4.74 **Bespoke licences.** Given the discretionary nature of the system, it should come as no great surprise to note that the OGA is empowered to grant licences on non-standard terms. Section 4(1)(e) of the Petroleum Act 1998 provides that the OGA shall make regulations prescribing the model clauses which shall be incorporated into petroleum licences "unless it thinks fit to modify or exclude them in any particular case". This discretion to tailor the terms of particular licences has been used on at least three occasions in recent times. First, when frontier area licences were granted in the May 1991 licensing round, this was done on an ad hoc basis, without the promulgation of a dedicated set of frontier area model clauses. Instead, the frontier area variations were advertised in advance[261] and incorporated into the six licences so awarded.[262] Second, as has already been noted, following consultation with the industry, all Seaward Production Licences issued in the 20th and 21st rounds were allocated on terms different from those contained in the model clauses then in force.[263] Third, at least one licence is known to have been issued which was tailored to accommodate the specific difficulties associated with re-developing a field which had already been relinquished and decommissioned by its previous operator.[264] The use of the power in the second given example has been criticised by Daintith and Hill as being "of doubtful legality" on the basis that the "in any particular case" criterion cannot be legitimately used

---

[260] Daintith, *Discretion*, para 3326.
[261] DECC, *Notice Inviting Applications under the Petroleum (Production) (Seaward Areas) Regulations 1988, London Gazette*, 29 June 1990, at 11243–11245. See particularly paras 2(b), 3(b) and 5–7.
[262] Daintith, Willoughby and Hill, para 1-602.
[263] See para I-4.52.
[264] The Argyll field began producing oil in 1975 and was the first productive oilfield in the UKCS. It was decommissioned by its then operators, BHP, in 1992 but was reopened by a joint venture led by Tuscan Energy (Scotland) Ltd in 2002.

to effect a change of general application.[265] It is submitted that this criticism is a valid one, and that, although Daintith and Hill direct no criticism towards it, even the first-mentioned example might be thought to lie at the border of acceptability. The third-mentioned example, however, would seem to be a wholly legitimate use of the power to issue a licence on modified terms. In the second edition of this work it was argued that "The fact that, post-2008, the model clauses no longer detail the licence's term and leave much key content to be stipulated by DECC in Sch 5 may make it less likely that the Department will have to issue bespoke licences in the future."[266] The point, in essence, was that when rules become flexible, there is less need to seek exceptions from them. The other way of looking at it might be to say that when a system becomes so flexible that no exceptional rules are necessary, are we still dealing with a system of rules at all? That is perhaps an appropriate point at which to commence a discussion of the Innovate Licence.

Innovate Licence. As we have already seen, the suite of licences which emerged in the early 2000s and which offered a standard licence and variations to meet the challenges posed by particular geographical areas or small-scale companies has not survived in the post-Wood era. The OGA's first significant act in its capacity as licence administrator was, for new licence grants, to discontinue the old system and replace it with one where only one form of Production Licence was on offer: the Innovate Licence.[267]   I-4.75

At first sight, this might seem a retrograde step; a sweeping away of the subtlety and flexibility offered by having multiple licence variants. This, however, would be a misleading impression. As OGA's Guidance on the licence states, its purpose is to offer "greater flexibility for each applicant to design a work programme around particular circumstances".[268] The OGA promises a speedier licensing process than existed previously,[269] and has relaxed the restrictions present in the traditional licence on the area that can be held under licence.[270] The radicalness of the new system is demonstrated by the   I-4.76

---

[265] Daintith, Willoughby and Hill, para 1-328/4; see also Daintith, *Discretion*, at para 3513.

[266] Gordon, Paterson and üşenmez, *Oil and Gas Law: Current Practice and Emerging Trends* (2nd edn, 2011), para 4.70.

[267] As was the case with the Promote Licence, the Innovate Licence was introduced somewhat irregularly, not by the issuance of revised model clauses but by the advance notification of new terms in the Official Journal announcement of the 29th round. See [2016] OJ C 244/05.

[268] OGA, Offshore Innovate Licence, available at www.ogauthority.co.uk/licensing-consents/offshore-licences/ (accessed 10 May 2017).

[269] See paras I-4.20 to I-4.21.

[270] OGA, *UKCS Frontier Basins and Licensing*, Presentation, APPEX Conference

fact that even the number of terms within the licence is not rigidly determined; while the expectation is that there will usually be three, each broadly fulfilling the same roles as in the existing three-term licences, the licensee is offered the option to skip the first term and proceed straight into the second. Although radical, this is perfectly logical, and may be an attractive option for licensees of areas which are being recycled into the system having previously been held and relinquished, and where the geology is well known; or perhaps in areas where state-funded public-access exploratory work has been conducted.

I-4.77   For the greater part, however, the flexibility has been achieved by building upon features that were already present in the Frontier and Promote Licences. The term in which the Innovate Licence most radically differs from prior practice – and where most flexibility is seen – is the first term. This is consistent with the OGA's statement that the licence is essentially concerned with optimising work programme design, since, as we have already seen,[271] "work programmes" really means "exploration programmes", and exploration is the purpose of the first term.

I-4.78   The first term will no longer be of a fixed length. It may be up to nine years in length – a figure arrived at as nine years was the longest of the first terms on offer among any of the previous licences (the nine-year Frontier Licence) – and can be subdivided into up to three phases (A, B and C), with the work package for each phase being set out separately in the work programme. OGA Guidance states that Phase A pertains to geotechnical studies and geophysical data reprocessing; Phase B is for undertaking seismic surveys and acquiring other geophysical data; and Phase C is for drilling.[272] The Guidance also makes it clear that phases A and B are optional and depend entirely on the applicant's plans. However, all licences having a first term must have a Phase C. The guidance envisages that each of the separate phases within the first term must meet certain success criteria if the licensee is to be permitted to move on to the next phase. As under the previous licence system, technical and financial capability is very significant.[273] However – as was the case with the Promote Licence[274] – it is not necessary for the licensee to demonstrate the technical and financial competence from the outset. As

2016, available at www.gov.uk/government/uploads/system/uploads/attachment_data/ file/505535/APPEX2016_-_EXTERNAL_-_Nick_Richardson_-_OGA_-_UKCS_Frontier_ Basins_and_Flexible_Licensing.pdf (accessed 10 May 2017).

[271] See para I-4.54.
[272] OGA, Offshore Innovate Licence.
[273] See para I-4.36.
[274] See para I-4.69.

Phases A and B do not involve breaking the ground, they are assessed as low-risk phases of operation and so the need to demonstrate financial and technical competencies is deferred until the licensee wishes to move into Phase C.

Landward licences

For a time, the main British[275] landward licensing regime differed **I-4.79** markedly from that applied in seaward areas. Three separate licences – exploration, appraisal and development – were awarded in succession as operations within the licensed area progressed from one phase to the next.[276] This regime was abandoned for new licences in 1995[277] to create the Petroleum, Exploration and Development Licence (PEDL), the current model clauses for which are to be found in the 2014 Regulations as amended.[278] The PEDL is the principal landward licence in current issue.[279] The PEDL shares the same basic structure and philosophy as the traditional Seaward Production Licence;[280] the flexible structure of the Innovate Licence has not been applied in the landward area. The Wood Review having been directed to how to maximise economic recovery upon the UKCS, that is to be expected; however, it may well be that, if the Innovate Licence is seen to have been a success, the model may creep onshore.

---

[275] The caveat given at para I-4.03 relative to Northern Ireland and the Isle of Man also applies here. Nor is this chapter concerned with methane drainage licences or the specialities associated with coal bed methane. For a short discussion of these issues, see Daintith, Willoughby and Hill, paras 1-312 and 1-313 respectively.

[276] Daintith, Willoughby and Hill, para 1-312. For a fuller discussion of the early onshore regime (which continues to be relevant for pre-1995 licences), see J Salter, *UK onshore oil and gas law: a practical guide to the legal regime relating to United Kingdom onshore oil and gas operations* (1986), Chapter 3.

[277] Petroleum (Production) (Landward Areas) Regulations 1995 (SI 1995/1436), Sch 3.

[278] The Petroleum Licensing (Exploration and Production) (Landward Areas) Regulations 2014 (SI2014/1686) Sch 2. The Petroleum Licensing (Exploration and Production) (Landward Areas) (Amendment) (England and Wales) Regulations 2016 (SI2016/1029) amend the 2014 Regulations by introducing (at Reg 2) a prohibition on hydraulic fracturing from a location within a protected area; a definition which includes Conservation Sites protected under European Law and Sites of Special Scientific Interest. The previous set of model clauses (which will continue to govern the majority of PEDLs for some time) were to be found in the Petroleum Licensing (Exploration and Production) (Seaward and Landward Areas) Regulations 2004, at Sch 6, as the same has been amended by the Petroleum Licensing (Exploration and Production) (Seaward and Landward Areas) (Amendment) Regulations 2006.

[279] The 2014 regulations in addition introduced a new Exploration Licence: The Petroleum Licensing (Exploration and Production) (Landward Areas) Regulations 2014 (SI2014/1686), Sch 3.

[280] The licences are divided into terms, and progress from one term to the next is contingent upon meeting certain conditions and relinquishing territory.

The seaward licence having initially evolved out of the landward one, there would be a nice circularity to this, were it to happen.

The duration of the terms of the licence differs from that of the traditional Production Licence;[281] the licences' Model Cl 2, although identical in function, are given different titles[282] and the PEDL omits as unnecessary several of the provisions which deal with the regulation of marine matters.

I-4.80 Like Seaward Production Licences, PEDLs are generally offered in rounds, although out-of-rounds applications are possible.[283] At the time of writing, there have been 14 landward licensing rounds in all. In England and Wales, the OGA is the relevant licensing authority. As we shall see in greater detail in Chapter I-9, the Scottish Government is to become the licensing authority in the Scottish onshore area, following the devolution of further powers under the Scotland Act 2016. It had already enjoyed devolved competence relative to planning matters and had used this to impose a precautionary moratorium relative to shale gas and underground gasification activities.

I-4.81 Much more oil has been produced under Seaward Production Licences than under landward licences. As can be seen by comparing the landward production to total production in BEIS's production statistics, landward production has never climbed above 5 per cent of total annual production,[284] and the overwhelming majority has come from one field: Wytch Farm, in Dorset. Onshore licensing will continue to have a role to play for years to come. Developments in directional and extended reach drilling mean that it is possible for a well commenced in an onshore borehole to be deviated a significant distance offshore. Thus oil and gas reserves which lie beneath environmentally sensitive areas within relatively close proximity to the coast may be reached by a rig located on an onshore location.[285] This approach may also appeal to a licensee for simple reasons of economics. Onshore drilling avoids the requirement for expensive

---

[281] The current model clauses provide for an initial term of five years, a second term of five years and a production period of 20 years, a slight variation upon the 2004 model clauses which provided for a six-year initial term.

[282] "Grant of licence" in the case of the traditional Seaward Production Licence and "right to search and bore for and get petroleum" in the case of the PEDL.

[283] OGA, *Licensing: Award of Licensing*.

[284] BEIS, "Crude oil and petroleum: production, imports and exports 1890 to 2015", available at www.gov.uk/government/uploads/system/uploads/attachment_data/file/548011/Oil_Production___Trade_since_1890.xls (accessed 9 May 2017).

[285] Rigzone, *Reach Exploration Awarded 16 Blocks In UK Licensing Round*, available at www.rigzone.com/news/article.asp?a_id=25174 (accessed 9 May 2017). At the time of writing, production is scheduled to commence from Reach Petroleum's Lybster field, an offshore discovery drilled from onshore.

offshore pipeline construction or tanker use and provides the licensee with the option either of constructing an onshore pipeline[286] or transporting oil by tanker, an option which may very well be the most economic when seeking to exploit a small field. And if the shale gas revolution takes hold in the UK with even a fraction of the success that it has enjoyed in the USA,[287] then – with offshore production expected to continue to tail off – we could, in the future, reasonably expect to see the number of landward licences increase and the volume of oil and gas produced thereunder get significantly closer to the volume of offshore production.[288]

CONCLUSION: CURRENT ISSUES AND EMERGING TRENDS IN THE UK PRODUCTION LICENCE

Reviewing the foregoing chapter, three themes or trends seem to stand out: the notion that MER UK's effect on the licence has been an indirect one; the idea of transition; and the concept of flexibility. Each shall be discussed in turn. **I-4.82**

Theme 1: The Wood Review – indirect effect and diluting significance

Judged by visible effects within its terms and conditions, the Wood Review and its subsequent programme of implementation have had very little effect upon the UK Production Licence. Given the way the UK licensing regime is structured that is hardly surprising. The UK Production Licence is contractual at least in form and possibly in nature. It contains no provision which would allow the state to make retrospective amendments to its terms. So – notwithstanding the fact that, as we shall see in Chapter I-5, the Wood Review initially saw retroactive amendment as a central plank of its amendment strategy – that was never a very realistic prospect. In the end, as we shall see **I-4.83**

---

[286] Onshore pipelines are cheaper to lay than offshore ones. However, the licensee may experience difficulties in overcoming the property law rights of the proprietors of onshore land: see Chapter II-13.

[287] The author is measuring success here simply by reference to volume of oil and contribution to the economy. Some US shale gas developments have undoubtedly had negative environmental consequences, by eg contaminating water: US EPA, *Hydraulic Fracturing for Oil and Gas: Impacts from the Hydraulic Fracturing Water Cycle on Drinking Water Resources in the United States (Final Report)*. US Environmental Protection Agency, Washington, DC, EPA/600/R-16/236F, 2016, available at https://cfpub.epa.gov/ncea/hfstudy/recordisplay.cfm?deid=332990 (accessed 9 May 2017). This, together with climate change considerations and loss of amenity for host communities, must be borne in mind when considering whether shale gas has been an overall success.

[288] The legal regulation of shale gas is further discussed at Chapter I-9.

in Chapter I-5, Wood was implemented by a strategy whereby the existing licences essentially glory on as before, but do so within the setting of an entirely new and more dynamic regulatory environment. Thus we can see straight away one indirect but extremely important effect that the Wood Review has had upon the licence; it has, quite simply, diluted its importance, making it a less significant resource management instrument than once it was. Having a licence is still important, of course; and so too is understanding the process of how to position yourself to have the best chance of obtaining one. But once it is received, the resource management clauses, MCs 16, 17 and 18, which previously contained so much of the state's power to tell the licensee what to do and when, are no longer determinative of these matters. As we shall see, different, more sophisticated and quite possibly more onerous demands may be made under the auspices of MER UK.

I-4.84   Another very significant indirect effect of MER UK – alluded to here and developed further in Chapter I-5 – is the fact that the sanction provisions that have been developed by Government primarily to give teeth to the MER UK obligation have also been applied to obligations under the licence. This is significant as the public law remedy for breach of licence was formerly revocation, and the licensing authority was reluctant to apply this remedy. So – apart from compliance secured by strategic considerations on the licensee's part, such as the desire to remain in good favour with the regulator – in all but the most egregious cases where revocation would be justified, the licence was fenced with little in the way of meaningful sanction. This situation has now changed quite radically, although the wording of the licence has not; a point which demonstrates *par excellence* the indirect effect of MER UK upon the licence.

Theme 2: Transition

I-4.85   The other major theme which seems to stand out in considering the licensing regime at the present time is the notion of transition. Much is changing, or on the verge of doing so. The notion that the Scottish Government might (without securing independence from the rest of the UK) become a petroleum licensing authority seemed more than far-fetched even a few years ago. It is now foreseeable that onshore development may become much more significant than it is at the present time. Brexit may, in time, have a bearing upon UK petroleum licensing law and practice. But by far the most significant transition currently in play in the field of UK petroleum licensing is the handing over of the reins from the light-touch, poorly resourced BEIS on to the better-resourced, more proactive OGA.

Theme 3: Flexibility

The OGA has been at the helm of the UK industry for only a short   I-4.86
period of time. It is already clear, however, that flexibility lies at the
heart of its proposed approach to licensing. In this it is continuing
a practice that was already to some extent present in the system,
and which was evidenced by, most obviously, the development of
promote and Frontier Licences. However, the abandonment, for
future licences, of the previous model in favour of the Innovate
Licence – a system which allows for significantly greater flexibility,
but at the cost of additional regulatory effort – says much about
the new era that we are embarking upon. It demonstrates that the
additional funding being devoted to the regulator has the potential
to fundamentally change the way in which the industry is governed.
The new approach will inevitably bring challenges; perhaps the most
obvious ones are ensuring consistency of approach and maintaining
transparency in an era of continual bespoking. It shows confidence:
a regulator willing to move out of the shelter provided by a set of
rigid rules and to come up with individual answers to individual
sets of circumstances is a regulator that is not afraid to take on
challenges. As we shall see, the Wood Review intended to establish
a more confident and assertive regulator, so this is encouraging. It
is perhaps unfortunate that this chapter has come to be written
while we remain within the transitional phase, before the OGA has
developed the model clauses that will allow it to give full expression
to the Innovate Licence. But already it is clear that the approach for
new licences is going to be markedly different from those which were
allocated to licensees over the first 50 years of the UKCS's history.

CHAPTER I-5

THE WOOD REVIEW AND MAXIMISING ECONOMIC RECOVERY UPON THE UKCS

Greg Gordon, John Paterson and Uisdean Vass[1]

INTRODUCTION

I-5.01   Aspects of the legal regulation of oil and gas exploration and production activities are endlessly adjusted and tinkered with. So the term structures of the standard UK Production Licences have been adjusted nine times since the first licences were awarded in 1964,[2] while the model clauses have gone through numerous iterative changes, but are fundamentally recognisable as the same kind of instrument as they were, say, 30 or 40 years ago.[3] Root-and-branch, or radical change – to the extent of introducing a new ethos to the UK's resource management and governance system – is, however, a much rarer event.

I-5.02   The Wood Review recommended just such a radical change. At the time of writing, the implementation process is almost complete: a new regulator, the Oil and Gas Authority (OGA), has been created; the implementing legislation has been passed and is in force; the MER UK Strategy[4] is in place, as are a host of more specific strat-

---

[1] Paras I-5.63 to I-5.70 are based upon preparatory work undertaken by James Cowie.

[2] DECC, *Oil and gas: licensing rounds*, available at www.gov.uk/guidance/oil-and-gas-licensing-rounds (accessed 11 May 2017): data taken from the "Past Licensing Rounds" table at the foot of the page. The table shows eight changes, but does not include the new 29th round, which (as discussed in Chapter I-4) has been offered on the Innovate Licence basis, which provides a revised term structure for new licences.

[3] The evolution of the UK model clauses can be traced in the Schedules to The Petroleum (Current Model Clauses) Order 1999 (SI1999/160).

[4] OGA, Maximising Economic Recovery for the UK (2016), available at www.ogauthority. co.uk/media/3229/mer-uk-strategy.pdf (accessed 11 May 2017). This is a rather bland name for what, as we shall see, is effectively the beating heart of new regime: the short

egies which provide guidance on how MER UK is to be secured in particular sectoral contexts;[5] and the OGA is rolling out the promulgation of a series of programmes which provide in more concrete detail how the OGA proposes to deliver on its near-term objectives within the particular sector.[6]

The system has still to bed in, and choices made by the new I-5.03 regulator, the OGA, about precisely how far it is prepared to go to dictate to the industry will have a bearing upon just how revolutionary the system is. But even if the OGA were to fight shy of using its powers of compulsion and instead adopt an essentially consultative and non-assertive approach, it is already clear that the effects of the change will be far-reaching and will transform the way in which the upstream oil and gas is governed by the state and managed by licensees.

This chapter will commence by locating the Wood Review within I-5.04 its broader historical context, noting previous key events in the development of the resource management system upon the United Kingdom Continental Shelf (UKCS). It then provides a detailed account of the Wood Review's findings and recommendations, and of the recommendations' legal implementation. Thereafter it will outline the essential features of the new system before concluding. Generally, the chapter's focus will be on Wood's effect upon high-level or systemic matters, not the fine detail of the system's application to particular issues or sectors;[7] however, the topic of asset management will be discussed in some detail, both because it is an example of particular importance and because the particular way in which Wood has been implemented means that, while this is an issue that would traditionally be dealt with under the licence, a broader view is now required.

BEFORE WOOD: HISTORICAL DEVELOPMENT AND THE UK'S PREVIOUS PHILOSOPHY OF GOVERNANCE

The Wood Review is by no means the first attempt at reforming the I-5.05 UK's system of licensing and governance. The legal framework for

---

document which sets out the means by which licensees and other relevant parties are to comply with the legal obligation to maximising the economic recovery of UK petroleum.
[5] See eg OGA, *Exploration Strategy* (2016), available at www.ogauthority.co.uk/media/2835/exploration_strategy_master.pdf; OGA, *Supply Chain Strategy* (2016), available at www.ogauthority.co.uk/media/2834/supply_chain_strategy_1016.pdf Decom Strategy (both accessed 11 May 2017).
[6] See eg OGA, EOR Delivery Programme (2017), available at www.ogauthority.co.uk/media/3171/eor_delivery_ver11.pdf (accessed 11 May 2017).
[7] These issues are dealt with throughout the remainder of this book. Virtually all chapters have at least a passing reference to Wood in general and/or MER UK in particular; but in Volume I, see in particular Chapters I-4, I-6 and I-12. In Volume II, see in particular Chapters II-3, II-5, II-11 and II-12.

offshore oil and gas operations emerged in the mid-1960s, when Section 1(3) of the Continental Shelf Act 1964 exported the substantially untested onshore petroleum regime into the offshore area.[8] In the licences granted under this regime, state companies participated alongside private investors, but in a rather unstructured way.[9] In the early 1970s, concerns emerged that the legal framework was not adequate to ensure that the UK obtained an appropriate share of its oil wealth. A Committee of Public Accounts report in 1973 noted deficiencies in the UK's fiscal system as well as weaknesses in the licence provisions which meant that the Government had little power to influence the licensee's operations.[10] In 1975, Parliament addressed these problems by passing the Petroleum and Submarine Pipe-lines Act and the Oil Taxation Act (hereinafter "the 1975 Acts"). In the public imagination, the 1975 Acts do not rank with landmark enactments such as the Representation of the People Act 1918 or the National Health Service Act 1946; however, given the significance that the oil and gas industry was to assume, they could fairly be described as among the most significant Acts of the 20th century. The 1975 Acts sought to rebalance the relationship between the oil companies and the state. This they did by establishing the British National Oil Corporation, a state national oil company which was intended to participate directly in the industry;[11] introducing new terms and conditions into the petroleum licence[12] intended to

---

[8] The onshore petroleum regime was contained in the Petroleum (Production) Act 1934. Although the Act had been in force for almost three decades, very few commercial finds of oil or gas had been made onshore. The Government therefore had little direct experience of the oil and gas industry prior to the development of the offshore area, although British companies such as BP had significant overseas experience. See further G Gordon, "British Hydrocarbon Policies and Legislation", in E Pereira and H Bjorneybe, *North Sea and Beyond* (2018) (hereinafter "Gordon, 'British Hydrocarbon Policies and Legislation'").

[9] See T Daintith, "Appendix A – State Participation", in T Daintith, G Willoughby and A Hill (eds), *United Kingdom Oil and Gas Law* (3rd edn, looseleaf, 2000–date) (hereinafter "Daintith, 'State Participation'"), para 1-A03; Gordon, "British Hydrocarbon Policies and Legislation".

[10] First Report from the Committee of Public Accounts, Session 1972–1973: North Sea Oil and Gas (HC 122), paras 89 and 98; A Kemp, *The Official History of North Sea Oil and Gas Volume 1: The Growing Dominance of the State* (2011), p 251.

[11] 1975 Act, Pt 1. This was not the beginning of direct state participation in the UK, but it placed the activity on a more systematic footing. See Daintith, "State Participation", paras 1-A12 to 1-A13.

[12] 1975 Act, s 17 and Sch 4. Not only were these conditions imposed prospectively, they were retrospectively incorporated into all existing offshore licences. The retroactive application of the provisions, in particular, proved controversial and led to a heated debate in Parliament. In seeking to defend their actions, the Labour Government contended that the problem that the provisions sought to address had been caused by the previous Conservative Government, who had failed to protect the nation's interests and had instead "given away" large tracts of the UKCS while retaining little or no operational control.

provide the state with at least some measure of control over the pace at which oil companies holding petroleum Production Licences undertook exploration and production activities;[13] providing the Government with at least some measure of control over the infra-structure that was being built by the oil companies to transport oil and gas to the shore; and introducing a new tax regime with rules particular to the oil and gas industry.[14] Taken together, these changes amounted to a radical reconstruction of the system of licensing and governance in the offshore area; to adopt the language of a computer operating system, it might be described as the launch of UKCS 2.0. This system was not, however, destined to last for long. Post-war consensus politics – which might have provided an environment suitable for the survival of a state national oil corporation[15] – came to an end with the election in 1979 of a Conservative Government led by Margaret Thatcher. Her administration advocated free market economics, private enterprise and a radically reduced role for the state,[16] and BNOC and the British Gas Corporation were among the first state-owned undertakings to be privatised. The end of the era of direct state participation in the upstream offshore oil and gas industry was a change sufficiently momentous to allow us to describe these reforms as the adoption of UKCS 3.0. Although this system has been patched and updated over time,[17] it was still essentially

---

For the Conservatives, Labour's reforms were socialism writ large: they were changing the rules of the game after it had commenced, and without offering compensation to the companies affected by the change, conduct, the Conservatives argued, that would be liable to undermine investor confidence and diminish the country's chances of developing a meaningful oil and gas industry.

[13] These clauses continue to be introduced into contemporary licences. Presently (ie in the model clauses contained in the Petroleum Licensing (Production) (Seaward Areas) Regulations 2008 (SI 2008/225), as amended by the Petroleum Licensing (Amendment) Regulations 2009 (SI 2009/3283) (hereinafter "the 2008 Regulations")) they are numbered model clauses 16, 17 and 18.

[14] Oil Taxation Act 1975. Part I thereof introduced Petroleum Revenue Tax, Part II Ring-fence Corporation Tax. For an account of these taxes, see Chapter I-7.

[15] One-nation Conservatism, the dominant strand of Conservative thought in the post-war era, had at its core the notion that the state could be a promoter of social cohesion, and was tolerant of at least some level of state participation in the UK's industrial sector: see eg H Bochel, "One Nation Conservatism and Social Policy, 1951–64", 18 (2010) *Journal of Poverty and Social Justice*, 123; P Dorey and M Garnett, "The weaker-willed, the craven-hearted: the decline of One Nation Conservatism", 5 (2015) *Global Discourse*, 69.

[16] See eg P Jackson, "Thatcherism and the Public Sector", 45 (2014) *Industrial Relations Journal*, 266.

[17] See eg the discussion of the Fallow and Stewardship initiatives in the Appendix below. These mature province initiatives were an attempt to address certain weaknesses in the resource management system that emerged in the early 2000s. Although they stood on a questionable legal foundation, in practical terms they enjoyed a considerable degree of success.

this structure that Wood sought to reform, creating UKCS 4.0. It is therefore worth briefly sketching the key features of this system before going on to consider the Wood Review and the manner in which its recommendations have been implemented.

I-5.06   Fundamentally, the pre-Wood system was one where oil companies (usually working in consortia) pursued their own field-level projects in the manner they calculated to be best suited to advance their economic interests against a background of "light-touch" (or "light-handed") regulation.[18] Although not without strategic aims and objectives, and willing to work with the industry in order to address specific issues identified as problematic,[19] the Government did not act in pursuit of a systematic development strategy of the type encountered in, for instance, Norway.[20] Instead, the Government cast itself "in a more passive role, restricted to approving or rejecting specific proposals submitted by the oil companies".[21] Many of the state's licence powers were negative in nature, involving the refusal of permission for operations of which it did not approve. It could refuse permission for individual developments that were not proposed to be carried out in a manner which the Government thought consistent with best industry practice[22] or refuse permission for drilling if dissatisfied with the operator's drilling plan.[23] Schemes like the Fallow Areas and Stewardship initiatives involved the state in asserting a right, supposedly founded in the licence terms,[24] to influence the manner in which certain exploration and production activities are carried out. These schemes probably constituted the high-water mark of the state's pre-Wood attempts to control and direct investment decisions; but even here, the focus of the schemes

---

[18] Sir I Wood, *UKCS Maximising Recovery Review: Final Report*, available at www.gov.uk/government/uploads/system/uploads/attachment_data/file/471452/UKCS_Maximising_Recovery_Review_FINAL_72pp_locked.pdf (accessed 11 May 2017) (hereinafter "Wood Review"), p 9. See also S Rush, *Access to Infrastructure on the UKCS: The Past, the Present and a Future* (2012), available at www.memerycrystal.com/uploaded/Articles/other%20files/Access%20to%20Infrastructure%20on%20the%20UKCS%20-%20SR%20-%20Feb%202012.pdf (accessed 11 May 2017), p 1. The term was there used specifically in relation to the Minister's regulation of third-party access to infrastructure, but it is apt to describe the Minister's approach more generally.

[19] See eg the discussion of PILOT PPWG at paras I-4.52 to I-4.54.

[20] For an account of Norwegian law in this regard, see E Nordveit, "Regulation of the Norwegian Upstream Petroleum Sector", in T Hunter (ed.), *Regulation of the Upstream Petroleum Sector* (2015), p 132.

[21] G Gordon, "Production Licensing on the UK Continental Shelf: Ministerial Powers and Controls" (hereinafter "Gordon, 'Production Licensing'"), 4 (2015) *LSU Journal of Energy Law and Resources* 75, at 81.

[22] Model Cl 23 (2008 numbering).

[23] Model Cl 19 (2008 numbering).

[24] See the Appendix below.

is tightly upon the individual asset. Thus these schemes fit into the "individual, field-specific mindset"[25] that, prior to Wood, generally characterised oil and gas operations upon the UKCS.

THE WOOD REVIEW

The various challenges of maturity that have already been adverted   I-5.07
to in previous chapters (declines in both production and production efficiency, increasingly marginal discoveries, ageing infrastructure, interdependency, increasing diversity among operators, and the like) prompted the then Secretary of State for Energy and Climate Change, Ed Davey, in June 2013, to commission Sir Ian Wood, the recently retired chairman of the Wood Group, to "undertake a comprehensive review of the regulation and stewardship of the UK's hydrocarbon reserves".[26] Given what has happened since, it is perhaps worth expressly mentioning that Wood was not commissioned as a result of low oil price:[27] price was still on a comparative high, standing at around $90 per barrel. It was the province's general performance that led to the review; and perhaps the most worrying aspect of the UKCS's performance at this time was that exploration and appraisal rates stood at historic lows. As Oil & Gas UK stated in their 2014 Activity Survey, "2011 and 2013 saw the lowest and second lowest numbers of exploration wells drilled, respectively, since drilling began on the UKCS in the 1960s".[28] This level of exploration and appraisal drilling was patently inadequate to inventory the remaining "yet-to-find" prize of the UKCS before much of the province's critical offshore infrastructure is lost to decommissioning. Neither was the UK alone in considering this to be an apposite time to review its oil and gas policy: Norway published a review in 2011[29] and Australia was to follow suit in 2015.[30] However, the UKCS was not in rude health.

---

[25] Gordon, "Production Licensing", 76.

[26] DECC, *Government Response to Sir Ian Wood's UKCS: Maximising Economic Recovery Review* (2014), available at www.gov.uk/government/uploads/system/uploads/attachment_data/file/330927/Wood_Review_Government_Response_Final.pdf (accessed 6 May 2017), Ministerial Foreword, p 4.

[27] Oil price was, however, a significant driver in the Fiscal Review that followed; see further Chapter I-7.

[28] OGUK, Activity Survey 2014, available for download from http://oilandgasuk.co.uk/wp-content/uploads/2015/05/EC040.pdf (accessed 12 May 2017), p 15.

[29] Norwegian Ministry for Petroleum and Energy White Paper, *An Industry for the future – Norway's petroleum activities* (2011), available at www.regjeringen.no/globalassets/upload/oed/petroleumsmeldingen_2011/oversettelse/2011-06_white-paper-on-petro-activities.pdf (accessed 12 May 2017). The Norwegian review is very detailed and specific, running to more than 200 pages.

[30] Australian Government Department of Industry, Innovation and Science, *Offshore*

I-5.08   Within a relatively short time (November 2013), the review team[31] produced an interim report which offered a concise and pithy assessment of some of the problems confronting the UKCS, with the responsibility being laid at the doors of both Government and industry. Wood's criticisms included poor stewardship by both parties; a lack of focus on maximising the economic recovery of the UK's hydrocarbons; historical fiscal instability; problematical legal and commercial behaviours which stood in the way of necessary co-operation between operators; and, despite what was seen as a high standard of strategic thinking between Government and industry, poor implementation of the strategies produced. There was no softening of the position when the final report was issued in February 2014.[32] The Wood Review made four principal recommendations, which we now address in turn, although, as we shall see, there is both significant overlap between these recommendations, and a great many sub-recommendations subsumed within them.

Recommendation 1 – "the tripartite approach"

I-5.09   Wood recommended that "Government (HM Treasury and the regulator) and Industry must adopt a cohesive tripartite[33] approach to develop and commit to a new, shared MER UK strategy to maximise the huge economic and energy security opportunity that still lies off the UK's shores".[34] It is important to note that the concept of maximising economic recovery does not have its roots in the Wood Review; this was already the stated policy of the Department of Energy and Climate Change (DECC) and its predecessors, albeit that it had no statutory or regulatory foundation.[35] By

---

Petroleum Resource Management Review Interim Report (2015), available at industry. gov.au/resource/UpstreamPetroleum/Documents/Offshore-Petroleum-Resources-Management-Review-Interim-Report.pdf (accessed 12 May 2017). This is a comparatively short (c. 50-page) document drawn at a similar level of abstraction to the Wood Review. At time of writing, Professor John Chandler of the University of Western Australia is finalising a book which will provide a valuable comparison of these three reviews.

[31] As well as Sir Ian, the team included representatives from the DECC and from OGUK.

[32] A great deal of work went into the preparation of the Review, including interviews with some 40 UKCS licensees representing some 95 per cent of all UKCS production.

[33] As the new regulator was not intended to wholly supplant the Government department, which retains a significant role in some areas (eg decommissioning), the relationship is probably more properly described as quadripartite.

[34] Wood Review, p 6. Wood considered that up to 12 billion boe remained to be developed, but only if industry behaviours and modes of governance were improved.

[35] Maximising economic recovery was, for instance, the policy aim that lay behind the Fallow Fields and Stewardship initiatives of the early 2000s, and of the Labour Government's depletion policy in the 1970s. See further Gordon, "British Hydrocarbon Policies and Legislation" and the sources cited therein.

contrast, the Wood Review envisaged this concept as playing a more elevated role, moving from a strategic aim to an enforceable legal obligation. Significantly, however, the Review offered no developed definition of what was understood as MER or any detail in relation to the strategy itself. Rather, it confined itself to setting out what may be understood as the principles upon which these could be developed and the broad areas that would require to be addressed. Thus strategy was seen as embodying a "holistic" approach to the regulation of exploration, development and production[36] and requiring the establishment of a new and more powerful regulator[37] which would work with Treasury and the industry to ensure that a "cohesive" approach was taken to the challenges of the mature basin. The involvement of the industry in this manner also implied that operators would commit to "much better collaboration" so that all may enjoy greater "opportunities and value". The achievement of MER was then identified as depending upon the regulator and industry developing strategies focused on key sectors of the value chain from exploration to decommissioning – the subject of the fourth recommendation, also discussed below.[38]

Recommendation 2 – a new, independent regulator

The second recommendation was that the DECC "should create a I-5.10 new independent Regulator responsible for operational regulation of the UKCS, focused on supervising the licensing process and maximising economic recovery of the UK's oil and gas reserves in the short, medium and long terms".[39] After setting out the wide range of regulatory tasks that require to be carried out in relation to the UKCS,[40] the Review concluded that having the regulator located within the DECC was "no longer adequate to meet the challenges of an increasingly complex basin".[41] In support of this assertion, the Review pointed to the dramatic decline in the number of personnel within the existing regulator throughout a time of increasing complexity of the regulatory task,[42] concluding that the

---

[36] Wood Review, p 15.

[37] The subject of the second and third recommendations, discussed below.

[38] Wood Review, p 15.

[39] *Ibid*, p 6.

[40] This included developing and delivering a coherent strategy; encouraging investment via a stable regulatory environment; promoting exploration; requiring the demonstration of sound stewardship and timely development; encouraging industry collaboration; ensuring retention of and access to key infrastructure; and overseeing planning for decommissioning.

[41] Wood Review, p 20.

[42] Staffing had declined from 90 in the 1990s, when there were 90 fields in production on

DECC was "significantly under-resourced and under-powered".[43] But the Review did not consider that simply adding more personnel to the DECC was sufficient; it called also for the creation of a wholly new body at arm's length from the DECC. It perceived many benefits of this type of arrangement, arguing that such a body operates "with a degree of autonomy from ministers" who "do not concern themselves with [its] day to day running".[44] Another major driver was the likely attitude of the sort of person whom Sir Ian wished to attract to the leadership role of the regulator: "it is likely that they will expect the freedom to run the organisation as they see fit, within a framework set out by minsters".[45] A variety of other justifications were advanced. It would demonstrate focus and maintain the momentum for reform, send a signal to industry regarding the more interventionist stance that the regulator would now take and facilitate the recruitment of high-quality candidates (including specialists from outside the Civil Service) by adopting a more flexible approach to remuneration.

I-5.11   It might have been expected that the prospect of a stronger regulator would be regarded with some dismay by an industry that had become used to light-handed regulation. Not so, according to Wood: "Industry is clearly saying they want a stronger Regulator, able to become proactively involved, minimise disruption and delays, and facilitate and accelerate progress." This surprising enthusiasm for interventionist regulation may be explained by the observation that the basin is now characterised by "increasing interdependence between operators" which in turn leads to an increased "number of disputes and disagreements over new field developments and access to infrastructure".[46] Whether industry was aware that fulfilling its wishes would come with a significant price tag is not clear, but the Wood Review in any case considered the model of an industry-funded regulator to be attractive.

Recommendation 3 – enhanced regulatory powers

I-5.12 The Review recommended that "[t]he Regulator should *take* additional powers to facilitate implementation of MER UK",[47] an activity which would involve scrutiny of licensee's activities;

---

the UKCS, to 50 in 2014, when there were 300 fields in production.

[43] Wood Review, p 20.

[44] *Ibid*, p 55.

[45] Wood Review, p 55.

[46] *Ibid*, p 21.

[47] *Ibid*, p 6, emphasis added. The wording is peculiar given that no regulator can *take* powers, only use those that have been devolved upon them.

playing a dispute resolution role; enforcing MER UK by means of informal and formal warnings and ultimately "loss of operatorship and then licence"; attending joint venture (JV) "Operating and Technical Management Committee meetings"; and ensuring "greater access to the timely and transparent data necessary for a competitive market".[48] To some extent, this could be achieved by the new regulator making greater use of under-utilised powers already contained within the licence. But additional powers were also clearly needed if the regulator was to achieve all of the functions set out within Wood. Meetings, for example, had previously been considered, at least in the UK, to be spaces of commercial freedom and confidentiality. But what did the Review envisage in relation to the source of the new Regulator's additional powers? How would these dovetail in with existing arrangements – particularly the licence, given that no mechanism exists for routine amendment thereto? And was it envisaged that the regulator would educate and persuade, or forcibly compel compliance with MER UK? Both those who wished for a continuation of the light-touch approach and those who looked forward to a more interventionist regulator could find material in Wood on which to hang their hopes. As regards the former approach, the Review noted that the new regulator "must have the capability to *facilitate and influence* greater collaboration between operators on exploration, field developments and infrastructure".[49] Similarly, in outlining the exploration strategy – which is part of the fourth recommendation, discussed below – the Review speaks of "*encouraging* appropriate data sharing".[50] One of the innovations emerging from the Review is the idea of the regulator playing a mediation role in disputes relating to the licence or the potential for collaboration. In such cases, which are to be brought to the regulator within six months, a non-binding opinion would be issued. The extent to which such an opinion would really be non-binding, however, is open to doubt insofar as the Review went on to suggest that where any party fails to accept it, "appropriate sanctions" may be applied, to the extent that the failure "is inconsistent with MER UK or other licence terms".[51] As a consequence, licensees may be less than reassured by the Review's suggestion that the "nonbinding dispute resolution procedure will not prejudice the normal legal rights of either party".[52] A tougher approach may also be discerned when the Review spoke, in outlining the asset stewardship strategy – again

---

[48] *Ibid*, p 6.
[49] *Ibid*, p 15, emphasis added.
[50] *Ibid*, p 15, emphasis added.
[51] *Ibid*, p 17.
[52] *Ibid*, p 17.

part of the fourth recommendation – of ensuring that "operators are *held to account* for the proper stewardship of their assets and infrastructure consistent with their *obligations* to maximise economic recovery from the fields under their licenses and with consideration to adjacent resources".[53] Further, the Review was unequivocal that "[a]ll licence holders will be *bound* to work within the *requirements* of MER UK".[54] As regards sanctions, the Review envisaged a graded response to a situation where a licensee is "not acting in accordance with the MER UK strategy", ranging from a public formal warning at one end of the spectrum to termination of licence at the other, by way of the intermediate options of facilitating a change of operator and suspension of the licence.[55]

I-5.13   It is the case, of course, that there is no necessary inconsistency inherent in a regulator exercising a generally light-touch approach, but always with the possibility of making use of stronger powers in the background. But how did the authors of the Review consider the MER UK obligation was to be implemented in law? The point was not developed as explicitly or fully as might have been wished; however, it would seem that (perhaps alongside a policy of persuasion)[56] the licence was initially seen as key to implementation. In listing early priority actions for the new Regulator, it was proposed that "terms for existing and new licenses should be reviewed to reflect the requirements of MER UK and the prevailing business environment".[57] More explicitly, the Review went on to state that "it would be appropriate for licenses to have conditions related to maximising economic recovery for the UK … to the extent such provisions are not already included".[58] We thus have an indication, although it is not spelt out clearly, that what was envisaged was the *retrospective amendment of existing licences* to include a new and more explicit clause compelling licensees to operate in compliance with the strategy. In support of this interpretation, one may point to the fact that, earlier in the document, there is one reference to "the new MER UK clause".[59] As we shall see, by the time of implemen-

---

[53] *Ibid*, p 16, emphasis added.
[54] *Ibid*, p 17, emphasis added.
[55] *Ibid*, p 17. As a result of these indications, Vass, in an early commentary on the Wood Review, concluded that "MER UK as a legal obligation would be included in every existing and future UKCS Production Licence": U Vass, *Sir Ian Wood's Review: A UKCS Game Changer?* (2014), available at www.bonddickinson.com/sites/default/files/bon_dic_1296_oil_and_gas_-_newsletter_jun14_v7.pdf (accessed 12 May 2017), at 2.
[56] The Review also regularly states that the regulator is to "seek commitments" from industry, language suggestive of a consultative approach.
[57] Wood Review, p 36.
[58] *Ibid*, p 36.
[59] *Ibid*, p 22.

tation, this approach had been abandoned in favour of MER UK as a statutory duty supplemented by a strategy document, a creature *sui generis* which would seem to operate at the boundary of guidance and regulation.

Recommendation 4 – sector strategies

The Wood Review's fourth and final recommendation was that   I-5.14 that the new regulator should, in collaboration with the industry, develop and implement "sector strategies" in the following areas: exploration, asset stewardship, regional development, infrastructure, technology and decommissioning.[60] Each shall be briefly introduced in turn.[61]

Exploration strategy
The key drivers for the formulation of an exploration strategy   I-5.15 were, firstly, the fact that the "rate of exploration drilling is totally inadequate to exploit the undiscovered resources of the UKCS within the lifespan of existing infrastructure"[62] and, secondly, that even if the rate of drilling and discovery manifest from 2000 to 2008 were to be recovered, the impact on the likely quantity of hydro-carbons discovered would not be significant. The Review called on the Government to consider how to stimulate exploration, pointing to the competition the UKCS confronted internationally. In this regard, the lack of fiscal incentives was highlighted,[63] and, while the Promote Licence was praised, the practical difficulties facing small companies in obtaining funding were pointed out. The Review also noted approvingly the approach in the Netherlands, where "the state-owned non-operating company routinely takes a 40 per cent share in each exploration well" and by sharing risk in this way has boosted exploration; however there was no express recommendation of a return to state ownership.

Secondly, the Review called on the regulator to facilitate regional   I-5.16 exploration plans, specifically mentioning the fact that this could be achieved by managing licensing rounds. This would seem to represent a departure from the manner in which the DECC has been

---

[60] *Ibid*, p 7.

[61] We shall return to the strategies when we come to discuss the implementation of the Wood Review and the activities of the newly created regulator, the OGA.

[62] Wood Review, p 32.

[63] The Review compared the position in the UK with that in Norway, where a company without production income, receives tax relief for exploration costs in cash. On fiscal matters more generally, although Wood did not discuss taxation in detail, it identified the practical importance of fiscal design and recommended that a review be undertaken as a matter of priority. This was duly done. See Chapter I-7.

used to dealing with licensing in recent years when in essence it has been a matter for the industry to decide where it wants to focus its attention, subject to restrictions related, for example, to environmental or defence considerations.

I-5.17   Among other issues highlighted as requiring attention were "[t]he size and shape of new licence blocks within new plays and less prospective areas should also be considered to avoid fragmentation and offer coherent opportunities to the market".[64] There was a recommendation that in facilitating the exploration of new plays, the regulator "should actively seek to create and encourage joint ventures to pursue such opportunities", although it is unclear what sort of role the regulator could play in practical terms in encouraging the creation of JVs. Involvement in commercial issues was also seen in the recommendation that: "[t]he Regulator, in consultation with Industry, should investigate what measures can be taken to increase the rate of exploration drilling, specifically concentrating on drilling costs, improving the supply of rigs to the UKCS, and companies' ability to access rigs".[65] Finally, it was recommended that the "Regulator should facilitate Industry and the seismic companies to carry out speculative seismic, particularly targeting new plays which lack up-to-date seismic coverage, and, if justified, *should support with Government funding*".[66] While such data would then be made available on a commercial basis, the idea of more direct state involvement is striking and demonstrates the extent to which Wood signals a shift away from the previous division of labour between industry and state.

Asset stewardship strategy

I-5.18   The motivation for the development of this particular strategy was the significant decline in production efficiency and overall production. The Review commenced its discussion of asset stewardship by indicating that the MER obligation would extend beyond the confines of the individual licence to encompass at least adjoining licences.[67] Whilst the Review noted the prior existence of the Stewardship Initiative, mentioned above, the idea that an operator could be "held to account for the proper stewardship of their assets and infrastructure consistent with their obligations to maximise

---

[64] Wood Review, p 34. This would appear to echo the approach evident in relation to Frontier Licences, where the DECC and its predecessors offered larger acreages to reflect the risk and uncertainty inherent in entering less well-explored and developed areas. See further the discussion in Chapter I-4.

[65] Wood Review, p 35.

[66] *Ibid*, p 35, emphasis added.

[67] *Ibid*, p 38.

economic recovery from their fields under their licenses *and with consideration to adjacent resources*"[68] would surely amount to a significant extension of the current, relatively opaque, arrangements. Moreover, it must be queried if such a discussion belongs to the rubric of asset development or regional development. The other elements discussed under the rubric of asset management focused more clearly on individual field issues: on greater clarity with regard to performance expectations and on ensuring that performance data are actually delivered to the regulator in a timely fashion.[69]

Regional development strategy
This strategy has as its objective "to ensure the development of UKCS resources on a regional, rather than solely a field basis". Given the individual, field-specific regulatory approach which had previously been utilised this was one of the more revolutionary ideas in the Wood Review, and it would seem that the authors were aware that such a radical change might require compulsion, rather than mere encouragement. The Review noted that "Operators should be *required* (emphasis added) where appropriate, to cooperate with the regulator and with other licence-holders in the wider adjacent area on all aspects of field and cluster development, from exploration through to decommissioning, with the overarching aim of maximising economic recovery from clusters as well as from individual fields".[70] But what concrete deliverables must be achieved as a result of that co-operation? What consequence will flow if co-operation comes to nothing? On this the Review was not clear. But a key part of achieving the Review's objective would seem to be solving the problems of third-party access to infrastructure, and the Review noted that licensees "should make their infrastructure and process facilities available, subject to their own capacity requirements and technical compatibility, at fair and economic commercial terms and rates to potential third party users".[71] What is not clear from this statement is whether something more than the then current arrangements are intended, as it could really be seen as an accurate account of what existed under the Petroleum Act 1998, as amended, together with the Infrastructure Code of Practice, and related documents. However, further light on this matter would seem to be shed by the discussion in the Review's infrastructure strategy section. There, the Review's focus was upon the new regulator making "full use of

I-5.19

---

[68] *Ibid*, p 38, emphasis added.
[69] *Ibid*, p 40, Actions 13 and 14.
[70] Presumably, that should be read to include regions as well as clusters and fields, given that the focus of this strategy is regional.
[71] Wood Review, p 41.

the current legal powers to resolve disputes and facilitate access to infrastructure". Given the long history of failure in relation to the exercise of ever increasing powers in the hands of the DECC and its predecessors, operators desperate to tie back smaller reservoirs to existing infrastructure could be forgiven for feeling a sense of disappointment. Nevertheless, the recommended action does perhaps open the way for new initiatives insofar as it suggests that "standard protocols should be established by the industry in conjunction with the regulator with set procedures, timetables and guidelines on issues such as co-mingling of liquids and other technical and commercial risks, with recourse to independent experts where appropriate". Cynics might suggest that we have been down this road before, although the Review did also suggest that the "protocol should take account of learning from past failures to agree",[72] a clear acknowledgement of the deficiencies of the previous way of doing things.

I-5.20   The Review further wanted the regulator and the industry to work together to "develop Regional Plans for each area and play across the UKCS".[73] It went on to list some nine issues to be considered in the fulfilment of this action, including an appropriate appreciation of the particular challenges facing gas as opposed to oil.[74] It is very interesting to note that in attempting to explain the significantly better position of gas development on the Netherlands Continental Shelf as compared to the UKCS, the Review noted that EBN, the state-owned company, "has a very strong influence" in relation to MER. It went on to observe that the "Dutch Government is an active owner of the infrastructure and regulates the industry in a more active manner facilitating a degree of transparency, fairness and an enhanced information flow which makes resolving disputes easier and achieves consistency across the region".[75] These are very striking observations which arguably raise the question of why earlier in the Review an enhanced regulator – rather than a state company – is the key recommendation.

Infrastructure strategy

I-5.21   The infrastructure strategy was expressed in the Review as being primarily about prolonging the life of existing equipment and encouraging new investment, with the focus particularly on critical hubs.[76] That said, and as we have seen in para I-5.19, quite a signif-

---

[72] *Ibid*, Action 18.

[73] *Ibid*, Action 15.

[74] *Ibid*, pp 42–43.

[75] *Ibid*, p 43. Indeed, the Review goes so far as to say that third-party access "is not an issue in the Netherlands".

[76] *Ibid*, p 44.

icant amount of the discussion under this heading was on the topic of third-party access. More radically, Action 20 recommended that the regulator should take measures to facilitate the development of new infrastructure business models either from new entrants or existing players. Insofar as the Netherlands was again referred to as a model, the Review could be seen as pointing towards direct state participation; however, at present, there seems to be little political appetite for this, and the principal trend in this regard has been the sale of infrastructure by upstream companies to a set of mid- and downstream specialists.[77]

Technology strategy
In the case of the technology strategy, the emphasis of the Review was on the fact that the success of the UKCS has been in no small measure due to the development and deployment of new technologies to cope with the challenges of the basin, and to ensure that this process continues.[78] Specifically, the Review called on industry to focus as a matter of priority on the most pressing challenges and to work with Government and research institutions to seek solutions. Among the priorities were identified the following: improving exploration outcomes; decommissioning cost reduction; production improvement efficiency; improved oil recovery; enhanced recovery; development of small fields; and extending the technological reach.[79] In terms of recommendation actions for the new regulator, the Review called for the establishment of dedicated groups to focus on achieving solutions to key challenges. In this regard, it is notable that the idea of mandating reviews of fields most suitable for the application of Enhanced Oil Recovery (EOR) was mooted, although the detail of how this would be achieved was not developed.[80] A similar uncertainty appeared in the second recommended action insofar as the Review called for an operator's deployment of existing technology and development of new technology to become "part of the annual stewardship review", but without indicating whether any sanctions could be imposed in the event that reported activity was felt by the regulator to be in some sense below par.[81] Noting, however, that operators may be averse to taking the risk of deploying new technologies, the Review also recommended that they should be encouraged and even incentivised to do so. It was also suggested that where trials of new technologies were undertaken, the results should

I-5.22

[77] See Chapter I-7.02.
[78] For a fuller discussion of law and technology in the oilfield, see Chapter II-12.
[79] Wood Review, pp 47–48.
[80] *Ibid*, Action 21.
[81] *Ibid*, Action 22.

be shared. No further detail for these suggestions was offered, leaving open the question of what sort of incentives might be required to persuade reluctant operators to take risks with perhaps ageing assets in a mature basin or the question of whether competitive pressures, not to mention competition law, might have a role to play.[82] Finally, the new regulator and the Office of Carbon Capture and Storage were recommended to continue working closely to look for opportunities not only in relation to the reuse of depleted reservoirs for carbon sequestration but also in relation to using CO_2 for EOR.[83]

Decommissioning strategy

I-5.23   The final strategy, focused on decommissioning, reflected the challenging balancing act that the MER UK strategy as a whole requires to perform. In this regard, the Review noted the need to ensure that premature decommissioning was avoided, whilst ensuring that decommissioning, when it did occur, was well planned and co-ordinated as well as allowing the UK to gain "a competitive industrial capability".[84] Noting the anticipated costs as well as the proportion to be borne by the tax payer, the Review also pointed to the benefits to be won from improved performance as well as the increased burden if costs were not controlled. Avoiding early decommissioning would also offer benefits in terms of additional hydrocarbon recovery. Whilst the Review noted industry initiatives in relation to decommissioning, it bemoaned the lack of DECC or Treasury involvement.[85] This last point provided the inspiration for the first recommendation for the new regulator, namely the establishment of a "single decommissioning forum".[86] The new regulator was called upon to set a target for the radical reduction of decommissioning costs "whilst respecting all current obligations", although there was no mention of what the consequences would be for failure to meet that target. Beyond that, the Review expected greater collaboration also with the supply chain to examine how industry could best share risks and costs.[87] This was bolstered by the following recommendation which called for the Industry Technology Strategy to have "decommissioning cost reduction as one of its key objectives".[88] Though tucked away on the final page of the main body of the report, Action 27 is perhaps one of the most

---

[82] *Ibid*, Action 23.
[83] *Ibid*, Action 24.
[84] *Ibid*, p 50.
[85] *Ibid*, p 51.
[86] *Ibid*, Action 25.
[87] *Ibid*, Action 25.
[88] *Ibid*, Action 26.

important parts of the Review insofar as it called on the regulator to ensure that "assets are not prematurely decommissioned, making the necessary linkage between decommissioning and access to infrastructure". Whilst there was recognition that this would involve a potentially complex balancing act, there was no indication in the Review of what the solution would be in a context where a hub is required for ongoing satellite production, but the hub operator has no continuing interest in producing from that piece of infrastructure. For less complex cases, the Review went on to recommend better business models for the transition between operator and decommissioning specialist to improve efficiency,[89] before concluding with a call for all parties to work together "to investigate game changing decommissioning concepts which could radically change the value proposition". Intriguingly, this last recommendation called on industry to be open "to considering decommissioning policies and initiatives in other countries and jurisdictions which achieve similar outcomes at less cost and/or less damage to the environment".[90] No indication of which countries or policies the Review had in mind are given, but insofar as the approach adopted on the UKCS is driven to a very considerable extent by the requirements of the OSPAR Convention and associated instruments,[91] it is difficult to see that the industry would have much room for manoeuvre in that respect without some fundamental change to that Regional Seas agreement.

THE WOOD REVIEW'S RECEPTION

The Wood Review rapidly received enthusiastic support from I-5.24 Government,[92] with the UK Government announcing it was "committed to implementing the recommendations contained in the Review as quickly as possible",[93] and from OGUK, the principal industry representative organisation.[94] The warmth of the reception

---

[89] *Ibid*, Action 28.

[90] *Ibid*, Action 29.

[91] See further Chapter I-12.

[92] Including the SNP-controlled Scottish Government, a point that is of some significance given how critical that party has at times been of UK oil and gas resource management policy: see eg para I-9.119 to I-9.122.

[93] DECC, *Government Response to Sir Ian Wood's UKCS: Maximising Economic Recovery Review* (2014), available at www.gov.uk/government/uploads/system/uploads/attachment_data/file/330927/Wood_Review_Government_Response_Final.pdf (accessed 6 May 2017), p 4.

[94] OGUK, Press Release, "Wood Review Final Recommendations Can be Game Changers for UK Continental Shelf", 24 February 2014, available at oilandgasuk.co.uk/wood-review-final-recommendations-can-be-game-changers-for-uk-continental-shelf (accessed 6 May 2017): "This is a seminal moment in the history of the UKCS. The report is a game changer. We have the opportunity to secure a bright future for our industry and

could no doubt be explained by Sir Ian's suggestion that a greater degree of regulatory oversight and fresh emphasis on collaborative behaviours would result in a win-win situation by "increasing the size of the pie"[95] – an outcome desirable for both Government and industry.[96] If, as Yergin so stirringly put it, upstream oil and gas operations involve the quest for the prize,[97] who having an interest in the sector could sensibly complain about the prize getting bigger? But while there was a warm welcome for the Wood Review from most, some who read the document entertained concerns, if not as to the aims and broad thrust of Wood, then as to how it could be implemented.[98] Wood advocated radical change, including, it would seem, the alteration of licence rights, in an environment that had seen 40 years' worth of accumulated licence grants and associated investment, much of it funded from abroad.[99] Moreover, in the mature province, it is misleading to talk of "the industry" as if it has one homogeneous group of companies all with closely aligned interests. Even if we ignore, for present purposes, the contracting community, the influx of new entrants over the last couple of decades means that there are about 120[100] different licensees on the UKCS. They range in size from the super-major to the smallest, locally based independent. They have differing levels of resource and

---

unlock at least a further £200 billion for the UK economy. Oil & Gas UK congratulates the Secretary of State for having the insight to commission the review and looks forward to the swift implementation of its recommendations."

[95] This expression, and the similar "increasing the diameter of the pipe", was used by a number of speakers at the OGUK conference held in Aberdeen on 11–12 June 2014.

[96] It may conceivably also have had shades of groupthink. See eg P t'Hart, *Groupthink in government: A study of small groups and policy failure* (1990).

[97] D Yergin, *The Prize* (1991), pp 11–13. Yergin's language was echoed by the Secretary of State in the Ministerial Foreword to the Government's Response to Wood. DECC, Government Response to Sir Ian Wood's UKCS: Maximising Economic Recovery Review (2014), p 4.

[98] See eg G Gordon, "First Thoughts on the Relationship Between the New 'MER UK Obligation' and the Existing Licence Terms". Paper presented at Oil & Gas UK Annual Conference 2014, Aberdeen, 11 June 2014; Deloitte, "Survey Report: Making the Most of the UKCS" (2014) (hereinafter "Deloitte, 'Survey Report'"), available at www2.deloitte.com/uk/en/pages/energy-and-resources/articles/cultural-shift-key-to-maximising-economic-recovery-of-ukcs-oil-and-gas.html (accessed 6 May 2017), at p 6.

[99] This dimension is relevant as, in addition to any features in domestic and/or EU law which could be utilised to protect investments, international investments may in addition attract protection as a result of bilateral trade agreements or the Energy Charter Treaty. The authors express no concluded view on the manner in which Wood has been implemented; they suspect that question cannot be answered at a macroscopic level, but only in the context of the facts of an individual case. But it is an issue to which consideration should be given in appropriate cases.

[100] Around 150 company groups are listed as licensees on DECC's website, of which around 20–30 hold onshore licences only.

differing appetites for risk. Some dislike collaboration and prefer to work on their own, at least some of the time.[101] Importantly, some own offshore infrastructure while others do not.[102] This lack of commonality in interest means that it would seem to be possible, after implementing Wood, for there to be net losers, even if the prize does indeed become bigger overall.

THE LEGAL IMPLEMENTATION OF THE WOOD REVIEW

The legal implementation of Wood – a process which is, at time of writing, just coming to an end, some three years after Wood first reported – has proved to be a difficult process. It has involved the establishment not just of a new regulator but of a new kind of regulator,[103] and of the introduction of the MER UK obligation not, as Sir Ian had suggested, by means of the introduction of retroactive licence condition, but the introduction of a whole new species of economic regulatory law. This has been designed as a complex interaction between two pieces of primary legislation, the MER UK Strategy and a number of other, more specific strategy documents and sectoral implementation programmes, and slots in alongside – or arguably, treads upon the toes of – the pre-existing features of the licensing and broader resource management system, creating what one of the authors has described as "a very crowded dance floor, necessitating very fancy footwork in constrained places to make the new [system] work".[104] **I-5.25**

In this section, we will first have regard to the establishment of the new regulator and will then move on to the issue of the implementation of the MER UK obligation. **I-5.26**

The establishment of the OGA

The Wood Review had recommended the establishment of a new, proactive regulator to take on the resource management functions of the understaffed, underfunded and underpowered DECC. The OGA **I-5.27**

---

[101] Some companies have a policy of buying a 100 per cent interest in already-producing fields on the basis that there is little or no geological risk associated with those fields and full ownership means that the company can control the pace and nature of activities within their licensed area more readily than if they have to secure the agreement of JOA partners.
[102] This factor is of particular importance because if Oilco A owns a significant volume of infrastructure while Oilco B owns none, in a third-party access scenario, A will often be an owner and B will always be an applicant (as noted in Chapter II-7).
[103] U Vass, "A Review of the New UK Energy Bill: Very Fancy Footwork!", 4 (2015) *LSU Journal of Energy Law and Resources* (hereinafter "Vass, 'Very Fancy Footwork'") 59, at 60.
[104] *Ibid*, at 73.

was established as that regulator. From 1 April 2015 it operated, as an interim measure, as an Executive Agency within the DECC. Following the entry into force of the Energy Act 2016, the OGA is now constituted as a Government Company. BEIS, the DECC's successor as the Government department having responsibility for energy, is the sole shareholder, a situation which would seem, at least theoretically, to allow the potential for the exertion of influence which could undercut the Wood Review's notion of the OGA as an "arm's length" regulator.

The implemention of MER UK

Conceptual confusion?

I-5.28 The primary legislation implementing what Wood termed the MER UK obligation was passed in two stages. The first involved the introduction, by amendment at committee stage, of two new clauses to the Infrastructure Bill 2014–2015.[105] The second stage involved the introduction of the range of sanctions and other provisions contained in the Energy Act 2016, as further discussed in paras I-5.35 to I-5.38, below. The provisions in the Infrastructure Bill would go on to become ss 41 and 42 of the Infrastructure Act 2015, which introduced amendments to the Petroleum Act 1998. Section 41 was introduced as the Government thought it important to put the "overriding principle [of maximising economic recovery] contained in Sir Ian Wood's report into statute".[106] Section 42 contained the levy arrangements through which the industry, in the form of certain licence holders, were to pay towards the new regulatory regime.[107]

I-5.29 We shall return to the detail of the statutory provisions below. It is first worth briefly noting that the introduction of the obligation to maximise economic recovery was not an altogether smooth process. In introducing the amendments by which the obligation was introduced into the Infrastructure Bill, Baroness Verma indicated that the Government had been working "at a furious pace" to bring forth these legislative measures.[108] The political desire to demonstrate swift

---

[105] Infrastructure Bill, Grand Committee [HL], 22 July 2014 (hereinafter "Infrastructure Bill, Grand Committee"), col GC409, available at www.publications.parliament.uk/pa/ld201415/ldhansrd/text/140722-gc0001.htm (accessed 14 May 2017).

[106] Baroness Verma, *ibid*. See also Baroness Kramer to the same effect in the Second Reading debate: Infrastructure Bill, Second Reading [HL], 18 July 2014, col 842, available at www.publications.parliament.uk/pa/ld201415/ldhansrd/text/140618-0001. htm#14061871000194 (accessed 14 May 2017).

[107] Although the fact that the industry is paying for the regulator by means of a levy is of considerable practical significance, we shall not discuss the mechanics of how the levy is charged. Our focus shall instead be upon the MER UK obligation.

[108] Infrastructure Bill, Grand Committee, col GC411. She also indicated that the policy intention of the Bill was "to reduce regulatory burden, empower a stronger, more capable

progress in implementation was understandable. The Government was under pressure from the industry to swiftly implement the Wood proposals; it also had other political reasons to demonstrate an ongoing interest in North Sea oil.[109] However, the rush towards implementation had a number of adverse consequences. Firstly, while the Government had been in discussions with the industry throughout the period between Wood's reporting and the introduction of the implementing measures in the Infrastructure Bill, there had been no time for specific consultation on the provisions as drafted.[110] As a result, the legislation required significant amendment in order to secure industry support. In particular, the industry was unwilling to accept the proposition that the Secretary of State would be empowered to interfere with commercial arrangements, including pre-existing commercial arrangements, having a significant adverse effect upon the objective of maximising economic recovery of UK petroleum. These provisions were removed by Government amendment during the Report stage.[111] That the industry and Government should find themselves in disagreement during the implementation phase of Wood, despite having each expressed themselves content with Wood's findings, provides further evidence for the authors' argument that the open-textured nature of the final report meant that there is significant ambiguity as to precisely what sort of regulatory interventions it proposed.

This ambiguity carried into the core question of what was meant I-5.30 by maximising economic recovery. When introducing the provisions in Committee, Baroness Verma stated that the Government would not offer a definition of "maximising economic recovery".

---

regulator that can mobilise and catalyse, and enhance the efficiency and co-ordination of activity in the UK Continental Shelf". Given that the OGA is to be significantly more interventionist than the DECC, the first element of this claim would seem to be optimistic. The other elements of her claim are more realistic and provide a useful summation of the anticipated benefits of the OGA.

[109] The Scottish independence referendum took place in September 2014. The Scottish Government contended that "successive Westminster Governments had failed to provide effective stewardship of Scotland's oil and gas assets" and, while welcoming the Wood Review, argued that only an independent Scotland would maximise the long-term value to the nation of the oil asset through eg the use of oil funds: Scottish Government, *White Paper: Scotland's Future* (2013), pp 300–302, available online at www.gov.scot/Publications/2013/11/9348/0 (accessed 27 August 2017). Demonstrating a strong political will to implement it may have seemed a useful means of undercutting the SNP's assertions of disinterest in the industry.

[110] This led to significant criticism of the Government from Lord Jenkin, a Conservative peer with significant experience of oil and gas matters: Infrastructure Bill, Grand Committee, col GC411.

[111] Infrastructure Bill, Report [HL], 10 November 2014, cols 35–38, available at www.publications.parliament.uk/pa/ld201415/ldhansrd/text/141110-0001.htm#1411109000477 (accessed 14 May 2017).

Although senior figures in the industry were very keen to see this term defined,[112] Baroness Verma defended the Government's position on the grounds that

> "[b]ecause of the continually changing nature of regulation, the developing needs of exploration and production in the North Sea, and changes in technology and approaches, we think that the concept of MER UK is something that itself is likely to change over time".[113]

I-5.31   One could observe first that that if MER truly is both incapable of definition and likely to change over time, it seems a fairly unpromising core concept for legally enforceable regulation. But it is submitted that in fact the difficulty arises because of a simple confusion of the concept with its means of implementation. Put shortly, Baroness Verma – and many others involved in the implementation process – have confused what MER UK is with the steps to be taken to bring MER UK about, a confusion which creates problems, but which, it is submitted, is not irresoluble. To expand a little on this point: although the industry is complex and fast-moving, the notion of MER UK is not. It is about producing the maximum petroleum that can be profitably produced from the UKCS as a whole. The one point of any complexity is whether one is interested in maximum by volume[114] or maximum by value.[115] That issue aside, the idea is both simple and stable, which is probably why it has cropped up repeatedly in policy initiatives of the past. The *idea* of MER UK does not vary depending upon advances in technology, variations in oil price, changes in cost base or as a result of the introduction of a new regulatory model. The *volume or value* of oil that is economically recoverable most certainly is affected by those factors; and as the various sections of the province move though their lifecycle from frontier areas towards cessation of production, the industry pushes the frontier of

---

[112] See eg Deloitte, "Survey Report", p 6.

[113] Infrastructure Bill, Grand Committee, col GC409.

[114] This is perfectly sensible, if eg the state's primary concern is energy security. It seemed to be what the Wood Review intended but, as we shall see, this seems to have changed, at least so far as strategy obligations are concerned.

[115] This makes sense if eg the state's focus is on maximising direct state tax take. As we shall see below, the Strategy provides, in its discussion of the Central Obligation, that relevant parties "must take the steps necessary to secure that the maximum *value* of economically recoverable petroleum is recovered": OGA, "Maximising economic recovery of UK petroleum: the MER UK strategy", available at www.ogauthority.co.uk/media/3229/mer-uk-strategy.pdf (accessed 24 August), para 10. This could raise questions of whether the OGA intends to return to the days of depletion policy management, a feature which was of significance to Government in the 1970s, and which can involve state intervention to manipulate production levels in order to optimise production during periods of high price. However it is understood that the OGA has no current intention of pursuing such a policy.

technology forward and is subject to the boom and bust cycles that have long characterised the economics of oil, the particular steps which need to be taken to promote MER UK will certainly change. But the concept does not. Thus while the achievement of MER UK is very much a moving target, what MER UK is should not be.

A question which arises when considering MER UK is whether it is a proper subject for enforceable legal obligation or is best left as a policy objective supported by a more specific set of discrete legal powers. Two of the authors of this chapter have argued that MER UK is a concept abstracted at too high a level of generality to serve as a useful basis for legal regulation; that trying to impose legal obligations at this level is productive of uncertainty and/or complexity; and that, as such, a less ambitious implementation strategy would be favourable, whereby MER UK retains a significant role within the system as a policy goal that should be legally implemented through a more specific set of regulations.[116] We continue to hold that view, but to some extent the point has become moot. Parliament has legislated; the die is cast.

I-5.32

MER UK as a legal obligation

As noted above, the legislative implementation of the Wood Review occurred in two tranches. The Infrastructure Act 2015 fulfilled the Government's intention to place "the principal objective of maximising economic recovery of petroleum upon the UKCS" on a legislative footing; however, it is important to note that the duty under the principal objective faces inwards towards the state, not outwards towards the industry. Thus the Act required the Secretary of State[117] to produce "one or more strategies for enabling the principal objective to be met".[118] It further provides that the principal objective is to be advanced, in particular, through:

I-5.33

"(a) development, construction, deployment and use of equipment used in the petroleum industry (including upstream petroleum infrastructure); and
(b) collaboration [among licensees, operators, infrastructure owners and those having responsibility for commissioning infrastructure]"[119]

---

[116] J Paterson and G Gordon, *AUCEL Response to Government Consultation on the Implementation of the Wood Review*, AUCEL Working Paper, available at www.abdn. ac.uk/law/research/centre-for-energy-law (accessed 10 May 2017).

[117] This function has not been transferred to the OGA by the Energy Act 2016 or The Petroleum (Transfer of Functions) Regulations 2016 (SI2016/898). At the time of writing, it therefore continues to be vested in the Secretary of State (of BEIS), not the OGA.

[118] Petroleum Act 1998, s 9A(2), inserted by Infrastructure Act 2015, s 41.

[119] Petroleum Act 1998, s 9(A)(1), as amended by the 2015 Act. As we shall see, collabo-

I-5.34   This provision has the potential to cause confusion. The reference to collaboration among industry actors may give the impression that this is an industry-facing obligation; however, it is not. As we shall see, the only obligation of the relevant industry actors is to obey "the strategy or strategies". What this section provides is a set of instructions to the Secretary of State as to matters which must be included within the strategy or strategies.[120] The Secretary of State was further obliged to "act in accordance with the current strategy or strategies"[121] while undertaking a range of regulatory and resource management functions.[122]

I-5.35   The Act did have an outward-facing dimension, in that it provided that a range of MER-relevant industry parties[123] "must act in accordance with the current strategy or strategies when planning and carrying out [specified] activities";[124] however, even after the provision had entered into force[125] that obligation was not active no such strategy had by that point been produced; neither had the sanctions regime for breach of the obligation been created.[126] It was not until the entry into force of the MER UK Strategy[127] that the industry-facing provisions of the Act came into existence, and not until the entry into force of Chapter 5 of the Energy Act 2016[128] that the obligations came to be fenced with meaningful sanctions for breach.[129]

---

ration, in particular, is seen as a key means of implementing MER UK: see further the discussion at paras I-5.46 to I-5.57.

[120] Alongside these mandatory elements, the Secretary of State is, in addition, entitled to include other matters within the stategy or strategies: Petroleum Act 1998, s 9(A)(3), as amended by the 2015 Act.

[121] Petroleum Act 1998, s 9B, inserted by the Infrastructure Act 2015, s 41.

[122] Petroleum Act 1998, s 9B(a)–(e), inserted by Infrastructure Act 2015, s 41. These include licensing, exercising any function or power under a petroleum licence, and utilising its powers relative to upstream infrastructure, contained in the Energy Act 2011 (on which see further paras I-6.45 to I-6.48).

[123] That is, licensees, operators, infrastructure owners and those who commission infrastructure.

[124] Petroleum Act 1998, s 9C, inserted by Infrastructure Act 2015, s 41.

[125] On 12 April 2015: The Infrastructure Act 2015 (Commencement No.1) Regulations 2015 (SI2015/481), Reg 3(b).

[126] The Act stated that the Secretary of State must regularly report on the industry's performance against the contents of the strategy or strategies; however even this rather soft form of enforcement was dependent upon the production of the strategy or strategies.

[127] Upon its publication on 18 March 2016: see OGA, Online Guidance: MER UK Strategy, available at www.ogauthority.co.uk/regulatory-framework/mer-uk-strategy (accessed 14 May 2017).

[128] On 1 October 2016: The Energy Act 2016 (Commencement No. 2 and Transitional Provisions) Regulations 2016 (SI2016/920), Reg 2(b).

[129] Energy Act 2016, s 42(3)(a). As we saw in Chapter I-4, the sanctions provisions also apply to breaches of the petroleum licence: s 42(3)(b). They also apply to "requirements imposed on a person by or under a provision of this Act which, by virtue of the provision, is sanctionable in accordance with this Chapter": s 42(3)(b). Such provisions include eg breaching the duty

The position, following the entry into force of the Energy Act I-5.36
2016 and the MER UK Strategy, is now that licensees, operators,
infrastructure owners and those who commission infrastructure are
obliged to act in accordance with the MER UK Strategy (and any
additional relevant sectoral strategy)[130] while undertaking relevant
petroleum-related activities.[131] The Secretary of State was obliged to
produce the first strategy within one year of the coming into force of
the Infrastructure Act, and this requirement was complied with when
the Strategy[132] came into force on 18 March 2016. Pursuant to the
Petroleum Act 1998 s 9(2) (as amended by the Energy Act Sch 1 para
7), the OGA (rather than BEIS, as formerly) has the responsibility to
revise an existing strategy or issue a new strategy, and to review the
strategy no later than within four years.

The sanctions for breach of the Strategy are set out in Section I-5.37
42(4) of the Energy Act 2016.[133] This provides for a suite of sanctions
provisions, made up of an enforcement notice,[134] a financial penalty
notice,[135] a revocation notice[136] and an operator removal notice.[137]

---

to provide the OGA with requested information and samples (made sanctionable by s 34(4))
and breaching the duty to inform the OGA of a forthcoming meeting (s 38(10)).

[130] A significant number of strategies have now been produced by the OGA and are
available for download from the "Publications" section of the OGA's website: see www.
ogauthority.co.uk/news-publications/publications (accessed 13 May 2017). These include
eg the Exploration Strategy: OGA, Exploration Strategy (2016), available at www.
ogauthority.co.uk/media/2835/exploration_strategy_master.pdf (accessed 14 May 2017).
Note, however, that unlike the MER UK Strategy, these sectoral strategies have been
produced by the OGA, not the DECC or BEIS. As was noted at para I-5.33, it would
seem that responsibility for the principal objective has not been formally transferred from
the DECC or BEIS to the OGA. Thus it would seem that this particular suite of strategies
do not impose any legally enforceable obligations upon the industry, although this could
easily be resolved on a prospective basis by reissuing them under the aegis of BEIS.

[131] Thus licensees must "act in accordance with the current strategy or strategies when
planning and carrying out activities as the licence holder": Petroleum Act 1998 (as
amended), s 9C(1); operators, "when planning and carrying out activities as the operator
under the licence": s 9(C)(2), and so forth.

[132] BEIS, The Maximising Economic Recovery Strategy for the UK (2016) (hereinafter "the
MER UK Strategy"), available at www.ogauthority.co.uk/media/3229/mer-uk-strategy.pdf
(accessed 14 May 2017). Although the version now published on the OGA website is
dual-badged as OGA and BEIS, it states (on the front cover) that it is "as reproduced by
the OGA". In truth, it is really a DECC document, dating from the period when the OGA
was an executive agency of the DECC and differing from the DECC only in its formatting
and design.

[133] These apply not just to the main MER UK Strategy, but to all MER-related strategies
produced by the Secretary of State, and, as was noted at para I-4.87, they also now apply
to breaches of licence terms and conditions.

[134] Energy Act 2016, s 43.

[135] Ibid, ss 44–46.

[136] Ibid, s 47.

[137] Ibid, s 48.

Prior to the imposition of a sanction, the OGA must serve notice under Section 49, intimating its intention to impose a sanction and providing the recipient with an opportunity to make representations.

I-5.38   Thus the legislation provides clarity on which parties must satisfy the principal objective and the consequences that may befall them if they do not. However, we need to have regard to the Strategy – a document that was not available, even in draft, at the time the Infrastructure Act 2015 was passed[138] – in order to get any real insight into the obligations that have been imposed upon the industry.

I-5.39   The Strategy is an unusual legal instrument, neither mere guidance nor one of the usual forms of delegated legislation, although draft Strategies are, however, subject to Parliamentary negative resolution procedure.[139] Perhaps the closest parallel to the use of the Strategy as a form of bespoke delegated legislation lies in the sphere of immigration law,[140] where the Rules and Policy process is utilised primarily to allow swift and frequent changes in practice. One would hope that that is not the reason why it has been selected for use in the present context, as fast and capricious change could have a disastrous effect upon investor confidence; and that was certainly not the purpose of Wood.

I-5.40   When it did appear, the MER Strategy was a relatively brief[141] document. It did not read like a strategy in the dictionary sense of the word,[142] but as a set of regulations, although it is respectfully submitted that it is not as tightly drafted as a regulatory device would ideally be.[143] Its structure is quite complex. At its heart is the Central Obligation. Alongside this sit the Supporting Obligations and Required Actions and Behaviours. The terminology used (specifically, the use of the word "supporting") might suggest a hierarchy of norms, where Supporting Obligations and Required Actions and Behaviours are binding only in so far as they promote the Central

---

[138] Thus, industry and the Government were to a very large extent being asked to buy sight unseen when they were consulted, or asked to vote, on the 2015 Act, a fact which led to disquiet within the industry: "[N]o-one presently knows what the strategy will look like ... [a]t this stage, experts are concerned that, vis-à-vis individual MER parties, MER will be an unquantifiable duty, partially or wholly unrelated to a particular MER party's licenses or assets": U Vass, "Very Fancy Footwork", 62.

[139] Section 9(G).

[140] See eg G McGill, J Vassiliou and D Stevenson, "Immigration", in *The Laws of Scotland: Stair Memorial Encyclopaedia* (2nd reissue, 2016), para 6: "They have a unique status, being neither delegated legislation nor merely statements of policy."

[141] Sixteen pages, many containing a fair amount of white space.

[142] The *Cambridge Dictionary*, for instance, defines strategy as "a detailed plan for achieving success in situations such as war, politics, business, industry, or sport, or the skill of planning for such situations".

[143] See eg the issue relating to whether the Safeguards apply to the Required Actions and Behaviours, discussed further at para I-5.45.

Obligation. However, this is not the case. The Introduction to the Strategy states that each of these elements are as binding as each other,[144] but that all of them are subject to a series of Safeguards, which – in the event of a clash – take precedence. Each of these elements of the Strategy shall be discussed in turn.

The Central Obligation

The Central Obligation is drawn at a very high level of abstraction. It provides: "Relevant persons[145] must, in the exercise of their relevant functions,[146] take the steps necessary to secure that the maximum value of economically recoverable[147] petroleum is recovered from the strata beneath relevant UK waters."   I-5.41

Supporting Obligations

In addition to the Central Obligation, there are 15 substantive[148] Supporting Obligations. These are arranged in headings: Exploration, Development, Asset Stewardship, Technology, Decommissioning and OGA Plans. The Supporting Obligations are all much more detailed and specific than the Central Obligation. For instance, paragraph 11 states: "The licensee of an offshore licence who has made a firm commitment to carrying out a work programme in respect of that licence must not relinquish the licence without first having completed the work programme as set out in the licence."   I-5.42

This provision helps to demonstrate how the Supporting Obligations put some flesh on the rather bare bones of the Central   I-5.43

---

[144] The MER UK Strategy, p 6. The main body of the document is less clear on this than one might have expected, but the drafting seems to be consistent with the intention expressed in the Introduction. For instance, in paras 8 and 9, the Central Obligation and Supporting Obligations are expressly stated to be subject to the Safeguards; there is no similar provision that would state that the Supporting Obligations are subject to the Central Obligation.

[145] These are the same as the categories of persons described at paras I-5.35 to I-5.36: the MER UK Strategy, p 14.

[146] "Relevant functions" is defined as "the functions which relevant persons are obliged by the Petroleum Act 1998 to exercise in accordance with the Strategy, but only insofar as those functions can affect the fulfilment of the principal objective." In its reference to the Petroleum Act at large, rather than the provisions relating to the principal objective in particular, the definition is somewhat open-textured, but this would seem to be a reference to the functions outlined at paras I-5.33 to I-5.36.

[147] "Economically recoverable" is defined as meaning "those resources which could be recovered at an expected (pre-tax) market value greater than the expected (pre-tax) resource cost of their extraction, where costs include both capital and operating costs but exclude sunk costs and costs (such as interest charges) which do not reflect current use of resources". *Ibid.*

[148] There are 16 paragraphs under the heading but para 9 simply repeats that they all fall to be read subject to the Safeguards.

Obligation. It also demonstrates how the obligations in the licence and those imposed by the MER UK Strategy can, on occasion, be so closely connected as to be intertwined.[149] Paragraph 11 is here stepping well into what has hitherto been the exclusive territory of the licence. Under the UK system, a licensee who failed to carry out its initial work programme would not find itself compelled to do so, nor subject to any kind of direct penalty for doing so. The licence would simply not be permitted to continue into the second term, and further consequences would be limited to loss of good standing with the licensing authority, meaning that the licensee might find itself at the back of the queue when applying for new acreage. The absence of a more immediate consequence may very well have been a lacuna within the UK system: to guard against the problem of failure to complete the work programme, some other systems demand performance, an obligation that is secured by bonds.[150] It may very well be a good idea, both for the state's interest and in the interests of fair play among licensees, to plug the gap. But it is certainly interesting that what is essentially a licence problem is now being fixed by having recourse to what can be seen as a new species of economic regulatory law. Will there be occasions when the operation of the Strategy is so restrictive of the licensee's freedoms as to operate as a derogation from the licence grant, or an indirect expropriation?[151]

I-5.44   Less obviously, but perhaps even more importantly, the relationship between the general Central Obligation and the express and specific provisions contained in the Supporting Obligations require further consideration. The Supporting Obligations deal with very specific situations, but it is submitted that it would be misleading to attach a great deal of emphasis to these. It must be remembered that the Central Obligation is very broad in its scope, and every bit as binding as are the Supporting Obligations. This can be tested by imagining that para 11 was not in the Strategy. Would that mean that the OGA was not empowered to insist upon completion of the work programme before relinquishment of the licence? It would seem to depend upon the circumstances. If relevant persons must, in

---

[149] Another strong example of this intertwining can be seen in the fact that the sanctions regime has been applied to breaches of the licence conditions. Thus a minor breach that would not have been previously been seen as serious enough to lead to revocation – and would therefore have escaped any practical result beyond an informal censure – may now result in eg a financial sanction. See further para I-4.13.

[150] A Marino and J Gower, "Oil and Gas Mineral Leasing and Development on the Outer Continental Shelf of the United States", 4 (2015) *LSU Journal of Energy Law and Resources* 1, at 19–21.

[151] See eg T Waelde and A Kolo, "Environmental regulation, investment protection and 'regulatory taking' in international law", 50 (2001) *ICLQ*, 811.

the exercise of their relevant functions, take the steps necessary to secure MER UK, then if it can be demonstrated that, on the facts of the individual case, the interests of MER UK would not be served by undertaking the work programme, there would be no obligation to undertake the work. If, however, the interests of MER UK would be served by undertaking the work programme, then it would seem that a duty to undertake the work could be founded on the Central Obligation. If this is correct, then there is a risk that the number of demands that OGA may legitimately make of the licensee are limited only by OGA's imagination.

Required Actions and Behaviours

Although the Required Actions and Behaviours are described, in the Introduction to the Strategy, as being "as binding as the Central Obligation",[152] it is not wholly clear that they are; or, perhaps more precisely, it is not wholly clear that they are obligations of the same form as the Central Obligation or the Supporting Obligations. Unlike the other categories of obligation, they are not expressly stated to be subject to the Safeguards;[153] given the significance that the Strategy attaches to the Safeguards, it is somewhat surprising that the relationship between is not set out expressly. There are, perhaps, two possible explanations for this. One is that the drafter of the Strategy considered the Required Actions and Behaviours to be in the nature of secondary obligations, parasitic upon the Central or a Supporting Obligation having been engaged and already found not to conflict with any of the Safeguards. This would be consistent with the character of the Required Actions and Behaviours; in essence, they are concerned not with the obligation to do things, but with the manner in which things should be done (eg in a timely fashion;[154] having given due consideration to collaboration;[155] with due regard to cost reduction[156]) or the consequences of a decision not to pursue MER UK (seek investment, divest or relinquish).[157] The other is that the omission of a paragraph clarifying the order of precedence was a straightforward drafting error. In that case, the wording of the

I-5.45

---

[152] The MER UK Strategy, p 6.
[153] The Central Obligation is made subject to the Safeguards by para 8 of the MER UK Strategy; the Supporting Obligations by para 9.
[154] The MER UK Strategy, para 27.
[155] *Ibid*, para 28.
[156] *Ibid*, para 29.
[157] *Ibid*, paras 30–34. The "invest, divest or relinquish model" set out in the Strategy would seem to owe much to the Fallow Areas Initiative, developed in the early 2000s. This Initiative has been superseded by the implementation of the Wood Review, but was a very significant staging post along the way to the new mode of regulation and is discussed further in the Appendix to this volume.

Introduction and the assertive nature of the Safeguards[158] themselves would strongly suggest that they are intended to take precedence over the Required Actions and Behaviours.

A key action of behaviour: collaboration

I-5.46   As we have seen, the Wood Review attached particular significance to the topic of collaboration, and collaboration is expressly referenced as part of the principal objective. It has been regularly identified by the OGA as one of the principal means by which MER UK will be achieved,[159] and it gives rise to some difficult legal and commercial questions. It is therefore worth sketching the extent of this particular obligation in some detail.

I-5.47   While collaboration in its broad sense runs through the whole MER UK Strategy, para 28 speaks specifically to collaboration, stating:

> "When considering how to comply with obligations arising from or under this Strategy relevant persons must:
> (a) where relevant, consider whether collaboration or co-operation with other relevant persons and those providing services relating to relevant functions in the region could reduce costs, increase recovery of economically recoverable petroleum or otherwise affect their compliance with the obligation in question;
> (b) where it is considered possible that such collaboration or co-operation might improve recovery, reduce costs or otherwise affect their compliance with obligations arising from or under this Strategy, relevant persons must give due consideration to such possibilities; and
> (c) co-operate with the OGA."

I-5.48   Sub-paragraphs (a) and (b) of para 28 are worded in a somewhat awkward, repetitive fashion. Sub-paragraph (a) requires MER parties to "consider whether collaboration or co-operation ... [with other MER parties] ... and those providing services [supply chain] ... could reduce costs, increase recovery of economically recoverable petroleum or otherwise affect their compliance with the obligation in question". Sub-paragraph (b) requires MER parties to give due consideration to collaboration and co-operation when the condi-

---

[158] Most of the Safeguards state that they apply to "any obligation". Thus it could be argued that they take precedence over any obligation-imposing provision in any event, and that paras 8 and 9 do no more than provide what is already implicit in the Strategy's schema. See further the discussion of the Safeguards at paras I-5.58 to I-5.62.

[159] See eg OGA, Press Release, "OGA Emphasises Collaboration as Key to Success", 20 April 2017, available at www.ogauthority.co.uk/news-publications/news/2017/the-oga-emphasises-collaboration-as-key-to-success/ (accessed 14 May 2017). This press release marked the launch of the OGA's new Collaborative Behaviour Quantification Tool, discussed further at paras I-5.54 to I-5.57.

tions of sub-paragraph (a) are satisfied. We do not believe that use of the word "consideration" can mean a mere formal "tick-the-box" consideration; in other words, para 28 envisages and requires actual collaboration.[160]

An MER party who does not wish to collaborate may, of course, argue one of the four substantive Safeguards which include illegality,[161] two discrete strands of lack of satisfactory expected commercial return,[162] beneficiary contribution[163] and overall damage to investors' confidence.[164] In this connection, para 2 on illegality assumes a special importance. It reads as follows: I-5.49

> "No obligation imposed by or under this Strategy permits or requires any conduct which would otherwise be prohibited by or under:
> (a) any legislation, including legislation relating to competition law, health, safety or environmental protection; or
> (b) the common law, including the OGA's duty to act reasonably."

There is always a possibility that collaboration as required by para 28 might be in violation of competition law.[165] The OGA addressed this issue in a paper published in November 2016.[166] This document refers to a letter written by the Competition and Markets Authority (CMA) (the UK competition regulatory authority) to Amber Rudd, then Secretary for State for Energy and Climate Change, on 3 December 2015. In the letter the CMA emphasised that the OGA should be careful not to encourage breaches of competition law. On the other hand the CMA recognised that collaboration can be beneficial and that concerns over competition law should not chill legitimate activity. The OGA notes that collaboration agreements may be horizontal or vertical in nature. In such cases, MER parties are urged to consider whether in any case of proposed collaboration: I-5.50

(1) there may be pro-competitive outcomes;
(2) the agreement may be *de minimis* for the relevant market;
(3) there may be an applicable block exemption.

---

[160] It is perhaps worth noting in passing that given that the Strategy is binding only upon MER parties, and that the oilfield service sector has not been included as an MER party, the Strategy could give rise to an asymmetrical situation whereby an MER party might be required to collaborate with an oil services company, but the service company could not be required to reciprocate.

[161] MER UK Strategy, para 2.

[162] *Ibid*, paras 3–4. The difference between these two strands is discussed at para I-5.60.

[163] *Ibid*, para 5.

[164] *Ibid*, para 6.

[165] See further the discussion in Chapter II-11.

[166] OGA, *Competition & Collaboration* (2016), available at www.ogauthority.co.uk/media/2952/oga_competitioncollaboration_ukcontshelf_16.pdf (accessed 14 May 2017).

I-5.51   In sum, while respecting the ambit of competition law, the OGA does not wish to see collaboration derailed because of overblown regulatory fears.

I-5.52   The OGA listed collaboration as one of its ten Asset Stewardship Expectations in a 2016 publication entitled "Asset Stewardship Expectations".[167] The Collaboration Expectation is that:

> "Licensees should build effective business relationships which aim to create more value than is possible alone, by embracing a culture of collaboration and utilising collaborative tools and processes. In particular, licensees should be able to demonstrate that collaboration forms a core part of their organisational culture, and that they are making use of appropriate collaborative behaviour tools."[168]

I-5.53   The goals are to "unlock value in Joint Ventures".[169] This type of objective has been present in the system since at least the days of the Stewardship initiative.[170] However the more overtly behaviour-focused concerns of the new system are illustrated by the further aims, to help parties to "develop self-awareness and self-help strategy";[171] and the system's educative function is made apparent by the references appended to the Collaboration requirement, which refers to relevant ISO standards and OGUK documentation, as well as to the OGA's own Guidance.[172] The Expectation also enjoins parties to recognise good examples of collaboration and encourage adherence to the Commercial Code of Practice (CCoP) and the Infrastructure Code of Practice (ICoP). Naturally, all this should promote compliance with MER Strategy, para 28. The Expectation makes clear that collaboration needs to be evidenced and measured.

I-5.54   In April 2017, the OGA issued the "Collaborative Behaviour Quantification Tool: Assessment Guidance Note"[173] (hereinafter "CBQT Note"), which outlines the methodology the OGA will use in measuring how collaborative operators are. The process will begin in 2017 and will be revisited every two years. Appendix A of the CBQT Note is a matrix listing eight headline collaborative values:

- reasonable;
- aligned;

---

[167] OGA, *Asset Stewardship Expectations*, (2016), available at www.ogauthority.co.uk/media/2849/asset_stewardship_expectations.pdf (accessed 14 May 2017).

[168] *Ibid*, p 14.

[169] *Ibid*, p 14.

[170] See the Appendix to this volume.

[171] OGA, *Asset Stewardship Expectations*, p 14.

[172] *Ibid*, p 14.

[173] OGA, *Collaborative Behaviour Quantification Tool: Assessment Guidance Note* (2017), available at www.ogauthority.co.uk/media/3596/420432-oga-cbqt-assessment-guidance-note_17.pdf (accessed 14 May 2017).

- learning;
- strategic;
- change;
- respect;
- accommodating;
- openness.

Opposite each value, the matrix describes negative, middling and I-5.55
positive descriptions of behaviour. To take one example, the first
"Learning" behaviour is about reactions to failure. The lowest
level score is "Failures are hidden"; the middling score is "Failures
are mostly shared when asked about them"; and the top score is
"Quickly and widely shares failures and learnings externally". No
fewer than 33 behaviours are thus described under the eight values.
Scoring is from one to five.

The operator and the OGA separately will analyse the operator's I-5.56
activity through the CBQT test. This will produce two test results.
The OGA and the operator will then take half a day to discuss the
two test results. The OGA team will need to be familiar with the
operator's activities. The teams will try to reach a synthesis and
agree identification of areas of good and bad practice. If necessary
the OGA can request that an operator submits an improvement
plan to address improvement of collaboration. An example of an
improvement plan is attached as Appendix C to the CBQT Note.

It is clear, therefore, that the OGA is taking this particular I-5.57
Required Behaviour and Action very seriously, and is building on the
legal mandate provided by para 28 in order to develop soft law tools
such as the CBQT.

Safeguards

Paragraphs 2–6 set out the Safeguards. Given the far-reaching scope I-5.58
of the Central Obligation, these Safeguards play the very significant
role of preventing the imposition of obligations in circumstances
where it would be undesirable or dangerous to do so.

Para 2 provides that no obligation imposed as a result of the I-5.59
strategy "permits or requires any conduct which would otherwise
be prohibited by statutory law or common law, including the OGA's
duty to act reasonably". This is a particularly important provision
as it means that eg the body of regulatory safety and environmental
law is elevated in status above the requirements of the MER UK
Strategy.[174] One might therefore argue that the objective is wrongly

---

[174] The effect of the provision is, of course, not restricted to those areas. Competition law,
for instance, would also be encompassed by this.

named, as the objective is not "primary" if it can be relegated to second in a competition for order of precedence. Be that as it may, it is manifestly the right result. We would enter upon very dangerous ground if the industry or its regulators perceived any obligation to balance MER with eg safety.

I-5.60   Paragraphs 3 and 4 are directed towards funding problems. Paragraph 3 provides that "no obligation is imposed by the Strategy if to do so would require investment or funding where no satisfactory expected commercial return will be made". Paragraph 4 is concerned with the different situation where a return would be made which is satisfactory for the purposes of the Strategy, but not sufficiently high to interest the particular investor, who, in consequence, decides to delay or discontinue the activity. Here, the Safeguard is restricted to a temporary stay of execution on enforcement: the OGA is obliged to discuss the situation with the defaulting party before proceeding towards enforcement.

I-5.61   Paragraph 5 is concerned with the situation where one relevant person is being required, as a result of the Strategy, to invest in infrastructure for the benefit of, or fund the activities of, another relevant person. It provides that the funder may require a contribution to costs which shall not exceed what is fair and reasonable in all of the circumstances, taking into account the importance of realising the benefited party's assets to meeting the Central Obligation. This is an interesting provision. In large measure it repeats what has been present in Government guidance for a long time: that it is appropriate to allow the host party to recover reasonable costs from the access-seeker. However, the addition, into the question of what is fair and reasonable, of the question of the importance of realising the benefited party's assets introduces a new element into the calculus. This would seem to suggest that the share of costs that can be recovered may be reduced if that is necessary to allow access. If so, access-seekers will be delighted at the limited scope of the Safeguard, while infrastructure owners may feel that their interests have not been adequately protected.

I-5.62   Finally, we come to paragraph 6, which provides that no obligation imposed by or under this Strategy requires any conduct where the benefits to the UK deriving from that conduct are outweighed by the damage to the confidence of investors in oil and gas exploration and production projects in relevant UK waters. This would clearly protect against eg uncompensated expropriation of assets, a state behaviour against which investors have long sought to protect themselves in politically volatile jurisdictions. How much further than that the Safeguard would extend is unclear. One could argue, for instance, that the uncertainties associated with the implementation of the whole MER UK process is damaging to the confidence of investors, who may be delaying investments until such time as they see the system in operation; but presumably the OGA would contend that the long-term benefits of MER UK will over time lead to an increase in investor confidence; and presumably Safeguard 6 was not included within the Strategy as an instrument of self-destruction.

Sector strategies and implementation programmes

Finally, we come to the documents which sit below the level of I-5.63 the MER UK Strategy: the Sector Strategies and Implementation Programmes. For reasons of space it is not possible to offer a commentary on all of these. However, the Asset Stewardship Strategy has been selected as an illustrative example of how this layer of the regulatory structure operates.

Previously, there has "been an absence of stewardship expectations I-5.64 to focus licensees and operators on delivery of existing licence and new MER UK obligations".[175] However, with the Asset Stewardship Strategy, the OGA aims to:

- ensure asset licensees fully identify opportunities and the means to realise them;
- increase the resource base through timely and efficient exploration, appraisal and development of resources;
- maximise recovery through optimising delivery efficiency, technology and collaboration;
- extend infrastructure life and ensure efficient late life management and decommissioning; and
- identify underperformance and best practices.[176]

It is envisaged that the Strategy, if followed, will help facilitate the I-5.65 delivery of the MER UK Strategy obligations. Furthermore, it is anticipated that the Strategy will deliver benefits to the industry, including: ensuring greater value overall; developing a stewardship process complementary to industry business management systems; and sharing best practices across the industry to support the Authority and improve performance.[177]

The Asset Stewardship Strategy
The Asset Stewardship Strategy supports the MER UK Strategy I-5.66 through four "complementary strategic elements", which are discussed below.

---

[175] OGA, *Asset Stewardship Strategy*, available at www.ogauthority.co.uk/exploration-production/asset-stewardship (accessed 11 May 2017), p 8.
[176] *Ibid*, p 5.
[177] *Ibid*, p 7.

I-5.67   **Stewardship expectations.** The OGA has developed, in consultation with the industry, ten good practice expectations aligned with the MER UK Strategy Supporting Obligations, covering the oil and gas lifecycle.[178] These expectations do not have "binding legal effect",[179] but, rather, set out practices which will help to facilitate the delivery of the MER UK Strategy. They cover areas of asset stewardship, including joint venture hub strategy, technology plans and collaboration.[180] Each expectation is broken down into sub-sections, explaining *what* the expectation to be achieved is; *how* indicators of behaviour can be achieved; and *why* the expectation is in place.[181]

I-5.68   **Rationalised Industry Survey.** The OGA has created a new, single UKCS Stewardship Survey. This replaces the nine previous industry surveys so as to streamline the way data is collected and to promote efficiency. The data collected will assist with asset stewardship reviews and economic modelling in the UKCS.[182]

I-5.69   **Benchmarking.** The OGA will use OGA data to analyse assets on factors such as production efficiency, recovery, operating cost and decommissioning cost. The OGA will work together with operators and licensees to improve performance based on the benchmarks created.[183]

I-5.70   **Tiered Stewardship Reviews.** The OGA has pledged to "undertake pro-active, structured and prioritised stewardship reviews with operators and licensees based on intelligence gathered, prioritising those with the greatest impact on MER UK".[184] The revised review system under the Asset Stewardship Strategy will see the OGA set the agenda for data-driven reviews, focusing on the whole upstream lifecycle of assets[185] "that offer the greatest MER UK impact".[186] There is a tiered structure to the review process, whereby reviews will be undertaken at different levels of management, ranging from

---

[178] *Ibid*, p 4.

[179] OGA, *Asset Stewardship Expectations*, available at www.ogauthority.co.uk/exploration-production/asset-stewardship/expectations (accessed 11 May 2017), p 4.

[180] Implementation Guides are made available on the OGA's website to assist operators and licensees in achieving the Expectations. See www.ogauthority.co.uk/exploration-production/asset-stewardship/expectations (accessed 11 May 2017).

[181] OGA, *Asset Stewardship Expectations*, p 5.

[182] OGA, *Asset Stewardship Strategy*, p 4. See www.ogauthority.co.uk/exploration-production/asset-stewardship/surveys (accessed 11 May 2017).

[183] *Ibid*, p 4. See www.ogauthority.co.uk/exploration-production/asset-stewardship/benchmarking (accessed 11 May 2017).

[184] *Ibid*, p 4.

[185] OGA, *Stewardship Review Guidance*, available at www.ogauthority.co.uk/exploration-production/asset-stewardship/reviews (accessed 11 May 2017) p 3.

[186] *Ibid*, p 5.

"operator group review at OGA Chief Executive and Managing Director (MD) level" to "individual operator technical and economic asset performance reviews at subject matter expert level".[187]

CONCLUSION

The Wood Review called for – and Government has delivered – the biggest shake-up in the management of the UKCS for at least three decades. The introduction of the MER UK requirement means that, in pursuit of the aim of maximising economic recovery over the UKCS as a whole, individual licensees and other key players in the upstream UKCS now have a host of additional obligations beyond those which were imposed by the previous system; new obligations which sit alongside – and in some cases, wrap around – the obligations of longer standing. The new obligations will be administered by a new regulator, better resourced than the previous ones and having a radically different mindset. Early in the operation of the new system as we are, it is already clear that the new regulator will not stand aloof from the sector, but will take a multi-faceted role, leading, encouraging, cajoling, teaching, facilitating and, perhaps, at times, requiring parties to move in the direction that it considers to be optimal. What is already clear is that in moving us from an essentially asset-based system to one which is concerned not just with assets, but with regions and indeed with the province as a whole, and with the regulation of behavioural processes as much as technical ones, Wood has wrought a revolution, not an evolution, in the governance of the UKCS.

I-5.71

As with any situation where radical change has been effected, it will take some time for its effect to become apparent. Depending on the extent to which the OGA feels it necessary to test the full extent of its powers to require certain behaviours, there is certainly potential for parties to disagree with the regulator's demands and for the new system to become mired in controversy and litigation. On the other hand, the collaborative approach may help to unlock the UKCS's potential and help to secure its future viability. Nothing in the oil industry is without risk, but that prize is certainly worth playing for.

I-5.72

---

[187] *Ibid*, p 3.

CHAPTER I-6

ACCESS TO INFRASTRUCTURE

Uisdean Vass

I-6.01 In earlier volumes of this work, the issue of third-party access (TPA) to infrastructure was noted to be of great significance.[1] In the new age of the Wood Review and MER (maximising economic recovery), the issue assumes even greater importance than before.[2] In a regime in which both the Oil and Gas Authority (OGA) and the MER parties are obligated to "take the steps necessary to secure … [the production of] … the maximum value of economically recoverable petroleum"[3] from the United Kingdom Continental Shelf (UKCS), it is vital that extensive existing infrastructure is utilised to best effect. Few new discoveries on the UKCS are large enough to justify the creation of entirely new infrastructure. The efficient use of existing infrastructure to produce new discoveries will postpone decommissioning, thus encouraging new exploration. In this chapter, owners of offshore petroleum infrastructure, whether oil companies or otherwise, are referred to as *owners* and the owners of commercial discoveries seeking access to such infrastructure are referred to as *applicants*.

I-6.02 The issue of TPA to offshore petroleum infrastructure provides a complex regulatory drama. In the context of the UKCS, all infrastructure, virtually without exception, has been built by oil and gas licensees seeking to produce their fields.[4] One could go further: most significant chains of infrastructure have been built to produce a big

---

[1] U Vass, "Access to Infrastructure", in G Gordon, J Paterson and E Üşenmez, *Oil and Gas Law: Current Practice and Emerging Trends* (2nd edn, 2011), Chapter 7.

[2] For an account of the "principal objective", MER obligation and the MER parties, see Chapter 1-5.

[3] *MER UK Strategy*, available at www.ogauthority.co.uk/regulatory-framework/mer-uk-strategy (accessed 14 May 2017), para 7 (Central Obligation).

[4] Arguably, the FPSO, which is sometimes owned by or leased from non-licensees and

field (eg Forties) or big fields. Independent infrastructure owners are only now entering the UKCS, and these companies have acquired existing infrastructure (CATS, Forties) as opposed to building it.

Therefore, the great majority of UKCS infrastructure has been I-6.03 built to service upstream discoveries. These said discoveries have occurred as a result of the award of Production Licences by the British state. Such licences are monopolistic grants (concessions, to use another name) which give the licensees an exclusive right (over a set area, for a set period) to carry out exploration and appraisal for petroleum and, if successful, produce and sell the hydrocarbons for (effectively) the life of the field. Clearly, where the licensee makes a commercial discovery, the licence proves to be an extremely valuable monopolistic grant.

While it is true that some infrastructure has been built outsize to I-6.04 cater for subsequent third-party production (or perhaps additional production from the original "discovering" licensee), the true driving money-making business of upstream oil companies is the production and sale of hydrocarbons. The building of production infrastructure is in essence a means to that end.

However, for the regulator and the other oil and gas players in I-6.05 the province, the existence and maintenance of an extensive web of offshore infrastructure (such as we have in the UKCS) is very far from being a matter of marginal interest. Infrastructure is built to accommodate early maximum flows from sponsoring fields and production levels soon fall off, and unused space (ullage) will sooner or later occur. It is probably true to say that the easy "big targets" in the North Sea have already been drilled, and what remains is medium-to-small fields which very often are not large enough or profitable enough to justify their own separate infrastructure.[5] It is therefore vital that the (often) small new entrant oil companies with (often) small discoveries are able to access the existing infrastructure of (often) large oil companies. The maintenance of this infrastructure is vital to encourage exploration and appraisal of new fields, and tariffs from new third-party input can, in turn, be vital to the continuing viability of old infrastructure. The continuing use of old infrastructure allows licensees to put off decommissioning, which further encourages exploration and appraisal. This virtuous circle depends on efficient TPA. TPA does not exist as an isolated regulatory issue.

A wide variety of factors come into play which makes efficient I-6.06 TPA in an unregulated situation problematic. First of all, the core

---

can be used to produce other fields in other locations on a subsequent basis, is a limited exception to the rule that only oil companies construct offshore infrastructure.

[5] It should be said that this modus operandi is less applicable West of Shetland and in the (much less explored) Atlantic Margin.

business of oil companies owning legacy offshore infrastructure is not third-party business. TPA may not factor high in the priorities of such oil companies. Accordingly, a degree of natural inertia can often be observed. Secondly, owners may conduct access negotiations in the context of wider considerations. Does the owner have a problem with the applicant in Kazakhstan? To the owner, would the applicant be a serious competitor elsewhere? Does the owner wish to do a deal with the applicant elsewhere? Additionally, the owner may wish (probably reasonably) to reserve ullage for its own subsequent production or that of its key joint venture partners. But inertia, or perhaps lack of finance, can also impede the would-be applicant from progressing its negotiation, especially where it is under no deadline to develop a small discovery. All of this occurs in a context where the interest of regulator and state is to facilitate rapid access because time is of the essence.[6] The regulator is not at all interested that the owner has a dispute with the applicant in Kazakhstan, or that the owner is prioritising other projects.

I-6.07   The North Sea segments of the UKCS (Northern, Central and Southern North Sea) are amongst the most heavily explored and produced offshore petroleum provinces in the world. There is a lot of generally old infrastructure with high levels of ullage in the gas-producing Southern North Sea (off East England). There is also extensive oil and gas infrastructure in the Central and Northern North Sea with some infrastructure (some of it new) lying in the challenging West of Shetland area. While parts of the UKCS (North Sea) are heavily developed, much of it on the west side (West of Shetland, Atlantic Margin) is frontier in nature. As we will see at para I-6.64, this creates challenges under competition law. If, for example, there is only one viable way to process and transport hydrocarbons from a small field West of Shetland, then an isolated "relevant market" may be created. Effectively, competition law becomes part of the TPA analysis.

I-6.08   In recent years, we have seen the advent of independent upstream infrastructure ownership in the UKCS. This is much to be encouraged. The business interest of independent infrastructure owners is not to discover and produce oil, but *solely* to gain tariff income through transporting (and perhaps processing) petroleum. Such companies want to maximise oil company throughput for as long as possible. However, it should be remembered that not all offshore infrastructure is suitable for independent ownership.[7]

---

[6] See Chapter I-5.

[7] An independent infrastructure owner is likely to be attracted to acquire major infrastructure such as a pipeline with existing throughput and a good economic potential to

It is also vital to understand the technical difficulties associated **I-6.09** with offshore access. Offshore infrastructure is created to deal with the volume and type of hydrocarbons and (crucially) *non-hydro-carbons* associated with the original big field or fields. We should note at the outset that it is very hard to really know how long a field will continue to produce. Much depends on an owner's technical view of the field and what priority it is willing to give the field. What are the genuine prospects for an owner to develop or acquire new production in the region of the infrastructure? The last thing an owner wants is to have its own production shut out of its own infrastructure. This can be a difficult judgement.

On the third-party side, production from a new petroleum field **I-6.10** may need additional processing before it meets the specifications of existing infrastructure. TPA is not merely access to pipelines and topsides but also access to services. It is the owner who typically carries out the construction work to connect a third-party field. What if taking one small field in year one will lose the owner the opportunity to take one larger field in year two? These types of problems are referred to as "sterilising capacity" problems.

Access to infrastructure tends to be easier onshore, eg with the **I-6.11** UK's National Transmission System (NTS) because there the work to prepare the third-party production to meet infrastructure speci-fication is much cheaper. This is also true where the *only* offshore access issue is merely inputting gas into a pipeline. Thus, there is a regulated tariff system for the Norwegian offshore gas transportation system.[8] One thing that shines through in the UKCS and Norwegian Continental Shelf oil systems is that access transactions are too complex for one-size-fits-all regulated systems. There are simply too many active factors for any top-down formulaic approach. There is therefore a system of negotiated access.

The fundamentals of negotiated access are the following: **I-6.12**

(a) Offshore infrastructure is built by, and at the risk of, private oil companies.

(b) In the first instance, it is up to the third-party owners of commercial discoveries (applicants) to approach owners to attempt to agree an acceptable access deal.

(c) However, the owner-applicant access transaction is of huge importance to the regulator, and if agreement cannot be reached between the parties, then the applicant can request the

---

attract new third-party throughput. These considerations may not exist in cases where eg an ageing fixed platform processes only a single field.

[8] See discussion in Grondalen and Lower, "Third Party Access to Infrastructure on the Norwegian Continental Shelf", 4 (2016) *LSU J. of Energy Law & Resources* 319.

regulator to consider the facts of the case, essentially write a fair deal, and compel access.

(d) The approach of the regulator is made public, and the owner and the applicant are made subject to legal and normative obligations in the private negotiation. This then becomes a directed and tensioned negotiation. A "sword of Damocles" is held over the head of owner and applicant, forcing them to play by the rules, or else ... or at least that is the thinking.

I-6.13   The purpose of this chapter is to examine how TPA, a complex regulatory issue of great importance, is handled in the UK. The recent advent of the Wood Review, MER and the creation of the OGA has, among many other things, greatly strengthened the regulator's hand in matters of TPA.

I-6.14   TPA on the UKCS has now become a complex regime with numerous laws, regulations, guidelines and industry codes. Accordingly, the student of, or legal advisor on, TPA must familiarise himself with the Energy Act 2011, Sections 82–91 (hereinafter "the 2011 TPA Provisions"), the Strategy (all of it, but particularly paras 15–17), the Infrastructure Code of Practice 2013 (hereinafter "ICOP") and the new 2016 Guidance on Disputes over Third Party Access to Upstream Oil & Gas Infrastructure (hereinafter "TPA Guidance").

I-6.15   It should be recognised that at the time of publication of this book, the TPA Guidance has been reviewed (autumn 2016) in the wake of the Strategy/Energy Act 2016, but ICOP is still a pre-MER document.

I-6.16   In broader terms, there will be a need to have a thorough understanding of the Energy Act 2016, the Infrastructure Act Section 41 (amending the Petroleum Act 1998 Section 9) and, of course, the Competition Act 1998. The latter three pieces of legislation are more extensively covered elsewhere in this work, but more than passing reference to them is made in this chapter.

I-6.17   The UK's regulatory TPA system is the most multi-faceted and complex in the world. Is the system effective? It is hard to know as honest feedback from regulator, owners and applicants is little in evidence, and very few of these players have any relevant comparative experience. However, the UK system does at least address in logical fashion most of the relevant questions.[9]

---

[9] To this author's knowledge, little work comparing distinct TPA systems has been done. The author himself has undertaken (2007–2008), an unpublished comparison of TPA systems in Brazil, the (federal offshore) USA, the UK, Norway and certain Canadian systems for the Province of Nova Scotia. A copy of this paper is on file with the University of Aberdeen. In March 2015 (Louisiana State University Law School, Baton Rouge, LA), the author participated in a review of key regulatory issues encountered in the established

As noted above in Chapter I-5, the OGA, as the principal UKCS **I-6.18** petroleum regulator, was created as an autonomous UK governmental body by virtue of Energy Act 2016 Section 1 with effect from 1 October 2016. BEIS, the successor to the DECC, continues to have significant offshore regulatory powers, particularly over decommissioning. However as far as TPA is concerned, the OGA is the exclusive regulatory authority.[10]

The next part of this chapter looks at the key legislation which **I-6.19** constitutes the heart of our TPA system. This is composed of the Energy Act 2011, the Infrastructure Act 2015 Section 41 (amending Section 9 of the Petroleum Act 1998), the Energy Act 2016, the Strategy and the Competition Act 1998.

Subsequently we will look at the ICOP and TPA Guidance. **I-6.20**

THE LEGISLATIVE FRAMEWORK

The Energy Act 2011

The mechanics of the UK system of TPA statutory regulation is **I-6.21** contained in the comparatively recent Energy Act 2011, Sections 82–91 (Part 2, Chapter 3 of the statute). The 2011 TPA Provisions were amended by the Energy Act 2016, but mainly to transfer the powers of the DECC to the OGA. References herein to the Energy Act 2011 are to that statute as amended. As a general observation, it may be said that there are a number of legislative innovations in the 2011 TPA Provisions which can be seen as precursors to the type of "new" legislative provisions which appear in Wood Review era legislation such as the Energy Act 2016.[11]

---

offshore producing jurisdictions of the USA (federal offshore), UK and Norway. Papers covering the USA (federal offshore), Norway and an overall comparative review are published, respectively, as Grauberger and Downer, "Third Party Access in the United States", 4 (2016) *LSU Journal of Energy Law and Resources* 293; Grondalen and Lower, "Third Party Access to Infrastructure on the Norwegian Continental Shelf" 4 (2016) *LSU Journal of Energy Law and Resources* 319; and Sweeney, "Introduction to Access to Third Party Infrastructure in Offshore Projects: A Comparative Approach" 4 (2016) *LSU Journal of Energy Law and Resources* 287. Sweeney's article includes a chart of key access issues affecting the three jurisdictions at p 291. Review of the chart leads one to conclude (with Sweeney) that open access is much less open in an enforceable sense in the USA than is the case with the UK and Norway. Norway's bifurcated system of regulated access for gas pipelines and negotiated access for everything else distinguishes it from the UK. Now the Strategy has been added to the mix in the UK. To this author, the UK and Norway are the most TPA-friendly regimes in the world.
[10] All of the DECC's regulatory powers over TPA previously stated in Energy Act 2011 ss 82–91 were transferred to the OGA by virtue of Energy Act 2016, Sch 1, paras 63–72.
[11] Compare Energy Act 2011 ss 83 (power of OGA to issue access order on own initiative) and 87 (power of OGA to obtain TPA related information), with Energy Act 2016 ss 22 (power of OGA to consider and decide qualifying dispute on own initiative), 24 (power

I-6.22   Foundational Section 82 of the Energy Act 2011 applies in cases where an applicant requests petroleum transportation and/or processing access to the owner of either "a relevant upstream petroleum pipeline", a "relevant oil processing facility" or a "relevant gas processing facility".[12] In this chapter, we sometimes collectively refer to such pipelines and facilities as "upstream petroleum infrastructure". In the context of Sections 82 and 83, an "owner" of "upstream petroleum pipeline", "oil processing facility" or "gas processing facility" means anyone who owns the pipeline or anyone leasing or controlling the pipeline or facility.[13] An "upstream petroleum pipeline", "oil processing facility" and "gas processing facility" are "relevant" for purposes of Section 82 if they are located in Great Britain, the Territorial Sea or the Continental Shelf.[14] However, an "upstream petroleum pipeline" is not "relevant" if it is subject to Section 17GA of the Petroleum Act 1998 (certain offshore infrastructure subject to Norwegian access legislation).[15]

I-6.23   The meaning of "pipeline" is defined rather broadly to mean "a pipe or system of pipes for the conveyance of anything", and "upstream petroleum pipeline" means a pipeline which is "operated or constructed as part of a petroleum production project", or which is used to convey petroleum from the site of one or more such petroleum projects directly to power generation or industrial projects, or directly or indirectly to a terminal, or to convey gas from a terminal to an onshore gas transportation system, or to a place outside the UK.[16] A "petroleum production project" means a project carried out under an offshore Production Licence.[17]

I-6.24   An "oil processing facility" means any offshore installation engaged in treating petroleum so as to produce stabilised crude oil and other hydrocarbon liquids to be ready for sale, or receiving or storing such stabilised crude or liquids prior to their further conveyance.[18] This covers FPSOs.

I-6.25   A "gas processing facility" means a facility used to process gas or prepare it for transport.[19] However, the 2011 TPA Provisions do not apply to an access application made to a gas processing facility for

---

of OGA to obtain dispute related information) and 34 (broad power of OGA to obtain information on any matter relevant to principal objective).

[12] Energy Act 2011, s 82 (1).

[13] *Ibid*, s 82 (10).

[14] *Ibid*, s 82 (3).

[15] *Ibid*, s 82 (3).

[16] *Ibid*, s 90 (1).

[17] *Ibid*, s 90 (1).

[18] *Ibid*, s 90 (1). See also the definition of "oil processing operations" in s 90 (2).

[19] *Ibid*, s 90 (1). See also the definition of "gas processing operations" in s 90 (2).

a downstream purpose.[20] Additionally "gas processing facility" does not cover onshore transportation facilities or liquefied natural gas (LNG) facilities.[21]

What is clear is that the definitions of "upstream petroleum I-6.26 pipeline", "oil processing facility" and "gas processing facility" definitions extend to subsea infrastructure, topsides, FPSO, pipelines and certain onshore facilities. TPA coverage is therefore very broad. One can say that any upstream petroleum infrastructure that can physically be shared is subject to the 2011 TPA Provisions.

If the applicant and owner cannot agree on an access application, I-6.27 then the applicant (not the owner) can, under Section 82 (4), request the OGA to issue an access notice. The access request is made under Section 82 (4): any access order is made under Section 82 (11).

However, the OGA cannot accept an access request under Section I-6.28 82 (4) if it is not satisfied that the parties have had a reasonable time in which to try to reach agreement.[22] In considering an access request under Section 82 (4), the OGA must either: (1) reject it; (2) adjourn it to allow further negotiation to take place between the parties; or (3) consider it further.[23] If the OGA decides to consider the application further, it must give the following persons (Consultation Persons) the right to be heard: the applicant and owner, any person having an existing throughput right through the pipeline or facility, and anyone else thought appropriate by the OGA.[24]

Section 82 (7) lists seven factors which, in so far as they are I-6.29 relevant, the OGA *must* take into account in reaching a decision on a Section 87 (4) access request. These are as follows:

(a) capacity which is or can reasonably be made available in the pipeline or at the facility;
(b) any incompatibilities of technical specification which cannot reasonably be overcome;
(c) difficulties which cannot reasonably be overcome and which could prejudice the efficient, current   and planned future production of petroleum;
(d) the reasonable needs of the owner and any associate of the owner for the conveying and processing of petroleum;
(e) the interests of all users and operators of the pipeline or facility;

---

[20] *Ibid*, s 82 (2). For downstream access applications to gas processing facilities, see s 12 of the Gas Act 1995.
[21] Energy Act, s 90.
[22] *Ibid*, s 82 (5).
[23] *Ibid*, s 82 (6)(a).
[24] Energy Act, s 82(6)(b).

(f) the need to maintain security and regularity of supplies of petroleum; and

(g) the number of parties involved in the dispute.

I-6.30　The seven factors cover a broad range of issues which may arise in an access decision. Note that the OGA is merely required to "take into account", not vindicate each particular factor. It is also important to note that Section 82 (7) was not amended by the Energy Act 2016 to require a consideration of either the Strategy or the principal objective. This issue is discussed below at para I-6.46.

I-6.31　Beyond Section 82 (7), the OGA may *only* grant an access notice under Section 82 (11) if *either* the condition stated in Section 82 (9) *or* (10) is met. The condition in Section 82 (9) is that an access notice *must not prejudice* the throughput of petroleum that the owner or associate of the owner "requires or may reasonably be expected to require", and/or must not prejudice the rights of a third party having existing throughput rights. The condition in Section 82 (10) is that if the throughput rights of owner or third party (as described in Section 82 (9)) *are* prejudiced by the access notice, then the person suffering loss must recover compensation from the applicant as determined by the Section 82 (11) access notice.

I-6.32　What Section 82 (9) and (10) read together mean is that there must be compensation provided by the applicant if existing owner or third party throughput is displaced. Note that nothing in Section 82 requires there to be available ullage in order for an access notice to be issued.[25] Admittedly, it would be a radical step to displace existing throughput to favour an applicant.

I-6.33　Section 82 (11) gives the OGA the power to: (1) grant the applicant the access requested in the Section 82 (4) request; (2) grant the applicant all necessary or expedient ancillary or incidental rights, which could include a right to have an applicant pipeline connected to a pipeline or facility of the owner; and (3) regulate the charges payable by the applicant for access or other granted rights. The access notice may also contain a provision which ensures that no one suffers loss as a result of the mixing of petroleum streams.[26]

I-6.34　It is clear that under Section 82 (11), the OGA has the power to lay down all the relevant terms and conditions necessary to facilitate access. However, the terms are not binding on the owner until the applicant accepts them.[27] An applicant might not find such terms to be sufficiently economically attractive to warrant acceptance.

---

[25] This position is consistent with Strategy, para 17 (b) and TPA Guidance, s 58.

[26] Energy Act, s 82 (12).

[27] *Ibid*, s 82 (17). TPA Guidance, para 8 refers to the terms of a s 82 (11) order as being "terms and conditions".

Under Section 83 of the Energy Act 2011, the OGA is given power **I-6.35** on its own initiative (ie without a Section 82 (4) request for access from the applicant) to make an access award under Section 82 (11). However, as a minimum, the applicant must have attempted to gain access to upstream petroleum infrastructure. Furthermore, the OGA may only issue an access award in such circumstances if the applicant and the owner have had reasonable time in which to come to an agreement and "there is no realistic prospect of them doing so".[28]

In reaching a Section 82 (11) decision on its own initiative, the **I-6.36** OGA must have regard to the factors listed in Section 82 (7) and the conditions of Section 82 (9) and (10) and give the Consultation Parties the opportunity to be heard. See discussion of Section 82 (6) above at para I-6.28.

The Section 83 procedure is there to allow the OGA to act in the **I-6.37** rather extreme circumstances in which the applicant may be intimidated by the owner or where it is obvious that no agreement will ever be reached. As discussed in Chapter I-5, in the wake of Section 22 of the Energy Act 2016, the OGA may of its own initiative raise a non-binding dispute resolution proceeding against an MER party on any issue relating to the principal objective. The referenced Section 83 procedure may therefore be seen as a precursor of the more recent general power.

Section 84 concerns the compulsory modification of upstream **I-6.38** petroleum infrastructure. The provision applies where an applicant has sought access to an upstream infrastructure under Section 82 (1) and has made an access request to the OGA under Section 82 (4). In order for Section 84 to be relevant, the OGA must be considering whether to make an access award under Section 82 (11).

If it appears to the OGA that the relevant pipeline or facility **I-6.39** should be modified so as to increase capacity or that it should be modified to allow a connection for the applicant's pipeline, then the OGA must issue a notice in terms of Section 84 (2).[29] The Section 84 (2) notice should provide for the following: (1) the technical modifications required; (2) the sums payable by the applicant to the owner in compensation; (3) any security required to be provided by the applicant; (4) the owner to be required to carry out the modifications in a certain period; and (5) the applicant to be required to make the payments.[30]

The OGA must make any decision on a Section 84 (2) notice **I-6.40** in the light of the seven factors mentioned in Section 82 (7) and

---

[28] *Ibid*, s 83 (3).
[29] *Ibid*, s 84 (2).
[30] *Ibid*, s 84 (3)

must allow the Consultation Persons to be heard. The provisions of Section 82 (9) and (10) are not relevant to this decision.

I-6.41   It is not absolutely clear that an applicant can specifically request a modification under Section 84 pursuant to a Section 82 (4) access request, but perhaps this is implicit.[31] Section 84 on mandatory modification is one of the few pre-Wood Review statutory instances where the UK Government *can* compel an oil and gas licensee to invest against its will. Even then, Section 84 envisages the applicant providing full compensation and security.

I-6.42   Section 87 (1) provides that where the OGA has reason to believe that a person has made or received a request for access under Section 82 (1), it may "require the person to confirm whether or not that is the case". This language comes from the pre-Wood Review era during which it was much more difficult for government to discover actual operating and commercial information. With specific reference to TPA, there was, prior to the adoption of Section 87, no way for the Government to know (absent confirmation by one of the parties) whether a Section 82 (1) access attempt had even been made. With the confirmation that a Section 82 (1) access attempt has been made, the OGA may request information which would assist it to make decisions under the Sections 82, 83 and 84 procedures.

I-6.43   Under the 2011 TPA Provisions, the OGA may compel an owner to grant access to its upstream petroleum infrastructure to an applicant's third-party petroleum throughput. It can also compel an owner to modify its infrastructure to accommodate the applicants' throughput. Once an access request under Section 82 (1) is made, the OGA can discover specified information to assist its decision-making. This broad power to obtain information relative to TPA issues is now in the era of MER extended to any information relevant to the principle objective pursuant to Energy Act 2016, Section 34.

I-6.44   However, under the 2011 TPA Provisions, the OGA can do nothing before a Section 82 (1) request is made (other than asking whether such a request has been made). Neither can it (under the 2011 TPA Provisions) address the situation where a third party producer feels that its *existing* access terms are unfair, or the circumstance where an applicant approaches a future infrastructure owner regarding a project which is at the design/construction stage.[32] Furthermore, the OGA cannot compel the applicant to accept an access award under Section 82 (11). We may note that the Strategy does give the OGA

---

[31] Perhaps the applicant under s 82 (4) would simply request that the relevant pipeline or facility be expanded or altered. The TPA Guidance provides a procedure for a s 82 (4) request but there is no mention of a request under s 84. However, TPA Guidance, paras 33 and 34 provide useful information on how the OGA will apply s 84.

[32] See supporting observation in TPA Guidance, para 12.

tools to handle these situations which verge on, but are not covered by, the 2011 TPA Provisions.[33]

Infrastructure Act 2015

We have seen how the Infrastructure Act 2015 created the statutory basis for the Strategy by enacting new sub-sections 9A–I of the Petroleum Act 1998. As noted, the Energy Act 2016 made some significant amendments to the newly extended Section 9 of the Petroleum Act 1998. While the impact of the Strategy on the 2011 TPA Provisions is discussed below at paras I-6.49 to I-6.63, let us briefly focus on a few important Infrastructure Act provisions.   I-6.45

Section 9B (c) states that the OGA must abide by the Strategy in exercising its "functions under Chapter 3 of Part 2 of the Energy Act 2011" (2011 TPA Provisions). Neither the Energy Act 2016 nor the Infrastructure Act 2015 otherwise make significant amendments to the 2011 TPA Provisions,[34] and, as we have seen, there is no   I-6.46

---

[33] Failure by a prospective applicant to seek access to upstream petroleum infrastructure within a reasonable time under s 82 (1), and so delaying the possible development of a commercial discovery, could be contrary to the Central Obligation (Strategy, para 7). Continued failure to act could lead to divestment under Strategy paras 30–34. Involved, no doubt, would be a discussion of Strategy, paras 3–4 (MER party does not need to make an investment or fund activity without a "satisfactory expected commercial return"). Also relevant to this situation might be the Fallow Fields Initiative. Failure by an owner to improve existing unreasonable contractual terms binding on an applicant, amounting to a violation of the Central Obligation (para 7), could be contrary to the Strategy. However, these terms would have to prejudice "the maximum value of economically reasonable petroleum ... beneath UK waters". In other words, the applicant would have to show that the terms seriously threatened the viability of its field. Such unreasonable terms may also be contrary to the Competition Act 1998. Though not covered by the 2011 TPA Provisions, Strategy, paras 13–14 require MER parties to "plan commission and construct production facilities" so as to maximise the value of economically recoverable petroleum in the region, including considering whether such totally new facilities could benefit producers or potential applicants. Lastly, it is true that an applicant is not obligated to accept an access award under s 82 (11) but if this results in failure to develop a commercial discovery, then there may be a violation of the Central Obligation (Strategy, para 7). Taken in conjunction with the OGA's huge new powers to obtain information on any matter relating to the principal objective (Energy Act 2016, s 34) the Strategy empowers the OGA to cover TPA from every angle.

[34] Energy Act 2016, ss 70–71 contain substantive amendments to the 2011 TPA Provisions. There are no changes in key ss 82–84 of the Energy Act 2011. Section 70 provides for an appeals process for any party being required to supply information under Energy Act 2011, section 87 (1)–(3) (see discussion above at para I-6.42). It goes on to provide that failure to provide such information is sanctionable by the OGA under Energy Act 2016, ss 42 *et seq*. However, failure in this regard cannot lead to the imposition of licence revocation or operator removal sanctions. Section 71 adds lengthy new sub-sections 89A and 89B to s 89 of the Energy Act 2011. New sub-section 89A covers the situation in which there has been an assignment or assignation of the applicant's interest, and new

mention of MER or anything akin to it in key Sections 82, 83 and 84 of the Energy Act 2011. On the contrary, access notices under Section 82 (11) and compulsory modification notices under Section 84 (2) must still be made by balancing the seven factors listed in Section 82 (7). However, our view is that the effect of (new) Section 9B (c) of the Petroleum Act 1998 is to import the Strategy into the 2011 TPA Provisions. The legal framework is not perfect and we would have preferred to see Section 82 (7) of the Energy Act 2011 amended to explicitly include the Strategy as a factor. But when all is said and done, we have no doubt that the Energy Act 2016 (though its amendment of Petroleum Act Section 9B(c)) has "added" the Strategy to TPA regulation.

I-6.47   Secondly, questions arise over the scope of the OGA's duty (Petroleum Act 1998, Section 9B (c)) to abide by the Strategy in TPA matters. Who is the duty owed to? Possible answers range from BEIS, to the MER parties, to any resident of the UK. The answer is not clear. One can ask a similar question regarding the MER parties. Petroleum Act Section 9C requires them to abide by the Strategy in their "relevant functions" which means the broad range of their offshore oil activities *including* TPA. Who is this duty owed to? BEIS, the OGA, all other MER parties or all residents of the UK? Again the answer is not clear. The Strategy itself says that all MER parties and the OGA are bound by all of its provisions.[35] The MER parties are clearly bound to the OGA to carry out the Strategy. But is the OGA duty-bound to the MER parties to comply with the Strategy by implementing it? The answer is probably in the affirmative.

I-6.48   The answers are important, for TPA and for broader reasons. If MER parties owe a duty to each other to comply with the Strategy, then this changes their commercial position. However, if the OGA is under a duty as to the MER party A to implement the Strategy relative to MER party B, then that again strengthens the commercial position of MER party A. If such duties are owed, the regular courts will be the arbiter of disputes.[36]

---

sub-section 89B covers the situation in which there has been an assignment or assignation of the owner's interest.

[35] Under Strategy, para 7, "relevant persons" must take the steps necessary to ensure that the maximum value of economically recoverable petroleum "is recovered from the UKCS". According to the Strategy Annex, "relevant persons means the OGA and the Petroleum Act 1998 Section 9C (MER) parties".

[36] The Energy Act 2016 provided that the OGA may bring sanctions proceedings against MER parties for, *inter alia*, failing to abide by the Strategy. Energy Act 2016, s 42. However, beyond this, Strategy issues are to be resolved by the regular courts.

STRATEGY

General

As we have seen, the MER parties and the OGA are obligated to   I-6.49
implement the Central Obligation (para 7) which is as follows:
"Relevant persons [MER parties and OGA] must, in the exercise of
their relevant functions, take the steps necessary to secure that the
maximum value of economically recoverable petroleum is recovered
from the strata beneath relevant UK waters."

The Central Obligation is always subject to the Safeguards (paras   I-6.50
2–6), as per the discussion above at paras I-5.58 to I-5.62.

Most of the Strategy is made up of Supporting Obligations   I-6.51
(paras 9–25) and Required Actions and Behaviours (paras 26–34).
Strategy, para 1 states: "The Supporting Obligations clarify how
the Central Obligation applies in certain circumstances and the
Required Actions and Behaviours are obligations which apply to
relevant persons [ie MER parties and OGA] when carrying out the
Central and Supporting Obligations."

After the enactment of the Strategy, any TPA decision must   I-6.52
be arrived at by utilising the statutory framework of the 2011
TPA Provisions, but also applying the whole Strategy (including
Introductory Principles and Safeguards) and taking cognisance of the
TPA Guidance, ICOP and any reported decisions. Any final decision
by the OGA or private arrangement between the parties must be
consistent with the Competition Act. The goal is to deliver "access to
infrastructure on fair and reasonable terms" (Strategy, para 17 (a)).
Reaching such a synthesis may be challenging.

Strategy, paras 16 and 17

Strategy, paras 16 and 17 are as follows: I-6.53

16. Owners and operators of infrastructure must ensure that it is
    operated in a way that facilitates the recovery of the maximum
    value of economically recoverable petroleum from (as applicable);
    a. the region in which it is situated; and
    b. where the infrastructure is used by or for the benefit of others,
       the regions in which those others are situated.
17. The obligation in paragraph 16 includes:
    a. allowing access to infrastructure on fair and reasonable
       terms; and
    b. where the infrastructure is not able to cope with demand
       for its use, prioritising access which maximises the value of
       economically recoverable petroleum.

Paras 16 and 17 of the Strategy are the second and third paragraphs   I-6.54
of a three-paragraph set of Supporting Obligations on "Asset

Stewardship".[3377] These Supporting Obligations are subject to the Safeguards (paras 2–6).

I-6.55   According to para 16, any owner or operator of infrastructure must operate it so as to maximise the value of economically recoverable petroleum (ERP) in its region or the region of third-party shippers or applicants. We have elsewhere drawn attention to the uncertain definition of ERP, but here that is less relevant as TPA applicants always seek to transport/process known reserves.

I-6.56   Para 17(a) elevates the "access to infrastructure on fair and reasonable terms" principle to statutory law. That particular principle does not explicitly appear in the 2011 TPA Provisions, though arguably the principle can be derived from the seven legal principles of Energy Act 2011, Section 82 (7) and the provisions of Section 82 (9) and (10). As we have noted, in circumstances when there is no ullage, para 17(b) still requires an owner to prioritise access so as to maximise the value of ERP. This is not inconsistent with prior law.

I-6.57   A Section 82 (11) access award can definitely be given in situations where an owner and/or third party is required to defer throughput. However, as Energy Act 2011, Section 82 (10) makes clear, there must be compensation for "any person who suffers loss" as a result of such a deferral. The Section 82 (10) compensation looks to be full compensatory payment.[38] Beyond compensation the difficulty here is that owner A and third party field owner B may be obligated under Strategy para 17(b) to defer production and amend (or suspend) their existing transportation and processing agreement to accommodate applicant C, who could be deemed to be more needful in order to accomplish the objectives of MER. The regulatory position is that although the OGA cannot tear up the existing commercial contract, it can sanction owner A and third party B if they fail to agree to amend (or suspend) such a commercial contract. Such a failure would violate the Strategy.[39] The methodology for valuing deferral losses is not clear.

---

[37] Para 15 relates more narrowly to infrastructure performance. It states: "The owners and operators of infrastructure must ensure that it is maintained in such a condition and operated in such a manner that it will achieve optimum levels of performance, including production efficiency and cost efficiency, for the expected duration of production, taking into consideration the stage of field and asset development, technology and geological constraints".

[38] See also TPA Guidance, para 50: "Existing users are given further protection by sections 82 (9) and (10) of the Energy Act 2011, which require that the reasonable expectations of owners and the rights of other users are not prejudiced unless they are compensated".

[39] See TPA Guidance, para 49, footnote 11: "There may be situations where existing contractual commitments [like our hypothetical transportation and processing agreement discussed above] work against maximising economic recovery and the OGA will consider

Now that TPA is made an explicit Strategy Supporting Obligation, I-6.58
it is extremely important to note the effect of the Safeguards and
the Required Actions and Behaviours. The Introductory Principles,
which serve to underlie, but are not strictly part of the Strategy, also
assume considerable importance.

Regarding Safeguards, an owner may object to a TPA decision I-6.59
under Section 82 (11) by arguing: (1) that the obligation would be
contrary to law (para 2); (2) that the obligation would require the
owner to make investments or fund activity upon which it would
not make a satisfactory expected commercial return (SECR) (paras
3–4); (3) that the obligation would require the owner to make investments or fund activity when the applicant is not required to make
a fair and reasonable contribution (para 5); and (4) that the access
decision if required to be carried out would result in more damage
than benefit to the UK because of its chilling impact on investors
(para 6).

Because Safeguards are effective on the crystallisation of a I-6.60
Strategy Obligation, we believe that Safeguards should be formally
raised by an owner after it has been handed a Section 82 (11) access
decision. Since an applicant cannot be forced to accept a Section 82
(11) access award, it is hard to see why an applicant would need to
invoke Safeguards.

It is unlikely that a Section 82 (11) award could be contrary to I-6.61
law. However, it is more likely that an access award could lead an
owner to declare a lack of SECR. As we have seen in Chapter I-5,
there is an objective/subjective test for SECR and this issue would
ultimately have to be resolved by a court. If a lack of SECR were
found to exist, the owner might ultimately have to divest/relinquish
its interest pursuant to Strategy, paras 30–34. The relationship of
para 5 to para 17 (a) needs careful thought. The principle of para 17
(a) is that of allowing "access to infrastructure on fair and reasonable
terms". However, para 5 adds the additional element: "taking into
account the importance of realising B's [ie applicant's] assets to
meeting the Central Obligation". This can be read to say that *better*
"fair and reasonable terms" can be approved for an applicant if
such favourable terms are needed for the fulfilment of the Central
Obligation. Little light is shone on the issue by the TPA Guidance.
It is entirely possible that a TPA award might have a chilling effect
on investors, which is intended to be safeguarded under the Strategy
(para 6). Such an award might be seen as being alarmingly "rich" for
applicants or, possibly, owners.

---

using its other powers such as those under the Energy Act 2016 to seek an improved
outcome".

I-6.62   We have already noted the issue regarding the status of the Introductory Principles (see para I-6.58). Introductory Principle (c), which is sometimes referred to as the "Allocation Principle", is quoted as follows:

> (c) compliance with the Strategy may oblige individual companies to allocate value between them, matching risk to reward. However, while the net result should deliver greater value overall, it will not be the case that all companies will always be individually better off.

I-6.63   In the context of TPA, this principle can be taken to suggest that owners may need to offer attractive terms to facilitate MER. The TPA Guidance incorporates the Allocation Principle at para 14.[40]

THE COMPETITION ACT 1998

I-6.64   Competition law in the form of the Competition Act 1998 is especially important for TPA. The so called Chapter I and Chapter II prohibitions in the Competition Act are based on Articles 101 and 102 of the Treaty on the Functioning of the EU (TFEU).[41] The relevant UK regulatory authority is the Competition and Markets Authority (CMA).

I-6.65   Under Chapter I, anti-competitive agreements between undertakings such as price fixing, market sharing and collusive tendering are prohibited. However, the prohibition will only apply where the agreement has an "appreciable" effect on the market. An "appreciable effect" exists where the parties combined share of the market is more than 10 per cent.

I-6.66   In the context of TPA, what is "the market"? If a company or its owner controls an "essential facility" which is a system through which third parties must send their product to reach market, then this can amount to a "relevant market". If it is, in practice, impossible to construct new infrastructure to compete with existing

---

[40] To conclude on the Strategy, we may note that TPA issues may, either solely or along with other matters, be covered in an OGA Plan, pursuant to Strategy, paras 23–25. Such an OGA Plan may require that "new and emerging technologies are deployed to their optimum effect", so as to maximise the value of ERP (Strategy, paras 18–19). Strategy, para 28 also requires MER parties to consider whether collaboration could "reduce costs [and] increase recovery of [ERP] …".

[41] EU competition law applies where there are anti-competitive effects across the borders of EU member states. Such law clearly applies to infrastructure straddling the UKCS and all other contiguous national continental shelf areas (including Norway), and may also apply where UKCS players headquartered in other EU states cause or suffer anti-competitive effects in the UKCS. However, there is very little difference between UK and EU competition law and this work will only discuss UK law, albeit referring to EU law where useful.

infrastructure, then that favours a finding of "essential facility". Under EU law, pipelines have been found to be "essential facilities" though there is no such holding under UK law.[42]

Clearly, owners need to be aware that their upstream petroleum I-6.67 infrastructure may be considered to be a "relevant market" for competition law purposes. Violation of either Chapter I or Chapter II prohibitions can lead to a fine of up to 10 per cent of the worldwide turnover of the offending company.

In many cases in the UKCS, an operator representing various I-6.68 owners may enter into a contract for joint services to be provided to an access applicant. Joint services are typically frowned on by UK and EU competition authorities because they are perceived as a mask for price fixing between companies (owners) who would otherwise compete. There are instances of UKCS infrastructure where joint owners enter separate contracts for access/service provision, though sometimes these separate services are provided subject to a single joint transportation and processing agreement.

It is possible to get an exemption for conduct which would I-6.69 otherwise violate Chapter I, if the conduct can be shown to have countervailing economic benefits. The CMA will not (as it once did) confirm such an exemption and companies must "self-assess" any exemption.

The analysis under Chapter II is different. Under this provision, I-6.70 anti-competitive conduct by undertakings wielding significant market power is prohibited. This is referred to as abuse of a dominant position. Generally, this position of dominance amounts to 40 per cent of a relevant market. This market position can be held singly or jointly with other companies. Abuses can include refusal to supply or discriminatory pricing. Unlike the case with Chapter I, there are no available exemptions for a Chapter II violation.

However, in order for there to be a Chapter II violation a company I-6.71 must not only hold a dominant position, it must abuse it. As the TPA Guidance states in para 2.2: "A dominant position essentially means that an owner is able to behave to an appreciable extent independently of competitive pressures, such as competitors, on that market".

Though the CMA has not taken any decisions on the application I-6.72 of competition law to TPA issues on the UKCS, it is unlikely that an owner might be found guilty of abuse of a dominant position if it complied with ICOP principles in favour of non-discriminatory access, unbundling and fair and reasonable tariffs and terms, which are discussed below at para I-6.79. See also TPA Guidance, para 24.

---

[42] See eg Gaz de France and Ruhrgas IP/04/573 of 30 April 2004, RWE Commission MEMO/07/186 of 11 May 2007 and ENI Commission MEMO/07/189 of 11 May 2007.

CODE OF PRACTICE ON ACCESS TO UPSTREAM OIL AND GAS INFRASTRUCTURE ON THE UKCS ("ICOP")

I-6.73 In 1994, the UK Government expressed concerns that a purely statutory approach to TPA was not working. The Government noted a "lack of transparency in the terms of gaining access to oil and gas infrastructure, particularly offshore, and the delays in new field developments and the increased costs that this imposed".[43] The D'Ancona Committee, composed of representatives of Government and industry, considered whether the solution might lie in enhanced regulation or in an industry code.[44] The Committee favoured the latter. Thus, in February 1996, the then United Kingdom Offshore Operators Association (UKOOA) (now Oil & Gas UK) accepted the "Rules and Procedures covering Access to Offshore Infrastructure" (RPGA). The RPGA included some of the key principles subsequently to appear in ICOP. In 2001, the then Department of Trade and Industry (DTI) produced a Guidance Note in an attempt to explain the considerations it might use to decide a TPA application should one be received.[45] Unfortunately, neither RPGA or Guidance Note produced any seismic change in TPA culture. Third parties declined to use their access rights, and the principles of the RPGA were honoured more in the breach than in the observance.

I-6.74   In 2004, UKOOA, working closely with Government, replaced RPGA with ICOP.[46] ICOP was revised by UKOOA's successor, Oil & Gas UK, in 2012. The purpose was to reflect the 2011 TPA Provisions (Energy Act 2011). At the date of writing, the ICOP, unlike the TPA Guidance, has not been updated to reflect the Strategy/Energy Act 2016.

I-6.75   ICOP provides a detailed programme of information provision/ exchange, states important TPA principles and embodies a structured negotiation system. This non-statutory industry code, supplemented by the TPA Guidance, lies over the separate legislative "backbone" of the 2011 TPA Provisions, the Strategy, the wider parameters of the Energy Act 2016 and the Competition Act 1998. While ICOP is, in strict terms, purely voluntary, the TPA Guidance promotes the use

---

[43] T Daintith, G Willoughby and A Hill, *United Kingdom Oil and Gas Law* (3rd edn, looseleaf, 2000–date) (hereinafter "Daintith"), para 1-733.

[44] *Ibid*, para 1-733.

[45] DTI Consideration of Applications for Resolution of Disputes over Third Party Access to Infrastructure: Guidance to Parties in Dispute (2001).

[46] In March 2005, the DTI published refreshed Guidance Notes. See DTI, "Guidance on Disputes over Third Party Access to Upstream Oil and Gas Infrastructure" (2005). These were in turn updated in 2009 and again in 2016 with the final result being the present TPA Guidance.

of ICOP (TPA Guidance, para 6). Whether or not the parties have attempted to use ICOP will be a factor in the OGA's Energy Act 2011, Section 82 (5) analysis as to whether the "applicant and the owner have had a reasonable time in which to reach agreement".[47] In sum, ICOP is not law, but it is very much part of the process.

The industry scope of ICOP covers the "processing and conveyance I-6.76 of all UK oil and gas throughout the hydrocarbon production and supply chain from wellhead through to receiving terminals and initial onshore processing facilities".[48] This covers all offshore and onshore infrastructure (including FPSOs) which is used to convey and process crude oil up until it is stabilised (ready for sale), and gas until such time as it enters the NTS. ICOP does not apply to the NTS, LNG facilities or interconnectors.[49]

The broad industry scope of ICOP thus precisely matches the I-6.77 broad industry scope of the 2011 TPA Provisions and the TPA Guidance. See discussion at paras I-6.22 to I-6.36 and para I-6.97.[50]

Like the 2011 TPA Provisions, ICOP applies to applicants and I-6.78 owners and also non-owner parties with capacity rights in relevant infrastructure. Again, like the 2011 TPA Provisions and the TPA Guidance, ICOP *only applies to new contracts* for access and infrastructure services. It is not relevant in the situation where existing contracting parties seek to renegotiate contractual terms. Going beyond the 2011 TPA Provisions, ICOP includes very important provisions for the publication and sharing of various kinds of access-related information on a pre-dispute and post-agreement basis.

The main principles of ICOP are as follows: parties will uphold I-6.79 infrastructure integrity and protect the environment; parties will follow the Commercial Code of Practice;[51] parties will provide meaningful information to one another prior to and during negotiations; parties will support negotiated access in a timely manner; parties undertake to ultimately settle disputes with an automatic

---

[47] TPA Guidance, para 28.

[48] ICOP, para 4 (1).

[49] *Ibid*, para 4 (1).

[50] It might also be noted that while the Strategy in para 7 has as its goal to "maximise the value of economically recoverable petroleum ... from the strata *beneath relevant UK waters*" (emphasis added) and "relevant UK waters" means the Territorial Sea and the UKCS, MER parties may have to take *onshore action* to maximise offshore production. Therefore, in our view, Strategy paras 16 and 17 have limited onshore effects in the same way as ICOP and the 2011 TPA Provisions.

[51] The Commercial Code of Practice, a short but influential document, is attached to ICOP as Annex C, comprising 12 bullet points. It states that its mission is to "Promote Co-operative Value Generation". It then lists six principles of best practice process (among them the wise injunction: "ensure personal issues do not become a barrier to progress") and five principles for senior management commitment, two of which are "ensure appropriate use of tactics" and "adopt a non-blocking approach".

referral to the OGA; parties will resolve conflicts of interest; owners will provide transparent and non-discriminatory access; owners will provide tariffs and terms for unlimited services where requested and practicable; parties will seek to agree fair and reasonable tariffs and terms where risks are matched by rewards; and parties will publish agreed commercial provisions.[52]

I-6.80   ICOP contains highly important provisions regarding information sharing. Firstly, every owner should ensure that it maintains basic public information reflecting "operational and ullage data", allowing a prospective applicant to "undertake basic economic screening of offtake options".[53] Owners are responsible to maintain such data at their cost. This work may be carried out by an operator for a number of relevant owners. High-level capacity information should be made publicly available through the owner's company website inter-linked with the DEAL portal.[54]

I-6.81   Secondly, further, more specific information should be available from the owner at the request of the applicant within a reasonable time eg within 14 days of the enquiry.[55] The owner/operator is not required to undertake material incremental work to pull together this enhanced information.[56]

I-6.82   Thirdly, when an applicant has reached the stage of desiring to negotiate an actual deal, it needs to provide the owner with significant information such as: name of field, licence, owners/operator; broad information on the proposed development including desired start-up date; and particular services required including production profiles, chemical specifications/other scientific information; and other services requested from other infrastructure owners.[57]

I-6.83   Having received this detailed information from the applicant, the owner should supply such further information as is needed to conclude a deal. While parties' data should be provided in good faith, it will also be provided on a no-liability basis.[58]

I-6.84   Lastly, owners should publish summaries of the key terms of access/services agreements within one month of these becoming unconditional. The information should appear on the corporate websites and/or the DEAL portal. Terms should remain noted until the relevant agreement terminates.[59]

---

[52] ICOP, para 5.
[53] *Ibid*, para 7.1 (1).
[54] *Ibid*, para 7.2 and Annex E, para 1.
[55] *Ibid*, para 7.3. See also Annex E, para 2.
[56] *Ibid*, para 7.3 (3).
[57] *Ibid*, para 7.4.
[58] *Ibid.*, para 7.4 (2)–(4).
[59] *Ibid*, para 14. See also Annex G.

ICOP paras 8 (Timelines); 10 (Minimising Conflicts of Interest), I-6.85
11 (Non-Discriminatory Negotiated Access), 12 (Separation of
Services) and 13 (Fair and Reasonable Tariffs and Terms) provide
considerable substance to the principles briefly stated in para I-6.79.

ICOP para 8 provides that access negotiations should proceed as I-6.86
quickly as possible so as to lead to a "fair, reasonable and technically
sound outcome" consistent with prudent corporate governance.[60] A
flowchart for an ideal negotiating schedule is included in para 6. At
the outset, the parties should agree a framework for negotiations
identifying, to the extent possible, the key technical, operational,
legal and commercial issues which will need agreement prior to
final commercial negotiation. The applicant should include in the
schedule the date upon which it proposed to submit the automatic
referral notice (ARN) (see further discussion below, in I-6.94).
Neither party should delay negotiations, but rapid negotiation to
obtain an unfair commercial advantage must also be avoided.[61]

ICOP para 10 provides that any company who may be both I-6.87
an owner and an applicant can only carry out negotiations in one
capacity (ie either as owner *or* applicant). Should a party elect to step
back from a negotiating role in any capacity, this will not affect its
rights to approve any final commercial settlement.

ICOP para 11 requires owners to respond to bona fide requests I-6.88
for access/services in good faith, without favouring or discriminating against any particular company or corporate group. As para
11 (1) says: "This principle of transparent and non-discriminatory
negotiated access applies to all infrastructure coming within the
scope of this code". Discrimination by an owner can be shown by
dissimilar conditions for similar transactions, but different terms
can be justified by variations in cost and risk of supply. Owners are
entitled to make reasonable provision for capacity for their own use
and prior third-party contracts should be respected.[62]

ICOP para 12 states that where it is "practical and requested" I-6.89
owners should offer terms and conditions for services on an unbundled
basis. This allows the applicant to better understand the owner's
commercial offer. Unbundling is also useful in cases where different
owners own different parts of the transportation/services chain.

---

[60] *Ibid*, para 8 (1).

[61] *Ibid*, para 8 (3)–(5).

[62] Para 11 (2)–(5) ICOP, Annex D para 6 speaks to "Reasonable Provision of Capacity".
Such reasonable capacity includes anticipated upsides and plateaux from currently
producing fields or new field developments expected within a reasonable period (five
years is cited) or relative to fields foreseen at the time the infrastructure was built. This
is precisely in line with the view expressed in the TPA Guidance para 49, footnote 9. See
discussion below at para I-6.102.

I-6.90   ICOP para 13 states that tariffs and terms should be "fair and reasonable, where risks taken are reflected by rewards".[63] Competition between systems, if possible, is most likely to lead to such fair and reasonable terms. As well as being entitled to reserve reasonable capacity for its own use, and respecting prior third-party contacts, the owner is entitled to take into account the system input effect of particular new third-party business.[64] By this, the Code envisages so-called "Sterilising Capacity" which covers the following kinds of situations: where taking in a new small field would prejudice negotiation with a subsequent large field; where a small field might, say, require all the de-propanising capacity in a particular facility (so prejudicing larger fields with less de-propanising needs) or where a sour gas field would prevent an owner from operating a gas pipeline sweet.[65]

I-6.91   If the capacity requested by an applicant does not exist, the owner is expected to provide the incremental capacity, but only if the applicant funds the investment in line with standard industry practice.[66]

I-6.92   A key precept of the Code is that liability and indemnity regimes can be as important an element in access and negotiations as tariffing. Liabilities and indemnities should be addressed early in any negotiation. In general terms, parties should accept a duty to mitigate loss when seeking recovery against another party.[67] ICOP para 13 then goes on to consider liabilities and indemnities for the Tie–in Phase and the Transportation and Production Phases separately. As a general principle, applicants must be willing to indemnify owners for liabilities and losses arising from tie-in or modification work, but such liability should be capped.[68] Parties should attempt to categorise and quantify covered losses and caps should be reasonable. The goal should be to cover defined risks with insurance.

I-6.93   With respect to the Transportation and Processing Phase, owners and applicants should abide by the familiar "mutual hold harmless" provisions subject to exclusions for wilful misconduct. Special provisions may be needed for delivery of off-spec production. In some cases, bespoke liability and indemnity regimes may attach to particular infrastructure and this may limit the discretion of owners.[69]

---

[63] ICOP, para 13.1 (1).

[64] *Ibid*, para 13.1 (2).

[65] *Ibid*, Annex D, para 7.

[66] This envisages the circumstances covered by Energy Act 2011, s 84. See discussion above at paras I-6.38 to I-6.41.

[67] ICOP, para 13.2.1 (1)–(4).

[68] *Ibid*, para 13.2.2.

[69] *Ibid*, para 13.2.3. In this connection, TPA Guidance para 61 states: "The L & I terms that would be determined by the OGA would have regard to the terms prevailing with existing users of a system ..."

ICOP's structured negotiation provision is contained in para 9. **I-6.94**
Where an applicant has obtained (and provided) the necessary information and has determined to begin serious commercial negotiations with a view to obtaining access/services, it should submit an ARN by completing the pro-forma in ICOP Annex E and sending it to the OGA and the owner(s). The ARN commits the applicant to approach the OGA in terms of Energy Act 2011, Section 82 (4) if negotiations are not complete within six months of the ARN submission. The ARN represents an irrevocable commitment by the applicant unless the applicant later confirms to the OGA that a deal has been done, or is no longer needed, or that more time is needed. If an applicant is engaged in negotiations with two or more owners of infrastructure, then multiple ARNs should be submitted. It is always up to the applicant when to submit an ARN, but having submitted an ARN, there is nothing that prevents an applicant from applying for access under Energy Act Section 82 (4) prior to the ARN deadline.

The ARN system has been in force since 2004. It seems to have **I-6.95**
been quite frequently used. However, anecdotally, parties appear to be trying to carry out full commercial negotiations and then (after failing) submitting an ARN. On this view, submission of an ARN can be viewed as hostile by an applicant, which is certainly not the intention of ICOP.

TPA GUIDANCE

TPA Guidance para 13 states the Government's main strategic goals **I-6.96**
for TPA:

> In summary, the Governments' main objective in operating its petroleum legislation is to ensure the recovery of all economical hydrocarbon resources taking into account the environmental impact of hydrocarbon development and the need to ensure secure diverse and sustainable supplies of energy for business and consumers at competitive prices.
>
> Access to infrastructure and associated services on fair and reasonable terms is crucial to maximising the economic recovery of the UK's oil and, particularly, gas because many fields on the UKCS do not contain sufficient reserves to justify their own infrastructure but are economic as satellite developments utilising existing infrastructure.

The TPA Guidance has the same broad infrastructure coverage, **I-6.97**
offshore, but also onshore where relevant to offshore operations, as the 2011 TPA Provisions and ICOP.[70] The TPA Guidance, which has been updated after the passage of the Energy Act 2016, is about

---

[70] See eg TPA Guidance, para 7, footnote 2.

how the OGA will handle infrastructure access requests under the 2011 TPA Provisions. The OGA hopes that "most issues related to infrastructure access ... [will] ... be resolved in timely commercial negotiations, and believes the potential use of its powers will act as an incentive to such an outcome".[71] As a fundamental principle, para 10 notes that in making any access award the OGA will need to identify "the relevant costs and risks and decide on fair and appropriate terms".[72] The TPA Guidance will be applied in a way that is consistent with the Strategy.[73]

I-6.98   TPA Guidance paras 18–25 discuss the importance of competition law to TPA. For further discussion see paras I-6.64 to I-6.72.

I-6.99   There is no standard format for making an access application under Energy Act 2011, Section 82(4), but TPA Guidance Annex 2 sets out a usable form of letter and information annexes.[74] The OGA will appoint a case manager for each separate application. TPA Guidance Annex 1 lists the milestones expected in any application consideration. Under Energy Act 2011, Section 82, the OGA must satisfy itself that there is a genuine dispute, and whether the parties have had reasonable time to reach agreement.[75] In deciding on the "reasonable time" issue, the OGA will look at the sufficiency of information exchange, whether the parties have negotiated in good faith and whether they have complied with ICOP.[76] The OGA can decide to allow an access request, defer such a request to allow further priority negotiation or reject the request.[77] Before using its Energy Act 2011, Section 83 power to intervene in a dispute unilaterally, the OGA will notify the relevant parties that it is considering doing so.[78]

I-6.100   Should the OGA opt to consider an access application, it will request further information from the owner. TPA Guidance Annex 3 provides guidance on what this information may consist of. The OGA will attempt to identify further necessary information at an early stage.[79] The OGA also expects the parties to copy their submissions to one another unless there is a good reason not to.[80] The OGA

---

[71] *Ibid*, para 8.

[72] *Ibid*, para 10.

[73] *Ibid*, para 14. On the OGA's duty to abide by the Strategy in TPA matters, see discussions at paras I-6.45 to I-6.48.

[74] TPA Guidance, para 27.

[75] See discussion of relevant TPA 2011 Provisions above at para I-6.28.

[76] TPA Guidance, para 28.

[77] *Ibid*, para 29. See also Energy Act 2011, s 82(6)(a).

[78] TPA Guidance, para 32. See also discussion of Energy Act 2011 Section 83 above at paras I-6.35 to I-6.37.

[79] TPA Guidance, paras 35–36.

[80] *Ibid*, para 37.

will seek to agree a decision-reaching timetable with the parties; if possible the schedule should not be longer than 16 weeks.[81] Any final decision will constitute a "comprehensive and detailed set of terms and conditions" including tariffs and liabilities and indemnities.[82]

TPA Guidance para 49 notes that the OGA, in deciding any access application, is bound by the seven statutory factors of Energy Act 2011 Section 82 (7). TPA Guidance para 50 goes on to say that the factors listed in Section 82 (7) are not exhaustive and that the OGA can take into account "any other material considerations, including financial information".[83] The OGA's decision will be influenced by the Strategy, including its Obligations and Safeguards. Para 50 also notes that Energy Act 2011 Section 82 (9) and (10) provide that the reasonable expectations of owners and other users cannot be prejudiced unless they are compensated. I-6.101

There are a number of very important footnotes to TPA Guidance para 49 which lists the seven mandatory criteria of Energy Act 2011 Section 82 (7). Footnote 9 discusses reasonable provision of future capacity for owners which would include: realistic upsides or plateau production from existing fields and new field developments where there is an existing development programme or one will be developed soon, or developments which were foreseen and were part of the reason for developing the original infrastructure. However, an owner cannot cite reasonable provision as a mere excuse to deny access to a competitor or for some other advantage. Nor can an owner claim reasonable provision for an unidentified purpose.[84] I-6.102

Footnote 10 to para 49 focuses on "sterilising capacity" which covers such situations as: where taking on a small field would mean that there would not be enough ullage for a prospective large field; where a small new field would consume all the de-propanising capacity in an oil facility thus prejudicing existing throughput; or where a new sour gas field would mean that a pipeline could not be operated sweet. The OGA stresses that every situation is unique.[85] I-6.103

Footnote 11 to para 49 is possibly the most controversial of all. Energy Act 2011 Section 82 (e) talks of "the interests of all users and operators of the pipeline or facility". Footnote 11 finishes with the statement: *"There may be situations where existing contractual* I-6.104

---

[81] *Ibid*, para 43.
[82] *Ibid*, para 44.
[83] *Ibid*, para 50.
[84] This observation relates to Energy Act 2011, ss 82(7)(a) ("capacity which is or can reasonably be made available") and 82(7)(d) ("reasonable needs of the owner …").
[85] Footnote 10 is relevant to Energy Act 2011, ss 82(7)(b) ("any incompatabilities of technical specification") and 82(7)(c) ("difficulties which cannot reasonably be overcome"). See also discussion of ICOP above at para I-6.90.

commitments work against maximising economic recovery and the OGA will consider using its other powers such as those under the Energy Act 2016 to seek an improved outcome" (emphasis added).

I-6.105   In other words, the OGA cannot unilaterally amend existing contracts but it could use sanctions powers against existing contractual parties if the "existing contractual commitments work against MER".

I-6.106   TPA Guidance paras 53–64 set out the OGA's guiding principles on setting transportation and processing terms. Para 53 reiterates that the OGA "supports the principle of non-discriminatory negotiated access to upstream infrastructure on the UKCS". This is consistent with the Strategy. Because there are so "many technical, economic and commercial variables", no formulaic approach is appropriate or possible.[86] A balance must be stuck between rewarding past infrastructure investment and encouraging the exploration and development of new fields. As para 54 states in highlighted prose: "The main issue is to identify the relevant costs and risks and to decide on fair and appropriate terms".

I-6.107   When considering an access request, the OGA:

(1) would not expect that the owner would be *worse off* as a consequence of granting access to the applicant;
(2) that the owner would be reimbursed for all associated costs including additional capital costs and any interruption of the owners' throughput;
(3) the applicant would have to bear a "fair share of the total running costs" post-connection; and
(4) unless the owner had already covered the basic capital costs of the infrastructure, the applicant would also bear a fair share of such outstanding capital costs.[87]

I-6.108   However, none of these considerations precludes the OGA from apportioning a share of total risks to the owner, in return for a corresponding reward.

I-6.109   Notwithstanding para 55's emphasis on the owner not being financially worse off and being able to recover reasonable running costs and capital costs, para 57 counters with the observation that the OGA will avoid an approach that is too owner-cost-reflective as this might reduce their incentive to bear risks, appetite to persevere with infrastructure, interest and desire to innovate and provide new services.

I-6.110   Generally, the terms likely to be offered by the OGA are similar to those which would be offered by owners if there was infrastructure competition.[88]

---

[86] TPA Guidance, para 54.

[87] *Ibid*, para 55.

[88] *Ibid*, para 56. The OGA goes on to emphasise that this statement does nothing to prejudice the OGA's rights to make an access decision.

Para 58 goes on to provide five very important bullet points. I-6.111

Bullet 1. If the applicant seeks access to infrastructure which has I-6.112 already paid off its original capital costs allowing for a reasonable return including risks,[89] then terms will generally reflect the owner's incremental costs and risks. This is a very important principle, because the message it sends to an owner of a "big old" field is that it should not seek to make a high return from depreciated assets.[90]

Bullet 2. If an applicant seeks access from facilities built oversize for I-6.113 the original field or maintained for the purpose of taking third-party throughput, then tariffs should include capital costs and a reasonable return. This tariff is likely to be higher than where there is actual competing infrastructure and should generally be higher than the case envisaged in Bullet 1 (incremental costs and risks). Thus, the message of Bullet 2 is that the OGA will reward owners who invest with the purpose of facilitating TPA. This bullet would broadly apply to access negotiations of the new independent UKCS infrastructure owners, such as NSMP and Antin. It is notable that there is very little specifically on these types of companies in the TPA Guidance.

Bullet 3. If the field or infrastructure is close to the end of its life, then I-6.114 the tariff may need to be higher than incremental costs level so as to facilitate maintenance of the infrastructure. In such circumstances, the OGA is likely to make provision for future cost-sharing. If it can be afforded, the applicant may be required to pay a higher than throughput share of costs, in order to maintain the infrastructure.[91]

Bullet 4. Where various potential applicants are competing for the I-6.115 infrastructure access, then the OGA is unlikely to require the owner to accept terms worse than the best offer from the potential applicants.[92] However, there could be an MER case for doing so.

Bullet 5. If an owner is required under an access order to accept the I-6.116 applicants' hydrocarbons ahead of the owners' own production,

---

[89] The OGA notes that it may often be difficult to determine when a system has been "paid out" under this analysis. *Ibid*, para 58, bullet 1.

[90] Exactly the same principle applies in Norway. See Grondalen and Lower, "Third Party Access to Infrastructure on the Norwegian Continental Shelf", 4 (2016) *LSU Journal of Energy Law and Resources* 324: "Tariffs are decisive as to whether the user field will be developed or not, which generally induces stipulation of low tariffs".

[91] Operating costs for this purpose would normally include replacing outdated metering equipment, necessary to accommodate the existing users of the facility. However, capital costs required to attract new third-party business/expand owner production would not be included. For further discussion on costs under this head, see TPA Guidance, para 58, footnote 14.

[92] Bullet 4 does not address the opposite situation ie where the applicant has the choice of competing lines of infrastructure.

or another third party's previously contracted production, then the owner would expect compensation for the costs of backing off such production. The compensation should be based on opportunity cost, but there is little guidance on how this should be computed.[93]

I-6.117   As for liabilities and indemnities, if there is to be a period of shutdown for tie-in or modification, the applicant should be required to pay reasonable liquidated damages subject to a cap.[94] Applicants will, absent wilful misconduct, be required to indemnify owners against liabilities and losses deriving from tie-ins or modifications. Such liabilities and losses should however be capped based on realistic exposure and a reasonable risk/reward balance. The OGA will be specific as to losses covered and would generally require the applicant to take out insurance for such losses.

I-6.118   Liabilities and indemnities during the transportation and processing purchase would typically be dealt with on a mutual hold harmless basis with an exception for wilful misconduct. Owner and applicant should mitigate their losses when trying to recover from each other. Liabilities and indemnities terms determined by the OGA would also have regard to pre-existing terms binding the users of any system. Each deal will be different.[95] The OGA will also supply terms relating to off-specification deliveries by the applicant during the transportation and processing phase. The applicant will normally have to indemnify the owner for damages caused by off-spec deliveries.[96] In determining this issue, the OGA will have regard to the following issues:

(1) whether the applicant indemnities are to be capped;
(2) the consequence to the owner/other users of an off-spec delivery;
(3) the existence (or not) of blending arrangements and who should be liable for blending failure;
(4) whether the off-spec delivery was known to occur or not, and whether the owner had consented prior to the event;

---

[93] Suppose that A is owner of Diamond field, an offshore oilfield with access to a gas pipeline and an FPSO for oil. In 2012, A agrees to take 10,000 barrels of oil per day from B who is owner of a small oilfield tied into A's field facilities. In 2017, at the request of C, an owner of a larger oilfield close to Diamond, the OGA directs A to allow C to have all the capacity in the Diamond field/pipeline facilities. What if A and B have no other workable export routes? How does one value the shut-in oil and gas of A and B which is all being produced at a spot price? What if there is a viable export route? Should A and B have to pay higher tariffs to access it?

[94] TPA Guidance, para 59.

[95] *Ibid*, para 61.

[96] *Ibid*, para 63.

(5) the availability of adequate clear production data, and the extent to which the owner can control the system;

(6) whether the identity of an off-spec user can be adequately proved in the context of a multi-user system;

(7) does a cross-user liability agreement (CULA) exist in the system – if so, the applicant would normally be required to adhere to it; and

(8) is the applicant desiring to deliver a contaminant into the system on a stipulated, long term basis?

CONCLUSION

The UKCS system of TPA is a highly complex and nuanced response I-6.119 to what is clearly an important regulatory dilemma. Efficient third-party access to existing infrastructure is key to the exploration and development of newer, smaller fields. Throughput from such fields will extend the life of the infrastructure, encourage regional development and put off the evil day of decommissioning. The regulator representing the state has a compelling interest in this, and not just the owner(s) and applicant.

The actual legislative TPA machinery is now contained in the I-6.120 Energy Act 2011. The 2011 TPA Provisions, which give broad regulatory power to the OGA, are well fit for purpose, even if the key points listed in Section 82 (7), (8) and (10) are potentially conflictive. However, the Strategy greatly expands the context of the 2011 TPA Provisions and firmly sets them in the demanding language of MER. Strategy Para 17 (a) clarifies that access to infrastructure must be offered "on fair and reasonable terms". However, TPA decisions do now involve consideration of Strategy Initial Principles (a)–(e) and the Safeguards (paras 2–6).

As noted, competition law plays a significant role in TPA analysis. I-6.121

While neither ICOP nor the TPA Guidance are law, these I-6.122 documents add useful flesh to the hard bones of the 2011 TPA Provisions, Strategy and Competition Act 1998. The ICOP, as an industry code, speaks for the need for publication and availability of relevant information, states key relevant principles and establishes a structured negotiation technique which, if not mandatory, does constitute an important standard. The TPA Guidance speaks to how the OGA will likely handle particular circumstances, offering more favourable terms to owners when they have built big to accommodate third-party throughput, and providing for opex-sharing as abandonment looms.

The negotiated access system envisages that the parties will do I-6.123 the right thing of their own accord. The body of law and principles is there to guide and encourage appropriate commercial behaviour but also to enforce the right result if all else fails. So many issues arise that a formulaic approach is impossible. Nonetheless, one sometimes fears that it will be difficult to know how the regulator would actually select governing principles and considerations from among what sometimes seems like multiple competing concepts.

CHAPTER I-7

THE UK'S OIL AND GAS FISCAL
REGIME: A RADICAL EVOLUTION

Claire Ralph

INTRODUCTION AND OVERVIEW

Given the rollercoaster that the UK's oil and gas fiscal regime[1] has **I-7.01** been on in the last five years, and its technical nature, the challenge of summarising its recent developments in a single chapter is significant. This is against a backdrop of major reform in other areas, such as the regulatory changes implementing many of the recommendations of Sir Ian Wood's Review discussed above in Chapter I-5.

Whilst similar in overall principles to the normal corporation tax **I-7.02** regime which applies to any corporate profits within the UK tax "net", including the activities of companies within the oil and gas supply chain, the taxation of the "acquisition, enjoyment or exploitation of oil rights"[2] is subject to different rules which distinguish it from non-oil exploitation activities, and is the focus of this chapter.

The taxation of oil and gas production is kept entirely separate **I-7.03** from the normal corporation tax regime which applies to all other types of economic activity within the UK by virtue of a ring-fence. This barrier ensures that profits arising from the "upstream" activities related to the exploitation of hydrocarbons from the UK mainland and the United Kingdom Continental Shelf (UKCS) are taxed separately, and the resultant profits cannot be reduced by offsetting non ring-fence losses from other activities that a company may be engaged in. The barrier only restricts movement in one direction; it is still possible for companies with a ring-fence trading loss to take

---

[1] This is commonly referred to, incorrectly, as the North Sea tax regime.
[2] Oil Taxation Act 1975 (hereinafter "OTA 1975"), s 13 or Corporation Tax Act 2010 (hereinafter "CTA 2010"), ss 274 and 279.

this across and set it against a non-ring-fence profit elsewhere within its corporate group, but this would restrict the rate of relief achieved to the (materially lower) non-ring-fence rate of corporation tax. The regime has evolved, at times in a relatively organic and haphazard way, and has attempted to keep pace with the commercial developments of the region and the speed at which the UKCS has matured in recent times.

I-7.04   The current fiscal regime is best described as a "cash flow" regime, meaning that it taxes profits which are broadly equivalent to the cash flow position that the company or field is in for the period. Income is compared to all categories of expenditure, whether operating or capital, exploration, development, production or decommissioning, broadly irrespective of quantum. Taxable profit, therefore, is the extent that total income exceeds total expenditure. Conversely, where total expenditure exceeds total income for the period, a loss arises and taxable profits are nil. Tax becomes payable in instalments on a payment on account basis, and historically on a quarterly payment profile for one aspect of the fiscal regime called Petroleum Revenue Tax. Interest and penalties apply to late payments and a range of other reporting and filing misdemeanours.

I-7.05   The regime contains three separate taxes, Ring Fence Corporation Tax (RFCT), Supplementary Charge (SC) and Petroleum Revenue Tax (PRT). The former two taxes are levied on company profits whereas for the latter the field is the taxable unit.

RECENT CHANGES IN THE FISCAL REGIME

I-7.06 The rates of the three taxes have fluctuated substantially in recent years (see Fig. I-7.1), upwards during the decade to 2011, and subsequently downwards between 2014 and 2016. Whilst comparing headline rates of tax in different international regions or across different stages of the exploration and production (E&P) lifecycle can offer a misleading impression of the international competitiveness of a regime, the proportion of taxable profits left in the hands of the UKCS investor has clearly increased substantially over the last 24 months.

I-7.07   The rate of RFCT has remained remarkably stable at a rate of 30 per cent for some considerable period.[3] Over the last decade, the rate of RFCT has diverged from the main rate of corporation tax charged on non-ring-fence profits, as the Government has sought to make the UK a more attractive destination for companies to be run from by reductions to the headline CT rate. The difference in rates

---

[3] Since 1999, Finance Act 1998, s 29.

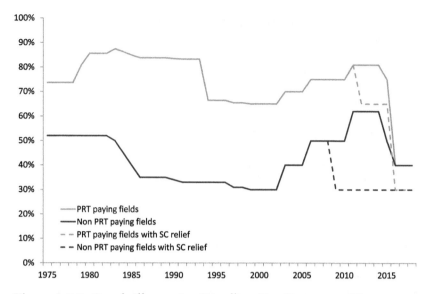

Figure I-7.1 Graph Illustrating Headline Tax Rates over Time

is presently[4] 11 percentage points[5], and set to increase further in future to 13 percentage points from April 2020.[6] The key differences between RFCT and the main corporation tax regime applying to all other companies in the UK will be discussed more fully later in this chapter.

SC has not enjoyed a similar level of consistency in rate and has   I-7.08 been flexed by successive governments. Firstly, upwards between 2002 (when it was first introduced) and 2011 when it peaked at 32 per cent.[7] Since 2011, the SC rate has been lowered three times, and is now at a 10 per cent rate, effective from 1 January 2016.[8] SC is applied to profits on top of the RFCT, meaning the combined headline rate stands at 40 per cent and is effectively another corporation tax which is exclusive to the profits from UK oil and gas production. The naming of this tax as a charge is misleading as the method of calculation is almost identical to that employed for RFCT, except no deduction is allowable for financing costs.[9] HMRC has long insisted that SC is not a separate tax in its own right, but arises

---

[4] As of 2017.

[5] 30 per cent vs 19 per cent.

[6] Finance (No 2) Act 2015, s 7 as amended by Finance Act 2016, s 46.

[7] CTA 2010, s 330 as amended by Finance Act 2011, s 7(1).

[8] Finance Act 2016, s 58, amended CTA 2010, s 330(1).

[9] CTA 2010, s 330(3)(a), whereas some allowance is given when computing a company's profits for RFCT, CTA 2010, s 286.

out of RFCT by virtue of adjustments – they therefore dispute the concept of an SC "loss", for example.

I-7.09   For over 20 years PRT was levied on chargeable field profits at a rate of 50 per cent[10] and was imposed first, with RFCT and SC charged on the remainder.[11] At the recent peak tax rates, when SC was at 32 per cent and RFCT at 30 per cent, fields paying PRT suffered an eye-watering marginal tax rate of 81 per cent.[12] In Budget 2015 the Government announced that the rate of PRT would fall to 35 per cent from 1 January 2016,[13] although this announcement was somewhat overtaken by the announcement in the Chancellor's 2016 Budget statement when PRT was set at a 0 per cent rate from 1 January 2016 on a permanent basis.[14] PRT was designed to only apply to the largest and most profitable fields by virtue of loss relief, Safeguard and Oil Allowance. These allowances were described in-depth in the previous (2nd) edition of this book, so for brevity's sake will not be repeated here. PRT was always a well-targeted tax, as can be seen in published HMRC statistics comparing the proportion of fields paying PRT to the number of fields theoretically liable to the tax.[15]

I-7.10   PRT was only applied to the oldest fields on the UKCS because it was turned off for fields given development consent after 16 March 1993.[16] The historical reasons for the "turning off" of PRT for new fields included the oil price crash in the early 1990s and the Government attempting to support the UKCS industry when cash flows temporarily became negative. Even as early as the 1990s, the Government of the day and those in the industry became increasingly aware that the future of the UKCS depended on smaller, more marginal fields which were not material enough to comfortably bear the burden of PRT as well as corporate taxation.

I-7.11   It is posited that that the Government would not have needed to develop the Supplementary Charge at all had the apparatus existed

---

[10] 1993–2015. OTA 1975, s 1(2).

[11] Although PRT was always deductible when calculating profits chargeable to RFCT and SC.

[12] Being 50 per cent + 50 per cent*(32 per cent + 30 per cent).

[13] PRT is set by reference to chargeable periods which run from 1 January to 30 June (1H) and 1 July to 31 December (2H). Legislation therefore refers to chargeable periods ending after 31 December 2015.

[14] Finance Act 2016, s 140(1) amended OTA 1975, s 1(2).

[15] Table 11.13 sets out, for the last chargeable period for which PRT was levied (1 July–31 December 2015) the number of fields within the charge to PRT ("Total number of fields assessed" was 57, but only 19 of those had taxable profits not covered by losses or Oil Allowance, and of this subset only five fields paid over £10 million in the period). Available at www.gov.uk/government/statistics/government-revenues-from-uk-oil-and-gas-production--2 (accessed 15 May 2017).

[16] Finance Act 1993, s 185.

to turn PRT back on for the newer, particularly profitable and productive fields when a recovery in the oil price was witnessed some eight years later. With hindsight, PRT should have been set to 0 per cent immediately after the oil price crash in the early 1990s and then could have been increased from this point when profitability returned to the sector in the early 2000s. Instead, with the rates of RFCT and normal CT aligned, and not wanting to increase the CT burden for companies outside the oil ring fence, the Government was left with little choice but to introduce an additional tax in the form of SC.

Using fields, rather than the corporate owners, as the taxable unit offers substantially more scope for policy design to vary across different levels of profitability. As it turned out, the additional layer of company taxation proved to be insufficiently flexible to cater for the increasingly heterogeneous nature of the UKCS in the mid to late 2000s, leading to the Field Allowance mechanism[17] which followed as a relief against SC. I-7.12

It is worth bearing in mind that whilst PRT has been permanently set at a 0 per cent rate since 1 January 2016, it still very much forms part of the regime overall, especially as regards to decommissioning which will be covered in more detail later in this chapter. I-7.13

There is almost no difference in the tax treatment of very different scales of profitability – the regime is neither progressive in effect (where the tax burden increases as income increases, such as UK income tax) nor regressive (where the tax burden decreases as income increases – a criticism often levied, somewhat unfairly, against most "sin" taxes such as alcohol and tobacco duty where consumption, and therefore duty paid, is higher for households further down the income scale). This is referred to as "neutrality" where the tax burden is a flat percentage of taxable profit.[18] I-7.14

THE POLICY RATIONALE FOR A SPECIFIC TAX REGIME FOR UK HYDROCARBON PRODUCTION – A BRIEF OUTLINE

The E&P of non-renewable mineral resources from a sovereign country's land mass and territorial waters is often subject to higher I-7.15

---

[17] CTA 2010, ss 351, 720, Pt 8, Chapter 7 and Sch 3, also CTA 2010, s 349A as inserted by Finance Act 2012, Sch 22.

[18] Theoretically there is a lower rate of RFCT (19 per cent) available for groups of companies with taxable profits for the year of below £300,000 but this rate has only been claimed by a few companies in each of the last five years and is almost irrelevant for the purposes of understanding the main characteristics of the regime overall. See www.gov. uk/government/publications/rates-and-allowances-corporation-tax/rates-and-allowances-corporation-tax (accessed 15 May 2017) for details. CTA 2010, s 279A.

than normal taxation, as well as onerous licensing conditions that govern the permitted activity of the companies engaged in the sector.

I-7.16   The nation's oil and gas, amongst other types of mineral wealth, can only be extracted once; and once produced, it is lost forever to the nation. Associated with this, oil and gas production often generates "supernormal" profits – profits which are higher than is required to compensate for the risks involved in the activity. The economic theory is that a substantial component of the supernormal profit (or resource rent) can be taxed away without materially diminishing the appetite within the investor community to (continue to) exploit the resource.

I-7.17   The level of resource rent varies enormously between different locations, and also over time, driven by factors such as underlying geology, commodity prices and the costs of operation. A number of key non-economic risk factors including political and fiscal volatility can also affect an investor's appetite for any particular region as well. Comparing headline tax rates to estimate whether a country is fiscally more attractive than another region is often misleading, and needs to be approached with extreme caution. The UK is often compared to Norway, for example, despite the fact that Norway's fields tend to be significantly larger than those on the UK's side of the median line, and the costs of operating in the different zones are often demonstrably different – the Norwegian operators benefiting from economies of scale that rarely exist in the UK segment.

I-7.18   The UK is referred to as a strict concession licence regime – the UK Government invites applications by international oil companies of all sizes and specialisms to competitively bid for licensed acreage on and offshore. The new Oil and Gas Authority (OGA) is responsible for awarding licences to a single operator per "bloc", and those companies suffer tax on all profits that arise from their "ring fence" profits from their E&P activities.[19]

I-7.19   Other countries have taken different routes to secure a financial return from their mineral resources – either via national oil companies who then return dividends to their home country's Exchequer; or establishing production sharing contracts (PSCs) with companies where the contract sets out the terms dividing up the wealth of the project between the Government and the company, which can vary project by project.

---

[19] See Chapter I-4 for a detailed discussion on OGA and the licensing regime in the UK.

PRINCIPLES OF GOOD OIL AND GAS FISCAL POLICY AND PRACTICE

Oil and gas investment is a capital-intensive, internationally competitive and high-risk investment. Depending on the availability of relevant infrastructure, the type of development and the individual economics of the field in question, the payback period[20] for a particular development can be a decade or two from initial exploration and discovery. Given this, and the amount of money required to be sunk into facilities and wells, a fiscal regime can aid or abet the risk profile of an oil and gas investment by providing protection from the downside of an investment when difficulties are encountered, in exchange for an increased tax cost for a good investment.   I-7.20

One aspect which is important to all investments is the degree of fiscal certainty and predictability that a Government builds up within the investor community. The long-term nature of the fiscal regime can often be vital in offering comfort that the fiscal regime modelled during the investment appraisal process, leading up to the final investment decision (FID) and will reflect the tax treatment the project actually receives. Where a regime develops a record of fiscal instability, investors require better returns than they would normally to sanction that particular project. That is what happened in the UK where there were three adverse rate rises within a decade and a number of other smaller changes affecting particular projects.   I-7.21

Between 2002 and 2011, the Government viewed the sector increasingly as a cash cow, a source of additional Exchequer revenues which had few political consequences, justified by an increase in Brent crude prices, which were taken as indicative of massive profitability. Not only did high prices boost the rates of return for all competitor regions of hydrocarbon production around the world (meaning that the UK, far from enjoying a premium related to the Brent blend, was simply benefiting from the worldwide supply/demand inequality), but also the cost base in relation to the operational and capital funding of the UK region was increasing[21] – many UK facilities were already beyond their original design lives by this point and the cost of running some of the worst-performing fields made them barely profitable at $100 oil. By July 2014, when the oil price began to slide, the problems the UKCS was suffering from had become very apparent indeed.   I-7.22

---

[20] This is the amount of time a project takes to recover the cumulative expenditures incurred to date.

[21] See Oil & Gas UK's annual Economic Report at www.oilandgasuk.co.uk (accessed 15 May 2017) for further background information.

I-7.23   Public data on the profitability of UK oil and gas production is not what it should be, but the Office for National Statistics does publish a dataset, and it shows profitability on a quarterly basis being depressed long before the oil price began to fall.[22]

DRIVING INVESTMENT – THE NEW BLUEPRINT FOR THE OIL AND GAS FISCAL REGIME IN THE UK

I-7.24   In December 2014, the coalition Government published a new roadmap setting out its approach to the taxation of the dwindling UK oil and gas province, called "Driving Investment: a plan to reform the oil and gas regime".[23] Whilst theoretically covering onshore and unconventional production as well, this document was very focused on offshore fiscal issues which had emerged during a period of consultation with the (offshore) industry over the preceding summer.

I-7.25   "Driving Investment" set out, in more explicit language than previously, the principles that were to guide Government policy in this area. Three key objectives stated at the time were:

(1) The overall tax burden will need to fall as the basin matures and as projects become ever more marginal.
(2) The government will consider the wider economic benefits of oil and gas production, rather than just the direct tax revenues the sector generates.
(3) The government's concept of a "fair return" will consider the global competitiveness of UK opportunities relative to international opportunities, the cost base and oil and gas revenues.[24]

I-7.26   The Government, rightly, wants to make sure that the nation captures some of the benefit of the nation's oil and gas resources, but there was an increasing acknowledgement that a smaller percentage of something amounted, in absolute terms, to more benefit than a higher percentage of a project which never took place. The goose that had laid the golden egg consistently over many years was at risk of being killed by punitive taxation. If that happened the wider economic benefits associated with the offshore industry, primarily in jobs, technological innovation and balance of trade benefits, risked being lost. Only time will tell if "Driving Investment" represents a

---

[22] Chapter 4 – UK continental shelf (UKCS) companies. Available at www.ons.gov.uk/economy/nationalaccounts/uksectoraccounts/bulletins/profitabilityofukcompanies/julytosept2016#uk-continental-shelf-ukcs-companies (accessed 15 May 2017).
[23] HM Treasury, "Driving Investment: a plan to reform the oil and gas fiscal regime" (Report, HM Treasury December 2014) (hereinafter "HM Treasury, 'Driving Investment'").
[24] *Ibid*, Executive Summary (pp 5–6).

genuinely new era of Government policy in this area, or a platitude offered long after the industry's glory days are over.

Another option for the Government to secure a return from I-7.27 mineral production is the use of a true royalty, where a portion of the final commodity value is paid over to the Government but with little or no allowance for the costs of extraction. Royalty regimes deliver more stable revenues compared to strict concession regimes, and accelerate cash flow into the Government compared to taxes as the project does not have to recover its initial upfront investment before the royalty becomes payable. It is worth bearing in mind that the UK had a royalty regime at one time, which was turned off for fields granted development consent after 1982,[25] and permanently abolished for all fields in 2002.[26] The main reason why the UK Government moved away from a royalty regime is that it can act in a regressive manner, meaning that the Government take proportionately increases as projects become more marginal with higher costs, damaging investor appetite as a region matures.

TECHNICAL CHARACTERISTICS OF THE RING-FENCE REGIME

The basics of how profits are taxed in the ring fence are the same as I-7.28 outside the ring fence, but some key differences are worth noting.

Capital Allowances

One of the major differences is the existence of 100 per cent First I-7.29 Year Capital Allowances for virtually all capital expenditure within the ring fence. Capital Allowances are the tax equivalent of depreciation in a company's financial accounts and therefore reduce profits chargeable to corporation tax. This means that all amounts of capital expenditure incurred in a period are immediately deductible against taxable profits. Outside the ring fence, most significant capital investment attracts 18 per cent or 8 per cent writing down allowances, and therefore takes at least eight years to achieve full relief for this investment.[27]

---

[25] Petroleum Royalties (Relief) Act 1983, s 1 read together with Finance Act 1983, s 36(2). This assumes that those development consents were granted for those licences that incorporated the 1982 model clauses. Petroleum Royalties (Relief) Act 1983, s 1(2)(a). The 1982 model clauses are set out in Petroleum (Production) Regulations 1982 SI 1982/1000, Sch 5.

[26] HM Treasury, "Budget 2002 The Strength to make long-term decisions: Investing in an enterprising, fairer Britain", Report, HM Treasury, 17 April 2002, at 5.82.

[27] Capital Allowances Act 2001, s 56(2).

I-7.30   Ring-fence activities are very heavily capital-intensive, and therefore the immediate tax relief for these expenditures makes the UK much more attractive than it would otherwise be, as any delay in achieving relief for investment is discounted in investment economics.[28] Capital expenditure has always been immediately relievable for PRT purposes; and for RFCT and SC, this feature was introduced at the point that SC was introduced in 2002.[29] It is extremely important that this feature is retained, and the Government seems to understand this: "Driving Investment" states: "100% first year capital allowances are available for virtually all capital expenditure, and it is clear that this remains a vital element of the regime".[30]

I-7.31   Where expenditure exceeds income for a period, a tax loss arises and profits are zero. This loss can be carried back a single year in line with the non-ring fence rules. It can also be carried forward and used to reduce future taxable trading profits – again in line with the current CT rules (which are changing outside the ring fence from April 2017). What is unique to the ring-fence regime is an allowance mechanism called the Ring Fence Expenditure Supplement (RFES), which was initially introduced in 2004 as a more restrictive Exploration Expenditure Supplement (EES).[31]

I-7.32   EES and its successor, RFES, were designed to help companies that were incurring expenditure but had limited/no production income upfront. EES originally targeted exploration expenditure driven losses but was then extended to cover all categories of ring-fence expenditure[32] in 2006.[33] The supplement increases the cash value of the tax loss carried forward so that it not only keeps up with inflation but also compensates investors who lack a tax-efficient business profile to gain immediate relief for capital expenditure and who would otherwise view UKCS projects less favourably to their tax-paying joint venture partners in making a Final Investment Decision. Initially EES and RFES were both "worth" an uplift of 6 per cent for six periods but the selection of these criteria seems not to be based on economic analysis. In 2011 and 2014, RFES was extended to 10 per cent[34] for ten periods[35] respectively as a better surrogate for the disadvantages such companies suffer.

---

[28] Also see Chapter I-2 on this.
[29] Capital Allowances Act 2001, Pts 2, 5, 6 as well as ss 162–165 as amended.
[30] HM Treasury, "Driving Investment", para 3.6.
[31] Income and Corporation Taxes Act 1988, Sch 19B, para 4.
[32] Except decommissioning expenditure.
[33] CTA 2010, s 312.
[34] CTA 2010, s 311.
[35] Finance Act 2015, s 47 and Sch 11.

RFES was originally envisaged to provide support only to a   I-7.33
minority of companies within the UKCS for a short period at the
start of their UKCS investment profile. However, with the collapse
in the oil price and the overhang of capital investment from the
2010–2014 period it is a reasonable assumption that in recent years
a greater proportion of companies have made at least one RFES
claim as oil prices have fallen dramatically and remained low for 18
months or more since June 2014.

The Investment Allowance

The Investment Allowance (IA) was introduced in 2015[36] to replace   I-7.34
the Field Allowance which had been in place previously as the
principal relief against a company's Supplementary Charge liability.
The allowance is based on the Onshore Allowance[37] which remains
in place for all fields developed from onshore drilling sites.[38]

The policy objective behind the development of the allowance was   I-7.35
to improve the post-tax return to the investor for UKCS projects
compared to other overseas projects within a company's portfolio.
Projects which were economically viable pre-tax were more likely
to be commercially sanctionable with the allowance on a post-tax
basis because the projects would only suffer SC on a portion of the
production income. It is a key part of incentivising life extension
projects and tie-ins of smaller fields upon which the future of the
UKCS increasingly depends.

The Investment Allowance is related to the amount of capital   I-7.36
expenditure incurred on a development after 1 April 2015 at a
rate of 62.5 per cent,[39] ie an allowance of £62.5 million for capital
expenditure of £100 million. The allowance is held on account
with the company which has incurred the expenditure, awaiting
"activation" by relevant income from that same development once
it reaches production. Without being activated the allowance is
not capable of being used against the company's SC liability in the
current or future accounting period. Once activated the company
requires an SC liability against which to offset the allowance.

Developments which require higher levels of capital investment   I-7.37
automatically attract more IA than smaller or less capital-intensive
projects. In some instances, replacing the prior Field Allowances with
the IA has reduced the amount of SC relief a development could

---

[36] Finance Act 2015 introduced amendments to CTA10, ss 322A onwards.
[37] CTA 2010, Pt 8, Chapter 8.
[38] See paras I-4.79 to I-4.82, including Wytch Farm in Dorset which is in the internal
waters of the UK coastline but treated as an onshore field for regulatory purposes.
[39] Finance Act 2015, Sch 12, s 2 332B.

attract. For example, a single well subsea tieback tapping a field of <45 mmbbl which used to generate the full £150 million small field allowance, which was likely to cover a substantial proportion of the production income from the field, would now require £240 million of qualifying expenditure to generate the same £150 million of Investment Allowance. The real benefit of the Investment Allowance as compared to the Field Allowances is that for the first time incremental investment in fields already in production are incentivised in the same way as greenfield developments.[40]

I-7.38   Capital expenditure across all stages of finding, appraising and developing an offshore field qualifies for Investment Allowance.[41] Whilst capital expenditure is defined for tax purposes differently than under accounting principles[42] most expenditures bringing into existence an asset with enduring economic benefit for the trade will qualify.[43]

I-7.39   Investment Allowance[44] is a value allowance[45] which gives a specific amount of relief in terms of production income. Value allowances have the benefit of being "used up" quicker in times of high commodity prices and more slowly for fields whose product was less valuable and/or in low price periods. As a relief against SC the cash value of the relief also flexes with the prevailing SC rate.[46]

I-7.40   For completeness, the Cluster Area Allowance (CAA)[47] is the other allowance which operates on the UKCS as a relief for SC profits but its relevance to the general reader is currently significantly limited, as only a single cluster has been determined around the central North Sea Culzean field. The only difference between IA and CAA is

---

[40] That is, without the administrative burden of compiling a Field Development Plan Addendum of the Brown Field Allowance or the specific field characteristics which were required to be met in the original Field Development Plan to qualify for the (greenfield) Field Allowances. See www.publications.parliament.uk/pa/cm201213/cmhansrd/cm120907/wmstext/120907m0001.htm (accessed 15 May 2017).

[41] CTA 2010, s 332B.

[42] Such as Generally Accepted Accounting Practice (GAAP) or International Financial Reporting Standards (IFRS).

[43] CTA 2009, s 931 together with Capital Allowances Act 2001, s 4. The test for determining capital expenditures for tax purposes comes from case law. See, for example, *ECC Quarries Ltd v Watkins (Inspector of Taxes)* [1975] 3 All ER 843 (Chapter D), (1977) 1 WLR 1386 and *Atherton (HM Inspector of Taxes) v British Insulated and Helsby Cables, Limited* [1925] KB 421, 10 TC 155 (CA).

[44] And the field allowances beforehand.

[45] As opposed to Oil Allowance for PRT which is a volume allowance.

[46] Profits chargeable to RFCT are unaffected by the investment or field allowances, meaning the Government still has the reassurance that hydrocarbon production income suffers a minimum 30 per cent tax rate, higher than the non-ring-fence corporation tax rate which began its downward trajectory in 2008 and has since accelerated.

[47] Finance Act 2015 introduced amendments to CTA 2010, s 356JC onwards.

that allowances generated by expenditures across the whole defined cluster area can be activated by production income from the same cluster area. Also, abortive exploration expenditure secures relief under the CAA where it would not under the IA.[48]

EXPLORATION TAX ISSUES

With no IA available for abortive exploration expenditures, and I-7.41 the delay between generating the allowance and its use against a company's SC liability, it is doubtful whether the investment allowance is a well-designed incentive for exploration-focused investors.

The other major area where the tax regime is not overly helpful I-7.42 towards exploration companies that do not have any production interests is the difficulty in raising the upfront finance to do the drilling in the first place. The comparison is routinely made with Norway, which has chosen to incentivise exploration by giving a cashable tax credit for companies that do not have a tax liability against which to achieve relief for exploration costs. Given the rates of tax in Norway,[49] the rate of the credit funds nearly four-fifths of the drilling costs, leaving a post-tax funding gap of only 22 per cent. The UK Government has consistently, and robustly, dismissed the often shrill requests for a similar exploration tax credit on the UKCS, preferring to directly fund new seismic data surveys of particular regions of the UKCS.[50]

TAX LOSSES, EXCEPT THOSE ARISING FROM DECOMMISSIONING ACTIVITIES

In a year where allowable expenditure exceeds taxable trading I-7.43 income in a company's corporation tax return, a trading loss arises, and a number of options exist to achieve tax relief, many of which are common to the regular, non-ring-fence corporation tax rules. The first step is to take the loss against other categories of taxable profits made by the company or other entity under common control within the same group in the same period.[51] Once a current year loss has been relieved, to the extent that there are further losses to relieve, the second route available allows the loss to be carried back against

---

[48] Because such allowance never gets activated and is otherwise permanently lost.
[49] Which are, at 78 per cent, not only very much higher than in the UK but have also remained remarkably stable over the last 20 years or so.
[50] See www.gov.uk/government/news/oga-awards-contracts-for-second-20m-uk-seismic-campaign (accessed 15 May 2017).
[51] CTA 2010, s 37.

the total taxable profits for the year immediately prior to the loss-making year.[52]

I-7.44   Once these two options have been exhausted, the only remaining opportunity to gain relief is to carry forward the loss into a future period to use against trading profits only, which happens automatically and so no election is required. Where a company is winding up its trade and ceasing permanently, an additional option for relief losses becomes available, terminal loss relief, which allows for enhanced carry back of losses against prior profits against the taxable profits of the preceding three years on a last in, first out (LIFO) basis.

LOSSES ARISING UPON DECOMMISSIONING

I-7.45   Obligations within the Energy Act 2008, and its predecessors,[53] ensure licensees pay for decommissioning activity. Such activity is expected to cost tens of billions of pounds to the end of the basin's productive life.[54] The tax regime attempts to cater for this phase via enhanced loss flexibilities, but an important aspect is that tax relief is only available when the process of decommissioning is actually carried out. The placing of funds into a decommissioning security agreement trust[55] is not, in itself, considered to be decommissioning activity and therefore does not attract tax relief at the time that the contribution is made,[56] no matter how far the money is alienated from the payee company, nor the restrictions upon its use.

I-7.46   For tax purposes, the principal definition of decommissioning relates to obligations to carry out approved decommissioning programmes.[57] HMRC has confirmed that this includes the major categories of "decommissioning" such as the plugging and abandoning of wells, removal of drill strings and other subsurface equipment, planning for, and eventual removal of topsides, ongoing monitoring of the marine environment and onshore recycling of all material.[58] This list is not exhaustive, and is simply to illustrate that decommissioning seems to include all processes which will be required to be covered under the definition.

---

[52] CTA 2010, s 39.

[53] See Chapter I-12 for discussions on decommissioning.

[54] See www.ogauthority.co.uk/media/1020/oga_decomm_strategy.pdf (accessed 15 May 2017).

[55] See Chapter I-12 for discussions on decommissioning security trusts.

[56] See www.gov.uk/hmrc-internal-manuals/oil-taxation-manual/ot28470 (accessed 15 May 2017).

[57] Capital Allowances Act 2001, s 163.

[58] See www.gov.uk/hmrc-internal-manuals/oil-taxation-manual/ot28050 (accessed 15 May 2017).

Depending on the portfolio of the assets within the decom- I-7.47
missioning company's group, there may well be other sources of
ring-fence income against which to claim relief for the decommissioning expenditure in the current year and/or the prior year. If this
is not the case, and there remains an amount of unrelieved loss,
there are other possibilities for achieving relief which are specific to
losses arising on decommissioning. Losses arising from decommissioning are capable of being carried back for many more periods
than normal tax rules otherwise allow, to 1975 for PRT and to 2002
for RFCT and SC purposes.[59] Carrying back a loss to a prior period
will set the profits for that year to nil and generate a repayment of
tax suffered from HMRC, plus interest, because the Exchequer, in
economic terms, has had monies not ultimately due for a period of
time, which could be considerable in many cases.

For the purposes of the company's cash flow, it would always be I-7.48
preferable to carry back the loss rather than carrying it forward, This
would generate a refund of the tax suffered, either by a direct remit-
tance or as a payment on account against a company's ongoing tax
liabilities.

UNCERTAINTY WITHIN THE DECOMMISSIONING TAX RELIEF REGIME AND THE DECOMMISSIONING RELIEF DEED

In the early 2000s, the industry had a deep-seated concern that I-7.49
at some future point, when tax revenues gradually fell away,
and decommissioning activity increased, that the Government's
commitment to tax relief for decommissioning could wane, tempting
the Government of whichever colour to consider restricting the loss
carry back available under legislation which became Corporation
Tax Act 2010.

Concern was focused upon PRT assets which typically have I-7.50
higher estimated abandonment costs and given that these were the
oldest fields on the UKCS their productive profiles, individually and
together, were falling quicker than their younger counterparts. The
prospect of the Government abolishing PRT at the time when PRT
revenues became negligible/negative, became the key concern because
it would have denied companies the accrued tax relief. This worry
only increased in the wake of Budget 2011 when the Government
unexpectedly increased the rate of SC to 32 per cent,[60] but restricted
tax relief for decommissioning losses to the 20 per cent rate.[61] This

---

[59] CTA 2010, s 42.
[60] HM Treasury, Budget 2011 (Report, HM Treasury, 23 March 2011) 1.146.
[61] HM Treasury, Budget 2011 (Report, HM Treasury, 23 March 2011) 1.149.

disconnect between the rate that income was being taxed at and the rate of relief for a particular subset of true business costs alarmed the industry. It also created some perverse incentives where the marginal rates of tax for incremental investment projects could get close to, or exceed, 100 per cent in certain cases due to the effective disallowance of costs for tax relief purposes.

I-7.51   A period of intense consultation between industry and the Government ensued[62] and the importance of the resultant policy development, the Decommissioning Relief Deed (DRD),[63] cannot be overstated. Whilst in many ways the DRD is designed to simply replicate the provisions of tax relief within the legislation in place at the date, there are instances where the DRD gives a more advantageous outcome for the company incurring "imposition expenditure" in the event of a default of a licensee. One example of this is that for imposition expenditure incurred upon the default of another entity extra tax capacity is created to ensure the company picking up the costs is not disadvantaged overall for having done so.[64] The Government's clear desire is to only pay out decommissioning relief via the Deed in extreme and exceptional cases.

I-7.52   One illustration of this desire to avoid DRD payouts was the decision in Budget 2016 to permanently set the rate of PRT to 0 per cent from 1 January 2016.[65] When the Government was seeking to remove PRT as an ongoing part of the regime it could have abolished PRT, but that would have triggered an adverse tax event under the DRD and would have led to relief accruing, in contrast to Government preferences. Relief available under the DRD, in particular circumstances, can be more generous than under the normal tax code, which is a further reason why Government concluded that it was unpalatable.

I-7.53   The zero rating of PRT has another economic effect which is worthy of note. PRT has not been abolished, rather taxable profits have continued to arise after 1 January 2016, albeit chargeable to tax at a 0 per cent rate, with no tax becoming payable to the Exchequer. There is a very significant difference between £0 being payable in a period and there being no tax due. Profits generated after 1 January 2016 will dilute the amount of decommissioning tax relief available when the field ceases production. This is because a proportion of decommissioning costs will be carried back and set

[62] HM Treasury, "Decommissioning Relief Deeds: increasing tax certainty for oil and gas investment in the UK continental shelf", Consultation, HM Treasury, 9 July 2012.

[63] See Chapter I-13 for a detailed discussion on the DRDs.

[64] See Chapter I-13 for further discussion on DRDs and their tax treatment.

[65] See www.gov.uk/government/publications/oil-and-gas-taxation-reduction-in-petroleum-revenue-tax-and-supplementary-charge (accessed 15 May 2017).

against taxable profits which have already benefitted from the 0 per cent rate, attracting no refund in the loss calculation on the normal LIFO basis. The result of this is that the cash value of the refund at the time decommissioning is carried out will be lower.

Given that the profitability of the older PRT paying fields has   I-7.54 declined considerably in recent years the beneficial incentive effect of zero rating PRT may have been taken too late in the basin's life to have a strong revitalising effect on many of the major hub infrastructure systems. However, at the margins, the impact of the overall tax rate on PRT fields halving in two years from 81 per cent to 40 per cent would otherwise be expected to improve their materiality for their owners' investment outlook.

THE TAXATION OF INFRASTRUCTURE

In the early days of the UKCS's development, the major trunklines   I-7.55 serving the main regions of the North Sea were instrumental and integral to the host field's production, and as such have been taxed within the ring fence as essentially extensions of the host field.[66] As the development of nearby satellite fields progressed, many of which were insufficiently profitable to support their own standalone infrastructure, these main systems saw an increasingly worthwhile business opportunity for third-party tariff business which could take up spare capacity, known as "ullage".[67]

Today, the UKCS has over 300 producing fields,[68] and fields are   I-7.56 increasingly interdependent, sharing relatively high fixed costs of operating this core infrastructure. This interdependency can create the platform for a domino effect where the cessation of production of a field means the proportion of the system costs it previously supported gets re-allocated among the remaining fields, and this eventually leads to the cessation of the next most vulnerable field, creating a vicious cycle.

From a tax perspective, the industry has argued for a number of   I-7.57 years that the taxation of infrastructure assets within the ring fence is wholly inappropriate given the much lower profitability enjoyed by the operators,[69] and the punitive tax charges if a piece of infra-

---

[66] Most of these fields would have been PRT liable and paying.

[67] See Chapter I-6 for a detailed discussion on Third Party Access to Infrastructure issues.

[68] Department of Energy and Climate Change, "UKCS Field Information", 3 August 2015, available at www.og.decc.gov.uk/fields/fields_index.htm (accessed 9 May 2017).

[69] Whilst the scope of the Wood Review did not include consideration of fiscal aspects one recommendation was explicitly tax focused – that companies owning and operating offshore infrastructure assets should no longer by taxed within the ringfence as their returns were not commensurate with the higher taxes on profits.

structure is sold to an entity outside the ring fence. The industry has, to date, not made a sufficiently convincing evidence-based case for reform of the tax treatment of infrastructure, and it is questionable whether there is enough political will for reform of this particularly technical area of the tax regime, partly related to the issues around mature asset transfers summarised below.

THE TAX-DRIVEN CHALLENGES OF TRADING MATURE UPSTREAM ASSETS

I-7.58 On the back of the much commended Wood Review,[70] which emphasised the importance and value of asset trading and ensuring the right assets found their way into the "right" hands (ie those companies with the strongest desire to drive their productive output and extend their operational lives), the market for mature UKCS assets came into sharp focus. The new OGA set this as one of its strategic aims when it took over responsibility for the upstream oil and gas industry as an empowered and better-resourced regulator.[71] Despite a few important deals announced very recently, for example the sale of Shell assets to Chrysaor[72] and Enquest's purchase of BP's Magnus field,[73] it is clearly still a buyer's market as there are many more assets up for sale than investors willing and financially strong enough to take on the challenge of mature asset ownership and operatorship on the UKCS. There is a key tax impediment to this challenge which is hard to negotiate around.

I-7.59   Typically, the owners of such assets are the companies who have owned and run the fields in their heydays when profitability was higher, productive capacity remained plentiful and the shadow of decommissioning was relatively small and far away on the horizon. Such companies paid substantial amounts of tax on these upstream profits against which to absorb the full costs of decommissioning when they eventually become due. But if such a company wishes to rationalise its asset ownership, perhaps refocusing on larger projects elsewhere on the UKCS or overseas, it may seek to sell a mature field to another, often smaller, company that is more focused on mature asset operations. This smaller company may not have any

---

[70] Sir I Wood, *UKCS Maximising Recovery Review: Final Report*, available at www.gov.uk/government/uploads/system/uploads/attachment_data/file/471452/UKCS_Maximising_Recovery_Review_FINAL_72pp_locked.pdf (accessed 15 May 2017).

[71] See Chapter I-5 for a detailed discussion on the OGA.

[72] See www.shell.com/media/news-and-media-releases/2017/shell-to-sell-package-of-uk-north-sea-assets-to-chrysaor.html (accessed 15 May 2017).

[73] See www.enquest.com/media-centre/press-releases/2017/24-01-2017.aspx (accessed 15 May 2017).

taxable history within the ring fence and, as such, except for the taxable profits it can build up between taking on the new asset and cessation of production, has no tax capacity in which to absorb the costs of decommissioning. Such a potential buyer, despite seeing a more positive future for the asset compared to the status quo ownership model, will suffer a loss value from the tax regime in any asset purchase because the RFCT and SC relief built up in the seller cannot move across with the decommissioning liability.

The only way around this issue from a strict tax perspective is to structure the deal as a share sale rather than an asset deal. This means the company with the licence interest and a portion of the tax capacity is sold instead.[74] However, this can be unpalatable for commercial reasons: for instance, the buying company may be acquiring a "dirty" asset – the company may have potential liabilities in fields it has long since exited that the acquirer has no interest or capacity to take over.[75] It is also difficult to structure such a deal and negotiate the various anti-avoidance clauses within the UK tax code, designed to stop abusive transactions from undermining the Exchequer yield in other areas. I-7.60

There have been recent discussions between the industry and HM Treasury about the potential for a portion of the tax capacity built up in the selling company to be transferred to the acquiring company alongside the decommissioning liability in future asset sales, subject to inevitable safeguards to prevent a market for tax capacity developing inappropriately. In the Chancellor's Spring Budget 2017, the Government announced its intention to publish a formal discussion document looking at this issue and establish a new expert panel to scrutinise options over summer 2017.[76] It would not be an understatement to point out that the UKCS's ability to survive and thrive in the wake of recent economically challenging times depends upon successfully extending the life of older fields which contain the best hope of securing near term value, and to keep areas open for future exploration and development projects. I-7.61

LOOKING FORWARD/EVALUATION

The UKCS fiscal regime has come a very long way in the last few years, and in early 2017 the tax regime is no longer the impediment I-7.62

---

[74] See Chapter II-9 for a detailed discussion on different ways in which acquisitions can be carried out.

[75] These liabilities may arise due to Secretary of State's powers under Petroleum Act 1998, s 34. See Chapter I-12 for a further discussion on this.

[76] HM Treasury, Budget 2017 (Report, HM Treasury 8 March 2017), s 4.6. See further para I-7.65.

to Maximising Economic Recovery from the UKCS that it was immediately after Budget 2011. The additional fiscal burden of PRT on older fields has, eventually, been removed. The allowance regime has been simplified, ensuring the regime is generally progressive and should support capital investment where such funding is available. The rates of tax at 40 per cent, with Investment Allowance against the SC, 100 per cent First Year Capital Allowances and RFES for loss-making companies, intuitively feels more appropriate for the opportunities available across the UKCS.

I-7.63 The challenge facing the vibrant and thriving future for UK oil and gas development is one of hard economic facts – smaller accumulations, with greater technical complexity, and ageing and expensive ancillary infrastructure. Without technological innovation breaking through the cost challenge, marginal UKCS projects will remain marginal compared to other more lucrative returns available overseas. Whilst the longer-term effects of the UK's exit from the European Union remain unclear, the important concerns of Britain's energy security and balance of trade mean that the fiscal regime offering as much stability and predictability as possible has never been so important.

I-7.64 With any luck this summary of the tax regime for the UK's oil and gas industry will remain current for longer than in previous editions. The future of many important facilities currently *in situ* on the UKCS depends on the future reform being supportive and targeted to the areas of most pressing need.

Afterword

I-7.65 In the Chancellor's Autumn Budget 2017, it was announced that transferable tax history (as discussed in para I-7.61 above) will be introduced from November 2018. The measure has been greeted with enthusiasm by OGUK and leading oil companies.

CHAPTER I-8

CONTINENTAL SHELF BOUNDARIES IN THE NORTH SEA AND THE NORTH ATLANTIC

Constantinos Yiallourides

INTRODUCTION TO THE LAW OF THE SEA

No discussion involving offshore petroleum exploration and exploi- I-8.01
tation would be complete without consideration of the international
legal framework governing the use of the oceans and the seas. The
international law of the sea, one of the most dynamic branches of
public international law, is the area of law that has made it possible
for the states to extend their control and jurisdiction into the sea
and that has opened the legal doors to exploration for, and exploitation of, natural resources lying on and under the seabed. The
progressive development of the law of the sea has occurred alongside
the increasing technological capability of states to exploit offshore
natural resources. No significant offshore hydrocarbon discoveries
were made until the later nineteenth century, before which use of the
ocean was largely limited to navigation and fishing.[1] Border security
and management and conservation of near-shore fisheries were the
coastal states' primary concerns with little attention paid to the
enactment of a legal regime that, although governing five-sevenths of
the globe, was of limited use.[2] However, when technology revealed
the presence of rich minerals lying beneath the ocean seabed, a
great interest arose in accessing and controlling these newly found
seabed resources, leading to a new era of extended claims over ocean
territory and resources and, ultimately, to increased possibilities
for conflicts among states. This situation provided the impetus for
consolidating the legal framework for the oceans and the seas and

---

[1] D Johnston, *Theory and History of Ocean Boundary-Making* (1988), pp 56–57.
[2] *Ibid*, pp 56–57.

for bringing some legal certainty to what was considered until then an utterly disordered situation.[3]

I-8.02   The North Sea is perhaps one of the most illustrative examples of the development of law and practice in this area. It is not by coincidence that, before the discovery in 1959 of a giant onshore gas field near Slochteren in the Dutch province of Groningen and the knowledge that the gas in the underlying reservoirs extended across the seabed of the North Sea, no serious effort was made by any of the regional governments to enact laws and regulations applicable to seabed petroleum exploration and exploitation.[4] However, once the huge mineral potentials of the North Sea became known following the Netherlands' success and subsequent geophysical systematic exploration,[5] the bordering states were quick to enact laws, at a domestic level, proclaiming exclusive control over petroleum resources in the adjacent areas of the seabed in an attempt to preserve what they considered as their sovereign rights in the North Sea seabed and subsoil.[6] Just as natural resources have motivated these states to extend their control and jurisdiction in and under the sea, natural resources have been instrumental in their efforts to determine the location of their boundaries offshore.[7] Indeed, as will be discussed further below, natural resources have been at the heart of national and international attention in maritime boundary negotiations, converting in a sense the classic "land

---

[3] For a detailed discussion on the progressive evolution of the international law of the sea, see J Harrison, *Making the Law of the Sea: A Study in the Development of International Law* (2011).

[4] B Nielsen, *The State Offshore: Petroleum, Politics, and State Intervention on the British and Norwegian Continental Shelves* (1991), pp 14–15.

[5] "In 1962 three seismological exploration parties were at work in the North Sea; in 1964 there were fifty". See R Young, "Offshore Claims and Problems in the North Sea", 59(3) (1965) *American Journal of International Law* 505, at 508 (hereinafter "Young, 'Offshore Claims'").

[6] Norway proclaimed sovereignty over the Norwegian continental shelf in May 1963 through a Royal Decree and passed an Act relating to the Exploration for and Exploitation of Submarine Natural Resources in June 1963. The 1963 Decree provided that "The seabed and its subsoil in the submarine areas outside the coast of the Kingdom of Norway are subject to Norwegian sovereignty in respect to the exploitation of and exploration for natural deposits, to such extent as the depth of the sea permits the utilisation of natural deposits ... but not beyond the median line in relation to other states", repr. in P Swan, *Ocean Oil and Gas Drilling* ([1979], 1996). The UK had also taken steps to establish its sovereign rights over its sector over the North Sea continental shelf by enacting in 1964 the Continental Shelf Act. The 1964 Act vested in the Crown any rights with respect to the sea bed and subsoil and their natural resources. Similar legislation was enacted by Denmark, the Netherlands and Germany in the mid-1960s; see Young, "Offshore Claims", 505; R B Clark, *The Waters Around the British Isles* (1987) (hereinafter "Clark, *Waters Around the British Isles*"), p 9.

[7] J Prescott, *The Geography of Frontiers and Boundaries* (1965), p 17.

dominates the sea" maxim into "oil dominates the land and the sea".[8]

CONTINENTAL SHELF BOUNDARIES IN THE NORTH SEA

North Sea: Geographical and legal context

The North Sea is a narrow, semi-enclosed sea of approximately 166,000 square nautical miles (570,000km²). It is bounded to the west by the east coasts of England and Scotland, to the south by the Netherlands, Belgium and Germany, and to the east by Denmark and Norway. It is strategically connected to the Atlantic Ocean from the south-west via the English Channel, and in the north across the boundary between Shetland and Norway. The North Sea is a relatively shallow sea with water depths less than 100m on average, although in the Norwegian Trench (a geological depression in the sea floor off the southern coasts of Norway) the depth reaches 700m.[9]   **I-8.03**

The early national claims of control and jurisdiction over offshore petroleum resources in the North Sea drew legal basis from the relevant rules of international law as was enshrined, at the time, in the 1958 Convention on the Continental Shelf (hereinafter "1958 CS Convention").[10] The 1958 CS Convention made clear that the rights of the coastal state over its continental shelf are "sovereign rights for exploiting its natural resources" and that "these rights are exclusive" in the sense that if the coastal state chooses not to explore the continental shelf or its natural resources, no one may undertake these activities without its express consent.[11] It is commonly accepted that the specific terminology of "sovereign rights" rather than "sovereignty" has been adopted to make it clear that the coastal state does not own the continental shelf, instead it has the rights necessary for and connected with the exploration and exploitation of seabed natural resources.[12]   **I-8.04**

With respect to the spatial extent of such rights, the 1958 CS Convention stipulated two key criteria: the "200 isobath"[13]   **I-8.05**

---

[8] *Maritime Delimitation and Territorial Questions (Qatar v Bahrain)* (Joint Dissenting Opinion of Judges Ranjeva, Bedjaoui and Koroma) [2001] 40 ICJ Rep 847, para 215.

[9] Clark, *Waters Around the British Isles*, pp 10–11.

[10] Convention on the Continental Shelf (adopted 29 April 1958, entered into force 10 June 1964) 499 UNTS 311 (hereinafter "1958 CS Convention").

[11] 1958 CS Convention, Art 2.

[12] S Jayakumar, "The Continental Shelf Regime under the UN Convention on the Law of the Sea: Reflections after Thirty Years", in M Nordquist et al. (eds), *Regulation of Continental Shelf Development* (2013).

[13] Simply put, "isobath" (also known as "depth curve" and "fathom line") is a line connecting points of equal water depth; see International Hydrographic Organisation, *Hydrographic Dictionary Part I Volume* (5th edn, 1994) pp 63, 85, 118.

or, beyond that limit, the "exploitability test".[14] While the first criterion was based on objective measurements of ocean depth, the second criterion was purely technological. Consequently shelf limits depended on general standards of marine technology concerning the exploitation of offshore natural resources. By the 1960s marine technological developments had made feasible the exploitation of the seabed at depths in excess of 1,000m.[15] Hence the *prima facie* entitlement of interested coastal states to exclusive control over the resources to be found in the continental shelf adjacent to their territory could not be easily challenged: if petroleum exploitation was technically feasible in water depths well beyond the maximum depth of the North Sea (700m) then, even under a restrictive definition of the 1958 CS Convention, continental shelf rights would stretch across the whole North Sea. Because of the narrow geographical configuration of the North Sea such rights overlapped substantially, triggering the need for maritime delimitation as a means to determine where the dividing line between two or more overlapping continental shelf entitlements should lie.[16]

I-8.06   The rules and principles of maritime delimitation have been subject to important refinements since the 1960s, many of which have had direct implications on the course of interstate negotiations and the resulting boundaries in the North Sea, but the bordering states of the North Sea have been quite successful in reaching agreement on their boundaries offshore. In part, this is due to the positive tenor of political relations among the states concerned.[17] But it is also largely due to the pressing need to gain access to seabed minerals and the realisation that, in the absence of clear and defined boundaries marking the jurisdictional limits of each state at sea, international oil companies (the funds and technical expertise of which were indispensable) would be quite reluctant to invest. As a matter of fact, shortly after the Groningen gas discovery, the UK Government was encouraged by several oil companies (including Shell, BP and ESSO) to reach boundary agreements with neighbouring countries before progressing with exploration work beyond seismic surveying.[18] The development of petroleum outside

---

[14] 1958 CS Convention, Art 1.
[15] P N Swan, *Oil and Gas Drilling and the Law* (1979), pp 11–22; M Nordquist et al., *Law, Science and Ocean Management* (2006), p 439.
[16] D Johnston and P Saunders, *Ocean Boundary Making: Regional Issues and Developments* (1988), p 17; Y Tanaka, *Predictability and Flexibility in the Law of Maritime Delimitation* (2006), pp 7–8.
[17] J Prescott and C Schofield, *Maritime Political Boundaries of the World* (2nd edn, 2004) (hereinafter "Prescott and Schofield, *Maritime Political Boundaries of the World*"), p 369.
[18] A Kemp, *The Official History of North Sea Oil and Gas: Vol. I: The Growing Dominance of the State* (2012) (hereinafter "Kemp, *The Official History, Vol. I*"), p 7.

of territorial waters was an unusual, expensive and incredibly risky venture, and the possibility that some of the offshore petroleum fields and associated infrastructure might be subject to competing jurisdictional claims by more than one sovereign government meant that companies would be quite unwilling to find themselves locked in disputes between states that might lead to inadvertent delays in their envisaged investment programme.[19] From the international oil companies' perspective, legal certainty and security of tenure were *sine qua non* for committing large sums of money and capital on risky exploration and exploitation operations.[20] Likewise, from the Governments' point of view, defining their maritime boundaries was important to ensure that exploration activity would not be delayed by uncertainty on the basis that oil companies would not be willing to take the risk that a dispute over the location of the maritime boundary led to the result that they had, in essence, obtained an exploration or Production Licence from the wrong coastal state. Thus, it behoved the Governments of the North Sea to move rapidly to define the limits of their maritime boundaries to deliver jurisdictional clarity and legal certainty to an area that would otherwise be characterised by legal uncertainty as to which of the two, or more, claimant states held exclusive sovereign rights and authority over petroleum operations in the area in question.[21]

Early continental shelf delimitations in the North Sea

The continental shelf boundaries in the North Sea were delimited I-8.07 progressively through multiple delimitation agreements that were achieved largely by way of interstate negotiations leading to bilateral treaties, including four treaties which followed directly from third-party adjudication of judicial nature.[22] The first North Sea boundaries were effected from 1965 to 1968, a period which followed the entry into force of the 1958 CS Convention in 1964 and coincided with the efforts of the UK to define the co-ordinates of the designated areas for its very first petroleum licensing round.[23] Further

---

[19] For an analysis, see C Yiallourides, "Oil and Gas Development in Disputed Waters Under UNCLOS", 5 (2016) *UCL Journal of Law and Jurisprudence* 59.

[20] *Ibid*; see also H Fox et al. (eds), *Joint Development of Offshore Oil and Gas* (1st edn, 1989), p 39.

[21] For a discussion, see M Pratt and D Smith, *How to Deal with Maritime Boundary Uncertainty in Oil and Gas Exploration and Production Areas* (2007).

[22] M C Wood, "Northern and Western European Maritime Boundaries", in J Charney et al. (eds), *International Maritime Boundaries Vol 5* (2005) (hereinafter "Wood, 'Northern and Western European Maritime Boundaries'"), p 3495.

[23] Continental Shelf (Designation of Areas) Order 1964 (UK), 3 (1964) *International Legal Materials* 640.

boundary agreements were subsequently concluded from 1969 to 1972, following the decision of the International Court of Justice (hereinafter "ICJ" or "Court") on the *North Sea Continental Shelf* cases of 1969 between Germany, the Netherlands and Denmark.[24] It has been suggested that, while the initial delimitations were shaped in accordance with the equidistance principle defined in Article 6 of the 1958 CS Convention, latter delimitations were inspired by the concept of "equitable principles" that emerged from the decision of the ICJ on the *North Sea CS Cases*.[25]

I-8.08   The first set of continental shelf delimitation agreements in the North Sea included six separate boundaries between the UK and Norway (1965),[26] the UK and Denmark (1966),[27] the UK and the Netherlands (1965),[28] Norway and Denmark (1965)[29] and Norway and Sweden (1968).[30] Interestingly, from the outset of bilateral diplomatic discussions on the determination of the above boundaries, the states concerned were in agreement that negotiations should commence with a view to agreeing boundaries based on the delimitation provisions of the 1958 CS Convention.[31] The Convention distinguished between two principal delimitation situations. For the situation where the continental shelf lay between two opposite coastal states, such as the UK and Norway, Article 6(1) provided that in the absence of interstate agreement or special circumstances, the boundary would be the median line, every point of which is equidistant from the coastal baselines of each state.[32] Under this solution, the boundary would simply be the halfway line between

---

[24] *North Sea Continental Shelf Cases (Germany v Denmark; Germany v Netherlands)* [1969] ICJ Rep 3 para 63 (hereinafter "*North Sea CS Cases*").

[25] *Ibid.*

[26] Agreement Relating to the Delimitation of the Continental Shelf between the Two Countries (UK/Norway) (adopted 10 March 1965, entered into force 29 June 1965) as amended by Protocol (adopted 22 December 1978, entered into force 20 February 1980).

[27] Agreement Relating to the Delimitation of the Continental Shelf between the Two Countries (UK/Denmark) (adopted 3 March 1966, entered into force 6 February 1967) as amended by Agreement (adopted 25 November 1971, entered into force 7 December 1972).

[28] Agreement Relating to the Delimitation of the Continental Shelf under the North Sea between the Two Countries (UK/Netherlands) (adopted 6 October 1965, entered into force 23 December 1966) as amended by Protocol (adopted 25 November 1971, entered into force 7 December 1972).

[29] Agreement Relating to the Delimitation of the Continental Shelf (Denmark/Norway) (adopted 8 December 1965, entered into force 22 June 1966) as amended by Agreement (adopted 24 April 1968, entered into force 24 April 1968) and Agreement (adopted 4 June 1974, entry into force 4 June 1974).

[30] Agreement Concerning the Delimitation of the Continental Shelf (Sweden/Norway) (adopted 24 July 1968, entered into force 18 March 1969).

[31] Kemp, *The Official History, Vol. I*, p 64.

[32] 1958 CS Convention, Art 6(1).

the opposite coastal states. Likewise, for situations where two states are adjacent to each other, such as Norway and Sweden, the 1958 CS Convention prescribed that, failing agreement to the contrary, "the boundary shall be determined by application of the principle of equidistance".[33] By this solution, the maritime boundary between two adjacent coastal states would simply be an extension of their land boundary in the sea.

The North Sea continental shelf cases of 1969 and subsequent continental shelf delimitations in the North Sea

Having close regard to the delimitation formula codified in the 1958 CS Convention, the UK Government approached its North Sea neighbours with proposals that the division of the maritime boundaries should be effected on the basis of equidistance-median line solutions. Such proposals were well received, particularly by Norway, Denmark, the Netherlands and Belgium. The states concerned were also in agreement that there were no special circumstances such as, for example, offshore insular features or deeply concave coasts that would justify a departure from equidistance.[34] I-8.09

The only notable opposition at the time was that of Germany, which was reluctant to accept that the use of equidistance would lead to a just and equitable division of the available continental shelf. Germany was of the view that, due to its concave (inwardly curved) coastline, a boundary based on equidistance would have given Germany an exceptionally small portion of continental shelf compared to those of to its neighbours on either side (Denmark and the Netherlands). According to Germany, in situations where a state's coastline is deeply concave as was to a moderate extent the coast of Germany, the effect of equidistance was to extend the land boundaries in the direction of concavity with the effect of depriving the coastal state of an important portion of the continental shelf appertaining to its territory. In contrast, in the case of convex (outwardly curved) coasts, like those of Denmark and the Netherlands, the application of equidistance would have a much wider effect on the areas allocated to those coasts. I-8.10

Discussions between Germany, Denmark and the Netherlands were protracted and eventually the matter was referred to the ICJ for resolution. The parties asked the Court to determine whether Germany was under a legal obligation to accept the equidistance principle defined in the 1958 CS Convention.[35] Of the three dispu- I-8.11

---

[33] 1958 CS Convention, Art 6(2).
[34] Wood, "Northern and Western European Maritime Boundaries", p 3496.
[35] *North Sea CS Cases*, para 21.

tants, Denmark and the Netherlands were parties to the 1958 CS Convention but Germany was not. It was necessary, therefore, to consider both the conventional rules on delimitation and those prescribed by international customary law. The Court considered that the equidistance principle could not be said to reflect or crystallise customary law, and was consequently not binding upon Germany.[36] The court emphasised that the continental shelf should be viewed as the natural extension of the coastal state's land territory under the sea and that delimitation should be effected in such a way as to leave as much as possible to each state "those parts of the continental shelf that constitute a natural prolongation of its land territory into and under the sea, without encroachment on the natural prolongation of the land territory of the other".[37] The Court also declared that in the course of determining their continental shelf boundaries, the parties had to take into consideration such factors as "the general configuration of the coasts", "the presence of any special or unusual features" and "the element of a reasonable degree of proportionality" whereby the portion of continental shelf allotted to each party should be in reasonable proportion to the linear extent of each party's coastline.[38] Based on the particular geographic circumstances of the case, the Court found that the application of equidistance, whilst "a very convenient one",[39] would unquestionably lead to an inequitable result. Ultimately, the Court rejected the mandatory application of equidistance and stressed the equitable outcome that needed to be achieved according to the particular geographical and geological circumstances at hand.[40]

I-8.12   While the Court refrained from drawing specific boundaries, the guiding principles it pronounced did help Germany to escape from a difficult settlement impasse with a far more favourable delimitation solution than equidistance would have offered. Importantly, it also gave rise to a new series of delimitation agreements either altering existing treaties or guiding the conclusion of new ones. The agreements that followed directly from the decision were those achieved between Germany, on the one side, and Denmark (1971),[41] the

---

[36] *Ibid*, para 21.

[37] *Ibid*, paras 19, 101.

[38] *Ibid*, para 101.

[39] *Ibid*, para 22.

[40] "Delimitation is to be effected by agreement in accordance with equitable principles, and taking account of all the relevant circumstances, in such a way as to leave as much as possible to each Party all those parts of the continental shelf that constitute a natural prolongation of its land territory into and under the sea", *North Sea CS Cases*, para 101.

[41] "Denmark-Federal Republic of Germany-Netherlands: Agreements Delimiting the Continental Shelf in the North Sea", 10(3) (1971) *International Legal Materials* 600.

Netherlands (1971)[42] and the UK (1971)[43] on the other. While a detailed assessment of the technical negotiations that predated the conclusion of the above agreements would go beyond the scope of the present chapter, it is interesting to note that all these agreements led, on the basis of attaining an equitable result, to the determination of boundaries which duly respected the special circumstances within the overall geographical context of the areas under delimitation and deviated from the strict application of equidistance.[44] The same result can be seen in the UK-France boundary in the southern end of the North Sea, determined partly by an arbitration tribunal and partly through an interstate agreement.[45] Equitable principles were applied to the effect that the British Scilly Isles and the Channel Islands, due to their proximity to France, were given a lesser effect than equidistance so as to produce an equitable boundary.[46]

UNRESOLVED CONTINENTAL SHELF ISSUES

Overlapping continental shelf rights in the Hatton-Rockall Plateau

Notwithstanding that the continental shelf boundaries between the I-8.13 UK and its regional neighbours in the North Sea are now clearly established, the same cannot be said for the maritime boundaries in the North Atlantic region. Indeed, in the Atlantic approaches to northern Europe the states concerned still remain unable to agree a number of issues despite several settlement attempts having been made. Such issues have revolved principally around the states' overlapping continental shelf rights over what is now commonly known as the Hatton-Rockall plateau.[47] The situation is further complicated by the presence of the North Atlantic rocky islet of Rockall. The latter

---

[42] *Ibid*, 600.

[43] Agreement Relating to the Delimitation of the Continental Shelf Under the North Sea Between the Two Countries (UK/Germany) (adopted 25 November 1971, entered into force 7 December 1972).

[44] For a detailed analysis of the North Sea continental shelf delimitations, see AG Oude Elferink, *The Delimitation of the Continental Shelf between Denmark, Germany and the Netherlands: Arguing Law, Practicing Politics?* (2014).

[45] *Delimitation of the Continental Shelf (UK/France)* (Award of 30 June 1977) Reports of International Arbitral Awards Vol XVIII; Agreement Relating to the Completion of the Delimitation of the Continental Shelf in the Southern North Sea (France/UK) (adopted 23 July 1991, entered into force 17 March 1992).

[46] Wood, "Northern and Western European Maritime Boundaries", p 3497.

[47] Denmark refers to the plateau as the "Faroe-Rockall microcontinent". Such terminology, also preferred by Iceland, implies that the plateau is an isolated piece of continental shelf that is geologically and geomorphologically disconnected from the UK and Ireland; see C R Symmons, "The Rockall Dispute Deepens: An Analysis of Recent Danish and Icelandic Actions", 35(2) (1986) *International and Comparative Law Quarterly* 344, at 345 (hereinafter "Symmons, 'The Rockall Dispute Deepens'"). See C Yiallourides, "It

was annexed by the UK in 1955 and subsequently incorporated into Scotland through the Island of Rockall Act of 1972, but its territorial status has been disputed by Ireland which rejects British sovereignty claims over the feature and surrounding resources.[48]

I-8.14   As far as the first of these issues is concerned, the conflicting continental shelf rights over the Hatton-Rockall Plateau, four states are involved: Denmark (on behalf of the Faroe Islands), Iceland, Ireland and the UK. These states have made multiple outer continental shelf claims across the Hatton-Rockall plateau area, many of which overlap to a considerable extent.[49] Both Denmark and Iceland expressed formally their claims over the area in 1985 by way of seabed designations in areas of the plateau which vastly overlapped with previous UK and Irish seabed designations in the period 1974 to 1977. Developments in the law of the sea have moved rapidly since these conflicting continental counter-designations were formalised; the most prominent being the conclusion of a new Law of the Sea Convention in 1982 (hereinafter "1982 LOS Convention" or "1982 Convention").[50]

I-8.15   The 1982 LOS Convention laid down a new and generally acceptable legal framework governing virtually all the uses of the oceans and the seas including the exploration for and exploitation of living and non-living natural resources.[51] While it goes beyond the scope of the present chapter to discuss in detail the 320 articles which make up the 1982 LOS Convention, it is nonetheless useful to consider some of the new features of the 1982 Convention, particularly those associated with the continental shelf regime, before turning to its legal implications on the delimitation of the continental shelf in the North Atlantic region. It should be noted that for states parties, the 1982 LOS Convention supersedes all pre-existing 1958 Geneva Conventions.[52]

I-8.16   While the 1958 CS Convention has not been rendered void after the emergence of the 1982 Convention, and "can still be referred to as

---

Takes Four to Tango: Quadrilateral Boundary Negotiations in the North East Atlantic", 87 (2018) *Marine Policy* 78–83.

[48] Prescott and Schofield, *Maritime Political Boundaries of the World*, p 374.

[49] At present, there are two large areas of trilateral overlap (Denmark/Iceland/UK and Denmark/Iceland/Ireland) and three areas of bilateral overlap (Denmark/Iceland (in two sections), Iceland/Ireland and Iceland/UK. For a discussion see Symmons, "The Rockall Dispute Deepens".

[50] United Nations Convention on the Law of the Sea (adopted 10 December 1982, entered into force 16 November 1994) 1833 UNTS 3.

[51] "A new constitution for the oceans", as stated by the President of the Third UN Conference on the Law of the Sea, Ambassador Koh, in 1982 when the Convention was adopted and opened for signature, cited in D Anderson, "The British Accession to the UN Convention on the Law of the Sea", 45(1) (1996) *International and Comparative Law Quarterly* 761.

[52] 1982 LOS Convention, Art 311(1). See also Yiallourides, "It Takes Four to Tango".

a source of international law to which in many respects the new LOS Convention is closely related", there are some significant changes which have rendered important parts of the 1958 CS Convention completely obsolete.[53] For example, while the substantive rights of states for exploring and exploiting the continental shelf established in the 1958 CS Convention remained largely unchanged, the 1982 Convention did prescribe a new method for the determination of outer limits of the continental shelf. Whereas Article 1 of the 1958 CS Convention had defined its outer limit in terms of bathymetry (measurement of depth) and exploitability beyond that limit, Article 76 of the 1982 Convention determines the continental shelf by reference to distance from the coasts (ie 200nm from the territorial sea baselines) or to criteria of geology and geomorphology (ie natural prolongation of the land territory to the outer edge of the continental margin), whichever lies further from the territorial sea baselines. As a consequence, coastal states are now able, on the basis of the physical characteristics of the seabed, to claim sovereign rights over a continental shelf extending up to the outer edge of the continental margin or, where the shelf does not physically prolong to that distance, to a continental shelf up to 200nm based on distance alone.

Article 76 of the 1982 Convention places the responsibility on claimant states to provide evidence based on concepts of geodesy, geology, geophysics and hydrography in order to prove that their continental margin physically extends beyond 200nm from their coasts.[54] Article 76(3) of the 1982 Convention illustrates the close interaction between the legal concept of the continental shelf and the physical properties of natural prolongation. It allows a coastal state to extend its seabed jurisdiction to the outer continental shelf but only where such shelf has been scientifically proven to form the submerged prolongation of the land territory to the outer edge of the continental margin where it reaches the deep ocean floor.[55] It has been observed that "although all states possess a continental shelf within 200 NM, regardless of the geological and geomorphologic elements, the entitlement of the area beyond 200 NM depends on

I-8.17

---

[53] Clark, *Waters Around the British Isles*, pp 282–283.

[54] For a discussion about the scientific methods and techniques involved in the determination of the continental shelf beyond 200nm, see B Kunoy, "The Rise of the Sun: Legal Arguments in Outer Continental Margin Delimitations", 53(2) (2006) *Netherlands International Law Review* 247, at 253; Z Wu et al., "Methods And Procedures to Determine the Outer Limits of the Continental Shelf Beyond 200 Nautical Miles", 32(12) (2013) *Acta Oceanologica Sinica* 126.

[55] R W Smith and G Taft, "Legal Aspects of the Continental Shelf", in P J Cook and C M Carleton (eds), *Continental Shelf Limits: The Scientific and Legal Interface* (2000), pp 91, 101.

such elements".[56] Article 84, which supplements Article 76, deals
with the question of publicising the designated outer lines of the
shelf by reference to charts or lists of geographical co-ordinates,
though neither it nor any other article offers any further guidance
"as to how the limit of the natural prolongation or continental
margin is to be determined".[57]

I-8.18   A highly complex and confusing legal situation has emerged as a
result of the new definition of the continental shelf prescribed in the
1982 LOS Convention. The crux of the controversy in the Hatton-
Rockall plateau lies in that each of the above states has offered
differing interpretations of the concept of natural prolongation and
has made use of different criteria to define the area of the continental
margin that naturally appertains to it.[58] It is important to note at
this point that all interested North Atlantic states are full parties to
the 1982 LOS Convention and its terms are binding upon them as a
matter of treaty law.[59]

I-8.19   While the UK and Ireland have based their continental shelf claims
on the concept of natural prolongation of their land mass offshore,
Denmark has contended that the Hatton-Rockall plateau is part of
a distinct "microcontinent" which is geologically and geomorphologically connected to the Faroe Islands and that, in the light of the
physical characteristics of the area (particularly the presence of a
steep geological depression known as "Rockall Trough") there is
a manifest break in the continuity or natural prolongation of the
UK's and Ireland's continental shelf so that neither of these two
countries have a natural prolongation connection with the Hatton-
Rockall plateau.[60] Similar to Denmark, Iceland also considers that its
continental margin naturally prolongs from the Icelandic mainland
until the "shelf-breaking" Rockall Trough – this being the natural
end-point of the plateau, implying that both Ireland and the UK

---

[56] B Kunoy, "The Rise Of The Sun: Legal Arguments in Outer Continental Margin
Delimitations", 53(2) 2006 *Netherlands International Law Review* 247, at 261.

[57] E D Brown, "Rockall and the Limits of National Jurisdiction of the UK Part 1",
2(3) (1978) *Marine Policy* (hereinafter "Brown, 'Rockall and the Limits of National
Jurisdiction of the UK Part 1'") 181, at 189.

[58] Prescott and Schofield, *Maritime Political Boundaries of the World*, pp 372–373.

[59] United Kingdom (25 July 1997), Republic of Ireland (21 June 1996), Iceland (21
June 1985), Denmark (19 December 2003), see "Chronological lists of Ratifications
of, Accessions and Successions to the 1982 LOS Convention" (UN Division for Ocean
Affairs and the Law of the Sea), available at https://treaties.un.org/pages/showDetails.
aspx?objid=0800000280043ad5 (accessed 22 April 2017).

[60] Symons cites several authors who have considered that the geological evolution of the
region indicates a separation of the Rockall microcontinent "wholly or partly" from the
European plate, thus evidencing a closer link to the Faroe Islands. See Symmons, "The
Rockall Dispute Deepens".

are physically cut off from any seabed rights in the area.[61] It is to be noted that while the average water depth in the Hatton-Rockall area is no more than 200m, the water depth of the Rockall Trough reaches down to 4,000m at its south-west end.[62]

From a geological and legal perspective, Iceland's position is I-8.20 closely related to that of Denmark's. This has not prevented Iceland from arguing that its claim is stronger "on the basis that the inter-position of the Faroes Channel weakens the Danish claim to a geological link to the Plateau and, indeed, constitutes a geomorpho-logical break with the Faroes".[63] On the opposite side, Denmark views the Hatton-Rockall plateau as being physically cut off not only from the UK and Ireland but also from Iceland, as evidenced by a "drop of the seafloor" similar to that of the Rockall Trough.[64]

The UK and Ireland were able to overcome their dispute regarding I-8.21 overlapping seabed rights in the Hatton-Rockall area through the Continental Shelf Agreement in 1988,[65] effectively turning the North Atlantic dispute from a quadrilateral to a trilateral one, between Ireland and the UK on the one hand and Denmark (on behalf of the Faroes) and Iceland on the other.[66] Despite several diplomatic discussions between the four countries, the dispute remains live with both Ireland and the UK rejecting the legal basis of the Danish and Icelandic continental shelf designations, and both Denmark and Iceland pointing out that the UK-Ireland continental shelf agreement

---

[61] *Ibid*, 363, 368.

[62] C Lysaght, "The Agreement on the Delimitation of the Continental Shelf between Ireland and the United Kingdom" 3(2) (1990) *Irish Studies in International Affairs* 81, at 82 (hereinafter "Lysaght, 'Agreement on the Delimitation of the Continental Shelf'").

[63] Symmons, "The Rockall Dispute Deepens", at 362.

[64] *Ibid.*

[65] The agreed Anglo-Irish continental shelf boundary, as illustrated on an annexed map to the 1988 agreement, is drawn in a zigzag manner so as to facilitate the granting of petroleum licences for rectangular blocks according to the common practice of both governments. The boundary extends in the south to over 300 miles from either coast to the edge of the continental margin. In the north, it extends to approximately 550 miles from either coast, crossing the Rockall Trough (which begins 100 miles offshore), and then extends across the whole Hatton-Rockall plateau. See Agreement Concerning the Delimitation of Areas of the Continental Shelf between the Two Countries (UK/Ireland) (adopted 7 November 1988, entered into force 11 January 1990); for a critical commentary see: Lysaght, "Agreement on the Delimitation of the Continental Shelf". It must also be noted that the two countries extended the 1980 boundary by 10nm in the north-northwest through a supplementary protocol (adopted 8 December 1992, entered into force 26 March 1993).

[66] C R Symmons, "The Irish Partial Submission to the Commission on the Limits of the Continental Shelf in 2005: A Precedent for Future Such Submissions in the Light of the 'Disputed Areas' Procedures of The Commission?", 37(3) (2006) *Ocean Development and International Law* 299, at 300.

has no binding effect upon them and does not prejudice their existing rights in the Hatton-Rockall plateau.[67]

I-8.22   The submissions of the UK, Ireland and Denmark to the Commission on the Limits of the Continental Shelf in 2009 and 2010 with respect to their outer continental shelf rights in the Hatton-Rockall plateau have been instrumental in clarifying the legal and scientific basis of the parties' perception of their continental shelf entitlements beyond 200nm. But the extent to which these submissions can influence the settlement of their maritime boundaries in the North Atlantic region is a different matter requiring further examination.

I-8.23   First of all, despite the provisions of Article 77(3) of the 1982 Convention which establishes that a state's title in the continental shelf exists *ipso facto* and *ab initio*,[68] Article 76 makes clear that the opposability of such title with regard to other states depends on the satisfaction of certain conditions, including the procedural requirement to submit to the Commission information on the limits of the continental shelf beyond 200nm and issuance by the Commission of relevant recommendations in this regard.[69] Article 76(8) of the 1982 Convention further provides that only the limits of the continental shelf established on the basis of the Commission's recommendations shall be binding and final. Notwithstanding the above, Article 76(10) provides that that submission of information on the limits of the continental shelf beyond 200nm is "without prejudice to the question of delimitation of the continental shelf between States with opposite or adjacent coasts". Therefore, the submission of such information in itself does not have any legal effect on unresolved delimitation issues. These unresolved issues remain subject to the terms stipulated in Articles 83 and 74 of the 1982 Convention concerning the delimitation of the continental shelf and the Exclusive Economic Zone (EEZ).

I-8.24   Secondly, "the Commission is a scientific and technical body with recommendatory functions entrusted by the Convention to consider scientific and technical issues arising in the implementation of Article 76".[70] The mandate to interpret and apply the provisions of Article 76 of the 1982 LOS Convention lies strictly with international courts and tribunals such as the ICJ and the International Tribunal

---

[67] C R Symmons, *Ireland and the Law of the Sea* (2nd edn, 2000), p 335; see also Verbal of Denmark Addressed to the UN Secretary-General (27 May 2009).

[68] As a matter of fact and by the very beginning.

[69] *Dispute Concerning Delimitation of the Maritime Boundary in the Bay of Bengal (Bangladesh/Myanmar)* (ITLOS Judgment) 51 (2012) *ILM* 840, para 411 (hereinafter "*Dispute Concerning Delimitation of the Maritime Boundary in the Bay of Bengal*").

[70] *Ibid*, para 407.

for the Law of the Sea (ITLOS).[71] Even so, given that international adjudicative bodies could be influenced by the Commission's technical recommendations, the 1982 LOS Convention restricts the competence of the Commission to consider submissions and make recommendations on the establishment of the outer limits of the continental shelf where outstanding delimitation disputes exist in areas which are the subject of the submission. Article 9 of Annex II to the 1982 LOS Convention provides that "The actions of the Commission shall not prejudice matters relating to the delimitation of boundaries between States with opposite or adjacent coasts". The Rules of Procedure of the Commission reiterate this requirement by mandating the Commission to refrain from considering and qualifying submissions made by any of the states concerned in a land or maritime dispute.[72] The practice of the Commission so far with respect to submissions involving unresolved land and maritime disputes indicates that the Commission will generally adhere to this rule.[73]

At this point it is also crucial to mention that Iceland has objected to the UK's, Ireland's and Denmark's submissions of preliminary information to the Commission. In a verbal note of 27 May 2009, addressed to the UN Secretariat, Iceland stated the following:

I-8.25

> The Hatton-Rockall Plateau area ... is part of the Icelandic continental shelf but is subject to overlapping claims by Denmark on behalf of the Faroe Islands, Ireland and the United Kingdom of Great Britain and Northern Ireland, and it is in dispute. In the view of the Government of Iceland, consideration and qualification of the submission by the

---

[71] *Ibid*, para 411.

[72] Rule 46(2) and Rule 5(a) Annex I, "Rules of Procedure of the Commission on the Limits of the Continental Shelf' (adopted 18 April 2008), available at www.un.org/Depts/los/clcs_new/commission_rules.htm_(accessed 28 March 2017): "The Commission may consider one or more submissions in the areas under dispute with prior consent given by all States that are parties to such a dispute".

[73] See eg the Commission's response to the submission made by Russia regarding the Sea of Okhotsk, in "Report of the Secretary-General, Oceans and the Law of the Sea" UNGA, 57th Session (8 October 2002), para 40, available at www.un.org/Depts/los/general_assembly/general_assembly_reports.htm (accessed 13 April 2017). Furthermore, in the East China Sea, in the light of Japan's objection, the Commission decided to "defer further consideration" of China's submission and instructed the parties to "take advantage of the avenues available to them, including provisional arrangements of a practical nature": R Lee and M Hayashi (eds), *New Directions in the Law of the Sea: Global Developments Vol 3* (2015), p 41; for a complete analysis, see B Kwiatkowska, "Submissions to the UN Commission on the Limits of the Continental Shelf: The Practice of Developing States in Cases of Disputed and Unresolved Maritime Boundary Delimitations or Other Land or Maritime Disputes. Part One" 28 (2013) *International Journal of Marine and Coastal Law*, 219 (hereinafter "Kwiatkowska, 'Submissions to the UN Commission on the Limits of the Continental Shelf'").

Commission on the Limits of the Continental Shelf would prejudice the rights of Iceland over the continental shelf in this area. The Government therefore does not give its consent to the consideration and qualification of the submission by the Commission.[74]

I-8.26   Likewise, Denmark requested the Commission to refrain from considering and qualifying the continental shelf submission of the UK, on the basis of Rule 46 of its Rules of Procedure. This stipulates that the Commission shall not consider and qualify a submission by a state involved in a maritime dispute in the area which is the subject of the submission, unless prior consent is given by all states parties to the dispute. The Commission has refrained from qualifying or making recommendations on the parties' submissions so far.

I-8.27   In view of this stalemate, in 2011 Denmark (Faroes) gave its consent to the Commission for considering the UK and Irish 2009 partial (Hatton-Rockall Plateau) submissions upon a strict condition that Denmark's 2010 (Faroes-Rockall Plateau) outer continental shelf submission will be considered by the Commission simultaneously.[75] Denmark's submission and "qualified consent" were, once again, met with the explicit objection of Iceland.[76]

I-8.28   Overall, by making a formal submission to the Commission, the states concerned have taken an important step to clarify the legal and scientific evidence underpinning the spatial limits of their outer continental shelf rights in the Hatton-Rockall plateau. Nonetheless, the submissions themselves do not exert any direct influence on the determination of the maritime boundaries in the North Atlantic region and the Commission is unlikely to consider them due to the constraints imposed by the Commission's Rules of Procedure and Iceland's objections.

I-8.29   The only alternative options to move this issue forward are for all parties to give their consent that the Commission review their submissions (for example by making a joint submission to the Commission, thereby avoiding future objections and internalising

---

[74] Verbal Note of Iceland Addressed to the UN Secretary-General (27 May 2009, 11 April 2011) Commission on the Limits of the Continental Shelf (UN Division for Ocean Affairs and the Law of the Sea), available at www.un.org/depts/los/clcs_new/commission_submissions.htm (accessed 22 April 2017).

[75] Verbal Note of Denmark Addressed to the UN Secretary-General (2 December 2010), Commission on the Limits of the Continental Shelf (UN Division for Ocean Affairs and the Law of the Sea), available at www.un.org/depts/los/clcs_new/commission_submissions.htm (accessed 22 April 2017). See also Kwiatkowska, "Submissions to the UN Commission on the Limits of the Continental Shelf", 219, 243.

[76] Verbal Note of Iceland Addressed to the UN Secretary-General (11 April 2011), Commission on the Limits of the Continental Shelf (UN Division for Ocean Affairs and the Law of the Sea), available at www.un.org/depts/los/clcs_new/commission_submissions.htm (accessed 22 April 2017).

the dispute within the group of submitting states)[77] or referring the matter to third-party settlement according to the dispute resolution procedures provided for in Part XV of the 1982 LOS Convention. The availability of the latter option begs the question as to whether an international court or tribunal might be reluctant to exercise its jurisdiction to delimit the continental shelf beyond 200nm until such time as the Commission has laid out recommendations to each party on its submission. Given that the Commission is precluded from considering submissions where unresolved delimitation issues exist or where one of the claimants has withheld its consent (such as Iceland in the present case), if an international court or tribunal were to decline to determine the continental shelf limits beyond 200nm, the issue would remain indefinitely: the court or tribunal would have to wait for the Commission to act and the Commission would have to wait for the court or tribunal to act.[78] International judicial practice provides strong indications that the absence of the Commission's recommendations cannot be said to deny jurisdiction to the ICJ, ITLOS or other ad hoc tribunal to determine the issue of delimitation between outer continental shelf designations beyond 200nm.[79]

The legal status of Rockall and its potential role in delimitation

A further notable controversy in the North Atlantic and an important source of friction that is often associated with the overlapping continental shelf claims in the Hatton-Rockall plateau concerns the potential effect of the disputed islet of Rockall in extending the UK continental shelf and fishing rights in the North Atlantic. Rockall is a remote rocky feature of approximately 624 square metres and rises to just 19m above water at high tide.[80] It lies at a distance of 402nm from Iceland, 322nm from the Faroes and 226 miles from the nearest point of the Republic of Ireland. It is about 289nm from

I-8.30

---

[77] It is quite interesting to note that the first case in state practice whereby two or more states made a Joint Submission to the Commission was the 2006 Joint Submission by France, Ireland, Spain and the United Kingdom in respect of the Celtic Sea and the Bay of Biscay. For a discussion, see C G Lathrop, "Continental Shelf Delimitation Beyond 200 Nautical Miles: Approaches Taken by Coastal States Before the CLCS", in D A Colson and R W Smith (eds), *International Maritime Boundaries Vol VI* (2011), pp 4139–4160; and H Llewellyn, "The Commission on the Limits of the Continental Shelf: Joint Submission by France, Ireland, Spain, and the United Kingdom", 56(3) (2007) *International and Comparative Law Quarterly* 677.

[78] *Dispute Concerning Delimitation of the Maritime Boundary in the Bay of* Bengal, para 390.

[79] B Kwiatkowska, "Submissions to the UN Commission on the Limits of the Continental Shelf", 219, 265–266.

[80] Brown, "Rockall and the Limits of National Jurisdiction of the UK Part 1", 275, 289.

the nearest point of the Scottish mainland and 165nm west of the Scottish Island of St Kilda.[81] The principal legal question is whether this remote, barren and minuscular feature can generate its own continental shelf and EEZ (the latter encompasses fishing rights in addition to seabed mining rights) and, if so, whether it can be used as a basepoint for drawing the delimitation lines between the UK and its regional neighbours.

I-8.31   The potential effect of an island on maritime boundary delimitation presupposes that the island is capable of generating full maritime projections to be delimited vis-à-vis other states with opposite coasts.[82] Under the previous 1958 CS Convention, a state's continental shelf rights extended "to the seabed and subsoil of similar submarine areas adjacent to the coasts of islands".[83] Article 10 of the 1958 Convention on the Territorial Sea and the Contiguous Zone (hereinafter "1958 Territorial Sea Convention") defined islands as "naturally formed" areas of land, "surrounded by water", which remain constantly "above water at high tide".[84] The indispensable condition that an island must be naturally formed and above water at high tide excluded by definition such offshore features as low-tide elevations and petroleum installations which are dealt with in other parts of the 1958 CS and Territorial Sea Conventions.[85] Because Rockall is a naturally formed marine feature that is above water at high tide, the UK contended that Rockall is technically an "island", and that consequently it can (1) generate continental shelf/fishery zone in its own right and (2) be given "full weight" in delimitation vis-à-vis other mainland coasts (in other words, be used as a base point in the construction of an equidistance line between the island and foreign territorial units).[86]

I-8.32   The validity of such claim was explicitly contested by Ireland, Denmark and Iceland, especially on account of the new legal regime

---

[81] *Ibid*, 275, 289.

[82] M Norquist et al. (eds), *United Nations Convention on the Law of the Sea, 1982: A Commentary*, vol 3 (1995), p 325.

[83] 1958 CS Convention, Art 1.

[84] Convention on the Territorial Sea and the Contiguous Zone (adopted 29 April 1958, entered into force 10 September 1964) 516 UNTS 205.

[85] *Ibid*, Art 11.

[86] In a statement of 30 January 1975, the British Minister of State, referring to the question of islands, said: "The British view is that the existing position under international law should be preserved, namely ... that islands should count in full for both the generation and the delimitation of zones of jurisdiction." Also, in the words of the Lord Advocate in a statement dated 23 January 1975: "Under present international law, in the Government's view, the island of Rockall generates its own continental shelf, and we are content to rely on that basis for the exploitation of oil and other purposes", cited in E D Brown, "Rockall and the Limits of National Jurisdiction of the UK Part 2", 2(4) (1978) *Marine Policy* 275, at 292.

of islands codified in the 1982 LOS Convention. The latter, though it adopted unchanged the main definition of islands provided in Article 10 of the 1958 Territorial Sea Convention and maintained the principle that islands are to be treated in the same way as applicable to continental territories, also introduced an important qualification. Article 121(3) of the 1982 Convention provides that "rocks which cannot sustain human habitation or economic life of their own" cannot generate continental shelf or EEZ areas but only a territorial sea and contiguous zone. In other words, any insular feature that is a natural formation and permanently above water at high tide is *prima facie* entitled to full maritime projections, but if such feature is classified as "rock" its capacity to generate a continental shelf (and EEZ) must be ascertained by examining two further conditions set forth in Article 121(3), that is "human habitability" and "economic viability".[87]

Rockall is both uninhabited and uninhabitable.[88] It lacks fresh   I-8.33 water and cultivable soil and, given its inaccessibility, is unsuitable as a base for a manned lighthouse, although it does have a light beacon placed on the rock in 1972 by the UK.[89] Therefore, it would appear that the present definition of an island provided by the 1982 LOS Convention, taken cumulatively with recent judicial precedents on the matter,[90] removed any legal basis for the UK's effort to employ Rockall as a basepoint for extended maritime zones in the area in question. Following the vigorous protests of affected neighbouring states, the UK acknowledged the legal status of Rockall as a rock within the meaning of Article 121(3) of the 1982 Convention and "rolled back"

---

[87] For a detailed analysis on the legal status of islands in international law, see C Yiallourides, "Senkaku/Diaoyu: Are They Islands?", 50(1) (2016) *The International Lawyer* 101.

[88] Back in 1985, a British citizen spent 40 days on Rockall in an attempt to validate its status as an island. In 1997, Greenpeace activists broke that record by two days, protesting against exploration, and in 2014 a British explorer was able to survive on Rockall for 45 days using rain-harvesting equipment. See "Rockall – a Timeline", *The Guardian*, 28 May 2013, available at www.theguardian.com/uk/2013/may/28/rockall-timeline (accessed 26 April 2017).

[89] Brown, "Rockall and the Limits of National Jurisdiction of the UK Part 1", 289.

[90] In the South China Arbitration, the Tribunal found *inter alia* that the capacity of a feature to sustain human habitation or an economic life of its own must be addressed by having regard to the principal factors indicating the natural capacity of any given feature to sustain human habitation and economic life (as opposed to mere human survival). Such factors may include inter alia "the presence of water, food, and shelter in sufficient quantities to enable a group of persons to live on the feature for an indeterminate period of time", *South China Sea Arbitration (Republic of Philippines v The People's Republic of China)* (award of 12 July 2016), available at https://pca-cpa.org/wp-content/uploads/sites/175/2016/07/PH-CN-20160712-Award.pdf (hereinafter "South China Award") (accessed 10 April 2017).

from its claim based on Rockall as a base point.[91] The UK outer continental shelf title beyond 200nm is now solely based on the natural prolongation of Scottish land mass. The UK's 200nm fishery limit in the area was redefined as being to the more easterly island of St Kilda. St Kilda arguably qualifies as a full territory- generating island given its rich history of stable human habitation, though currently it is not permanently inhabited.[92] While the UK's outer continental shelf title in the Hatton-Rockall plateau remains unaffected, the implication of the UK's acceptance that Rockall cannot be used as a base point has been the reduction in the area of the British fishery zone by 164nm, that is an estimated 60,000 square miles.[93] Further, while the UK still claims a 12nm territorial sea around Rockall consistent with its status as a rock under Article 121(3) of the 1982 Convention, Ireland rejects Rockall's capacity to generate even a territorial sea.[94] Nonetheless, the issue of Rockall has in recent years receded in importance. It is to be noted that the UK and Ireland gave no effect to Rockall for the purpose of delimiting their continental shelf boundaries in 1988 and one may reasonably expect that the same limited effect would be accorded to Rockall in future delimitations between the UK, Denmark and Iceland in the North Atlantic region.[95]

CONCLUSION

I-8.34   The North Sea stands out as an area with the most comprehensive pattern of established maritime boundaries worldwide. Early continental shelf boundaries in the region were negotiated on the basis of the 1958 CS Convention which provided for equidistance-based solutions, while subsequent boundaries were heavily influenced by the 1969 decision of the ICJ in the *North Sea CS Cases* which rejected the mandatory character of equidistance and rather stressed the aim of achieving an equitable delimitation outcome in the light

---

[91] In a press communiqué of May 1985, the Danish Foreign Affairs Ministry stated that "Denmark and others plead that [this] uninhabited skerry cannot be granted the status of an island in the sense of international law", and that "the new UN Convention on the Law of the Sea clearly confirms this view", cited in Symmons "The Rockall Dispute Deepens", at 348.

[92] H Haswell-Smith, *The Scottish Islands* (2004), pp 314–326.

[93] According to Carleton, this represents the first time that any country worldwide has pulled back from a 200nm maritime zone claim on the basis that the island concerned cannot sustain human habitation or economic life of its own. See C Carleton, "The Development of Maritime Zones and Boundaries from 1964 to the Present", in C Schofield et al. (eds), *The Razor's Edge: International Boundaries and Political Geography* (2002), pp 127, 147.

[94] C Symmons, "Ireland and the Rockall Dispute: An Analysis of Recent Developments", 6(1) (1998) *Boundary and Security Bulletin* 78, at 80–81.

[95] Prescott and Schofield, *Maritime Political Boundaries of the World*, pp 372–373.

of the overall geographical and geological context of the areas to be delimited. In the Atlantic approaches to Northern Europe, regional neighbours remain in dispute regarding overlapping continental shelf rights in the Hatton-Rockall plateau despite several rounds of talks. Interest in oil and gas operations has undoubtedly driven boundary negotiations in these regions, but the legal and technical complexities surrounding the newly acknowledged definition of the continental shelf beyond 200nm in the 1982 LOS Convention have added to the controversy with respect to the establishment of boundaries extending beyond 200nm from the baselines of each state. Maritime boundary developments are further complicated by the presence of the sovereignty disputed islet of Rockall, thought the British "pull back" from using this feature as a generating point for its vast continental shelf and fisheries zone claims in the North Atlantic has meant that much of the discussion on the legal status and effect of Rockall is now of academic rather than practical significance.

Notwithstanding that the far-reaching British, Irish and Danish I-8.35 claims have recently translated into formal outer continental shelf submissions to the Commission on the Limits of the Continental Shelf, in view of Iceland's objections, such moves need to be re-appraised. The Commission is restricted by its Rules of Procedure from considering and qualifying submissions with respect to areas where delimitation issues remain unresolved unless all claimants have given their consent. In any event, the option of resorting to the dispute settlement procedures provided in Part XV of the 1982 Convention still remains. Not only do these procedures involve adjudicative processes but there is also an opportunity for states to refer their disputes to a conciliation commission whose decisions are not binding upon the parties, allowing them to negotiate a solution based on the conciliation commission's report.[96] As matters now stand, it appears that quadrilateral action is indispensable to any future settlement attempt, whether by negotiation or through a judicial process. Evidently this is so because parties to a bilateral delimitation agreement (such as Ireland and the UK) cannot create an opposable boundary with regard to the other two affected claimants. Neither would a trilateral agreement be binding upon the fourth claimant state. Likewise, any court or tribunal looking at the Hatton-Rockall plateau continental shelf dimension would inevitably have to assess all overlapping claims so as to ascertain the "effects, actual or prospective of any other continental shelf delimitation" in the same area.[97]

---

[96] 1982 LOS Convention, Art 298(1)(a).
[97] *North Sea CS Cases*, para 101(D)(3).

REGULATORY LAW

CHAPTER I-9

CURRENT PRACTICE AND EMERGING TRENDS IN REGULATING ONSHORE EXPLORATION AND PRODUCTION IN GREAT BRITAIN

Tina Hunter, Steven Latta and Greg Gordon

INTRODUCTION

There has been a long and sustained interest in the development of onshore petroleum resources, which commenced much earlier than the exploitation of offshore hydrocarbons. At least as far back as the First World War, as a consequence of Churchill's decision to convert the British Naval Fleet fuel source from coal to oil,[1] there has been onshore extraction of petroleum.[2] Initial attempts to develop a significant onshore oil and gas industry were held back by poor prospectivity and until the advent of the offshore oil and gas industry in the 1970s, the UK was heavily dependent upon imports.[3] By far the largest development onshore has been that of the conventional[4] Wytch Farm oil and gas field in Dorset, which was discovered in 1973 and has been producing since 1979 at much the same rate as the larger offshore fields.[5] In these conventional onshore fields,

I-9.01

---

[1] D Yergin, *The Prize* (1991), pp 11–13.

[2] UKOOG, "History" (2015), available at www.ukoog.org.uk/onshore-extraction/history (accessed 12 March 2017). Even before then, a range of oils, including paraffin, were produced from oil shale by a process of destructive distillation. See eg J McKay, *Scotland's First Oil Boom: The Scottish Shale-Oil Industry, 1851–1914* (2012).

[3] See further G Gordon, "British Hydrocarbon Policies and Legislation", in E Pereiea and H Bjorneyebe, *North Sea and Beyond* (2018).

[4] Here, conventional refers to the extraction of oil and gas (petroleum) from "conventional" reservoirs, which are usually sandstone or limestone reservoirs. This is in contrast to petroleum extraction from "unconventional" reservoirs, usually shale or coal, of which shale oil and shale gas are the most common and well known.

[5] Originally operated by BP, Wytch Farm was sold to Perenco in 2011. See www.perenco-

well stimulation techniques such as hydraulic fracturing (HF) have frequently taken place, with Cuadrilla reporting that it undertook HF at its Elswick field in 1993.[6]

I-9.02   Since 2010, onshore petroleum activities have focused on the exploration for unconventional shale oil and gas (for convenience, and on the basis that most current activity relates to gas, collectively called "shale gas" in this chapter). The Westminster Government has actively encouraged such activity, as shale gas is seen as an excellent source of gas to balance dwindling North Sea gas resources and to promote energy security.[7] The huge success of shale gas in the USA has undoubtedly served to stimulate this activity. To date, exploration for shale gas in Great Britain[8] has been minimal, largely in part due to the public concerns that have been raised over the use of HF in the development of such shale gas reservoirs. Shale gas activities initially caught the attention of the public in the spring of 2011, when Cuadrilla resources undertook HF of an exploration well at Preese Hall, near Blackpool. This triggered two mild seismic events that resulted in the UK Government imposing a temporary moratorium on shale gas activities while a review of the risks relating to HF was undertaken. Subsequent investigation and report by the Royal Society and the Royal Academy of Engineering (RS & RAE Review),[9] as well as a report commissioned by the Department of Energy and Climate Change (DECC), now BEIS, to review Cuadrilla's reports of the incident,[10] concluded that the activity was safe and that any uncertainties relating to shale gas extraction can be addressed through robust monitoring systems and a goal-based regulatory approach. The RS & RAE Review made a variety of recommendations designed to ensure that the current petroleum regulatory framework is fit for purpose in relation to onshore shale gas activities, with a particular focus on the integrity of shale gas wells.[11] In response to such recommendations, a review

---

uk.com/about-us/wytch-farm.html (accessed 22 March 2017). See also para 1-4.82.

[6] See Cuadrilla, *What is Fracking?* at https://cuadrillaresources.com/about-fracking/what-is-fracking (accessed 12 March 2017).

[7] See further the discussion at I-3.52.

[8] "Great Britain" is used in preference to "the United Kingdom" throughout this chapter as – as was noted at para I-4.03 – Northern Ireland enjoys significant legislative competence over onshore oil and gas, including its own long-standing (but to date little-used) licensing system. This chapter will not discuss Northern Irish law in detail, although some reference will be made to it in the discussion of the Scottish position at paras I-9.111 to I-9.126.

[9] The Royal Society and the Royal Academy of Engineering (RS & RAE), *Shale Gas Extraction in the UK: A Review of Hydraulic Fracturing* (2012).

[10] C Green, P Styles and B Baptie, *Preese Hall Shale Gas Fracturing: Review & Recommendations for Induced Seismic Mitigation* (2012).

[11] *Ibid.*

of the regulatory framework was undertaken by the DECC, implementing several changes to the existing regulatory framework which also incorporated applicable EU legislation relating to unconventional petroleum extraction, including those directives covering the environment, water and mining waste.

Aside from the review and regulatory changes in response to the RS & RAE Review, in September 2014 the UK Government established the Task Force on Shale Gas that was tasked with a number of functions: I-9.03

- to undertake an independent scrutiny of the benefits and concerns associated with shale gas exploration and production;
- to provide a trusted and transparent source of accurate online information for the public and policymakers;
- to host discussions and meetings on issues and research into shale gas exploration and production;
- to commission and peer-review research into questions related to shale gas exploration and production;
- to communicate the results of research and meetings to interested stakeholder groups; and
- to publish a final comprehensive report of its research and findings.[12]

The Task Force released a number of interim reports, with the final report released in December 2015.[13] As part of the process, all stakeholders were invited to submit evidence to the Task Force. It is expected that the final recommendations and conclusions will be utilised to inform government policy and regulation in the years ahead.   I-9.04

The Preese Hall incident, and its link to HF, triggered a significant volume of community concern and activism, thrusting onshore petroleum activities, both conventional and unconventional (shale gas), into the spotlight. To date, the extraction of petroleum in general, and the use of HF in particular, has cast a pall over the extraction of oil and gas onshore, and created a battleground between local authorities, the community and companies intending to undertake shale gas extraction. The epicentre of the battle is generally when a company seeks a permit to undertake drilling operations for a well (which may be an exploration, appraisal or production well).[14]   I-9.05

---

[12] Task Force on Shale Gas, *Our Mission* (2015), available at www.taskforceonshalegas. uk/mission (accessed 4 May 2017).

[13] Task Force on Shale Gas, *Final Conclusions and Recommendations* (2015), available at www.taskforceonshalegas.uk/reports (accessed 1 April 2017).

[14] At present, there are no production wells for shale gas in the UK. The Preese Hall well that was undergoing HF at the time of the seismicity was an appraisal well that was undertaking stimulation in order to determine the flow capabilities of the shale formation.

However, prior to an application for consent to drill, a company is required to gain a licence to undertake activities. Therefore, this chapter will discuss both the licensing framework that grants access to the petroleum and the consent system that grants a licence holder permission to drill a well for exploration, appraisal or production.

I-9.06   Following this introduction, this chapter will be divided into three main sections. The first will consider the UK-wide legal framework with respect to the granting of a licence and consent to drill in the UK, focusing on health and safety and environmental aspects of the consent system. The second will consider, in detail, the planning issues and restrictions related to the extraction of petroleum onshore, and, in particular, recent issues in shale gas consent applications. Finally, there will be a consideration of the legal, policy and political impact of the Scottish devolution process to this particular area.

REGULATION OF ONSHORE PETROLEUM ACTIVITIES IN THE UK – NATIONAL APPROACHES

I-9.07   In regulating onshore petroleum activities, the Oil and Gas Authority (OGA) undertakes the role as the lead regulator, the role of which is twofold. Firstly, the OGA grants licences to applicants through the onshore licensing rounds, thereby granting access to petroleum resources. Secondly, as a consenting authority, the OGA grants the licence holder permission to undertake drilling activities (whether for exploration, appraisal or production).

Access to petroleum – licensing

I-9.08   As companies seek to explore for, and ultimately produce, petroleum onshore, there is a legal requirement to hold a licence to search for petroleum, as stipulated under ss 3 and 4 of the Petroleum Act 1998. The OGA acts under powers delegated to the relevant Minister through the Petroleum Act 1998 (hereinafter "the Act") to license oil and gas production. The Act enables "[t]he Secretary of State, on behalf of Her Majesty, [to] grant to such persons as he thinks fit licences to search and bore for and get petroleum in a landward area",[15] and the state can grant a licence with terms and conditions that the Secretary of State thinks fit.[16] The Crown has the right to grant

---

To date, most applications for consent to drill have been for exploration wells, and not all have included HF in the well plan – see eg the consent to drill at Balcombe. See cuadrillaresources.com/site/balcombe (accessed 1 April 2017).

[15] Petroleum Act 1998, s 2(1).

[16] Petroleum Act 1998, ss 3(3) and 4, and Sch 1, which sets out the model clauses for petroleum licences.

such access to petroleum as a vestige of the common law doctrine of tenure, where the Crown owns all land in England and Wales. As such, no landowner has absolute ownership. Rather, the landowner holds the land "of" the Crown as tenants (and therefore the doctrine is known as the doctrine of tenure).[17] Although abolished in Scotland by Section 1 of the Abolition of Feudal Tenure etc. (Scotland) Act 2000, all Crown property rights in Scotland belong to Scotland as a sovereign territory, with the Crown's ownership of petroleum in Scotland administered by the UK Government.[18]

Under the Act, the OGA (as delegated by the Secretary of State) administers the licensing of oil and gas production. There is a distinction in the licensing regime between *seaward* and *landward* licences, with landward licences being the focus of this chapter. The landward licence is known as a Petroleum Exploration and Development Licence (PEDL), granted pursuant to Section 4 of the Act and The Petroleum Licensing (Exploration and Production) (Landward Areas) Regulations 2014. The award of a PEDL grants the licence holder permission to explore for all forms of petroleum, including shale gas, but excludes underground coal gasification and CO_2 sequestration.[19] The PEDL does not, however, confer the requisite permissions required for shale gas exploration and production, such as land access rights and planning consent. Licences are generally issued during designated licensing "rounds" that occur on a regular basis.[20] The 14th licensing round was the most recent round, with 93 new licences awarded over 159 blocks. The onshore blocks are 100km². As part of the licence application, the applicant must carry out a work programme, which outlines the minimum amount of work the applicant must undertake in order to retain the licence, and forms a key part of the licence.

I-9.09

When granted, all PEDLs run for three successive terms:

I-9.10

- Initial term: the licence may continue into a second term if the agreed work programme has been completed and a minimum amount of acreage has been relinquished. The initial term length under the 14th licensing round is five years.
- Second term: the term length here is five years. This may continue into a third term if a development plan has been

---

[17] C Harpum, S Bridge and M. Dixon, *Megarry and Wade: The Law of Real Property* (8th edn, 2012).

[18] Land Reform Review Group, *The Land of Scotland and the Common Good* (2014), p 56.

[19] DECC, *The Unconventional Hydrocarbon Resources of Britain's Onshore Basins – Shale Gas* (2013), p 2.

[20] The 13th, 12th, 11th and 10th licensing rounds took place in 2008, 2004, 2003 and 2001 respectively.

approved and all acreage outside the development has been relinquished.

- Third term: runs for an extended period to allow production. The model clause term length is 20 years.

I-9.11 The format and terms and conditions of the PEDL are contained in the model clauses for Landward Petroleum Exploration Licences, as set out in Sch 2 of the *Petroleum Licensing (Exploration and Production) (Landward Areas) Regulations 2014*. However, under s4 of the Act, the OGA is free to offer licences on other terms it sees fit.

Regulation of petroleum exploration and production: consent to drill

I-9.12 In order to undertake drilling operations to either explore for or produce petroleum, there is a requirement for a licence holder to gain consent to drill. In addition, the UK Government has developed *Guidance on fracking: developing shale gas in the UK*.[21]

I-9.13 When undertaking this permissioning role, the OGA undertakes a co-ordinating function,[22] granting consent to drilling only when all the following conditions have been satisfied:

  (a) the *Planning Authority* has granted permission and the relevant planning conditions have been discharged; [23]
  (b) the requisite environmental permits from the relevant *Environment Agency* are in place;
  (c) the *Health and Safety Executive* has had notice of, and is satisfied with, the well design;

---

[21] BEIS, *Guidance on fracking: developing shale gas in the UK* (2017), available at www.gov.uk/government/publications/about-shale-gas-and-hydraulic-fracturing-fracking/developing-shale-oil-and-gas-in-the-uk#regulation. In 2013, the UK Government also developed a Regulatory Roadmap for best practice in onshore oil and gas exploration for England, Wales, Scotland and Northern Ireland, available at www.gov.uk/government/publications/regulatory-roadmap-onshore-oil-and-gas-exploration-in-the-uk-regulation-and-best-practice (accessed 9 April 2017). The Roadmap serves as a point of reference and information for all stakeholders, but particularly for those who seek to understand the permitting and permissioning process for onshore oil and gas. It provides stakeholders with a number of documents, including an overview of onshore oil and gas exploration regulation and best practice for England, Wales, Scotland and Northern Ireland. However, these roadmaps are out of date and do not contain much of the legislative reform regarding hydraulic fracturing that has occurred since 2015.

[22] The Office of Unconventional Gas and Oil (OUGO) was established within DECC to formally coordinate the activities of the various agencies involved in onshore unconventional gas development.

[23] Planning requirements will be considered in paras I-9.29 to I-9.110.

(d) the well design has been assessed by an *independent competent well examiner*; and

(e) the *British Geological Survey* has been notified of the intent to drill.[24]

The regulation of all petroleum activities in the UK is undertaken I-9.14 using a goal-setting, risk-based approach that requires operators to identify the hazards specific to their activity, assess the risks and then specify the measures they will implement to reduce those risks to a level that is as low as reasonably practicable (ALARP).[25] In the United Kingdom, different agencies regulate the assessment of risk in relation to petroleum activities. Under the Act and associated regulations, there is a requirement for a risk assessment to be undertaken for safety (personal safety of the worker and the safety of the facility – known as process safety),[26] the integrity of the well itself and environmental impacts/harm associated with the activity. Each of these will be considered in turn.

Safety

The primary legislation protecting workers from harm when under- I-9.15 taking shale gas exploration and production is the Health and Safety at Work etc Act 1974 (HSWA) which works in conjunction with the primary petroleum legislation, the Petroleum Act 1998 (PA).[27]

The current regulatory framework for worker and process safety I-9.16 was implemented in 1992 in response to the 1988 Piper Alpha offshore platform disaster in which 167 men lost their lives. The subsequent Cullen Inquiry recommended a move from prescriptive regulation to the objective or goal-based regulatory framework.[28] A detailed overview and analysis of both the Piper Alpha disaster

---

[24] T Hunter and J Paterson, "Offshore Petroleum Facility Integrity in Australia and the United Kingdom: A Comparative Study of Two Countries Utilising the Safety Case Regime", 9(6) (2011) *OGEL* 1–27, at 7–8.

[25] *Ibid*, at 4.

[26] Personal safety can be defined as safety from hazards that give rise to incidents affecting individuals, such as slips, trips and falls. Process safety hazards involve the process of running a facility, and can involve leaks, spills, equipment malfunctions, corrosion, metal fatigue and overpressure. It is important to note that process safety is not just about equipment failure, but it may also include poor decision-making. Process safety incidents can have catastrophic consequences and result in multiple fatalities and injuries, as well as substantial economic, property and environmental damage. Examples of process safety failure include the Piper Alpha platform disaster in 1988 and the BP Texas City Refinery fire in 2005. J Baker et al., *The Report of the BP US Refineries Independent Safety Review Panel* (2007).

[27] The Secretary of State has the capacity to make further regulations under s 4 of the Petroleum Act 1998.

[28] Reform was implemented through the Offshore Safety Act 1992 (UK) and the Offshore Installations (Safety Case) Regulations 1992 (UK).

and the subsequent changes to the legal regime is found in Chapter I-9. The agency responsible for the regulation of health and safety persons, and the safety of facilities and processes on the facilities is the Health and Safety Executive (HSE). With regards to the risks arising from petroleum activities, the HSE also plays an important role in regulating other areas of petroleum activities such as wells (see paras I-9.19 to I-9.24) and pipelines.[29]

I-9.17   General safety of the onshore drilling site is regulated under the Borehole Sites and Operations Regulations 1995. These Regulations require the licence holder to produce, prior to operations, a health and safety document that demonstrates that risks have been assessed and that health and safety will be safeguarded. The regulations also specify the response plans the document must contain, including:

(a) an escape plan with a view to providing employees with adequate opportunities for leaving work places promptly and safely in the event of danger and an associated rescue plan with a view to providing assistance where necessary;

(b) a plan for the prevention of fire and explosions including in particular provisions for preventing blowouts and any uncontrolled escape of flammable gases and for detecting the presence of flammable atmospheres;

(c) a fire protection plan detailing the likely sources of fire and the precautions to be taken to protect against, detect and combat the outbreak and spread of fire; and

(d) in the case of a borehole site where hydrogen sulphide or other harmful gases are or may be present, a plan for the detection and control of such gases and for the protection of employees from them.[30]

I-9.18   It is important to note that the general safety legislative regime for onshore oil and gas activities should be read in conjunction with other regulations and guidance documents from the HSE, including those for well integrity.

Well integrity

I-9.19   The regulation of the integrity of onshore oil and gas wells is undertaken by the HSE. The primary legislation regulating safety of the site and well activities are the Borehole Site and Operation Regulations 1995 (BSOR) and the Offshore Installations and Wells (Design and Construction, etc) Regulations 1996 (DCR). To reduce the risk of well blowout, with the possible consequential harm

---

[29] Pipeline safety is regulated under the Pipelines Safety Regulations 1996.
[30] Borehole Sites and Operations Regulations 1995, Reg 7 (2).

to both persons and facilities, and to ensure that well integrity is maintained, the HSE undertakes the scrutiny and approval of well design and construction. In addition, Guidance documents explain how the HSE implements these regulations.[31]

When a well is submitted for approval, the HSE considers the **I-9.20** well design from the point of view of safety. The overall approach is focused on risk assessment and on ensuring that the operator is managing identified risks on an ongoing basis. The HSE, in conjunction with other parties such as independent well verifiers, focuses on assuring well casing quality and integrity, maintaining well integrity, cement bond logging and monitoring the integrity of the cement bonding, and testing the well casings. As part of the UK oil and gas health and safety regime, the onus is on the operator to identify hazards, assess risk and reduce those risks.[32] Wells are inspected by experienced inspectors throughout construction and operation to ensure that operators fulfil their regulatory obligations. Such obligations include well notifications relating to initial design, weekly operations reports while construction is ongoing, and meetings and inspections during operations.

A Guide to the Well Aspects of the Offshore Installations and **I-9.21** *Wells (Design and Construction, etc) Regulations 1996* sets out, in detail, the requirements relating to the design, construction and operation of a well, requiring the operator to assure the regulator at every step that safety and well integrity is assured. Importantly, this guidance explicitly recognises the need to establish public confidence.

The importance of well integrity has also been recognised by the **I-9.22** industry peak body, the UK Onshore Operators Group (UKOOG), who developed guidelines for shale gas wells.[33] These comprehensive guidelines, written by a working group comprising the DECC, HSE and the applicable environment agency, address the following areas:

- safety and environmental management;
- disclosure and transparency;
- regulatory compliance;
- well design and construction;
- well integrity;
- fracturing and flow-back operations;

---

[31] HSE, *A Guide to the Well Aspects of the Offshore Installations and Wells (Design and Construction, etc) Regulations 1996 – Guidance on Regulations*, L84 (2nd edn, 2008); HSE, *Shale Gas and Oil Guidance for Planners* (2015).

[32] As set out in the Cullen Inquiry and adopted as part of the offshore safety regime. Refer to Chapter I-10 for an analysis and discussion of this regime.

[33] UKOOG, "UK Onshore Shale Gas Well Guidelines: Exploration and Appraisal Phase" 3 (March 2015), available at www.ukoog.org.uk/images/ukoog/pdfs/ShaleGasWellGuidelinesIssue3.pdf (accessed 1 March 2017).

- environmental management;
- fracturing fluids and water management; and
- fugitive emissions management.

I-9.23   The UKOOG guidelines were to be applicable to wells designed and constructed for well stimulation, including HF,[34] and are recognised as good industry practice, referencing relevant legislation standards and practices. The guidelines were developed to assure operators who adopt the guidelines that they have complied with all relevant regulations and established standards for shale gas well design and operations, including all matters relating to fracturing operations, fracturing fluids and flow-back fluids.[35]

I-9.24   The HSE works closely with the relevant environment agency and BEIS, sharing relevant information relating to well integrity and worker/workplace safety and integrity, to ensure that there are no gaps between the safety, environmental protection and planning considerations, and that all material concerns are addressed. Given the combined safety and environmental considerations, the HSE works in conjunction with environmental regulators as appropriate on the basis of a formal agreement, and undertakes monitoring throughout well construction and operation.

Environmental impacts and harm arising from the activity

I-9.25   Whilst the well site and the design, construction and operation of wells are regulated by the HSE, environmental permits are also required in order for onshore well operations to be undertaken. Such environmental regulation is undertaken by the relevant environment agency. In the UK there is no single environment agency. Rather, the regulation of environmental impacts from activities (including oil and gas activities) is regulated by the relevant environment agency. The Environment Agency (EA) in England and Wales and the Scottish Environmental Protection Agency (SEPA) in Scotland regulate the assessment of risk related to the environment. The Local Planning Authority (LPA), which is generally responsible for the granting or otherwise of planning permission, considers some aspects of environmental impact such as noise, traffic and the like, and will be considered at paras I-9.29 to I-9.110. Environmental regulation of Scotland will be considered at paras I-9.111 to I-9.126.

I-9.26   The HSE and the EA work in partnership to assess environmental hazards and risks from the manufacture, use and disposal of industrial chemicals, and waste materials. The HSE retains the role as the competent authority regarding such chemicals, with the

---

[34] *Ibid*, 5.
[35] *Ibid*.

EA is the enforcing authority for REACH (Registration, Evaluation, Authorisation and Restriction of Chemicals) Regulations, which regulate the chemicals used in the process of hydraulic fracturing. To effectively regulate onshore well operations, the EA and HSE have a Memorandum of Understanding in place that details the day-to-day arrangements for regulating petroleum activities. Under the Environmental Permitting Regulations 2010, the Environment Agency has responsibility for:

(a) the management of water resources;
(b) protecting communities from the risk of flooding;
(c) issuing permits for certain industrial, farming, waste management, surface water; and
(d) groundwater discharge activities.[36]

In response to the public concern regarding the environmental impacts of onshore petroleum activities, in particular the impact of HF (also known as fracking) on water, the Environmental Audit Committee published its report entitled *Environmental Risks of Fracking* in January 2015, proposing an outright ban on fracking in the United Kingdom.[37] The proposed ban on fracking was rejected by the Government, who instead opted for a number of restrictions on HF. These restrictions were introduced by Section 50 of the Infrastructure Act 2015 and defined in the Onshore Hydraulic Fracturing (Protected Areas) Regulations 2016 which, *inter alia*, include the following:

I-9.27

(a) The requirement that the level of methane in groundwater be monitored for a period of 12 months before the start of fracking operations.
(b) A ban from HF within:
   (i) protected groundwater source areas (any land at a depth of less than 1,200m or within 50m of a surface water supply);[38] and
   (ii) within other protected areas (designed to protect National Parks, the Norfolk and Suffolk Broads, other defined "areas of outstanding natural beauty" and World Heritage sites).[39]

---

[36] HSE and Environment Agency, *Working Together to Regulate Unconventional Oil and Gas Developments* (hereinafter "*Working Together* agreement"), 2012.

[37] The report can be found at www.publications.parliament.uk/pa/cm201415/cmselect/cmenvaud/856/85602.htm (accessed 1 March 2017).

[38] Protected groundwater resources are defined in Reg 2 of the Onshore Hydraulic Fracturing (Protected Areas) Regulations 2016.

[39] Other protected areas are defined in Reg 3 of the Onshore Hydraulic Fracturing (Protected Areas) Regulations 2016.

I-9.28   Section 50 of the Infrastructure Act 2015 inserts Section 4A
         (Onshore Hydraulic Fracturing: Safeguards) and supplementary
         provisions into the Petroleum Act 1998 but, as at April 2017, is yet
         to come into force. Consequently, the Onshore Hydraulic Fracturing
         (Protected Areas) Regulations 2016, which enter into force when
         Section 4A (3) of the Petroleum Act 1998 enters into force, are not
         yet in force.

THE ROLE OF PLANNING IN REGULATING ONSHORE PETROLEUM ACTIVITIES

I-9.29   To understand town planning and the laws and policies which regulate
         this area, it is useful to appreciate the context in which the concept
         of town planning has evolved. The history of modern planning in the
         UK, relating to *spatial* planning or *land use*, and generally referred
         to as *town planning*, as a Governmental function dates back almost
         200 years and emerged as a consequence of the impact of the indus-
         trial revolution at the turn of the nineteenth century.[40] The term
         "town planning" itself is one which didn't enter into the law in the
         UK until 1909 with the Housing & Town Planning Act 1909.

I-9.30   The twentieth-century development of British town planning has
         a background which reflects many different dimensions and perspec-
         tives. Ostensibly for community change, town planning policy and
         legislation quickly became an important weapon in the twentieth
         century armoury of state powers. Town planning has now come
         to embrace towns, cities, rural areas and countryside. Initially it
         was an activity which, broadly, sought to regulate the use of land
         by establishing control over the process of the development of
         the physical environment. Town planning has changed over time,
         however, and modern town planning is a comprehensive exercise of
         analysis and prescription covering the regulation of environmental
         and community affairs far beyond its early remit.[41]

I-9.31   The Town and Country Planning Act 1947 saw the creation of
         Local Planning Authorities (LPAs) (being predominantly the local
         authority or National Park Authority that is empowered by law to
         exercise statutory town planning functions for a particular area of
         the United Kingdom) and the bestowal (upon these LPAs) of wide
         powers to deal with development control, and the requirement to
         produce Local Development Plans to set out detailed policies and
         specific proposals for the development and use of land as well as
         making land subject to the control of the planning regime.

---

[40] P Hall, *Urban & Regional Planning* (5th edn, 2010).
[41] G Cherry, *The Politics of Town Planning* (1982).

Currently, the key pieces of planning legislation include the I-9.32 following:

In England

- The Town and Country Planning Act 1990 which consoli- I-9.33 dated previous town and country planning legislation and sets out how development is regulated.
- The Planning and Compulsory Purchase Act 2004 which made changes to development control, compulsory purchase and application of the Planning Acts to Crown Land.
- The Planning Act 2008 which sets out the framework for the planning process for nationally significant infrastructure projects and provided for the community infrastructure levy.
- The Localism Act 2011 which provides the legal framework for the neighbourhood planning powers and the duty to co-operate with neighbouring authorities.

In Scotland

In Scotland there are two key pieces of legislation that govern the I-9.34 operation of the Scottish planning system:

- The Town and Country Planning (Scotland) Act 1997 is the basis for the planning system and sets out the roles of the Scottish Ministers and local authorities with regard to development plans, development management and enforcement. This Act was substantially amended by the Planning etc. (Scotland) Act 2006.
- The Planning (Listed Buildings and Conservation Areas) (Scotland) Act 1997 is mainly concerned with the designation and protection of listed buildings and conservation areas. This Act was amended by the Historic Environment (Amendment) Scotland Act 2011 and the Historic Environment (Scotland) Act 2014.

In the context of fracking-related activities, the UK-wide Infrastructure I-9.35 Act 2015 also came into force in February 2015 and impacted directly on the planning system with measures to cut delays and fast-track planning applications, streamlining aspects of planning procedure and the facilitation of energy exploitation.

Development control

Development control/development management (the terms are I-9.36 used interchangeably) is the specific role of the UK's system of town and country planning which regulates land use and development. Development Management is given specific legal consideration in Part 3 of the 1997 Act in Scotland and Control over Development in Part 3 of the Town and Country Planning Act 1990 for England.

I-9.37   Section 28 of the 1997 Act and Section 57 of the 1990 Act state that "planning permission is required for the carrying out of any development of land".

I-9.38   The definition of "Development' lies at the heart of town planning law and is statutorily defined in both the 1990 (Section 55) and 1997 (Section 26) Acts as "the carrying out of building, engineering, mining or other operations in, on, over or under land, or the making of any material change in the use of any buildings or other land".[42]

I-9.39   Planning law in the UK is, therefore, a long-established and mature area of the law. The nature of the operations involved in onshore shale gas operations dictate that planning permission is one of the approvals required before any activity may start on site.

I-9.40   The planning systems in all four UK countries is "plan-led", in that planning policy at both a national and local level is set out in formal development plans which describe the favoured land use and the types of development which typically should (and shouldn't) get approved. The four planning systems are fundamentally similar to each other in structure; however, there has been a greater divergence between them in recent years.

I-9.41   In Scotland, Local Development Plans cover the whole of the country and provide the vision for how communities will grow and develop in the future. Each Planning Authority (ie Local Authority or National Park Authority) is required to publish and then update Local Development Plan(s) covering their area at least once every five years, which are subject to detailed public consultation.

I-9.42   In England, the National Planning Policy Framework (NPPF) directs that each LPA should produce a Local Plan for its area (there is no *legal* requirement to do so, however).

Planning conditions

I-9.43 Planning conditions are a feature of the planning regime in both Scotland and England. The power to impose conditions upon the grant of permission enables the LPA to go beyond the adjudicative "yes/no" role by using conditions to control detailed aspects of

---

[42] Planning (Listed Buildings and Conservation Areas) (Scotland) Act 1997, s 26(1).

developments. In determining an application, the power to impose conditions in Scotland is conferred by the 1997 Act which empowers a LP "to such conditions as they think fit".[43] As a matter of policy, conditions should only be imposed where they are:

(a) necessary;
(b) relevant to planning;
(c) relevant to the development to be permitted;
(d) enforceable;
(e) precise; and
(f) reasonable in all other respects.[44]

In England the Town and Country Planning Act 1990 is the **I-9.44** authority which enables the LPA to impose "such conditions as they think fit"[45] and this must be done in line with the NPPF, which directs that planning conditions should only be imposed where they meet the "six tests".[46]

The appeals process

In terms of the appeals process, in England the applicant has the right **I-9.45** of appeal to the Secretary of State. In practice, the normal procedure is for the appeal to be decided by a planning inspector in the name of the Secretary of State, after either considering written representations, holding an informal hearing or holding a full inquiry. The Secretary of State also has powers to "recover" an appeal and to take the decision himself.

In Scotland, if a planning decision was taken by the Planning **I-9.46** Committee (as opposed to being delegated to planning officers) then any appeal against that decision will be made to Scottish Ministers. Planning appeals made to Scottish Ministers are considered by a Reporter (a Senior Planning Officer), appointed by the Directorate for Planning and Environmental Appeals (DPEA). In most instances the appeal decision is made by the Reporter on behalf of the Scottish Ministers. However, in a small number of cases the Reporter does not issue the decision, but instead submits a report with a recommendation to the Scottish Ministers, who make the final decision.

---

[43] 1997 Act, s 37(1).

[44] See www.gov.scot/Publications/1998/02/circular-4-1998/circular-4-1998-a (accessed 11 April 2017).

[45] Town and Country Planning Act 1990, s 70(1)(a).

[46] See www.gov.uk/guidance/national-planning-policy-framework/decision-taking#para206 (accessed 11 April 2017).

The planning application process

I-9.47   The planning application process is often formally commenced with informal pre-application discussions between the applicant (or his agent) and a representative, typically a Chartered Town Planner, of the relevant LPA.

I-9.48   Pre-application discussions are generally considered to be good practice in that they assist the applicant in understanding the specific requirements of the LPA and, consequently, reduce the risk of submitting invalid or inadequate applications.

The development plan

I-9.49   The role of the Local Development Plan, which sets out the development strategy for an area, is something which is fundamentally important in relation to the planning decision-making process; s70(2) of the 1990 Act and s37(2) of the 1997 Act states that "in dealing with such an application the authority shall have regard to the provisions of the development plan". Pre-application discussions are an early opportunity for an applicant to both demonstrate his proposed application is in harmony with the development plan and present the proposed development in a positive light more generally.

The LPA considers the application

I-9.50   When the applicant is at the stage where he is ready to submit the planning application, it should be submitted together with the necessary supporting information and requisite fee.[47]

I-9.51   Planning applications will be accompanied by information such as site and locational plans and any other supporting information which is required for a given application and which will be dictated in large part by the nature and location of the development. This information may include such reports as traffic impact assessments, environmental and ecological statements, geophysical surveys and noise impact studies.

I-9.52   When a planning application has been received by the LPA, it should be validated as soon as is reasonably practicable (as an example, Scottish Planning Series Circular 4 2009: Development Management Procedures states that: "The administrative checking of applications in this regard should be carried out as soon as possible but certainly within five working days of receiving the application"). The LPA should then start the determination process.

I-9.53   The LPA will consult with various statutory and non-statutory consultees. The identity of consultees will be determined by the

---

[47] An indication of current fees is available at https://ecab.planningportal.co.uk/uploads/english_application_fees.pdf (accessed 11 April 2017).

nature and location of the proposed development but, by way of illustration, in Scotland[48] statutory consultees can typically include:

- Transport Scotland
- Scottish Water
- Scottish Environment Protection Agency (SEPA)
- Scottish Natural Heritage (SNH)
- Historic Scotland

Unless the development is of minor significance (where the decision making will be delegated to planning officers), the application is decided by the Planning Committee which is made up of local councillors (elected members of the Local Authority).   I-9.54

A complicating issue with regard to fracking-related applications is that most development plans typically do not address unconventional gas exploration, due to its relative novelty.[49] In any event, the LPA must take into account all material considerations when determining a planning application. These considerations provide the grounds for the reasons both for objecting to proposed developments and for the granting of/refusing to grant planning consent.   I-9.55

Development plans/material considerations
In terms of the status of the Development Plan, Scottish planning law dictates that "in making any determination under the planning acts, regard is to be had to the development plan and the determination shall be made in accordance with the development plan *unless material considerations indicate otherwise*".[50]   I-9.56

National Planning Policy Framework

Similarly, the NPPF sets out the Government's planning policies for England (and Wales) and how these are expected to be applied. Planning law dictates that planning applications "must be determined in accordance with the development plan, *unless material considerations indicate otherwise*".[51]   I-9.57

The issue of material considerations is one of the most debated in planning law, one in which there is no settled legal definition and   I-9.58

---

[48] Guidance on statutory consultees in England is available at www.gov.uk/guidance/consultation-and-pre-decision-matters#Statutory-consultees-on-applications (accessed 11 April 2017).
[49] R Turney, "Fracking and the National Planning Policy Framework", Landmark Chambers, 17 October 2013.
[50] Town and Country Planning Act 1997, s 25.
[51] Town and Country Planning Act 1990, s 70(2).

one which is worthy of specific consideration in the context of shale gas operations.

I-9.59   *Stringer* v *Minister of Health for Housing and Local Government*[52] provides, as a starting point, that any consideration which relates to the use and development of land is capable of being regarded as planning considerations. "Whether a particular consideration falling within that broad class is material in any given case will depend on the circumstances."[53]

I-9.60   The Royal Town Planning Institute also provides guidance on "Material Planning Considerations" and sets out some more, specific, examples of what might merit the "material" classification.[54]

I-9.61   Amongst the examples set out, those which have a particular resonance with fracking operations include:

- Government and Planning Inspectorate requirements – circulars, orders, statutory instruments, guidance etc;
- Principles of case law held through the Courts;
- Overshadowing/loss of outlook to the detriment of residential amenity;
- Highway issues: traffic generation, vehicular access, highway safety;
- Noise or disturbance resulting from use, including proposed hours of operation;
- Capacity of physical infrastructure, eg in the public drainage or water systems;
- Adverse impact on nature conservation interests and biodiversity opportunities;
- Incompatible or unacceptable uses.

I-9.62   The extent to which an issue may constitute a material consideration can be very wide. In general, however, the broad view taken is that planning applications are concerned with land use in the public interest, so that the protection of purely private interests such as the impact of a development on the value of a neighbouring property or loss of private rights to light should not be considered a "material" consideration.[55]

I-9.63   Since *Stringer*, the courts have not been entirely silent on the matter of defining material considerations. Lord Scarman, in the 1985 case

---

[52] *Stringer* v *Minister of Health for Housing and Local Government* [1970] 1 W.L.R. 1281.
[53] *Ibid*, per Cooke J, at 1294.
[54] See www.rtpI-org.uk/media/686895/Material-Planning-Considerations.pdf (accessed 4 May 2017).
[55] Planning Practice Guidance, "How must decisions on applications for planning permission be made?", available at www.gov.uk/guidance/determining-a-planning-application (accessed 4 May 2017).

of *Westminster City Council* v *Great Portland Estates plc*,[56] defined a
material consideration by whether it served a planning purpose and
whether that planning purpose was related to the *use and character*
of the land. He also, significantly, added:

> "Personal circumstances of an occupier, personal hardship, the diffi-
> culties of businesses which are of value to the character of a community
> are not to be ignored in the administration of planning control, and
> it would be inhuman pedantry to exclude from the control of our
> environment the human factor. The human factor is always present,
> of course, indirectly as a background to the consideration of the
> character of the land use. It can, however, and sometimes should, be
> given direct effect as an exceptional or special circumstance ... such
> circumstances, when they arise, fall to be considered not as a general
> rule but as exceptions to a general rule to be met in special cases. If a
> planning authority is to give effect to them, a specific case has to be
> made and the planning authority must give reasons for accepting it".[57]

He went on to note that: "If a planning authority is to give effect   I-9.64
to them [the special cases], a specific case has to be made and the
planning authority must give reasons for accepting it".[58]

In so doing, this would appear to have opened the door for an   I-9.65
LPA to exercise discretion as to what constitutes a "material consideration" in any given case.

In recent years, concerns such as promoting social objectives were   I-9.66
deemed to be material considerations in the planning context. In *R.
(on the application of Copeland)* v *Tower Hamlets LBC*[59] a school's
attempt to encourage healthy eating by its pupils was considered.
Here, the proximity of a fast-food takeaway to a school was deemed
capable of constituting a material planning consideration.

In practice, the LPA has a fairly broad licence to: I-9.67

(a) interpret its own development plan policies;
(b) give weight to certain policies over others; and
(c) decide whether particular material considerations outweigh
    the provisions of the development plan.

This effectively provides significant scope for LPAs to refuse appli-   I-9.68
cations on the basis of a broad range of impacts and concerns,
including temporary increases in the levels of traffic or visual impacts
(which may be no greater than for other local developments which
are granted consent without controversy).[60]

---

[56] *Westminster City Council* v *Great Portland Estates plc* [1985] A.C. 661.
[57] *Ibid*, per Lord Scarman, at 670.
[58] *Ibid*.
[59] *R. (on the application of Copeland)* v *Tower Hamlets LBC* [2010] L.L.R. 654.
[60] C Howard, "Fit to Frack" (2014), *JPL* 13, 43.

I-9.69   Significantly, the *weight* attached to material considerations in reaching a decision is also a matter of judgement for the Planning Authority. Ultimately, the decision makers will have to consider the advantages and disadvantages of any given proposed development. *Tesco Stores* v *Secretary of State*[61] is the leading authority in this area:

> "The law has always made a clear distinction between the question of whether something is a material consideration and the weight which it should be given. The former is a question of law and the latter is a question of planning judgement, which is entirely a matter for the planning authority. Provided that the planning authority has regard to all material considerations it is at liberty to give them whatever weight the planning authority thinks fit (or no weight at all). The fact that the law regards something as a material consideration therefore involves no view about the part, if any, which it should play in the decision making process".[62]

The issue of "public concern" and planning applications

I-9.70   In the UK context, fracking operations have been characterised by their controversy and public concern. There has been specific consideration around the issue of whether public concern, by way of opposition to a development, constitutes a material consideration in its own right.

I-9.71   On the face of it, there is an obvious connection between land use and public concern. As one commentator put it: "It is the public in part, or as a whole, who are affected by the consequences of any development, whether those consequences are good, bad or neutral".[63]

I-9.72   This was also the approach taken by the Court of Appeal in *West Midlands Probation Committee* v *Secretary of State for the Environment*.[64] Here, the Planning Inspector found that the apprehensiveness of the residents living in the vicinity of the proposed development "would be capable of being a material consideration provided, of course, that there were reasonable grounds for entertaining them".[65]

I-9.73   At the Court of Appeal, it was submitted that apprehension and fear are not material planning considerations since they do not relate to the character of the use of the land. However, it was noted that "justified public concern in the locality about emanations from

---

[61] *Tesco Stores* v *Secretary of State for Environment* [1995] 1 W.L.R. 759.

[62] *Ibid*, per Lord Hoffman, at 780.

[63] A Piatt, "Public concern – a material consideration?" (1997), *JPL* 397, at 399 (hereinafter "Piatt, 'Public concern'").

[64] *West Midlands Probation Committee* v *Secretary of State for the Environment* (1998) 76 P. & C.R. 589.

[65] *Ibid*, at 591.

land as a result of its proposed development may be a material consideration".[66]

This position, in turn, appeared to be aligned with the earlier authority in the Court of Appeal case of *Gateshead MBC v Secretary of State for the Environment*.[67] It was stated that: I-9.74

> "Public concern is, of course, and must be recognised by the Secretary of State to be, a material consideration for him to take into account. But if in the end that public concern is not justified, it cannot be conclusive. If it were, no industrial development – indeed very little development of any kind – would ever be permitted."[68]

The "very substantial public opposition to the proposal" was one of the reasons given by the Planning Authority for the refusal of planning permission for the construction of a chemical waste treatment plant in the subsequent case of *Newport County Borough Council v Secretary of State for Wales and Browning and Ferris Environmental Services Ltd*.[69] I-9.75

In *Newport*, Aldous L.J. noted, however, in contrast to the previous decisions that: I-9.76

> "A perceived fear by the public can in appropriate (perhaps rare) occasions be a reason for refusing planning permission, whether or not that has caused local opposition. It follows that the Circular contemplates that planning reasons such as public perception can (again, perhaps rarely) warrant refusal, even though the factual basis for that fear has no scientific or logical reason."[70]

The waters are muddied further in this area when we consider that a decision which turns upon the weight attached to public concern could be open to legal challenge if the decision maker has not demonstrated that they have an adequate understanding of the factors influencing the level of opposition to (*or support for*) a particular development proposal.[71] I-9.77

There would appear scope to argue that, for planning decision makers, an understanding of the factors which influence a given level of public concern (and the consequent weight assigned to these issues) may be considered to be analogous to an understanding of a given planning policy; and it is established that: "it is essential that the policy is properly understood by the decision-maker. If the body I-9.78

---

[66] *Ibid*, per Pill L.J., at 597.
[67] *Gateshead MBC v Secretary of State for the Environment* [1995] Env. L.R. 37.
[68] *Ibid*, per Glidewell L.J., at 49.
[69] *Newport County Borough Council v Secretary of State for Wales and Browning and Ferris Environmental Services Ltd* [1998] Env. L.R. 174.
[70] *Ibid*, at 185.
[71] Piatt, "Public concern", 397.

making the decision fails to properly understand the policy, then the decision will be as defective as it would be if no regard had been paid to the policy."[72]

I-9.79   Post-*Newport*, we were left with the situation of having two conflicting Court of Appeal decisions, *Gateshead* and *Newport*, the former representing a rationalist position and the latter a populist one.

I-9.80   The law since does not appear to be entirely settled on this matter. The 2003 High Court Case of *Phillips* v *First Secretary of State*[73] was not predominantly focused on public concern issues. However, there was an important public concern element to it. In his judgment, Richards J. noted that:

> "if there were two alternative sites each of which was otherwise acceptable in environmental terms, it would be open to a decision-maker to refuse approval for one of those sites if the location of a mast on that site would give rise to substantially greater public concerns than its location on the alternative site".[74]

I-9.81   This would appear to be in line with the 2002 decision in *Trevett* v *Secretary of State for Transport, Local Government and the Regions*,[75] where, in relation to public concerns, it was held that they should be given such weight as might be appropriate in the particular circumstances of the case. This is somewhere between the strict rationalist approach which entirely discounts public concerns that are not objectively justified, and a strict populist approach: in as much as *Trevett* appears to be stating that local authorities are entitled to have regard to the extent to which the perceived health risks (the public concern) are objectively justified, in deciding what weight to attribute to them.

I-9.82   The issues of what constitutes a material consideration have provided much debate and little consensus, with the consequence being that there is considerable scope for the application of discretion by LPAs when having such matters in their contemplation.

Town planning; as much a political function as a legal one?

I-9.83   Town planning in the UK is undertaken by a number of different actors. There is a planning profession, the Royal Town Planning

---

[72] E. C. Gransden & Co. Ltd. and Another v Secretary of State for the Environment and Another (1987) 54 P. & C.R. 361, per Woolf J, at 94.

[73] Phillips v First Secretary of State [2004] J.P.L. 613.

[74] Ibid, at 614.

[75] Trevett v Secretary of State for Transport, Local Government and the Region [2002] EWHC 2696 (Admin).

Institute (RTPI), and various other professions and professionals with varying degrees of involvement depending on the nature of the planning application under consideration. These professions have been trained with their own set of ideologies and perspectives and they have their own objectives and ideas of what they consider desirable or undesirable.[76] In the context of a commercial property development, a property developer's perspective on a proposed development will often be at odds with an environmentalist's whose view, in turn, may not be aligned with that of a conservationist.

As discussed, in the British town planning system, it is local politi- **I-9.84** cians, sitting as members of the local Planning Committee, who ultimately decide on the outcome of any planning applications.

In the context of LPAs in both Scotland and England, politicians **I-9.85** and professional planners perpetuate the notion that planning can be, and often is, distinct from politics.[77] Certainly, the planning system aspires to be, and aspires to be perceived as, a quasi-judicial process where applications are considered on their merits in line with law and policy. Furthermore, in the guidance to the local politicians that make up LPAs: "planning decisions cannot be made on a party political basis or in response to lobbying".[78]

Inevitably, though, democratically elected decision makers will **I-9.86** hold political affiliations and personal interests which will, at least, have the potential to influence their decisions. A study by the British Chamber of Commerce in 2011 found that seven out of ten British companies believe planning decisions are taken on political grounds rather than on the quality of the application.[79] This is something which also does not appear to be lost on the media.[80]

Town planning can also be described as a political activity in **I-9.87** as much as it is a function of government and the "players" who operate the system. Those who influence decisions and policies within it are motivated, at least to some extent, by their own values and actions or by their self-serving interests.[81]

---

[76] Cherry, *The Politics of Town Planning*.

[77] A Blowers, *The Limits of Power: The Politics of Local Planning Policy* (1980).

[78] Local Government Association and Planning Advisory Service, Probity in Planning for Councillors and Officers, April 2013, available at www.local.gov.uk/sites/default/files/documents/probity-planning-councill-d92.pdf (accessed 4 May 2017).

[79] British Chambers of Commerce, "The Planning System" (2011), available at www.britishchambers.org.uk/assets/downloads/policy_reports_2012/12-03-02%20Planning%20Factsheet%20-%20Consistency.pdf (accessed 4 May 2017).

[80] C Hope, "Planning system driven by 'increasing political pressure to help builders'", *The Telegraph* (25 March 2013), available at www.telegraph.co.uk/news/earth/greenpolitics/planning/9953217/Planning-system-driven-by-increasing-political-pressure-to-help-builders.html (accessed 4 May 2017).

[81] Cherry, *The Politics of Town Planning*.

I-9.88   Planning is also a political activity in the sense that it is an alloc-
ative activity: in granting planning consent it, in effect, distributes
rewards, resources and opportunities throughout the community
and, consequently, there will be both winners and losers resulting
from planning decisions. It is not difficult to imagine that, in arriving
at planning decisions, the "influencers" will again be motivated by
their own values in terms of who should be winning and losing in
this context.

I-9.89   The decision-making process in town planning is a complex web
of interaction between institutional frameworks of government and
external pressures from the community and other stakeholders.
The various interfaces between the public and the planning system
through pressure groups, consultees, political groups and the climate
of public opinion creates significant pressures on LPAs,[82] which they
rarely fail to be responsive to.[83] While the duty of the LPA is to
decide each case on its planning merits, the political risk associated
with making a decision around a particularly controversial proposal
is obvious, particularly given the discretion afforded to the LPA.

I-9.90   Recent "fracking" planning applications in the UK would suggest
that this wide discretion is being exercised.

Lancashire County Council: Planning Application LCC/2014/0096

I-9.91   Dated 29 March 2014. This was a planning application submitted
by Cuadrilla Bowland Limited and related to the "construction and
operation of a site for drilling up to four exploratory wells, hydraulic
fracturing of the wells ... and a connection to the gas grid network
and associated infrastructure to land to the north of Preston New
Road, Little Plumpton".

I-9.92   When the application was originally submitted, Friends Of the
Earth, on behalf of Preston New Road Action Group, initially
expressed concerns regarding the consultation period of 21 days for
consideration of the Environmental Statement accompanying the
planning application. Lancashire County Council took account of
these concerns and extended the formal consultation period to 12
weeks.

I-9.93   The Committee, sitting on 14 September 2014, was informed
that the planning application had generated a considerable amount
of interest with several thousand representations having been
received objecting to the proposal, primarily in respect of visual,
traffic, habitats, noise, pollution and seismic activity. It was
therefore proposed that the Committee visit the site before deter-

---

[82] See A Joyner, "Current trends in local planning and political discretion", J.P.L. (1995),
Occ Pap 23, 24–38.
[83] Cherry, *The Politics of Town Planning.*

mining the application. This would have the benefit of members being familiar with the site and environs before determining the application, thereby having a clear understanding of the issues associated with the proposal.[84] This application was deferred at the Development Control Committee meeting of 28 January 2015 to enable "further and other information" submitted by the applicant in respect of noise, air quality and landscape and visual amenity to be considered.

At the Develop Control Committee Meeting held on 23 June 2015, **I-9.94** it was recommended by the planning officers that the application should be "granted subject to conditions controlling time limits, working programme, restriction on permitted development rights, highway matters, soil management, hours of working, safeguarding of water courses, control of noise, dust, lighting, security, ecology, archaeology, landscaping, restoration and aftercare".[85]

The application was decided on meetings between 23 and 29 **I-9.95** June, and, despite planning officers maintaining that there were no substantive planning reasons which justified the Committee reaching a decision to refuse the application, after taking legal advice, the application was refused on the grounds relating to issues of visual amenity and unacceptable noise.[86]

Interestingly, the associated planning application (LCC/2014/0097) **I-9.96** relating to monitoring works and associated with LCC/2014/0096 was also decided on 29 June, and was refused planning consent. A video of the planning decision shows the Senior Planning Officer advising the Committee that:

> "we need to know, very clearly, the reasons why you have decided that this should be refused, particularly given the decision last week on Roseacre Wood (a very similar application) where, completely conversely of course, planning permission was resolved to be granted".

According to the minutes, at this point: "The meeting was adjourned **I-9.97** to enable the Committee members to consider the grounds for refusal".[87] It appeared that in the period between 25 and 29 June, the Council had changed their view on this, almost identical, fracking-related application. There were 36,000 objections to the

[84] See http://planningregister.lancashire.gov.uk/PlanAppDisp.aspx?recno=6586 (accessed 4 May 2017).

[85] See http://council.lancashire.gov.uk/documents/s64253/PNR%20Report%200096.pdf (accessed 4 May 2017).

[86] See http://council.lancashire.gov.uk/documents/s66389/Minutes.pdf (accessed 4 May 2017).

[87] The Roseacre Wood application (LCC/2014/0102) referred to an almost identical application to that of LCC/2014/0097 for monitoring works and consent was granted, subject to conditions, on 25 June 2015.

proposed development and just 400, predominantly from businesses, supporting the application.

I-9.98   It appeared that, overwhelmingly, the will of the public was against the approval of the application and there was considerable pressure on the Planning Committee to refuse the application.

I-9.99   The LPA took legal advice from David Manley QC "as a matter of utmost urgency" on the validity of a proposed refusal (of the application) on 24 June. There was a hiatus to request more legal information on the application and the application was deferred until 29 June. The weekend of 27–28 June saw anti-fracking group Preston New Road Action Group instruct legal counsel Dr Ashley Bowes and Friends of the Earth instruct Richard Harwood QC to comment and respond to David Manley's advice, which they both did and circulated the same weekend. That weekend, in effect, saw Members bombarded with conflicting last-minute legal advice from three senior barristers which they had to make sense of and factor into their decision-making process, all against the backdrop of considerable anti-fracking protesting at Lancashire Council's offices and UK-wide media interest. Ultimately, the Council refused the application for drilling, to the considerable consternation of Cuadrilla.

I-9.100   The decision by Lancashire Council was challenged by Cuadrilla, who appealed to the Secretary of State under Section 78 of the Town and Country Planning Act for planning permission to be granted. The Department for Communities and Local Government granted such approval on 6 October 2016. In granting the approval, controversy arose as to whether the planning system was undermined and subverted.[88]

Falkirk Council: Planning application P/12/0521/FUL

I-9.101   Dated 29 August 2012, this application related to the "Development for Coal Bed Methane Production, Including Drilling, Well Site Establishment at fourteen locations, Inter-Site Connection Services, Site Access Tracks, a Gas Delivery and Water Treatment Facility, Ancillary Facilities, Infrastructure and Associated Water Outfall Point" at Letham Moss, Falkirk.[89]

I-9.102   This application was received by Falkirk Council in August 2012 and, while the initial time period for the determination of the application was four months, the Council sought to extend the period for

---

[88] BBC News, "Fracking in Lancashire given go-ahead by government", available at www.bbc.co.uk/news/uk-england-lancashire-37567866 (accessed 4 May 2017).

[89] For information and details relating to this application, see http://edevelopment.falkirk.gov.uk/online/applicationDetails.do?activeTab=summary&keyVal=M9I1L9HC4X000 (accessed 4 May 2017).

further consultation as the application was of a scale and type that had not been widely experienced or considered previously.

The matter quickly became very contentious and every member I-9.103 of one prominent political party within the local authority signed a community charter which opposed the extraction of unconventional gas in the area, co-ordinated by local action group Falkirk Against Unconventional Gas. This effectively meant that these politicians could not objectively participate on the Planning Committee in relation to this application. The Convener of the Planning Committee noted that, with regards to the handling of this application, "certain politicians used the controversial nature of the application to put scoring political points before their planning committee responsibilities".

On 20 December 2012, correspondence from the Council to the I-9.104 Applicant stated that a decision would not be made by 7 January 2013, as initially anticipated, as the Council had decided to request that a hearing take place in relation to the application, based largely on the grounds that "a large number of representations seek clarification of technical and legislative aspects of the proposals" (the application shows a total of 2,474 comments on the application with 2,446 (99 per cent) objections to the application and just seven supporting it).

A subsequent letter from the Planning Authority to the applicant's I-9.105 agent on 1 March 2013 stated that "no date has yet been set for the hearing" and explained that "Falkirk Council is in the process of commissioning an external consultant to peer review technical aspects of the proposals relating to the technically novel nature of the application". Certain elements of the process had to be assessed in greater detail and a consultant was commissioned to evaluate the technical aspects of the proposal.

In correspondence dated 3 May 2013, it was noted that there was I-9.106 still no date set for the public hearing and that, while a technical consultant had been appointed, they had not reported back to the committee. The letter, consequently, requested an extension of time of a further two months, until 7 July 2013, to allow "full consideration and assessment of the application".

On 5 June 2013 the Applicant lodged an appeal to the Directorate I-9.107 for Planning and Environmental Appeals (DPEA) on the basis that the Council had failed to determine the application within the statutory timescale and that it was therefore deemed to have refused the application.

The DPEA subsequently set about the process of considering I-9.108 the appeal (there were almost 3,000 documents associated with this case). Prior to the DPEA making a decision on the appeal, an announcement was made in the Scottish Parliament on 28 January

2015 by Mr Fergus Ewing, Minister for Business, Energy and Tourism, that: "there is to be a moratorium on granting consents for unconventional oil and gas developments in Scotland while further research and a public consultation is carried out", and that remains the position at the time of writing. The prospects for this moratorium and its political implications will be further discussed below.

I-9.109 With regards to both these applications, there is at least a strong suggestion that political expediency had a role to play in the (lack of) decision making at both a local and Central Government level.

I-9.110 In summary, the legal planning framework throughout the UK should, in theory, be robust enough to deal with hydraulic fracturing as a land use in the same way that it has dealt with various new and emerging land uses throughout the last century. It appears, however, that for as long as this particular use remains as contentious as it is, achieving planning consent will not be without its difficulties.

THE CHALLENGE AND PROMISE OF DEVOLUTION – THE EMERGENCE OF DIFFERENTIAL REGIONAL POLICY

I-9.111 Having set out the regulatory framework for unconventional oil and gas activities in Great Britain as a whole, we turn now to consider the implications of devolution, and in particular the emergence of policy positions within the devolved institutions that differ from or challenge the policies of the Westminster Government and Parliament. Although the story could be told through the prism of, eg, Stormont-Westminster relations,[90] we will focus in particular upon the example of Scotland, not least of all because the position is in evolution and is of relevance to the broader Scottish independence debate.

Scottish Devolution

I-9.112 Scotland was an independent nation until 1707, when it became part of the Kingdom of Great Britain and in due course the United Kingdom of Great Britain and Northern Ireland. The idea that Scotland should enjoy some measure of autonomy while remaining part of the UK has a long history: the Scottish Home Rule Association, for instance, was highly active in the period between its formation in 1886 and the First

---

[90] Northern Ireland enjoys broader competences than the other devolved assemblies, largely due to the legacy of historic home rule arrangements. Thus – unlike its Scottish or Welsh counterparts – the Northern Irish Assembly enjoyed broad devolved powers over energy from the time of its creation. However there is no current production of oil or gas in Northern Ireland.

World War.[91] By the latter part of the twentieth century, home rule –
now rebadged as devolution – again enjoyed considerable support.
The Scottish Constitutional Convention, a broad-based civic organisation, drafted a framework for devolution[92] which was adopted by
the majority of the main Scottish political parties.[93] In the General
Election of 1997, the Labour Party made the offer of a referendum
on devolution a key manifesto commitment. At the referendum, the
Scottish electorate voted overwhelmingly in favour of the creation
of a devolved Scottish Parliament and accompanying Executive,[94]
institutions which were duly established by the Scotland Act 1998.

Under the devolution settlement put in place by the Scotland   **I-9.113**
Act, Westminster continues to enjoy parliamentary sovereignty.
The Scottish Parliament is provided with a general competence
to legislate, subject to a number of restrictions. Of these restrictions, the most relevant for present purposes is that the Scottish
Parliament has no power to legislate on matters which have been
reserved to Westminster.[95] The UK Parliament continues to have
the power to legislate on matters which – as they have not been
reserved – must be taken to have been devolved to the Scottish
Parliament;[96] however, this is subject to a constitutional convention
("the Sewel Convention") that it "[will] not normally legislate
with regard to devolved matters ... without the consent of the
Scottish Parliament".[97] This Convention – initially instituted by
announcement by the Government Minister for whom it is named –
was placed on a statutory footing by Section 2 of the Scotland Act
1998. It should be noted, however, that even in its statutory form, the
Convention only recognises what will *normally* happen. While there
is a strong expectation that Westminster will respect Holyrood's
wishes in areas within its legislative competence, it remains open to
Westminster to legislate on devolved matters without the consent of
the Scottish Parliament.

---

[91] See eg The Scottish Home Rule Association, *Statement of Scotland's Claim for Home
Rule* (1888). See also Sir R H B Lockhart, *Home Rule for Scotland*, 24 (1945) *Foreign
Affairs* 678.

[92] Scottish Constitutional Convention, *Scotland's Parliament, Scotland's Right* (1995).

[93] The Labour Party, the Scottish National Party and the Liberal Democrats all adopted
the Scottish Constitutional Convention's proposals. The Scottish Conservative and
Unionist Party opposed devolution.

[94] The Executive was renamed the Scottish Government in 2008.

[95] Scotland Act 1998, s 29(2)(b).

[96] The UK Parliament has done so several times, but only in compliance with the terms of
the Sewel Convention, discussed below.

[97] HL Deb 21 Jul 1998 Vol 592 c 791. The Scottish Parliament has developed a Legislative
Consent procedure which is used when it is content for Westminster to legislate on a
devolved issue: see www.gov.scot/About/Government/Sewel (accessed 7 May 2017).

The road to the 2014 independence referendum

I-9.114 By the time of the creation of the devolved Scottish Parliament in 1999, Labour had been the dominant force in Scottish politics for more than a decade.[98] However, by the end of the 1980s, the Scottish National Party (SNP) had succeeded in forging a clear political identity as a left-of-centre party which left it well placed to capitalise upon an increase in popular nationalist sentiment following devolution.[99] By 2007, the SNP had become the largest party in the Scottish Parliament. In the Scottish election of 2011, the SNP obtained both an overall parliamentary majority and a clear mandate for holding a referendum on whether Scotland should leave the UK.

The independence referendum and the Smith Commission

I-9.115 The (first)[100] Scottish independence referendum took place on 18 September 2014. Of the votes cast, 55 per cent supported Scotland remaining within the UK; 45 per cent favoured independence. Thus the result showed a clear (but by no means overwhelming) majority in favour of retaining the Union. Prior to the referendum, almost all opinion polls showed that a majority of Scots were against leaving the UK. However, in the immediate run-up to the referendum date, the lead narrowed, and a poll held on 6 September 2014 showed a narrow majority in favour of independence. Shortly before the vote, the leaders of the pro-Unionist parties committed themselves to the devolution of further powers to Scotland in the event of a vote against independence.[101]

I-9.116   The pledge to devolve further powers to Scotland led to the formation of a Commission comprised of representatives of the Conservative, Green, Labour, Liberal Democrat and Scottish National parties, and chaired by Lord Smith of Kelvin ("the Smith Commission"). The Commission reported in November 2014 and

---

[98] Prior to this, Labour and the Conservatives (known, prior to 1965, as the Unionist Party) enjoyed closely matched levels of popular support. Throughout the 1960s and 1970s, the Conservative Party lost ground to Labour, a trend that accelerated under the leadership of Margaret Thatcher.

[99] J Mitchell, "From Breakthrough to Mainstream", in G Hassan (ed.), *The Modern SNP: From Protest to Power* (2009), p 38.

[100] In the aftermath of the UK's decision to leave the European Union, the Scottish Government – noting that a significant majority of Scottish voters in the Brexit referendum voted to remain in the UK – has called for a second referendum on Scottish independence. At the time of writing, it remains to be seen if and when this will occur.

[101] This came to be known as "the pledge" – see eg BBC, "Scottish independence: Campaigns seize on Scotland powers pledge", 16 September 2014, available at www.bbc.co.uk/news/uk-scotland-scotland-politics-29219212 (accessed 7 May 2017) or, more dramatically, "the vow": D Clegg, "The Vow", *Daily Record*, 16 September 2014.

recommended a range of new powers and competences.[102] Among these were proposals relative to onshore oil and gas. We shall return to these in due course, but first we need to consider the devolved Scottish institutions' competences relative to energy as these existed at the time of the initial devolution settlement.

The devolved Scottish institutions' initial competences relative to energy

The matters reserved to the UK Parliament are set out in Schedule   **I-9.117**
5 to the Scotland Act. The Scottish Constitutional Convention had envisaged that the Scottish Parliament would enjoy a wide competence relative to energy, naming it as among "one of the principal areas which will fall within the [Parliament's] powers".[103] The Convention's approach was, however, criticised for a lack of information and detail,[104] and it was noted that electricity and gas were regulated on a UK-wide basis and operated within a UK-wide market.[105] By the time that devolution came to be implemented, a significant change in policy had emerged. The range of areas of energy policy reserved to Westminster by the Scotland Act was very extensive,[106] and included:

> "Oil and gas, including—
> (a) the ownership of, exploration for and exploitation of deposits of oil and natural gas
> ...
> (c) offshore installations and pipelines."[107]

Other key matters of energy policy such as nuclear energy and the   **I-9.118**
generation, transmission and distribution of electricity were also reserved to Westminster.[108] The Scottish Parliament and Government

---

[102] The Smith Commission, *Report of the Smith Commission for Further Devolution of Powers to the Scottish Parliament*, 27 November 2014, available for download from http://webarchive.nationalarchives.gov.uk/20151202171017/http://www.smith-commission.scot (accessed 28 August 2017) (hereinafter "*Report of the Smith Commission*").

[103] Scottish Constitutional Convention, *Scotland's Parliament, Scotland's Right* (1995), p. 32. Among the areas for devolution there listed is "Energy, including electricity generation and supply".

[104] The Scottish Council for Development and Industry, *The Scottish Energy Sector and the Scottish Constitutional Convention's Blueprint for a Scottish Parliament* (1997), p 25.

[105] The Constitution Unit, University College London, *Scotland's Parliament: Fundamentals for a New Scotland Act* (1996), p 140.

[106] As well as containing a reservation relative to oil and gas, coal, nuclear energy and matters relating to the generation, transmission, distribution and supply of electricity are also reserved: Scotland Act 1998 Sch 5 Pt II Head D.

[107] Scotland Act 1998 Sch 5 Pt II Head D2.

[108] Scotland Act 1998 Sch 5 Pt II Heads D1 and D4.

were not, however, left wholly without competence relative to energy policy: energy efficiency, fuel poverty and the encouragement of renewables were all devolved.[109]

The emergence of a distinct energy policy in Scotland

I-9.119   These competences were thought sufficient to merit the appointment of a Scottish Energy Minister.[110] But of even greater significance has been the indirect influence that the Scottish Parliament (and, in particular, Scottish Government) has been able to exert over energy policy by using powers relative to other areas of devolved competence. The fact that town and country planning, in particular, is a devolved matter has given the Scottish Government a *de facto* veto over any new energy projects requiring an element of construction or land-use in Scotland. Since the SNP emerged as the party of government in Scotland, it has used this veto in order to assist it in forging an energy policy distinct from that pursued at UK level.[111] Thus when the UK Government in 2008 announced an intention to take steps to encourage energy companies to invest in new nuclear power stations, the SNP-led Scottish Government made it clear that its focus was on renewables and that it would not consent to the siting of new nuclear plants in Scotland.[112] "No to new nuclear", allied to a strong support for renewables, remains a key element of the Scottish Government's energy policy. In 2013, in opposing new nuclear, which he characterised as expensive and susceptible to interruption,[113] the Scottish Energy Minister stated that the Scottish Government had "an ambitious but achievable target to generate the equivalent of 100% of electricity from renewable sources by 2020".[114] The Scottish Government's opposition to new nuclear

---

[109] G Gordon, A McHarg and J Paterson, "Energy Law in the United Kingdom", in M Roggenkamp et al., *Energy Law in Europe* (2016), para 14.11.

[110] Currently Paul Wheelhouse, who is in addition Minister for Business and Innovation.

[111] It should be noted that devolution is specifically about empowering a particular region or community to set its own policy agenda. The mere fact that the Scottish Government (or Northern Irish Government) has chosen to formulate its own policy distinct from that of the rest of Great Britain provides no grounds for criticism. Criticism might, however, properly be directed towards the architects of the devolution process for not legislating more explicitly on the extent to which it is legitimate to use powers granted to the devolved Scottish institutions in order to frustrate the execution of UK policy within a reserved area.

[112] *The Scotsman*, "MSPs Vote No to New Nuclear Stations", 17 January 2008.

[113] A previous Scottish Energy Minister in addition characterised nuclear as unsafe. In 2007, Jim Mather MSP stated, "We completely reject the development of dangerous, unnecessary and costly new nuclear power stations in Scotland": R Milne, *Utility Week*, 19 October 2007, 15.

[114] See eg F Ewing, Minister for Business, Energy and Tourism, "Scotland's Response on

continues to impact upon the Westminster Government's ability to implement UK-wide policy. The Energy Act 2013 introduced reforms and market arrangements to support an increase in the capacity of low-carbon generation. These measures were in large part introduced to facilitate new-build nuclear; however, as Heffron and Nuttall note, the practical reality is that while the Energy Act applies in Scotland, "its powerful arrangements intended to make possible new nuclear build will have no direct impact [there]".[115]

Unconventional oil and gas is another area where the SNP-led **I-9.120** Scottish Government has pursued a policy which is at odds with that adopted at UK level. Although the SNP has been a powerful advocate for the offshore oil and gas industry, taking an active interest in the offshore tax regime and arguing in favour of better governance in order to extend the life of the offshore industry,[116] it has adopted a different approach to the fledgling onshore industry. While, as we have already seen, the UK Government is making concerted efforts to stimulate the development of shale gas,[117] the Scottish Government's policy has been cautious,[118] and avowedly so.[119] When the Infrastructure Bill 2015 was passing through Parliament, the Scottish Government formally opposed the introduction to Scotland of the provisions relative to the right to use deep-level land. Following this, the UK Government agreed to amend the Bill so that these provisions would not apply in Scotland.[120] In January 2015, the Scottish Energy Minister announced a moratorium on the granting of consents for unconventional oil and gas developments in Scotland pending the outcome of further research and public consul-

---

Nuclear Energy" (2013). Statement no longer available online but quoted in R Heffron and W Nuttall, "Scotland, Nuclear Energy Policy and Independence", University of Cambridge Energy Policy Research Group Series Working Paper 1409 (2014), 9 (hereinafter "Heffron and Nuttall, 'Scotland, Nuclear Energy Policy and Independence'").

[115] *Ibid*, 3.

[116] Former First Minister Alex Salmond could be particularly trenchant in his criticism: see eg *The Herald*, "Salmond tells Alexander: listen to man who wrote the book on oil when you were just three", 13 June 2014.

[117] See the earlier sections of this chapter and para I-3.52.

[118] A McHarg, "Energy Policy Devolution and the Smith Commission", 2 December 2014, available at www.scottishconstitutionalfutures.org/OpinionandAnalysis/ViewBlogPost/tabid/1767/articleType/ArticleView/articleId/4766/Aileen-McHarg-Energy-Policy-Devolution-and-the-Smith-Commission.aspx (accessed 7 May 2017).

[119] The Scottish Government, *Talking Fracking: A Consultation on Unconventional Oil and Gas* (2017) (hereinafter "Scottish Government, *Talking Fracking*"), available at https://consult.scotland.gov.uk/energy-and-climate-change-directorate/fracking-unconventional-oil-and-gas/supporting_documents/00513575.pdf (accessed 7 May 2017), p 9: "The Scottish Government's approach to unconventional oil and gas is therefore one of caution …". The expression "cautious and evidence-based" approach appears five times within the document.

[120] Infrastructure Act 2015, ss 43–48 and 56.

tation. The press release announcing this decision specifically noted the difference in policy between Scotland and England and Wales, by stating it came "days after the UK Government voted against a moratorium".[121] The purpose of the moratorium was to permit the Scottish Government to:

- undertake a full public consultation on unconventional oil and gas extraction;
- commission a full public health impact assessment; and
- conduct further work into strengthen planning guidance [sic].[122]

I-9.121   In October 2015, the Scottish Government announced that the moratorium would remain in place while research was undertaken into a further range of work packages including seismic monitoring, decommissioning and aftercare, climate change and economic impact. A similar moratorium was put in place relative to underground coal gasification. The Scottish Energy Minister stated that the Scottish Government's approach was to take "a precautionary, robust and evidence-based approach" to unconventional oil and gas, in contradistinction to the "gung-ho approach of the UK Government".[123] It is noteworthy that both of these announcements expressly drew attention to the difference between the policy of the Scottish Government and that of the UK Government. The latter announcement, in particular, seeks to draw a comparison between what it presents as the prudence of the Scottish Government and the impatience of the UK Government. It is also worth noting, however, that the Scottish Government's position is one of provisional caution, not determined proscription. Comparison with the case of new nuclear may help to illustrate this point. As noted above, the Scottish Government conclusively ruled out the construction of new nuclear capacity. In so doing, it repeatedly adverted to its commitment to renewables. In the context of its statements on unconventional oil and gas, the Government's language has been more equivocal. The statements speak of "potential negative impacts" but also of the "potential opportunities of new technologies". Perhaps surprisingly, neither they nor the Government's 63-page Consultation document make a single reference to Scotland's ambitions for the renewables sector.

---

[121] The Scottish Government, Press Release, "Moratorium Called on Fracking", 28 January 2015.

[122] Ibid.

[123] The Scottish Government, Press Release, "Moratorium on Underground Coal Gasification", 8 October 2015. The moratoria have been continued while the Scottish Government engages in public consultation on the development of unconventional oil and gas: see Scottish Government, Talking Fracking.

In announcing the initial moratorium, the Energy Minister stated **I-9.122** that he hoped it would be "widely welcomed as proportionate and responsible".[124] Reaction has, however, been mixed. One leading industry figure, while acknowledging that it would be "frustrating" if the moratorium were to go "on and on", has accepted that it is appropriate for the Scottish Government to proceed with caution while it collates and analyses information.[125] Others have indicated that they will not invest further in Scotland until the moratorium is lifted.[126] The reaction of environmental groups has generally been negative. One campaigner has expressed the view that the moratorium is nothing more than "political posturing – a temporary fix to help get the SNP through the next two elections" and inappropriate for a "progressive, environmentally-responsible government".[127] The Green Party[128] consider the moratorium to be weak and advocate a permanent ban. Some degree of division is also apparent within the SNP itself. An anti-shale group of SNP members has been established. This group contends that shale gas development runs contrary to the SNP's strong climate change, environmental and community credentials.[129] Although the group's attempt to strengthen the current moratorium with a view towards ultimately introducing a complete ban was defeated at the SNP's 2015 party conference, a significant number of members voted in favour of the motion, suggesting that the proper approach to unconventional oil and gas development is a contentious issue within the party.[130] The moratorium was extended in a decision of 3 October 2017. Announcing the decision, the Scottish Energy Minister stated that it was substantially based on overwhelming public opposition, but declined to issue a legislative ban. Thus the moratorium could be recalled at any time.

---

[124] The Scottish Government, Press Release, "Moratorium Called on Fracking".

[125] *The National*, "Row as Fergus Ewing fails to rule out fracking in Scotland", 29 April 2015.

[126] BBC, "Cluff stops spending on Forth underground coal gasification plant", 7 January 2016, available at www.bbc.co.uk/news/uk-scotland-scotland-business-35251793 (accessed 7 May 2017).

[127] *The National*, "Row as Fergus Ewing fails to rule out fracking in Scotland", 29 April 2015.

[128] The Green Party, Press Release, "Greens call on public to push government towards fracking ban", 30 March 2017, available at greens.scot/news/greens-call-on-public-to-push-government-towards-fracking-ban (accessed 7 May 2017).

[129] *The Scotsman*, "SNP members form anti-fracking group called Smaug", 17 September 2015, available at www.scotsman.com/news/environment/snp-members-form-anti-fracking-group-called-smaug-1-3889468 (accessed 7 May 2017).

[130] BBC, "SNP conference 2015: Members reject fracking ban call", 16 October 2015, available at www.bbc.co.uk/news/uk-scotland-scotland-politics-34552962 (accessed 7 May 2017).

Further devolution: opportunities and challenges

I-9.123   As has already been noted, in November 2014, the Smith Commission reported on the additional powers that should be devolved to the Scottish Parliament, including powers relative to onshore oil and gas. Offshore oil and gas was not affected. The Commission did not propose to devolve the power to determine the financial consideration to be received.

I-9.124   Provisions implementing these proposals were included within the Scotland Act 2016, although at time of writing these have not been brought into force. Section 47(1) of the Act devolves "the granting and regulation of licences to search and bore for and get petroleum that, at the time of the grant of the licence, within the Scottish onshore area". Matters of financial consideration are expressly excluded and therefore this matter remains in the hands of the Westminster Parliament. Section 47(1) further devolves the issue of "access to land for the purpose of searching or boring for or getting petroleum under such a licence", thus heading off the possibility for further conflict between the Scottish and Westminster Governments on this particular issue.

I-9.125   The 2016 Act in addition empowers the UK Secretary of State to amend the model clauses applicable to existing licences or to individual existing licence terms, making "any amendment that appears to the Secretary of State to be necessary or expedient"[131] as a result of the Scottish Government's new devolved power. Although this will involve changing the terms of existing licences, a matter which has, at times, been controversial in the past,[132] as this is likely to involve only mechanistic changes such as changing the identity of the licensing authority, it would seem, on this occasion, to be uncontroversial. A more interesting issue is the precise territorial scope of the powers to have been devolved. The Smith Commission proposed the devolution of "the licensing of onshore oil and gas extraction underlying Scotland". Two issues arise which have given rise to a degree of complexity in the drafting of the implanting legislation. Firstly, the reference to "onshore" would tend to give the impression the Commission was here concerned solely with oil beneath the territorial land mass of Scotland. However there are at least two issues that need to be further considered. Firstly, the PEDL is often colloquially described as the "onshore" form of petroleum licence.[133] It is, however, more properly described as the

---

[131] Scotland Act 2016, s 49(1).
[132] See para I-4.17.
[133] See eg the OGA Guidance, *Onshore Licenses*, available at www.ogauthority.co.uk/licensing-consents/onshore-licences (accessed 7 May 2017).

"landward" licence[134] – as being the appropriate licence for developments on the *landward* side of the line described in Schedule 1 of The Petroleum Licensing (Exploration and Production) (Landward Areas) Regulations 2014,[135] an area which includes a significant amount of inland waters, including the area lying between the outer points of the Hebrides and the Scottish mainland. This raises the question of whether the Commission's intention was truly to restrict the grant of devolved power to oil and gas beneath Scotland's land mass, whether it was intended to include deposits lying on the landward side of the relevant line; or whether it simply did not apply itself to that question. The legislation addresses this issue by devolving powers relative to the "Scottish onshore area". This is then defined as "the area of Scotland that is within the baselines established by any Order in Council under Section 1(1)(b) of the Territorial Sea Act 1987 (extension of territorial sea)".[136] As at the date of drafting, no such baselines have yet been established. Secondly, oil and gas – more obviously when within a conventional reservoir, but to some extent within shale, too – is fugacious in nature: it is capable of movement within its reservoir rocks. It is therefore possible that oil that did not lie directly underneath the land mass of Scotland, might, by a process of drainage, be drawn there either from an offshore area or from under the land mass of England. The fugacious nature of oil and gas would seem to lie behind the legislation's reference to "the granting and regulation of licences to search and bore for and get petroleum that, *at the time of the grant of the licence*, is within the Scottish onshore area".[137] Thus the Scottish Parliament has no power to grant licences which, although located within the Scottish onshore area, would permit drainage into the Scottish onshore area from outside.

When the provisions enter into force, the Scottish Government I-9.126 will, for the first time, become a licensing authority for oil and gas operations. The Scottish Government will no longer be limited to the above-discussed veto conferred by its competence relative to land-use and planning, but will be empowered to exert a broader level of policy control. The unification of licensing and planning competence in the same government is to be welcomed. As has already been noted, it will reduce the potential for tension between Westminster and Holyrood; it does not, however, remove it entirely. At least for as long as the moratorium or more permanent ban is in place, and

---

[134] See The Petroleum Licensing (Exploration and Production) (Landward Areas) Regulations 2014 (SI2014/1686).

[135] In the same way, "offshore" licences are more properly described as seaward licences as being seaward of the same line.

[136] Scotland Act 2016, s 47(3).

[137] Scotland Act 2016, s 47(2).

there is a marked difference between the policy followed at UK level and that adopted in Scotland, the Scottish Government will be able to capitalise on that difference, casting itself as more sceptical, more responsible and less impetuous than the UK Government. This would fit with the well-established narrative of the SNP-led Government as a responsible steward and protector of Scotland, and would also enhance the SNP's green and progressive credentials. But depending on the economic benefits projected to accrue as a result of development, there may in due course come to be very significant pressure to permit development. What, then, if the moratorium is removed, and development goes ahead? The fact that full control over all aspects of law and policy has not been devolved means that new areas for disagreement may emerge. The Smith Commission provided, and the implementing legislation confirms, that corporation tax and "all aspects of the taxation of oil and gas receipts" will continue to be reserved to Westminster.[138] Thus the Scottish Parliament will not be able to implement its own fiscal regime relative to onshore oil and gas operations. The terms of the fiscal regime is an essential element borne in mind by oil and gas companies when deciding whether or not to invest in a particular oil and gas province. As we have already seen, in the context of offshore oil and gas operations, the SNP has frequently asserted – often with considerable justification – that the UK's tax regime is not well-designed to encourage investment. Thus – although this matter is far removed from the issues on which Holyrood and Westminster have thus far locked horns – disputes about the optimal design of the tax system could yet prove to be a fertile source of disagreement between Holyrood and Westminster.

CONCLUSION

I-9.127 To date, the path of shale gas exploration and development within Great Britain has been difficult. The UK Government has clearly articulated that shale gas is a valuable, even essential part of the energy mix for the UK, and that the development of such resources is not just necessary but inevitable. In promoting the development of shale gas, the UK Government pointed to the robust nature of the petroleum regulation framework, and its effectiveness in regulating petroleum activities to date. Furthermore, confidence in the planning system was cited as a reason for shale development, with a reliance on planning permission to approve the process in the community. Such confidence has prompted Westminster to pursue a shale gas development policy, where Exploration Licences have been granted.

---

[138] *Report of the Smith Commission*, paras 81 and 82.

In England there has been a sustained resistance to shale gas I-9.128 development since 2012, both in the Midlands and the High Wealds. In Scotland, such resistance has crystallised into a government moratorium on the granting of planning and environmental permission for the activity, effectively placing a moratorium on shale gas extraction. In particular the Scottish Government decided that a precautionary, robust and evidence-based approach to shale gas development was warranted. Whether it has adopted this approach as a result of a genuine belief that caution is merited, out of a desire to be seen to adopt a different approach to that adopted by the UK Parliament, or to allow it to demonstrate its green and progressive credentials is not wholly clear. Whatever the reason, the same restraint has not been exhibited in England. Instead, the UK Government has stepped in to override planning rejection of shale gas exploration applications, such as Cuadrilla's application for drilling at Little Plumpton in Lancashire. Although planning permission was initially rejected by Lancashire Council, Cuadrilla appealed to the Secretary of State. After a lengthy assessment, planning permission was granted. Such government action is a stark contrast to the Scottish position.

CHAPTER I-10

HEALTH AND SAFETY AT WORK OFFSHORE

John Paterson

INTRODUCTION

I-10.01   The United Kingdom Continental Shelf (UKCS) is among the most hostile environments in which the offshore oil and gas industry operates anywhere in the world. The substances it produces are naturally volatile. The nature of the activity means that the workforce lives in very close proximity to the workplace. Travel to and from the workplace is by helicopter. The industry has been characterised by a constant striving for newer and better technology to allow it to operate in deeper water or to cope with higher-temperature and higher-pressure reservoirs. All of these issues very obviously impact upon health and safety at work. They also raise profound questions for those charged with regulating occupational health and safety in the offshore industry. How should a regulator do the things that it would traditionally be expected to do? How does it draft regulations for a novel setting that is then characterised by constant change? How does it monitor and inspect when the regulated area is remote and relatively inaccessible?

I-10.02   It is not surprising, then, that the regulatory approach to health and safety at work offshore has seen a constant evolution throughout a history that is now into its fifth decade. There have been severe setbacks along the way – serious accidents that have revealed regulatory shortcomings – and it is perhaps only in recent years that it could be said with any confidence that the regulatory regime is well adapted to the challenges presented by the industry. Even as that position is reached, however, the realisation dawns that the character of the regulated area is changing again. The UKCS is now described, in the main, as a mature province, albeit that especially to the west of Scotland it may also be said to still possess the characteristics of the frontier, all of which raises a continuously changing set of challenges: the gradual withdrawal of the major oil companies and the arrival of new entrants with innovative ideas but perhaps less experience; infrastructure at or beyond its design life which the Government is keen to take full advantage of to maximise the recovery of the hydrocarbon resources, not least via the intervention of the new regulator, the Oil and Gas Authority (OGA); the decommissioning of platforms, pipelines and other equipment that has become redundant; the search for new reservoirs in deeper water that has previously been less attractive due to the difficulties and costs involved – and all of this in the context of oil prices substantially below those which subsisted between 2011 and the middle of 2014. All of these features of the UKCS have the potential to impact upon occupational health and safety.

This chapter traces the evolution of the regulatory approach to health and safety at work offshore. The evolution is divided into four phases. The first phase covers the period when health and safety at work was dealt with under the licence and focuses upon, first, the Continental Shelf Act 1964 and, then, the findings of the Inquiry into the Sea Gem accident. The second phase covers the period during which the detailed prescriptive regime recommended by the Sea Gem Inquiry was gradually developed and implemented. It accordingly considers: the Mineral Workings (Offshore Installations) Act 1971; the tension in due course between this Act and the Health and Safety at Work, etc. Act 1974; and the findings of the Burgoyne Committee on Offshore Safety. The third phase covers the period during which the permissioning approach now in place was developed. Thus, it considers the Piper Alpha disaster and the findings of the subsequent public inquiry, the Offshore Safety Act 1992 and the subsequent safety case and goal-setting regulations; the revised Offshore Installations (Safety Case) Regulations 2005. The fourth phase runs from the Macondo disaster in the US to the present, including the Offshore Safety Directive and the impact it has had upon the architecture and orientation of the UK's regulatory approach. This treatment of the issue of health and safety at work offshore serves a number of purposes. By contrasting the current permissioning approach with the foregoing approaches, it highlights the particularity of the means by which health and safety at work offshore is now regulated. This is of interest not only to those based in the UK, but also to the increasing number from other jurisdictions who are involved in the development of the regulation of health and safety in the offshore industry and who are looking to the UK for inspiration. Secondly, it reveals the extent to which the problems that the current approach seeks to deal with were evident from relatively early on but never adequately dealt with in the recommendations

I-10.03

of Inquiries or in the discussions of legislators. This serves also to bolster the new approach in the face of criticism from those who regret the passing of the prescriptive regime. Finally, it allows an appraisal of the ability of the permissioning approach to respond to the emergent challenges posed by the UKCS. Selected developments in criminal law are then considered before the concluding remarks to this chapter ask the question whether the troubling findings that have periodically emerged in relation to the implementation of the safety case approach in the context, variously, of an investigation by the regulator, a Select Committee inquiry and the criminal courts will finally be addressed by the latest developments under the Offshore Safety Directive or whether, if anything, the challenges are only becoming more intense such that a concerted effort to deal with what appears to be a continuing weakness is now required.

FIRST PHASE: THE LICENSING APPROACH

The Continental Shelf Act 1964

I-10.04 The early days of the oil and gas industry in the North Sea coincided with a period of economic difficulty for the UK. As a result, the possibility that there may be reserves of oil and gas on the UKCS almost seemed an answer to prayer as the country struggled with a crippling balance of payments deficit.[1] It has been suggested that this pressing need goes a long way to explaining the speed with which the Government then acted to ensure that the legal framework was in place to allow the industry to explore for and in due course to exploit oil and gas.[2] The presence of developed and politically stable markets around a geologically interesting location made the North Sea particularly attractive to the industry, but the financial investment involved meant that, without a firm legal framework, work would not proceed. The Continental Shelf Act 1964 was accordingly passed, based on the United Nations Convention on the Continental Shelf of 1958 which had conferred "sovereign rights" in the continental shelf on coastal states.[3] It would be going too far to

---

[1] For an indication of the extent to which this issue dominated government thinking at the time, see Sir A Cairncross, "Devaluing the Pound: the Lessons of 1967", *The Economist*, 14 November 1992, 25–28.

[2] See eg W G Carson, *The Other Price of Britain's Oil: Safety and Control in the North Sea* (1981) (hereinafter "Carson, *The Other Price of Britain's Oil*"), pp 141*ff*; see also L Turner, *Oil Companies in the International System* (3rd edn, 1983); G.-P. Levy, "The Relationship between Oil Companies and Consumer State Governments in Europe, 1973–1982", 19 (1984) *Journal of Energy and Natural Resources Law* 9.

[3] The Convention granted states "sovereign rights for the purpose of exploring [the continental shelf] and exploiting its natural resources" (Art 2(1)) and entitled them "to

say that at this time no attention was paid to the question of occupational health and safety, but it would nevertheless be true to say that this issue was by no means uppermost in the minds of lawmakers. Insofar as the whole point of the 1964 Act was to assert the UK's sovereign rights over the continental shelf, the assumption was that the general law, including statute law, could be extended offshore, which Section 3 purported to do. This, however, fell foul of the accepted canon of statutory interpretation that holds that statutes are assumed to extend only to Great Britain unless otherwise stated.[4] Accordingly, this section could not have the effect of extending, for example, existing factory legislation to cover offshore installations – even if the definitions in such onshore legislation could be stretched to accommodate these novel structures, which was by no means certain.

With the option of simply extending onshore occupational safety legislation offshore thus effectively ruled out, the treatment of this issue at the time under the legal regime explicitly dedicated to the offshore industry looks, with the benefit of hindsight, decidedly inadequate. In this regard, the Government, again perhaps motivated by the concern to allow rapid exploration and development in the North Sea, drew extensively upon the approach taken some 30 years earlier when the legal framework had been put in place for the nascent (and never more than marginally important) *onshore* oil and gas industry under the Petroleum (Production) Act 1934. This Act vested mineral resources in the Crown and required that those who wished to exploit them must obtain a licence. Regulations governing the grant of licences were promulgated in 1935 under the authority of the 1934 Act, with model licence clauses annexed.[5] These regulations and model clauses were essentially repeated with minimal modification for offshore licences in the Petroleum (Production) (Continental Shelf and Territorial Sea) Regulations 1964.[6] The whole issue of occupational health and safety then fell to be dealt with in one clause of the licence where it was provided that: "[t]he Licensee shall comply with any instructions from time to time given by the Minister in writing for securing the health, safety and welfare of persons employed in or about the licensed area".[7] In practice, the

I-10.05

---

construct and maintain or operate ... installations" to that end (Art 5(2)). See also para I-4.09.

[4] For a discussion of this point, see T Daintith, G Willoughby and A Hill, *United Kingdom Oil and Gas Law* (3rd edn, looseleaf, 2000–date) (hereinafter "Daintith, Willoughby and Hill"), para 1-510.

[5] Petroleum (Production) Regulations 1935 (SR and O 1935/426).

[6] SI 1964/708.

[7] Sch 2, Cl 18.

Minister wrote to each licensee instructing them in this regard to carry out operations "in accordance with such provisions of ... the Institute of Petroleum Model Code of Safe Practice in the Petroleum Industry issued in October, 1964, as relate to the safety, health and welfare of persons employed".[8] It is important to realise, therefore, that although a licensing approach may be regarded as a very inter-ventionist form of regulation,[9] this was by no means the case in the context of the licences issued under the 1964 Regulations insofar as they applied to health and safety. This issue was simply not subject to any detailed regulatory stipulations or inspections, but rather was left to the industry itself to deal with. Even if the Minister did become aware of a shortcoming in the treatment of health and safety, the contractual nature of the licence limited both the remedies open to him and the individuals against whom they could be applied. The only remedy available would be a revocation of the licence, which would presumably only be countenanced in the case of the most egregious behaviour, while the only party whom the Minister could in any event take action against in this regard would be the licensee – a significant limitation in the context of an industry so characterised by subcontracting.[10]

The Sea Gem Inquiry

I-10.06   It was only a short time before the deficiencies of this approach to health and safety offshore became evident. In April of 1965, the first commercial gas field on the UKCS was discovered by the jack-up drilling rig Sea Gem, justifying the Government's and the industry's optimism with regard to the North Sea.[11] In December 1965, the same rig collapsed and sank with the loss of 13 lives.[12] This tragedy did not necessarily mean, of course, that there was a problem with the law. The fact that the Minister of Power discovered that he had no statutory authority to set up an inquiry into the accident because the rig fell into no category recognised by law did, however, set alarm bells ringing. An inquiry was, nevertheless, established but it "was without statutory authority,

---

[8] For details of this letter of instruction and the Institute of Petroleum Code, see Ministry of Power, *Report of the Inquiry into the Causes of the Accident to the Drilling Rig Sea Gem* (Cmnd. 3409, 1967) (hereinafter "Sea Gem Inquiry"), pp 17–22.

[9] See eg A I Ogus, *Regulation: Legal Form and Economic Theory* (1994) (hereinafter "Ogus, *Regulation*"), pp 214*ff*.

[10] For more detail see R W Bentham, "The United Kingdom Offshore Safety Regime: Before and After Piper Alpha", (1991) *Journal of Energy and Natural Resources Law* 273 (hereinafter "Bentham, 'The United Kingdom Offshore Safety Regime'").

[11] See *Petroleum Press Service* April 1965, 127.

[12] See "Triumph and Tragedy in the North Sea", *Petroleum Press Service* 1966, 5.

had no power to compel the attendance of witnesses, nor was it empowered to administer oaths".[13] The Inquiry was presided over by a lawyer sitting with two assessors with engineering expertise and its recommendations led in due course to the establishment of the prescriptive regulatory approach that characterised the middle phase of the evolution described below. It is worth considering the approach of the Sea Gem Inquiry in a little more detail, however, in order to understand the reasoning that lay behind these recommendations.

While the Inquiry's ultimate recommendations certainly dealt with the difficulties attending the licensing approach to health and safety outlined above, its motivations were actually quite different. The Inquiry focused in no small measure upon the difficulties posed for *the law* by a code of practice that had been drafted by *the industry*. For example, the inquiry considered the following paragraph of the Institute of Petroleum Code: I-10.07

> "[o]nly persons, who, in the opinion of the supervisor, are essential to the safe raising and lowering of a self-elevating type of mobile drilling platform, should be on the unit when this is being done. They should wear a lifejacket at all times while any phase of the raising and lowering operation is being carried out."

The Institute of Petroleum clearly had no problem with this wording, but the Sea Gem Inquiry was not impressed. "This paragraph would, if it had legislative force, be a prescription for unlimited litigation." Furthermore, a "court of law might have considerable difficulty in determining what" was meant.[14] That said, however, the Inquiry had earlier acknowledged that the "authors of the code emphasise that it deals with recommendations regarding safe practice only and does not necessarily include anything issued in the form of regulation or instructions by the appropriate national authorities".[15] Nevertheless, concern with the imprecision of the language in the Code led to the Inquiry recommending "a code of [statutory] authority with credible sanctions".[16] I-10.08

This recommendation is entirely justified given the clear shortcomings of the licensing approach generally, but it is surprising that it was not qualified in any way given the Inquiry's own findings about the difficulty of legislating for such a complex area. "The field over which the Inquiry ranged is so large and the evidential material so complex" that "generalisations could well be both inapt and I-10.09

---

13 See Sea Gem Inquiry, p 1.
14 *Ibid*, para 8.8.
15 *Ibid*, para 8.2.
16 *Ibid*, para 10.2(i).

dangerous".[17] And, indeed, it could be suggested that the Inquiry had implicitly recognised the impossibility (or at least the difficulty) of trying to produce a comprehensive regulatory code for such a complex and rapidly changing field. Most of the deaths on the Sea Gem had actually been caused by strict adherence to the Code where it was unequivocal in its terms. The Code required personnel to muster on the helicopter deck in the event of an emergency, but this instruction had relevance only in the event of a fire and when helicopter evacuation had been arranged, not in the circumstances of a structural collapse when escape to the sea via lifeboats or rafts (of which the Sea Gem had plenty for all the crew on board) was the appropriate response.[18] It is by no means clear that "a code of [statutory] authority with credible sanctions" would have avoided this problem.

I-10.10   Furthermore, it could also be suggested that the Inquiry accurately identified the extent to which the operation of the industry in the North Sea would inevitably be a learning process. As the Inquiry put it: "In terms of North Sea drilling the SEA GEM represented a pioneering, not to say an experimental project".[19] Insofar as in 1965 it was already clear that what had taken place to date was only the first tentative step into uncharted territory, the idea that this would stop being a learning process anytime soon was surely a naïve one. That it was appropriate to call for a detailed regulatory code in such circumstances must accordingly be questioned. But that is what the Inquiry did. It thus revealed a fundamental belief, firstly, that responsibility for ensuring safety lay with the Government and, secondly, that the law could come to terms with the problems identified so as to provide a comprehensive regulatory code which, if enforced, could ensure safety in the industry.

I-10.11   It might thus be said that the Inquiry actually had quite profound insights into the nature of the problems confronting regulation in this area, but did not follow through on those insights when it came to making its recommendations. It could be offered in the Inquiry's defence that it had few, if any, regulatory alternatives to turn to in 1965. It could also be suggested that the Inquiry's confidence in the appropriateness of its recommendations was unduly bolstered by the reliance it placed on the comparison it drew between an offshore installation and a ship to which the Merchant Shipping Acts applied. In this last regard, it is significant that of the six recommendations made, no fewer than five drew a direct comparison with those Acts.[20]

---

[17] *Ibid*, para 10.1.

[18] *Ibid*, para 9.2.

[19] *Ibid*, para 10.1.

[20] In their other recommendations the Inquiry called for an accepted discipline and

SECOND PHASE: THE PRESCRIPTIVE APPROACH

The Mineral Workings (Offshore Installations) Act 1971

Despite the urgency of the Sea Gem Inquiry's recommendations   I-10.12
when it reported in July 1967, it was another four years before any
kind of legislative response was in place. The lapse of time is perhaps
an indication of the difficulty confronted by the Labour Government
and the subsequent Conservative Government in drafting the sort of
statutory code with credible sanctions that would meet the requirements of the Inquiry.

The Mineral Workings (Offshore Installations) Bill began life   I-10.13
in the Lords and moved in due course to the Commons. In both
Houses, the influence of the Sea Gem Inquiry can be seen, with
legislators broadly accepting its assessment of the problem and of
the appropriate solution. Thus, Earl Ferrers in the Lords acknowledged the problem of the lack of any sanction short of revocation
of the licence.[21] The Under Secretary of State for Trade and Industry
in the Commons, Nicholas Ridley, noted the problems arising from
the fact that the regime under the 1964 Act was contractual rather
than mandatory.[22] In place of this approach, a detailed code was
to be established which would set out clear requirements for the
industry, enforceable by the regulator and attracting graded penalties.[23] Flexibility was, however, to be a feature of this new approach,
not least in recognition of the fact that the industry was new and
developing rapidly.[24] The starting point for the new approach would
be the registration of every installation. The approach would then be
characterised by three means by which control could be exercised:
certification as fit for use of all installations by Certifying Authorities,
with those already active in the certification of shipping envisaged
in this role; the *appointment of masters*, later designated installation

---

chain of command similar to the Merchant Navy (Sea Gem Inquiry, paras 10.2(ii), (iii));
loudspeakers so that orders could be communicated (para 10.2(iv)); the keeping of records
of eg increases and decreases of loading on the rig and other matters affecting the rig as a
structure (para 10.2(v)); and a daily round equivalent to that of a ship's master designed
to keep "everybody up to scratch" (para 10.2(vi)). "Interestingly, however, while this new
regulatory code was directly related to the UK experience of legislating for the control of
merchant shipping (and installations had obvious similarities to ships), no attempt was
made to apply merchant shipping laws directly to installations", B Barrett and R Howells,
"Safe Systems for Exploiting the Petroleum Resources of the North Sea", 33 (1984)
International and Comparative Law Quarterly 811, at 817.

[21] Earl Ferrers, *Hansard* HL (series 5), vol 315, cols 741–746, 742–743 (18 February
1971).

[22] Hon. N Ridley (Under Secretary of State for Trade and Industry), *Hansard* HC (series
5), vol 816, cols 645–649, 647 (28 April 1971).

[23] Ferrers, at col 743; Ridley, at col 647.

[24] Ferrers, at col 743; Ridley, at col 648.

managers, as the focal point of responsibility; and *regulations* to be made in due course within the framework of the Act.[25]

I-10.14   The regulations in particular were seen to be the way in which the regime could keep pace with technological change. This flexibility and adaptability to change did not in any way mean, however, that the regulations would not be comprehensive. It was foreseen that they would cover the safety both of the installation itself and of the operations on it and provide the basis for detailed inspection and enforcement.[26] Furthermore, the framework Act would allow coverage of both existing installations and those not yet designed or built.[27] The need for the regulations to keep pace with change was clearly understood: concern was expressed that they should neither cramp development nor cause waste and extravagance for no good reason.[28] Furthermore, Parliament envisaged that regulators would not adopt a heavy-handed approach to their task: enforcement was foreseen as being "benevolent" and "advisory" with prosecution as a last resort.[29] In this regard, there was recognition from both Government and opposition of the fact that the industry had offered ready co-operation.[30]

I-10.15   While there was, then, little disagreement between Government and opposition about the nature of the new regime, a tension was evident between, on one hand, a desire to allow flexibility by leaving more to the regulators to determine and, on the other, a desire to restrict that flexibility by laying down more rules at the level of the Act. Thus, there was concern that matters seen to be fundamental, such as the provision of radio and radar, were not being set out in the Act but rather being left to regulations.[31] Equally, there was concern that the terms in which authority was delegated to the Secretary of State to make regulations assumed a great deal and lacked specification.[32] The Government position in each case was to insist on the need for flexibility in view of the uncertainty surrounding future developments.[33]

I-10.16   The Mineral Workings (Offshore Installations) Act 1971 was fully in force by 31 August 1972,[34] but the history of the eventual

---

[25] Ferrers, at cols 744–745.

[26] *Ibid*, at col 744.

[27] Ferrers, at col 744; Ridley, at col 648.

[28] Ferrers, at col 746; Ridley, at col 648.

[29] Ferrers, at col 746.

[30] Ridley, at col 648.

[31] Lord Brown, *Hansard* HL (series 5), vol 315, col 748 (18 February 1971).

[32] Mr D Mudd, *Hansard* HC (series 5), vol 816, col 654 (28 April 1971).

[33] Ferrers, at col 750; Ridley, at col 648.

[34] Mineral Workings (Offshore Installations) Act 1971 (Commencement) Order 1972, SI 1972/644.

promulgation of regulations under it appears somewhat different to what was envisaged by Parliament. Far from there emerging quickly a comprehensive range of regulations which were then updated as and when required by the rapid pace of change, it was actually 1980 before the full set of regulations – in the form of eleven Statutory Instruments – was in place.[35] While the issue of registration was swiftly dealt with (regulations in this regard appeared in 1972), it was 1976 before regulations relating to what might be regarded as core issues of health and safety at work were produced.[36] Furthermore, issues relating to emergency response were not covered by regulations until after that point. It is not insignificant that some of the North Sea's major fields began producing in the period before 1980, that is, before all the regulations were actually in place.[37] That said, however, by the time the last Statutory Instrument was promulgated, the regulations did present the appearance of the sort of comprehensive code that the Sea Gem Inquiry had envisaged some 13 years before, covering as they did everything from the construction and survey of installations and well control, through fire-fighting and life-saving equipment. Furthermore, while the regulations under the 1971 Act were centred on the installation itself, regulations were also promulgated under the Petroleum and Submarine Pipelines Act 1975 in relation to offshore pipe-laying operations.[38]

As important as the regulations, however, was the identity of the I-10.17 *regulator* itself. Although the responsibility for safety in the offshore industry under the 1971 Act passed between several different Government departments, it is regarded as significant by many

---

[35] The regulations introduced under the 1971 Act in this period were: Offshore Installations (Registration) Regulations 1972 (SI 1972/702); Offshore Installations (Managers) Regulations 1972 (SI 1972/703); Offshore Installations (Logbooks and Registration of Death) Regulations 1972 (SI 1972/1542); Offshore Installations (Inspectors and Casualties) Regulations 1973 (SI 1973/1842); Offshore Installations (Construction and Survey) Regulations 1974 (SI 1974/289); Offshore Installations (Public Inquiries) Regulations 1974 (SI 1974/338); Offshore Installations (Operational Safety, Health and Welfare) Regulations 1976 (SI 1976/1019); Offshore Installations (Emergency Procedures) Regulations 1976 (SI 1976/1542); Offshore Installations (Life-saving Appliances) Regulations 1977 (SI 1977/486); Offshore Installations (Fire-Fighting Equipment) Regulations 1978 (SI 1978/611); Offshore Installations (Well Control) Regulations 1980 (SI 1980/1759).

[36] It later transpired that even this delay did not preclude the 1976 Regulations having been introduced "hurriedly" and with "inadequate consultations". See the Department of Energy's submission to the Burgoyne Committee, J H Burgoyne, *Offshore Safety: Report of the Committee* (Cmnd 7866, 1980) (hereinafter "the Burgoyne Report"), Submission 37, para 7.

[37] For example, Forties in September 1975, Brent in November 1976, Piper in December 1976 and Ninian in December 1978.

[38] Submarine Pipe-lines (Diving Operations) Regulations 1976 (SI 1976/923); Submarine Pipe-lines (Inspectors, etc.) Regulations 1977 (SI 1977/835).

commentators that the body responsible for health and safety was always the same as that which was responsible for licensing and thus for ensuring the efficient exploitation of the nation's hydrocarbon resources.[39] For the majority of the period from 1971 to the end of what is referred to in this chapter as the middle phase, the responsibility for safety lay with the Petroleum Engineering Division (PED) of the Department of Energy (DEn).[40] This location of responsibility for safety within the industry's "sponsoring" department very soon set offshore oil and gas at odds with the general trend in health and safety regulation in the UK. No regulatory regime is likely to last forever,[41] but few are already profoundly in question before they are even in force. That, however, was essentially the position of the 1971 Act.

Tension with the Health and Safety at Work, etc. Act 1974

I-10.18   While the 1971 Act was passing through Parliament, a committee set up by the Government under the chairmanship of Lord Robens was examining the whole question of the regulation of health and safety at work. In its report,[42] which was published in June 1972 (and thus before the 1971 Act was fully in force at the end of August 1972), the Committee came down firmly against precisely the sort of prescriptive regulatory approach that had just been mandated for the offshore industry. The Robens Committee had three main problems with this sort of approach. Firstly, it took the view that there existed too much law relating to health and safety at work and the detailed

---

[39] For example, Carson, *The Other Price of Britain's Oil*, p 163; Bentham, "The United Kingdom Offshore Safety Regime", 276; K Miller, "Piper Alpha and the Cullen Report" (1991) *Industrial Law Journal* 176, at 178–9.

[40] The administrative history of the regulators responsible for offshore health and safety is somewhat complicated. Initially, responsibility lay with the Petroleum Division of the Ministry of Power. In 1969, responsibility was transferred to the Ministry of Technology and a Petroleum Production Inspectorate was set up as a subdivision of the Petroleum Division. Only a year later, however, the function was moved to the Department of Trade and Industry where the responsible division became the Petroleum Production Division. In 1974, the Department of Energy was founded and the Petroleum Production Division became a part of that Department. It was this move in particular which concerned critics who saw safety subordinated in a Department "for which energy production was the primary concern". In 1977, the Petroleum Engineering Division was established and most safety functions became its responsibility. See Carson, *The Other Price of Britain's Oil*, pp 161–163; Burgoyne Report, Appendix 7.

[41] Though even an opposition Member of Parliament at the time of the passage of the 1971 Act believed its inherent flexibility meant it would last 100 years. Mr E Ogden, *Hansard* HC (series 5), vol 821, col 678 (14 July 1971).

[42] Lord Robens, *Safety and Health at Work: Report of the Robens Committee* (Cmnd 5034, 1972) (hereinafter "Robens Report").

prescription of every aspect of work had the effect of persuading people that health and safety was purely a matter of Government regulation and not of individual responsibility. Secondly, it believed that too much of the existing law was irrelevant to real problems. Finally, it contended that there was a major disadvantage in attempting to address the problem of health and safety with the wide array of administrative agencies then engaged in the field.[43] Summing all of this up, the Committee concluded that:

> "[t]here are severe practical limits on the extent to which progressively better standards of safety and health at work can be brought about through negative regulation by external agencies. We need a more effectively self-regulating system. This calls for the acceptance and exercise of appropriate responsibilities at all levels within industry and commerce. It calls for better systems of safety organisation, for more management initiatives, and for more involvement of work people themselves. The objectives of future policy must therefore include not only increasing the effectiveness of the state's contribution to health and safety at work but also, and more importantly, creating conditions for more effective self-regulation".[44]

The tension between this approach and that of the Sea Gem Inquiry some five years earlier is abundantly clear. And the two reports produced quite different results. As has been seen, the Sea Gem Inquiry led to the 1971 Act and in due course to a comprehensive set of detailed regulations specific to the offshore oil and gas industry. In stark contrast, the result of the Robens Report was the Health and Safety at Work, etc. Act 1974 (HSWA 1974) which: **I-10.19**

> "introduced a broad goalsetting, non-prescriptive model, based on the view that 'those that create risk are best placed to manage it'. In place of existing detailed and prescriptive industry regulations, it created a flexible system whereby regulations express goals and principles, and are supported by codes of practice and guidance. Based on consultation and engagement, the new regime was designed to deliver a proportionate, targeted and risk-based approach."[45]

Furthermore, the regulatory function was centralised in the Health and Safety Commission (HSC) and the Health and Safety Executive (HSE) instead of the broad array of industry-specific agencies that had existed previously. Perhaps because the 1971 Act was just emerging as the Committee deliberated, Robens did not discuss offshore safety in any depth. Instead, the 1971 Act was noted as one of a category of statutes that, though not considered in detail,   **I-10.20**

---

[43] *Ibid*, paras 28, 30 and 41.
[44] *Ibid*, para 41.
[45] HSE, *Thirty Years On and Looking Forward* (2004), p 3.

the Committee thought capable of, on the face of it, being brought within the proposed unified system perhaps after the main arrangements had been made.[46] But that did not happen.

I-10.21   It would be going too far to say that the Robens approach was greeted uniformly with open arms at the time,[47] or that the subsequent HSWA 1974 was thereafter regarded by all as an unalloyed good.[48] But a measure of the perceived differences in the value of this approach as opposed to detailed regulation enforced by the offshore industry's "sponsoring" department may be gained from the numerous calls there were in the succeeding years for responsibility for offshore safety to be transferred to the HSE.[49]

I-10.22   Even if the offshore industry was not, then, brought within the ambit of the 1974 Act at this time in the way that Robens anticipated, that Act did have an impact. Responsibility for offshore safety remained with the DEn and was not transferred to the HSE, but the general duties contained in the HSWA 1974 were in due course expressly extended offshore,[50] including the duty on the employer to ensure, so far as is reasonably practicable, the health, safety and welfare of employees.[51] The provisions relating to workforce involvement, such as safety committees and safety representatives, were not, however, similarly extended. The point has been made that the two different regimes (that under the 1971 and 1975 Acts and that under the 1974 Act) operated on the basis of different enforcement procedures. The former relied on criminal penalties and, in the ultimate, the power to suspend operations, while the

---

[46] Robens Report, para 109.

[47] See eg A D Woolf, "Robens Report: The Wrong Approach?", 2 (1973) *Industrial Law Journal* 88.

[48] See eg R Kinnersley, *The Hazards of Work: How to Fight Them* (1973), pp 228–230; R Baldwin, "Health and Safety at Work: Consensus and Self-Regulation", in Baldwin and McCrudden (eds), *Regulation and Public Law* (1987) (hereinafter "Baldwin, 'Health and Safety at Work'"), pp 132–158; S Dawson, P Willman, A Clinton and M. Bamford, *Safety at Work: The Limits of Self-regulation* (1988); P James, "Reforming British Health and Safety Law: a framework for discussion", (1992) *Industrial Law Journal* 83 (hereinafter "James, 'Reforming British Health and Safety Law'"); Ogus, *Regulation*, p 188; S Tombs, "Law, Resistance and Reform: 'Regulating' Safety Crimes in the UK", 4 (1995) *Social and Legal Studies* 411.

[49] For example, by the dissenting members of the Burgoyne Committee (see Note of Dissent, para 13, Burgoyne Report, p 60) and by the Trades Union Congress (see Submission 62, para II (b), Burgoyne Report, p 292), in contrast to the opposition of the industry put forward by UKOOA (see Submission 43, Burgoyne Report, pp 241, 247–249).

[50] Health and Safety at Work, etc. Act 1974 (Application Outside Great Britain) Order 1977 (SI 1977/1232). The PED carried out the HSE's inspection function under an agency agreement between the HSC and the DEn. For the agency agreement see Burgoyne Report, Appendix 11.

[51] Section 2.

latter relied on a more flexible system of improvement notices, prohibition notices and lastly criminal penalties. It has also been emphasised that the two regimes each envisaged different inspectorates and that the industry was concerned that the HSE would not understand the special problems faced in the offshore situation, including the extreme cost of delays.[52] Questions accordingly arose as to how the spirit of the 1974 Act could survive *without* the workforce involvement envisaged by that Act and precisely *with* the sort of inspectorate that the Robens Committee had criticised.

The Burgoyne Committee

If, however, the regulatory situation for the offshore oil and gas   I-10.23
industry was, to put it at its most charitable, extremely complicated by the end of the 1970s, an opportunity arose at that time to review it in detail and to recommend any necessary changes. This occurred as a consequence of the blowout on the Ekofisk Bravo Platform in the Norwegian Sector of the North Sea in April 1977. Recognising that this could have been a major disaster resulting in significant loss of life, the UK Government established a further committee, under the chairmanship of Dr J H Burgoyne, to consider, among other things: "so far as they are concerned with safety, the nature, coverage and effectiveness of the Department of Energy's regulations governing the exploration, development and production of oil and gas offshore and their administration and enforcement".[53] The Report of this Committee, published in March 1980, is an important document that took on renewed significance in the context of events some eight years later. It follows the line adopted by the Sea Gem Inquiry in some respects but departs from it, significantly, in others. It too, however, can be read, admittedly with the benefit of hindsight, as having had profound insights into the nature of the problems confronting the regulation of health and safety offshore but then as having failed to follow through on those in the form of recommendations.

Burgoyne agreed with the Sea Gem Inquiry that the ultimate   I-10.24
responsibility for ensuring safety lay with the Government, and also that this end could be achieved through monitoring and enforcement by the regulator.[54] But Burgoyne departed from its predecessor when it came to the role that the law would play in this regard. The Sea Gem Inquiry wanted "a code of statutory authority with credible sanctions". Parliament in passing the 1971 Act was clearly of the

---

[52] Daintith, Willoughby and Hill, para 1-857.
[53] Burgoyne Report, para 1.1.
[54] *Ibid*, para 6.5.

view that the framework statute would allow regulations to keep pace with developments in the industry. But Burgoyne basically withdrew from the idea that law could ever be flexible enough to provide detailed safety regulations for the totality of this complex and evolving industry. The Committee's proposal was that the role of Government was "to *set objectives* designed to achieve a uniformly high standard of safety throughout the Industry".[55] It is possible to detect resonances of the Robens Committee in this statement, resonances that only become clearer when Burgoyne states that "[t]he Government shall discharge its responsibility for offshore safety via a single Government agency whose task it is to *set standards* and to ensure their achievement",[56] and that "[m]ethods of implementation should be advised as fully and flexibly as possible in guidance notes, which should be recognised as being non-mandatory".[57] But if those resonances seemed to indicate that Burgoyne was minded to place the offshore industry squarely under the 1974 Act, then they were misleading.

I-10.25   It seems instead that the Committee's apparent rejection of prescription at the level of the regulations stemmed from its recognition that a tension existed between the law and the regulated area. But whereas the Sea Gem Inquiry was troubled by the inability of the law to understand a Code of Practice which had been drafted by the industry, the Burgoyne Committee was confronted by evidence that the converse problem had now arisen and that difficulties were being encountered *by the industry* in dealing with regulations drafted in accordance with the needs *of the legal system*. As the PED expressed it in its submission to Burgoyne:

> "[i]t is ... difficult to draft regulations which can be readily understood by a person without legal training. We accept, however, that legal conventions must be respected and that, in the ultimate, regulations must be able to stand up in a court of law. We think that the regulations under the 1971 Act are now understood by the offshore oil industry, largely through explanation and interpretation from the Inspectorates to educate the industry through guidance notes".[58]

I-10.26 Burgoyne thus appears to have accepted the PED's contention that it could make the regulations understandable to the industry. While

---

[55] *Ibid*, para 6.2; emphasis added.

[56] *Ibid*, para 6.5; emphasis added.

[57] *Ibid*, para 6.15.

[58] Submission by the Department of Energy, Petroleum Engineering Directorate to the Burgoyne Committee. See Burgoyne Report, Submission 37, para 8, *Ibid*, p 228. It is worth noting that lawyers themselves were not enthusiastic about the comprehensibility of the 1971 Act. See W Dale, "Statutory Reform: The Draftsman and the Judge", 30 (1980) *ICLQ* 141, at 147.

the union members of the Committee expressed strong opposition, the majority decided in favour of leaving the PED in charge of safety rather than handing responsibility to the HSE.[59] It would be wrong to say, however, that the Committee was sanguine about the PED's abilities in this regard or that it was unaware of the extent to which the offshore industry had fallen behind developments onshore under the 1974 Act. Accordingly, it recommended that the PED should strengthen its relations with the HSE in order to benefit from the latter's expertise in occupational health and safety,[60] and it made a series of recommendations in relation to workforce involvement.[61] These latter reflected to some extent the terms of the Safety Representatives and Safety Committees Regulations 1977 applying onshore,[62] but there was no call for their extension offshore nor did Burgoyne "consider it essential to embody these principles in mandatory regulations".[63]

These issues contributed to the union members of the Committee issuing a Note of Dissent. As regards the question of safety representatives and safety committees, they pointed out that the tripartite Offshore Industry Advisory Committee had recently reached agreement in principle regarding the extension of the onshore regulations.[64] Whatever the actual reasons for the Burgoyne Committee's position, its refusal to recommend the extension of the onshore regulations may have been an acknowledgement that these operated so as to allow employers who did not recognise trade unions to avoid the mandatory involvement of safety representatives,[65] a very relevant point in the context of an industry with a very low level of unionisation.[66] With regard to the role of the PED, the Note of Dissent mentioned fears of the "possibility of shared values and membership of closed groups"[67] as between industry and regulator. But that assessment must be read in the light of the difficulties in communication perceived by the PED mentioned above.

I-10.27

---

[59] Burgoyne Report, para 6.6.

[60] *Ibid*, para 4.24

[61] *Ibid*, para 5.94.

[62] SI 1977/500 made under the HSWA 1974.

[63] Burgoyne Report, para 5.97.

[64] See the Note of Dissent, para 25, Burgoyne Report, p 63.

[65] See D M Kloss, *Occupational Health Law* (3rd edn, 1998), p 137.

[66] See S S Andersen, *British and Norwegian Offshore Industrial Relations: Pluralism and Neo-Corporatism as Contexts of Strategic Adaptation* (1987); C Woolfson, J Foster and M Beck, *Paying for the Piper: Capital and Labour in Britain's Offshore Oil Industry* (1996) (hereinafter "Woolfson, Foster and Beck"), pp 44*ff*. It is also worth noting that the original intention was to allow safety representatives to be appointed by either the workforce or trade unions. This provision in the HSWA 1974 was, however, repealed by the Employment Protection Act 1975 as part of the Social Contract between the then Labour Government and the TUC (see James, "Reforming British Health and Safety Law", 90).

[67] See the Note of Dissent, para 8, Burgoyne Report, p 59.

I-10.28   Burgoyne revealed that the orientation of the law envisaged by Parliament at the time of the passing of the 1971 Act was in fact difficult to implement. Regulators were unable to keep pace with developments in the offshore industry at the level even of secondary legislation and had instead to resort to guidance in order to achieve the necessary flexibility. They were also experiencing difficulties in communicating legal requirements to the industry. In view of the fundamental nature of these issues, it might be expected that these would have figured largely in the extensive parliamentary debate that followed publication of the Burgoyne Report.[68] In fact, these issues were not mentioned at all. Instead, the focus was mainly on where responsibility for safety offshore should ultimately lie, with the DEn or the HSE. The Conservative Government took the view that it should remain with the DEn on the basis that those administering safety should have an intimate knowledge of this complex field, and indeed have immediate access to the related expertise within other units of the Department.[69] Nevertheless, it was explicitly recognised that there was a danger in isolating the offshore from health and safety developments onshore and, therefore, the Secretary of State for Energy was to take over the responsibilities of the Employment Secretary under the HSWA 1974 – the matters which had previously been the subject of the agency agreement between the two Departments.[70] The former would still seek the advice of the HSE but ultimate responsibility would now rest with him. Whereas unions were concerned that this gave rise to an unacceptable conflict of interest, the Government's view was that the risks actually lay in splitting the various responsibilities of the Department of Energy given the limited number of experts at its disposal.[71] The Government acknowledged that there was room for new safety initiatives in the offshore industry, but maintained that the new administrative arrangements would be able to respond to this need efficiently and effectively.[72]

I-10.29   The Government's confidence in an arrangement that ran counter to practically every other industry was not shared by the opposition, who regarded it as a "dog's breakfast".[73] It was also concerned, firstly, by the Burgoyne Committee's finding that none of the safety committees visited during their investigations had ever been

---

[68] *Hansard* HC (series 5), vol 991, cols 1472–1546 (November 1980).

[69] Mr H Gray (Minister of State, Department of Energy) (hereinafter "Gray"), at cols 1476–1480.

[70] Gray, at col 1479.

[71] *Ibid*, at cols 1479, 1482.

[72] *Ibid*, at col 1484.

[73] Dr D Owen, at col 1492.

contacted by the PED and, secondly, by the PED's view that they had never had the need to speak to such a committee.[74] The Opposition was accordingly clear that the responsibility for offshore safety should be transferred to the HSE.[75]

Whatever the difference of view between Conservative and Labour on this issue, what is more significant in view of the issues highlighted in this chapter is the fact that legislators of all persuasions remained convinced of a direct correlation between the promulgation and enforcement of prescriptive regulation, on the one hand, and improved occupational health and safety, on the other. Labour saw the conflict of interest within the DEn between safety and production as having the potential to reduce overall safety, whereas the Conservatives saw any increase in the distance between the technical experts of the DEn and the safety regulators as having the potential to reduce overall safety. But, crucially, all legislators appear to have adhered to the rationale they deployed in 1971 as regards the appropriate regulatory approach. It could be objected that the Robens approach upon which the 1974 Act is based would have resulted in a move away from prescriptive regulation and hence that the Labour programme by this time must be seen as fundamentally different,[76] but it is significant that neither side once mentioned this point during the course of the debate. The *orientation* of the law (prescriptive or goal-setting) was not an issue for legislators whereas the location of responsibility was. Whatever the Burgoyne Committee had discovered about difficulties of communication and about the limitations of prescription, this was not apparently visible to legislators. It is ironic that if the Opposition had focused on these aspects of the Burgoyne Committee's findings, the Government would have found it much more difficult to resist the calls for a transfer of responsibility to the HSE.

I-10.30

THIRD PHASE: THE PERMISSIONING APPROACH

The Piper Alpha disaster and the Cullen Inquiry

The extent to which the Burgoyne Report and the subsequent parliamentary debate represented a missed opportunity to bring offshore health and safety into line with the most advanced thinking on the subject became all too clear eight years later. On 6 July 1988 the

I-10.31

---

[74] Mr H Walker, at col 1533.

[75] Dr D Owen, at cols 1489–1492.

[76] This point can, however, be overstated. There is evidence that even up to the mid-1980s the development of the approach to regulation onshore proposed by Robens and provided for by the HSWA 1974 was still relatively in its infancy. See Baldwin, "Health and Safety at Work", p 145.

Piper Alpha production platform was almost entirely destroyed by a series of explosions and subsequent fires. A total of 167 men lost their lives, making this by far the worst accident in the history of the offshore industry. A Public Inquiry was set up[77] under the chairmanship of a senior Scottish judge, Lord Cullen. This was charged principally with providing answers to two questions: "[w]hat were the causes and circumstances of the disaster ...? and What should be recommended with a view to the preservation of life and the avoidance of similar accidents in the future?"[78]

I-10.32   The Inquiry became, at the time, the longest and most thorough ever seen in the UK. Its two-volume Report is a damning indictment of the state of safety in the UK sector of the North Sea in the late 1980s. The main cause of the disaster was a failure of the permit to work (PTW) system, especially as it related to communication between the different shifts on the platform. On the night in question, this led to equipment being used that was in fact subject to maintenance. A gas escape and explosion resulted.[79] What was already a serious incident was made much worse by a number of other factors, each one sufficient to raise profound questions about the adequacy of the regulatory regime in place at this time. Firstly, the Claymore platform, to which the Piper Alpha was connected, continued to pump oil, thus feeding the fires that followed the initial explosion.[80] Secondly, the Offshore Installation Manager, such a focus of attention in the Sea Gem Inquiry's recommendations, "took no initiative in an attempt to save life".[81] Thirdly, emergency systems, for which specific regulations had finally been passed in the late 1970s, failed as a result of the intensity of the explosion. Fourthly, the platform's Stand-by Vessel, the *Silver Pit*, proved ineffective in the circumstances[82] as did the *Tharos* fire-fighting vessel, which was actually on hand at the time of the disaster.[83] The platform's owners, Occidental Petroleum, were severely criticised by Lord Cullen for this state of affairs. They were said to be unprepared for this sort of emergency and to have adopted a superficial attitude to such risks. Adequate safety arrangements were frequently not in place and where they were they were often ignored, as exemplified by the PTW system.[84]

---

[77] By the Secretary of State for Energy under the Offshore Installations (Public Inquiries) Regulations 1974 (SI 1974/338).

[78] Lord Cullen, *The Public Inquiry into the Piper Alpha Disaster* (Cm 1310, 1990), (hereinafter "the Cullen Report"), para 1.1.

[79] *Ibid*, Chapter 11.

[80] *Ibid*, paras 7.37 to 7.40.

[81] *Ibid*, para 8.35.

[82] *Ibid*, para 9.42.

[83] *Ibid*, paras 9.49 to 9.57.

[84] *Ibid*, Chapter 14.

Lord Cullen's criticism did not, however, stop with the operator. **I-10.33**
The PED was also singled out. In Lord Cullen's view, the inspections carried out by the PED were "superficial to the point of being
of little use as a test of safety on the platform"[85] and not really an
effective means of assessing the *management* of safety.[86] This litany
of criticism went to the core of the existing regulatory approach
and completely undermined existing assumptions: the Government's
faith in the regulation of safety by the "sponsoring department",
specifically reiterated despite the opportunity to revisit the issue in
the aftermath of the Burgoyne Report; the regulator's adherence to
a mode of regulation and inspection that had already been called
into question by Robens in 1972, and its questionable grasp of core
occupational health and safety, so tellingly already exposed in its
evidence to the Burgoyne Committee;[87] and the industry's ability to
prepare for and respond to emergencies, which had been expressed
so confidently in its submission to Burgoyne.[88]

Lord Cullen's trenchant criticism of both the operator and the **I-10.34**
regulator were translated into no fewer than 106 recommendations
which, taken together, constituted a radical reorientation of occupational health and safety regulation offshore. Some of the issues
that had emerged already at the time of the Burgoyne Committee
a decade earlier were again apparent in the Cullen Report. Now,
however, they were followed through to their logical conclusion
and worked into a comprehensive and coherent system. As regards
the regulator, Cullen adopted the proposal made by the authors of
the Note of Dissent to the Burgoyne Report – albeit for different
reasons – in recommending that responsibility for health and safety
offshore should be transferred from the "sponsoring department" to
the Health and Safety Executive. The approach to regulation would
similarly see radical reform. The proposal was that the operator
of each and every installation on the UKCS should submit to the
regulator a Safety Case, that is, a document making the case that the
installation is safe both in its construction and operation. The Safety
Case would demonstrate that certain objectives have been met,
including the following: that the Safety Management System (SMS)
of the company and that of the installation are adequate to ensure
that the design and the operation of the installation and its equipment
are safe; that the potential major hazards to the installation have

---

[85] *Ibid*, para 15.48.
[86] *Ibid*, para 15.50.
[87] See note 36 above.
[88] "UKOOA has full confidence in the ability of the industry ... to cope with any
emergencies that may arise", Burgoyne Report, Submission 43, para 5.3. For the full Code
of Practice and Plan for Offshore Emergencies, see Appendix A to Submission 43.

been identified and appropriate controls provided; and that adequate provision is made for ensuring, in the event of a major emergency affecting the installation, a temporary safe refuge for personnel and their safe and full evacuation, escape and rescue.[89] Superficially, it may be wondered whether there was anything dramatically different here from the recommendations made by the Sea Gem Inquiry in 1967. But whereas it too had called for a comprehensive approach, what Cullen had in mind was quite distinct.

I-10.35   As troubling as it may have appeared to lawyers and regulators, Cullen essentially abandoned any idea that law, even in its most flexible forms such as secondary legislation, could provide a detailed and comprehensive code covering all the aspects of the industry. Rather he placed a considerable degree of responsibility on *the operator*: to identify risks to occupational health and safety, whether at the level of issues with a catastrophic potential or at that of the more mundane slips, trips and falls; to demonstrate, by means of Quantified Risk Assessment (QRA) where appropriate, that these risks had been minimised; and to show how this risk minimisation had been (or was to be) achieved. He thus very significantly departed from the views on the location of responsibility expressed by both the Sea Gem Inquiry and the Burgoyne Committee, stating that "a regulator cannot be expected to assume direct responsibility for the on-going management of safety ... this is and remains in the hands of the operator".[90] As far as he was concerned, an operator may depart from procedures outlined either in regulations or in official guidance provided that this was justified in the Safety Case.

I-10.36   Such a position accordingly implied a reorientation of the regulations themselves. Thus, Cullen called for prescriptive regulations as far as possible to be replaced with goal-setting regulations.[91] He argued, indeed, that prescriptive regulation could actually be part of the problem rather than a solution, as this approach encouraged a

---

[89] Cullen Report, para 23.2.

[90] *Ibid*, para 21.4.

[91] The recommendation was that the Construction and Survey Regulations, the Fire Fighting Regulations, the Life-Saving Appliances Regulations and the Emergency Procedures Regulations should be revoked and replaced by: (1) Construction Regulations, covering *inter alia* the structure and layout of the installation and its accommodation; (2) Plant and Equipment Regulations, covering *inter alia* plant and equipment on the installation and in particular those handling hydrocarbons; (3) Fire and Explosion Protection Regulations, covering *inter alia* both active and passive fire protection and explosion protection; and (4) Evacuation, Escape and Rescue Regulations, covering *inter alia* emergency procedures, life-saving appliances, evacuation, escape and rescue. Each of these sets of regulations should include goal-setting regulations as their main or primary provisions and should be supported by guidance notes giving advice which is non-mandatory. Cullen Report, para 21.69.

compliance mentality rather than a wider consideration of safety[92] and was unable to cope with the overall interaction of components.[93] The extent of his departure from the position both of the Sea Gem Inquiry and indeed of the legislators at the time of the passing of the Mineral Workings (Offshore Installations) Act 1971 can perhaps best be seen when he associated himself with the remarks of one witness to the Inquiry who claimed that safety could not be legislated.[94]

This change in the orientation of the regulations naturally **I-10.37** implied a different notion of compliance as well as a change in the role of the regulator. Lord Cullen's view in this regard was that the operator under the new regime must satisfy itself by means of audits that the Safety Management System was being adhered to and that the regulator should review the operator's audit and ensure that the output from the SMS was satisfactory.[95] This shift in responsibility away from the regulator and towards the operator was also evident in the degree of freedom that the latter would have in the regime envisaged by Cullen with regard to the specification of the standards to be used to demonstrate compliance with the goal-setting regulations.[96] This fitted in, however, with Cullen's view that the primary function of the Safety Case was to ensure that every operator produces a Formal Safety Assessment (FSA) to assure itself that its operations are safe. Only secondarily would it be a matter of demonstrating this to the regulators, albeit that this would meet the legitimate expectations of the workforce and the public and provide a sound basis for regulation.[97] There could perhaps be no clearer indication than this of the extent of Cullen's departure from a traditional view of the process and function of regulation.

The recommendations regarding the use of FSA, QRA and other **I-10.38** such procedures are another interesting feature of Cullen's approach. He was explicitly impressed by the HSE's use of such methods in relation to onshore industries under the Control of Industrial Major

---

[92] *Ibid*, para 21.51.

[93] *Ibid*, para 21.42.

[94] *Ibid*, para 21.4. The witness was Mr R E McKee, the Chairman and Managing Director of Conoco (UK) Ltd.

[95] *Ibid*, para 21.60.

[96] *Ibid*, para 21.70.

[97] *Ibid*, para 17.35. It is worth noting that the idea of a formal safety assessment was mentioned at the second reading stage of the Mineral Workings (Offshore Installations) Bill in 1971. Mr David Watkins raised the issue or "damage control ... a technique whereby it is possible for experts to examine thoroughly any place where people are subject to any possible form of hazard arising from their work and, by reporting fully, enable steps to be taken to eliminate many of the causes of accident". *Hansard* HC (series 5), vol 816, col 663 (28 April 1971).

Accident Hazard (CIMAH) Regulations 1984[98] – indeed this was one of the factors that helped him decide which agency should be given responsibility for regulating offshore safety.[99] His assessment of the value of this approach appeared to be based in part on its contribution to successful communication between industry and regulator, which, as has been seen, was a concern to both the Sea Gem Inquiry and the Burgoyne Committee. Cullen found no evidence to support the concerns expressed by the authors of the Note of Dissent to the Burgoyne Report regarding the closeness of the PED and the industry or a conflict of interest between safety and production in the DEn.[100] Rather, he seems to have shared the PED's concerns as expressed in the evidence to the Burgoyne Committee as regards the difficulty in achieving communication between law and the regulated area. Whereas the Burgoyne Committee took some steps towards addressing this problem in its recommendations of non-mandatory guidance and flexibility, Cullen met the problem head-on and recommended an approach that would allow the regulators to speak to the industry in a language that it could understand. In evidence to the Inquiry, the HSE explained that it used QRA as a means of founding "legal or political judgements as firmly as possible on a rigorous scrutiny of the facts" because the "technologically based industries or scientifically numerate organisations" it dealt with expected "a structured and logical approach".[101] Cullen was clearly impressed by this.

I-10.39   The issue that caused such contention for the Burgoyne Committee and indeed for Parliament thereafter was, of course, the involvement of the workforce in health and safety. While the Inquiry was ongoing, the DEn brought in a set of regulations that provided for workforce involvement.[102] This was not an extension of the corresponding onshore regulations but rather reflected the approach recommended by the Burgoyne Committee. As a result, in the course of the Piper Alpha Inquiry, there were calls for these to be amended to incorporate trade union involvement.[103] While Cullen was sympa-

---

[98] SI 1984/1902 implementing the so-called Seveso Directive (Council directive 82/501/EEC of 24 June 1982 on the major-accident hazards of certain industrial activities) and now replaced by the Control of Major Accident Hazard (COMAH) Regulations 1999 (SI 1999/743) which in turn implement the Seveso II Directive (Council Directive 96/82/EC of 9 December 1996 on the control of major-accident hazards involving dangerous substances, as amended by Directive 2003/105/EC).

[99] Cullen Report, paras 22.28 and 22.34.

[100] Ibid, para 22.38.

[101] Ibid, para 17.53.

[102] Offshore Installation (Safety Representatives and Safety Committees) Regulations 1989 (SI 1989/971).

[103] Details of trade union evidence to the Cullen Inquiry can be found in the Cullen

thetic to the view that the appointment of representatives by trade unions could be beneficial with regard to credibility and the ability to resist pressure, he concluded that the particular circumstances of the offshore workforce in terms of its low level of unionisation and its fragmentation meant that such a change would have the effect of removing representation from a large number of workers.[104]

The Offshore Safety Act 1992 and the Safety Case Regulations 1992

Just as the recommendations of the Sea Gem Inquiry and the Burgoyne Committee had been almost unequivocally accepted, so did the Government accept without demur Cullen's 106 proposals for change.[105] The foundation was laid with the passing of the Offshore Safety Act 1992 in March of that year which served to extended fully the Health and Safety at Work, etc. Act 1974 offshore and allowed regulations to be made repealing those made under the Mineral Workings (Offshore Installations) Act 1971 and the Submarine Pipelines Act 1975. The Health and Safety Commission announced that it would set about the reform of offshore safety on three fronts: the development of new regulations specific to the offshore environment in line with Cullen's vision for a goal-setting approach; ensuring that the offshore environment was subject to new regulations being made to implement European directives on health and safety at work; and the extension of offshore of existing health and safety regulations as appropriate.[106]

I-10.40

In November, the Offshore Installations (Safety Case) Regulations 1992 were made,[107] coming into force on 31 May 1993. These required operators of fixed installations to prepare a safety case prior to completion of the design and to send it to the HSE so as to allow time for any concerns raised to be taken account of in the design.[108] Further, operation of a fixed installation was not permitted unless the safety case had been accepted by the HSE.[109] Owners of mobile installations were not permitted to move such installations into UK waters (as defined) with a view to operation without prior acceptance of a safety case.[110] Where combined opera-

I-10.41

Report, para 21.78–80).

[104] *Ibid*, para 21.85

[105] For the parliamentary debate, see *Hansard* HC (series 6), vol 180, cols 329–345 (12 November 1990); vol 187, cols 472–567 (7 March 1991).

[106] For details of the HSC's statement see Daintith, Willoughby and Hill, para 1-932.

[107] SI 2885/1992.

[108] Reg 4(1).

[109] Reg 4(2).

[110] Reg 5.

tions involving a fixed and a mobile installation were to be carried out, a separate safety case would be required.[111] Finally, the 1992 Regulations required an operator of a fixed installation to prepare a safety case in relation to the proposed abandonment of an installation.[112] In each case (except combined operations), the operator or owner had to include sufficient particulars to demonstrate that (1) the management system was adequate to ensure that relevant statutory provisions would be complied with; (2) adequate arrangements were in place for auditing and reporting; (3) all hazards with the potential to cause a major accident had been identified; and (4) risks had been evaluated and measures taken to reduce them to the lowest level that is reasonably practicable.[113] In each case, schedules to the regulations provided further detail on what the safety case was to contain.[114] Regulation 9 embodied Cullen's vision that the safety case should be a "living document" insofar as it provided that an operator or owner should revise it as often as required. Where any such proposed revision would render the safety case "materially different" from the version last seen by the HSE, the regulation required that the regulator should accept it before it be made. Regulation 9 also required the triennial resubmission of the safety case. Without renewed acceptance by the HSE, continued operation was not permitted. An indication of the radical nature of the safety case approach was to be found in regulation 10 which imposed a duty on the operator or owner to ensure that health and safety procedures and arrangements contained in the safety case were actually followed. Criminal liability could arise from a breach of that duty. It was also incumbent on every person to whom the regulation applied to co-operate with the operator or owner so as to enable the latter to comply with the Regulations.[115] Transitional provisions allowed the 200 or so installations already operating on the UKCS to continue to do so after the coming into force of the Regulations, provided that a safety case was submitted within six months of that date and provided the HSE had accepted the case within 30 months of that date.[116] The HSE announced in November 1995 that it had achieved the successful assessment and acceptance of all safety cases for existing installations before the deadline. The HSE at this time also indicated just how far the understanding of

---

[111] Reg 6.

[112] Reg 7.

[113] Reg 8. The ALARP standard (As Low As Reasonably Practicable).

[114] Sch 2 for fixed installations, Sch 3 for mobile installations, Sch 4 for combined operations and Sch 5 for abandonment.

[115] Reg 14.

[116] Reg 13.

the locus of responsibility for safety had changed since the Sea Gem Inquiry and the Burgoyne Committee. It stated that acceptance of a safety case "cannot guarantee" that safety management systems are working effectively but rather allows inspectors to "target their continuing intervention".[117]

The goal-setting regulations

The process of introducing the new goal-setting regulations was also **I-10.42** efficiently completed. The first of these, the Offshore Installations (Management and Administration) Regulations 1995,[118] deals with the following matters: notification of entry into or departure from relevant waters of an installation and of any change in the duty holder of an installation;[119] the appointment of an installation manager, his powers of restraint and putting ashore and the duty of others to co-operate with him;[120] the keeping of records of those on the installation;[121] the use of permit to work systems;[122] the availability and comprehensibility of health and safety instructions;[123] arrangements for communications with the shore, other vessels, aircraft and other installations;[124] helicopter operations;[125] gathering of meteorological and related information;[126] availability of contact details for the HSE;[127] health surveillance of workers;[128] supply of drinking water and provisions;[129] identification of the installation;[130] the possibility of exemptions;[131] and the application of the Employers' Liability (Compulsory Insurance) Act 1969.[132] Reading through this list, the level of detail may give the impression that these are really prescriptive rather than goal-setting regulations. Closer inspection

---

[117] See HSE Press Release E180:95, 22 November 1995.

[118] SI 1995/738 as amended by the Offshore Safety (Miscellaneous Amendments) Regulations 2002 (SI 2002/2175).

[119] Reg 5. "Duty holder" is defined by Reg 2(1) as meaning the operator in relation to a fixed installation and the owner in relation to a mobile installation.

[120] Regs 6, 7 and 8.

[121] Reg 9.

[122] Reg 10.

[123] Reg 11.

[124] Reg 12. There are particular requirements in respect of helicopter landings on not-normally-manned installations.

[125] Reg 13.

[126] Reg 14.

[127] Reg 15.

[128] Reg 16.

[129] Regs 17 and 18.

[130] Reg 19.

[131] Reg 20.

[132] Reg 21.

of the specific provisions reveals, however, that, while some elements of prescription remain, important issues are indeed dealt with on the basis of goal-setting. For example, regulation 6(b) requires that the duty holder ensure that "the installation manager is provided with appropriate resources to be able to carry out effectively his function", but offers no further specification of what these resources might be. Similarly, regulation 10 requires that a permit to work system be utilised where this is required by the nature of the work or the circumstances in which work may be carried out, but provides no detail on the work or circumstances envisaged. The duty holder's response to these regulations will be determined by the Formal Safety Assessment they carry out under the Safety Case regulations.

I-10.43   The same approach is found in the second set of regulations usually referred to under the goal-setting heading, the Offshore Installations (Prevention of Fire and Explosion, and Emergency Response) Regulations 1995.[133] These begin by imposing a general duty on the duty holder to take appropriate measures to protect persons on the installation from fire and explosion and to secure effective emergency response.[134] The means by which this duty is fulfilled is principally via an assessment covering: the identification of events which could cause a major accident involving fire or explosion or the need for evacuation, escape or rescue to avoid such an accident; "the evaluation of the likelihood and consequences of such events" (that is, a risk assessment); the establishment of performance standards for evacuation, escape, recovery and rescue equipment; and "the selection of appropriate measures".[135] Thereafter a series of more specific requirements are listed in regulations 6 to 21, but none of these affects the generality of the duty in regulation 4(1).[136] The more specific provisions relate to preparation for emergencies (including helicopter emergencies) and preparation of an emergency response plan;[137] prevention of fire and explosion;[138] detection of incidents;[139] communication in, and the control of, emergencies;[140] mitigation of fire and explosion;[141] muster areas, arrangements for evacuation, means of escape and arrangements for recovery and rescue;[142] suitability of personal protective equipment

---

[133] SI 1995/743.
[134] Reg 4(1).
[135] Reg 5.
[136] Reg 4(2).
[137] Regs 6, 7 and 8.
[138] Reg 9.
[139] Reg 10.
[140] Regs 11 and 12.
[141] Reg 13.
[142] Regs 14, 15, 16 and 17.

for use in an emergency;[143] suitability and condition of plant;[144] life-saving appliances;[145] and information regarding plant.[146] Once again, while there are elements of prescription in these more specific provisions, most make reference to the duty holder taking "appropriate measures" or making "appropriate provision" without specifying in detail what these measures or provisions should be.

The third set of "goal-setting" regulations is the Offshore Installations and Wells (Design and Construction, etc.) Regulations 1996.[147] The principal parts of this statutory instrument (relating respectively to integrity of installations[148] and wells[149]) each start by imposing a general duty. In the case of integrity of installations, the duty is imposed on the duty holder to ensure that "an installation at all times possesses such integrity as is reasonably practicable".[150] In the case of wells, the duty is imposed on the well operator[151] to ensure that a well is utilised at all stages of its life in such a way that "so far as is reasonably practicable, there can be no unplanned escape of fluids" and that risks to health and safety are as low as is reasonably practicable.[152] In each case, more specific provisions follow which are expressly without prejudice to the generality of these duties.[153] In the case of integrity of installations, the more specific provisions relate to the design of, and work to, an installation;[154] operation of an installation;[155] maintenance of integrity;[156] reporting of danger to an installation;[157] and decommissioning and dismantlement.[158] In the case of wells, the more specific provisions relate to assessment

I-10.44

---

[143] Reg 18.

[144] Reg 19.

[145] Reg 20.

[146] Reg 21.

[147] SI 1996/913.

[148] Pt II.

[149] Pt IV.

[150] Reg 4(1).

[151] Defined by Reg 2(1) as the person appointed by the concession owner to organise and supervise well operations, or in the absence of such an appointment the concession owner.

[152] Reg 13(1).

[153] Regs 4(2) and 13(2).

[154] Regs 5 and 6. Note that, uniquely among the regulations in this instrument, a defence is provided by Reg 22 to a person charged with contravention of these provisions. The defence is to the effect that the commission of the offence was the fault of another person (not being one of his employees) and that he took all reasonable care and exercised all due diligence to avoid the commission of the offence.

[155] Reg 7.

[156] Reg 8.

[157] Reg 9.

[158] Reg 10. Note that Part III of and Schedule I to the Design and Construction Regulations impose further requirements relating to installations, for example, with regard to the design and construction of helicopter landing areas.

of conditions below ground;[159] design with a view to suspension and abandonment;[160] fitness for purpose of materials;[161] well control;[162] arrangements for examination by independent persons of the well;[163] provision of drilling and other information relating to a well;[164] imposition of a duty of co-operation with the well operator on persons concerned in an operation relating to a well;[165] information, instruction, training and supervision.[166]

I-10.45   Three other sets of regulations follow the same goal-setting approach as that contained in the instruments described above – the Pipelines Safety Regulations 1996;[167] the Diving at Work Regulations 1997;[168] and the Lifting Operations and Lifting Equipment Regulations 1998[169] – but these will not be discussed in further detail here.

I-10.46   The goal-setting regulations discussed above also served in part to give effect to the requirements of European law in the shape of the Framework Directive on Safety at Work[170] and the Extractive Industries Directive.[171] Further general regulations covering both onshore and offshore industries were also introduced in 1992 to implement European requirements.[172]

---

[159] Reg 14.

[160] Reg 15.

[161] Reg 16.

[162] Reg 17.

[163] Reg 18.

[164] Reg 19.

[165] Reg 20.

[166] Reg 21.

[167] SI 1996/825. These revoke the Offshore Installations (Emergency Pipe-line Valve) Regulations 1989 (SI 1989/680), the regulations requiring the installation of emergency valves in the immediate aftermath of the Piper Alpha disaster.

[168] SI 1997/2776.

[169] SI 1998/2307.

[170] Council Directive 89/391/EEC of 12 June 1989 on the introduction of measures to encourage improvements in the safety and health of workers at work.

[171] Council Directive 92/91/EEC of 3 November 1992 concerning the minimum requirements for improving the safety and health protection of workers in the mineral-extracting industries through drilling (eleventh individual Directive within the meaning of Art 16 (1) of Directive 89/391/EEC).

[172] Management of Health and Safety at Work Regulations 1992 (SI 2051/1992) (now replaced by SI 3242/1999) implementing Council Directive 89/391/EEC of 12 June 1989, Council Directive 92/85/EEC of 19 October 1992 on the introduction of measures to encourage improvements in the safety and health at work of pregnant workers and workers who have recently given birth or are breastfeeding (tenth individual Directive within the meaning of Art 16 (1) of Directive 89/391/EEC) and Council Directive 94/33/EC of 22 June 1994 on the protection of young people at work; Provision and Use of Work Equipment Regulations 1992 (SI 2932/1992) (now replaced by SI 2306/1998) implementing Council Directive 89/655/EEC of 30 November 1989 concerning the minimum safety and health requirements for the use of work equipment by workers at

All of this activity clearly indicates that, if nothing else, the I-10.47
regulation of health and safety offshore during the 1990s changed out
of all recognition. Now referred to as a *permissioning* approach,[173]
it appeared to many to represent a significant step forward for what
had been a problematic industry sector. And, indeed, during those
years there was a tendency to suggest very much that a corner had
been turned and that a cultural change had occurred in the industry's approach to safety. Some commentators, however, were less
enthusiastic. Woolfson, Foster and Beck, for example, writing in
1996, suggested that this perception was due more to efforts on the
part of the offshore industry to project a new safer image and to
divert attention from ongoing health and safety problems.[174] These
authors contended, indeed, that the new permissioning regime was
inherently flawed. They had three principal concerns. Firstly, Safety
Cases placed the major burden of responsibility on line management
in a highly technocratic way that few understood, especially because
external consultants had often produced them. Secondly, the handling
of compliance was also technocratic because the HSE now audited
processes rather than checking to see if its regulations were being
observed. The authors believed that goal-setting regulations could
only be effective if anchored in genuine workforce involvement, and
this was affected by their third concern, namely that there was a lack
of trade union support for the safety representatives who formed the
cornerstone of workforce involvement.[175]

---

work (second individual Directive within the meaning of Art 16(1) of Directive 89/391/
EEC) itself amended by Directive 95/63; Manual Handling Operations Regulations
1992 (SI 2793/1992) implementing Council Directive 90/269/EEC of 29 May 1990 on
the minimum health and safety requirements for the manual handling of loads where
there is a risk particularly of back injury to workers (fourth individual Directive within
the meaning of Art 16 (1) of Directive 89/391/EEC); Personal Protective Equipment at
Work Regulations 1992 (SI 2966/1992) implementing Council Directive of 30 November
1989 on the minimum health and safety requirements for the use by workers of personal
protective equipment at the workplace (third individual directive within the meaning
of Art 16 (1) of Directive 89/391/EEC) (89/656/EEC); Personal Protective Equipment
Regulations 1992 (SI 3139/1992) implementing Council Directive 89/686/EEC of 21
December 1989 on the approximation of the laws of the member states relating to personal
protective equipment; Health and Safety (Display Screen Equipment) Regulations 1992
(SI 2792/1992) implementing Council Directive 90/270/EEC of 29 May 1990 on the
minimum safety and health requirements for work with display screen equipment (fifth
individual Directive within the meaning of Art 16 (1) of Directive 89/391/EEC).

[173] See, for example, Health and Safety Commission, *Policy Statement: Our Approach to
Permissioning Regimes* (2003).

[174] Woolfson, Foster and Beck, pp 360–361.

[175] *Ibid*, p 346. For further critical comment see D Whyte, "Moving the Goalposts:
the Deregulation of Safety in the Post-Piper Alpha Offshore Oil Industry" (1997)
Contemporary Political Studies 1148.

The Offshore Installations (Safety Case) Regulations 2005

I-10.48   As evidence both that these criticisms may carry some weight and equally that the HSE adopts a proactive stance to the regulation of health and safety offshore, the 1992 Safety Regulations were repealed and replaced with an updated set in 2005.[176] That said, however, the appearance of new regulations might seem surprising given that the 1992 Regulations explicitly envisaged change, with the safety case being understood as a living document that would evolve through the lifetime of an installation. The HSE noted, however, that the system in place under the 1992 Regulations was increasingly perceived to be excessively bureaucratic. Equally, while it was seen to have produced a significant change in the health and safety situation offshore in the early years, each subsequent three-year cycle of safety case resubmissions appeared to have produced less in the way of improvement.[177]

I-10.49   Thus, as regards the focus of the new regulations on bureaucracy, a triennial resubmission of the safety case is no longer required, with this requirement replaced by a five-yearly "thorough review" (or as directed by HSE).[178] Providing that specific safety cases are no longer required for combined operations, design or decommissioning also reduces the burden of bureaucracy. In the first two instances, simpler notification procedures are in place,[179] while in the last a modification of the existing safety case is all that is required, albeit that this must be accepted by the HSE.[180] Note also that a right of appeal to the Secretary of State against a refusal to accept is introduced.[181]

I-10.50   Turning to the focus on diminishing returns, there are changes to the requirements relating to workforce involvement and the demonstration of compliance. As regards workforce involvement,

---

[176] Offshore Installations (Safety Case) Regulations 2005 (SI 3117/2005), which came into force on 6 April 2006.

[177] See generally: Health and Safety Commission, *Proposals to Replace the Offshore Installations (Safety Case) Regulations 1992* (2004); and Health and Safety Commission, *A Strategy for Workplace Health and Safety in Great Britain to 2010 and Beyond* (2004).

[178] 2005 Regulations, Reg 13. Note that there is an exception to this rule where there are "material changes" which will still require to be accepted (Reg 14). For further detail see HSE, *Offshore Installations (Safety Case) Regulations 2005 Regulation 13: Thorough Review of a Safety Case*, Offshore Information Sheet No. 4/2006; also HSE, *Procedure for dealing with thorough review summaries submitted under regulation 13 of the Offshore Installations (Safety Case) Regulations 2005* (2007).

[179] Reg 10 (Notification of combined operations), Reg 6 (Design and relocation notification for production installation) and Reg 9(1) (Design notification in respect of a non-production installation). Note that the terms "fixed" and "mobile" used in the 1992 Regulations are thus replaced in the 2005 Regulations by "production" and "non-production" respectively.

[180] Reg 11 (Safety case for dismantling fixed installation).

[181] Reg 24.

while this was already a feature of the 1992 Regulations, the revised approach requires that the safety case summarises consultation with the workforce not only with regard to its preparation but also its revision and review.[182] In other words, by these means, the regulator seeks to ensure that the workforce is directly engaged in the safety case process on an ongoing basis. As regards the demonstration of compliance, while the 1992 Regulations required a demonstration that major hazard risks had been reduced to a level as low as reasonably practicable (ALARP), the 2005 Regulations require that they are identified and evaluated, and that relevant statutory provisions will be complied with.[183] At first sight, this can appear to be a retrograde step. In fact, the change reflects the recognition that the safety case regulations were not the appropriate location for the setting of standards. Insofar as Lord Cullen had called for the replacement of the previous prescriptive regulations with goal-setting regulations, it is in these latter that standards should appear. Furthermore, insofar as there continue to be some prescriptive elements setting absolute standards,[184] there was actually a contradiction between these and the 1992 Regulations.[185] ALARP has not, therefore, disappeared from the offshore health and safety environment; rather it is now to be found only in the goal-setting regulations.[186]

In recognition of the changing character of the UKCS as a maturing province with the appearance of new and perhaps less experienced entrants,[187] the 2005 Regulations place a new duty on the licensee in respect of health and safety – a curious echo of the position in the early phase of the evolution of health and safety regulation offshore, albeit of a quite different nature. Thus, regulation 5 provides that the licensee must ensure that "any operator appointed by him is capable of satisfactorily carrying out his functions and discharging his duties under the relevant statutory provisions" and then "take all reasonable steps to ensure" that the operator does indeed behave in this way. As the HSE's guidance makes clear, if in its opinion "the operator appointed by the licensee is unable to discharge the management and control functions, the duty to submit the safety case and other

I-10.51

---

[182] Sch 2, para 3. The Offshore Installations (Safety Representatives and Safety Committees Regulations 1989 (SI 1989/971) are consequently amended.

[183] Reg 12.

[184] As discussed previously; see paras I-10.42 to I-10.43.

[185] See HSE, *Offshore Installations (Safety Case) Regulations 2005 Regulation 12: Demonstrating compliance with the relevant statutory provisions*, Offshore Information Sheet No. 2/2006.

[186] As discussed previously, see paras I-10.42 to I-10.45.

[187] See further in Chapter I-4 above.

related duties will revert to the licensee".[188] The licensee will accordingly have to take particular care in the appointment of the operator if it is not to be faced with a considerable regulatory burden that it itself may struggle to cope with.

I-10.52   Similar concerns arising from the changing character of the maturing province lie behind the alteration in the definition of "operator" in the 2005 Regulations. Whereas in the 1992 Regulations, this was defined in relation to a fixed installation as "the person appointed by a concession owner to execute any function of organising or supervising any operation to be carried out by such installation", this is now defined in relation to a production installation as "the person appointed by the licensee to manage and control directly *or by any other person* the execution of the main functions" of such an installation.[189] The net effect is to emphasise to operators that while they can delegate functions to contractors, this does not absolve them of their responsibilities under health and safety legislation and regulations.

I-10.53   A further change is to be found in new guidance issued by the HSE with regard to risk assessment.[190] Here the regulator notes that the 1992 Regulations focused attention on QRA, which often required specialist consultants to be involved.[191] While this is seen to have been useful in the post-Piper Alpha era, the regulator believes that the understanding of offshore risks is now mature. Accordingly, it believes that risk assessment should focus on adding value and be management-owned rather than consultant-owned.[192] The guidance notes that risk assessment should be proportionate to the complexity of the problem in hand and the magnitude of risk. QRA thus features only where the risk level and the complexity of a problem are high, with qualitative and semi-quantitative approaches being identified as appropriate for lower-level situations.[193] Lest this should be read as a lessening of the level of responsibility rather than just a lessening of the regulatory burden, the guidance makes clear, firstly, that the

---

[188] HSE, *Offshore Installations (Safety Case) Regulations: Guidance*, Operations Notice 71, May 2006. See also the definition of "operator" in Reg 2.

[189] Reg 2 (emphasis added).

[190] HSE, *Guidance on Risk Assessment for Offshore Installations*, Offshore Information Sheet No. 3/2006 (hereinafter "*Guidance on Risk Assessment*").

[191] It is a question whether this is a problem unique to the offshore oil and gas industry. Recall that the reasoning behind Lord Cullen's recommendation of QRA was evidence from the HSE that it used this approach as a means of founding legal or political judgements as much as possible on a rigorous scrutiny of the facts because the technologically based industries or scientifically numerate organisations it dealt with expected a structured and logical approach. See para I-10.38.

[192] *Guidance on Risk Assessment*, p 2.

[193] *Ibid*, p 3.

main purpose of risk assessment is to decide whether more needs to be done to reduce risk and, secondly, that the duty holder must demonstrate that risks are controlled and are not intolerable.[194]

With the appearance of the 2005 Regulations, therefore, it could   I-10.54 have been assumed that the regulatory regime for health and safety at work offshore had reached a stage of maturity commensurate with the challenge of the maturing province. As the concluding sentence of this chapter in the first edition of this book warned, however, the characteristics of the mature province might only have been beginning to test the ability of the permissioning approach. And, within a very short time, significant doubts were indeed raised.

Key Programme 3: asset integrity

Whilst conducting a major effort from 2000 to 2004 to reduce hydro-   I-10.55 carbon releases offshore, the HSE "became increasingly concerned about an apparent general decline in the condition of fabric and plant on installations".[195] As a consequence, a further initiative was set up to run between 2004 and 2007 focused on the issue of asset integrity, designated Key Programme 3 (KP3). Around 100 installations (or 40 per cent of the total on the UKCS) were inspected with the concentration on the maintenance management of safety critical elements. The HSE defined *asset integrity* as "the ability of an asset to perform its required function effectively and efficiently whilst protecting health, safety and the environment" and *safety critical elements* as "the parts of an installation and its plant ... whose purpose is to prevent, control or mitigate major accident hazards ... and the failure of which could cause or contribute substantially to a major accident", whilst *maintenance management* in relation to safety critical elements was understood to be "the management systems and processes which should ensure that [such elements] would be available when required".[196]

Given the very close connection between these issues and the idea of   I-10.56 the safety case as a "living document" designed to ensure the ongoing safe operation of an installation, KP3's findings are sobering to say the least. With regard to maintenance management, there was considerable variation both across the industry and between installations within the same company. Problems arose because there was difficulty in keeping track of which equipment was defective or overdue

---

[194] *Ibid*, pp 8–9.
[195] Health and Safety Executive, *Key Programme 3: Asset Integrity Programme: A Report of the Offshore Division of the HSE's Hazardous Installations Directorate*, November 2007 (hereinafter "KP3"), p 5.
[196] *Ibid*, p 5.

for maintenance.[197] Very strikingly, given the claimed advances in the years following the Piper Alpha disaster, the HSE found that there was "a poor understanding across the industry of [the] potential impact of degraded, non-safety-critical plant and utility systems on safety-critical elements in the event of a major accident"[198] and that "the role of asset integrity and [the] concept of barriers in major hazard risk control" was "not well understood".[199] A finding that might have cheered observers of safety on the UKCS in the seventies, to the effect that monitoring by management tended to focus on occupational safety, was nevertheless problematical insofar as it thus served to mask the significance of "major accident precursors".[200]

I-10.57   As regards the condition of the infrastructure as a whole, the HSE was heartened to find that structural integrity was "well controlled" and the main hydrocarbon boundary "reasonably well controlled", but concerned that other parts of the hydrocarbon infrastructure such as pipes and valves were in decline – again a very striking finding given the resonance with the circumstances of Piper Alpha disaster.[201] There was also evidence of the low oil price having prompted deferrals in maintenance that had not been reversed, especially where there were plans to sell on assets in due course. Not surprisingly, this state of affairs was having an adverse effect on workforce morale.[202] Again calling into question the ongoing significance of the safety case as a "living document", the HSE also found that there was insufficient testing of safety critical elements leading to diminished reliability.[203]

I-10.58   By way of an explanation for these deficiencies, the HSE identified three underlying problems relating to *learning*, the *engineering function* and *leadership*. As regards the first of these, there was seen to be a problem both of inadequate auditing and monitoring and of a lack of processes to allow learning to be embedded.[204] Insofar as auditing and monitoring are supposed to be integral parts of the sort of safety management system prioritised in the setting of the safety case, this appears to indicate a significant failure of the permissioning approach to safety offshore. As regards the second underlying problem, the issue here was the relative strength of the engineering function in companies which was seen to have declined

---

[197] *Ibid*, pp 11–13.
[198] *Ibid*, p 6. See also p 13.
[199] *Ibid*, p 6.
[200] *Ibid*, p 6.
[201] *Ibid*, p 6.
[202] *Ibid*, p 6.
[203] *Ibid*, p 7.
[204] *Ibid*, p 8.

"to a worrying level".[205] While the report was not explicit as to which other functions engineering had lost out to, it may be inferred that these are related to finance – a conclusion borne out by the third underlying problem identified above. With regard to leadership, the HSE noted that senior management in setting priorities for spending had to balance safety and financial risks and often did not properly understand the impact on these risks of operating with "degraded [safety critical elements] and safety-related equipment".[206] This lack of understanding, coupled with the HSE's findings that well-publicised findings during KP3 were not being acted upon even as the programme proceeded,[207] does raise the possibility that cost factors were predominant in a way that the safety case approach might have been supposed to have made impossible given its emphasis on risk and safety assessment on an ongoing basis.[208] A review of KP3 in 2009 reported good progress, but indicated that more needed to be done and that the challenges would remain and indeed become more acute.[209]

FOURTH PHASE: THE MACONDO DISASTER AND THE OFFSHORE SAFETY DIRECTIVE

The Macondo disaster and the emerging interest of the EU institutions

Up to this point, the development of health and safety at work regulation for the offshore industry on the UKCS had very much been in the hands of the UK authorities, even if on at least one occasion their deliberations had been driven by events elsewhere.[210]   I-10.59

---

[205] *Ibid*, p 8.

[206] *Ibid*, p 8.

[207] *Ibid*, p 6.

[208] That such an interpretation is by no means unreasonable may be inferred from the comments of the Chief Executive of Petrofac, Ayman Asfari, at the Oil and Money Conference in October 2008, where he indicated that "his company had seen installations which were in bad need of repair", that he feared "firms will fail to spend enough in improvements" and that he was "concerned that the industry would end up in a situation where budgets were curtailed, leading to more risk of accidents". Oil & Gas UK disagreed with this sentiment whereas the offshore arm of the RMT Union indicated that this served to confirm their warnings in this regard. See BBC News online at http://news.bbc.co.uk/1/hi/scotland/north_east/7696232.stm (accessed 11 May 2017). The fact that the offshore industry was criticised by the Chair of the HSE at this time in relation to its accident statistics for problems relating to the "control of potential major incident risks" would tend to suggest that the Chairman of Petrofac and the RMT had a point. See HSE Press Release E039:08, 13 August 2008.

[209] HSE, *Key Programme 3 – Asset Integrity: a review of industry's progress*, July 2009, pp 3–5.

[210] The Burgoyne Committee in the aftermath of the Ekofisk Bravo blowout discussed

That situation changed in April 2010 with the Macondo disaster in the Gulf of Mexico, resulting in the loss of the Deepwater Horizon drilling rig, the deaths of 11 men and the industry's worst ever environmental disaster. Such was the scale of the accident that policymakers and regulators across the globe – and not just in the United States – immediately began to re-examine their assumptions about how offshore oil and gas operations were regulated. The UK was no different, but here the EU institutions also became involved to an unprecedented degree, in no small measure because of their fears that an oil spill in the Black Sea or the Mediterranean would have very serious consequences indeed. The European Parliament[211] and the European Commission[212] by the end of the year had signalled their belief that new legislation in addition to the general environmental law and the Extractive Industries Directive[213] would be required, even suggesting that this should apply also to the operations of EU companies in other parts of the world, and with the Commission ultimately proposing a new Regulation.[214] Both the UK and Norway (which, while not a member of the EU, as a member of the European Economic Area, the EU also considered would be affected by the Regulation[215]) immediately expressed concern that the proposal would contradict their approach to health and safety regulation, even if the Commission stressed that the draft Regulation was inspired by best practice in member states with oil and gas production. The UK was so concerned because if the Commission legislated by way of a Regulation, then under EU law this would have direct effect without transposition into domestic law and supersede any contrary domestic provisions. The Commission's perceived lack of expertise in the field in contrast to that evident in the UK or Norway, often gained the hard way, meant that the

---

above at paras I-10.23*ff.*

[211] European Parliament resolution of 7 October 2010 on EU action on oil exploration and extraction in Europe.

[212] European Commission, Communication from the Commission to the European Parliament and the Council: Facing the challenge of the safety of offshore oil and gas activities, SEC(2010) 1193 final.

[213] Directive 92/91/EEC.

[214] Proposal for a Regulation of the European Parliament and of the Council on safety of offshore oil and gas prospection, exploration and production activities, COM(2011) 688 final, 27 October 2011.

[215] In the event, Norway declined to implement the Regulation on the basis that the EEA treaty did not apply offshore: see E Nordveidt, "Regulation of the Norwegian Petroleum Sector", in T Hunter (ed.), *Regulation of the Upstream Petroleum Sector* (2015), p 132. The EU does not accept Norway's position in this regard and proceedings may follow to attempt to compel compliance: see eg C Oliver and R Milne, "Norway's offshore drilling fight with EU a cautionary tale for UK", *Financial Times*, 18 January 2016, available at www.ft.com/content/9ed984b0-bab0-11e5-b151-8e15c9a029fb (accessed 11 May 2017).

intervention was all the more unwelcome. Consequently, when the Commission indicated that it would be willing to consider legislating by way of a Directive,[216] industry and government in producing states affected were somewhat reassured, insofar as a Directive both requires transposition into domestic law and allows the member state greater freedom as to how it will achieve the overall objective rather than imposing a particular course of action.

The Offshore Safety Directive and the immediate challenges for the UK

The Directive, widely known as the Offshore Safety Directive (OSD), was passed in June 2013[217] and although there was relief that the problems that a Regulation would have caused had been avoided, there were nevertheless questions as to how the UK would transpose it that were not necessarily straightforward to answer. Two issues in particular had to be resolved: the need to avoid any conflict of interest between economic development, on one hand, and health, safety and environmental regulation, on the other; and the need to deal holistically with health and safety and environmental protection. I-10.60

The first of these issues had implications for the UK's regulatory architecture, that is, the question of who regulates. The UK had, of course, separated health and safety from licensing in accordance with Lord Cullen's recommendations after the Piper Alpha disaster. Now, however, there was also a need to separate environmental protection from licensing: what would be the role of the Department of Energy and Climate Change (DECC) in the context of the Directive given that it was both the licensing authority and the key environmental regulator for the offshore industry. One option, given that environmental regulatory functions were shared with the Environment Agency (EA) (SEPA in Scotland) and the Marine and Coastguard Agency (MCA), would have been to transfer the DECC's responsibilities to those other bodies. It was not clear, however, that such a transfer would have made it straightforward then to meet the Directive's requirement for an holistic approach to health and safety and environmental regulation. With that last thought in mind, the next option that presented itself was to transfer environmental responsibilities to the HSE. But that was not a straightforward matter either, as the HSE had, by definition, no existing environmental I-10.61

---

[216] See www.alynsmith.eu/index.php?mact=News,cntnt01,detail,0&cntnt01articleid=109 2&cntnt01returnid=153 (accessed 29 August 2017).

[217] Directive 2013/30/EU of the European Parliament and of the Council of 12 June 2013 on safety of offshore oil and gas operations and amending Directive 2004/35/EC Text with EEA relevance.

expertise. Furthermore, even were that solution to be chosen, it would produce an uncomfortable anomaly insofar as the offshore oil and gas industry would alone be subject to the HSE dealing with both regulatory dimensions and would lead to questions of consistency of approach. That solution might nevertheless have been countenanced for as long as offshore health and safety was the responsibility of a specialist division within the HSE, but it is less obvious where that responsibility has been subsumed within a division responsible for high-risk industries more generally. The Directive itself provides for a more creative solution, however, in that it is open to the possibility that the competent authority responsible for the health, safety and environmental regulation of the offshore oil and gas industry may be composed of more than one body, with the proviso that one of them takes the lead role and there is a coherent approach among them. It would thus have been possible for the UK to meet its obligations under the Directive by transferring environmental regulatory functions from DECC to the EA (SEPA) and the MCA and then those bodies working with the HSE to form the competent authority (with the HSE presumably taking the lead role).

I-10.62   Even if a solution to the question of who regulates so as to comply with the Directive could, therefore, be envisaged, this would also need to take account of the implications of that document for the means of regulation – specifically the need for an holistic approach to the regulation of health, safety *and* the environment. The Safety Case since its inception in the offshore setting in 1995 has required a focus on the identification of hazards affecting health and safety and thereafter a quantification of risks to health and safety and a detailing of measures employed or to be employed to reduce those risks to a level as low as is reasonably practicable. The equivalent document in the Directive, however, the Major Hazard Report, calls for health and safety hazards as well as *environmental* hazards to be considered simultaneously and then for appropriate risk assessment and management. Given this additional requirement, would the UK's implementation of the Directive imply wholesale reform of the Safety Case Regulations or would it be possible to meet the overall objective of the MHR by creatively combining existing documents, such as the Safety Case and the Oil Spill Response Plan, relating to an installation? And would the answer to that question depend upon or have implications for the earlier questions relating to regulatory architecture?

The Offshore Safety Directive Regulator

I-10.63 Whilst the complications facing the UK in relation to the implementation of the Directive were very real, they were reduced to some extent by developments in a parallel and unrelated policy review, which led to fairly fundamental change in the regulatory architecture relating to the offshore hydrocarbon industry. The then Secretary of State for Energy and Climate Change, Ed Davey, in 2013 established an independent review under the leadership of Sir Ian Wood into how the UK could maximise the economic recovery of remaining offshore oil and gas reserves.[218] Among the review's recommendations, the most relevant to the present discussion is that a more powerful and better-resourced regulator should take over the majority of the Secretary of State's functions in relation to petroleum licences.[219] Accepting this recommendation, the Government moved quickly to establish the Oil and Gas Authority,[220] the net effect of which is to remove the conflict within first the DECC and now BEIS between licensing and environmental regulation. Accordingly, it became a relatively straightforward matter to solve the problem raised by the Directive by essentially copying the approach the UK has adopted in relation to COMAH:[221] the competent authority under the OSD, described as the Offshore Safety Directive Regulator (OSDR), is a partnership between the HSE's Energy Division and the successor within BEIS to the Oil and Gas Environment and Decommissioning unit within the DECC.[222]

The Offshore Installations (Offshore Safety Directive) (Safety Case etc) Regulations 2015

The regulatory changes required by the OSD are implemented in the   I-10.64
main by the Offshore Installations (Offshore Safety Directive) (Safety Case etc) Regulations 2015.[223] The 2005 Regulations are repealed in relation to operations in external waters, that is, in the territorial sea or on the continental shelf, but they remain in force in relation to operations in

---

[218] For details, see www.gov.uk/government/groups/wood-review-implementation-team (accessed 11 May 2017).

[219] Sir I Wood, *UKCS Maximising Recovery Review: Final Report*, available at www.gov.uk/government/uploads/system/uploads/attachment_data/file/471452/UKCS_Maximising_Recovery_Review_FINAL_72pp_locked.pdf (accessed 15 May 2017).

[220] See DECC and OGA, "Oil and Gas Authority: Framework Document", April 2015.

[221] See now the Control of Major Accident Hazard Regulations 2015, SI 2015/483. For details of the Competent Authority arrangements, see www.hse.gov.uk/comah/authorityindex.htm (accessed 11 May 2017).

[222] See the Memorandum of Understanding between DECC and HSE, 15 July 2015. At the time of writing, in early May 2017, the precise details of the new arrangements within BEIS were not entirely clear. Given the brief statement on the HSE's website to the effect that "[t]he changes do not materially affect the operation of OSDR and OSDR documents are being revised to reflect BEIS", it appears likely that the Oil and Gas Environment and Decommissioning unit will simply continue as before within BEIS. See www.hse.gov.uk/osdr (accessed 7 May 2017).

[223] SI 2015/398.

internal waters.[224] The 2015 Regulations refer to new definitions of key actors, including "offshore licensee", "operator" and "well operator", contained in the Offshore Petroleum Licensing (Offshore Safety Directive) Regulations 2015.[225] There are also new definitions relative to the physical infrastructure and operations including, for example, "installation", "major accident" and "production installation".[226]

I-10.65   Perhaps the key change introduced by the OSD is the need for environmental major accident hazards to be considered in the safety case in addition to those related to occupational health and safety. In practice this means that one new document is needed whilst some existing documents and processes have their scope expanded. The new document is the Corporate Major Accident Prevention Policy (CMAPP), a written policy which "establishes the overall aims and arrangements for controlling the risk of a major accident and how those aims are to be achieved and those arrangements put into effect by the officers of the duty holder".[227] Consistent with the EU institutions' original desire to attempt to influence the behaviour of the industry globally as well as within Europe, the CMAPP must cover the duty-holder's installations outside the EU as well as those covered by the Regulations.[228] The matters the CMAPP must address are set out in Schedule 1 to the Regulations. They include details of where responsibility lies, the command and control systems involved (including senior management), how competency is ensured, how desirable behaviours are incentivised, what evaluation and audit arrangements are in place, how standards are maintained and the extent to which all of these arrangements are applied beyond the EU. The CMAPP must be prepared in accordance with certain matters, which are set out in Schedule 2. These include: taking measures to prevent unplanned escapes of hydrocarbons; ensuring no single containment barrier failure could lead to a major accident; paying "particular attention to evaluation of the reliability and integrity requirements of all safety and environmental-critical systems and base inspection and maintenance systems on achieving the required level of safety and environmental integrity";[229] ensuring the relia-

---

[224] SI 2015/398, Reg 4 and Sch 13, Pt 1, paras 33–40.

[225] SI 2015/385, Reg 1. These regulations are substantially concerned with transposing the OSD's requirements in relation to the environmental criteria the Secretary of State must now apply when considering the grant of seaward petroleum licences.

[226] *Ibid.* The first two of these are now defined by specific reference to the OSD.

[227] 2015 Regulations, Reg 7(2)(a).

[228] Reg 7(2)(b). This is backed up by a new requirement for a UK-registered licensee, operator or well operator to report when requested by the OSDR on the circumstances of any accident they or a subsidiary have been involved in outside the EU. See Reg 34(1).

[229] 2015 Regulations, Sch 2, para 2. As will be seen in due course, this is a particularly important development.

bility of data collection and preventing data manipulation; ensuring the monitoring of compliance with statutory duties by incorporating them into standard operating procedures; creating and sustaining a strong safety culture, especially through appropriate arrangements in relation to employees and contractors.

Whereas it was previously necessary to have in place a Safety Management System (SMS), the new requirement is for a Safety and Environmental Management System (SEMS).[230] This "must include the organisational structure, responsibilities, practices, procedures, processes and resources for determining and implementing the corporate major accident prevention policy".[231] Additionally, the SEMS "is to be integrated with the overall management system of the duty holder",[232] and must address the following particulars: "organisational structure and personnel roles and responsibilities; identification and evaluation of major hazards as well as their likelihood and potential consequences; integration of environmental impact into major accident risk assessments in the safety case; controls of the major hazards during normal operations; emergency planning and response; limitation of damage to the environment; management of change; monitoring of performance; audit and review arrangements; [and] the measures in place for participating in tripartite consultations and how actions resulting from those consultations are put into effect".[233] As with the CMAPP, the SEMS must be prepared in accordance with the matters set out in Schedule 2, discussed in the foregoing paragraph.

I-10.66

Transposition of the OSD requires that the verification scheme that was already a feature of the safety case regime under the 2005 Regulations be extended to include environmental as well as safety-critical elements.[234] The scheme must ensure that such elements are "suitable" and "remain in good repair and condition".[235] The means by which this is to be achieved are then set out and include: examination and reporting by a verifier;[236] and taking and noting of action

I-10.67

---

[230] 2015 Regulations, Reg 8(1).

[231] Reg 8(2).

[232] Reg 8(3).

[233] Sch 3.

[234] Reg 9(1).

[235] Reg 9(1)(a) and (b).

[236] Defined as "independent and competent person ... who performs functions in relation to a verification scheme" (Reg 2(1)). Such a person is only to be regarded as "independent" insofar as there are essentially no circumstances where their objectivity would be compromised by their having any responsibility for the issues they are verifying (Reg 2(7)). He or she is only to be regarded as competent insofar as they possess "reasonable technical competence ... sufficient" for the task (Reg 2(8)).

by the duty holder following a report.[237] In addition, the verification scheme must be drawn up in consultation with the verifier and any reservations on the part of the latter must be noted.[238] Similarly, the duty holder must also prepare a list of the environmental and safety critical elements, seek the verifier's comment and note any reservations on the part of the latter.[239] The matters to be included in a verification scheme include: the selection principles employed by the duty holder in relation to a verifier; arrangements for communicating appropriately with the verifier; "the nature and frequency of examination and testing"; arrangements for the making and keeping of records relating to the verification scheme, as well as communicating relevant information "to the appropriate level" in the duty holder's management system.[240]

I-10.68   Insofar as well design and construction were central issues in the Macondo disaster, the OSD unsurprisingly requires enhancements to the UK's existing arrangements relating to well examination, particularly in relation to the competence of well examiners and communication between well examiners and well operators. This is achieved by way of a well examination scheme, which operates according to similar arrangements as for a verification scheme.[241] As regards well operations, well notifications to the competent authority remain a feature of the system but must now also include "a statement, made after considering reports by the well examiner under regulation 11(2)(b), that the risk management relating to well design and its barriers to loss of control are suitable for all anticipated conditions and circumstances".[242] Where the well operator plans a material change to a well operation, he must consult the well examiner and include a copy of the latter's report with the notification to the competent authority.[243]

I-10.69   Transposition of the OSD also implies changes to the duty holder's obligations and new powers in the hands of the competent authority. As regards the former, while the duty holder has always had to ensure that "procedures and arrangements" in the safety case are followed, there is now also a similar duty to ensure that operations set out in notifications of combined operations or in well notifications are conducted "in pursuance of" the relevant plans.[244] Whilst

---

[237] Reg 9(2).
[238] Reg 9(3).
[239] Reg 9(4).
[240] Reg 10(1) and Sch 4, Pt 1.
[241] See Regs 10–13 and Sch 4, Pt 2.
[242] Reg 21(3).
[243] Reg 21(4) and (6).
[244] Reg 28(3) and (4).

the meaning here is relatively clear, and the accompanying guidance (in draft form at the time of writing) specifies that the relevant plans are required "to be followed",[245] it might be thought that the wording in the Regulations might more usefully and accurately read "in conformity with" the relevant plans.

The circumstances of the Macondo disaster also explain a further   I-10.70 new duty imposed on the duty holder requiring him to ensure that, where "an activity ... significantly increases the risk of a major accident ... suitable measures" are taken "to ensure that the risk is reduced as low as is reasonably practicable". Furthermore, the Regulation goes on to spell out that that those measures "include, where necessary, suspending the relevant activity until the risk is adequately controlled".[246]

The new Regulations also include more detail with regard to   I-10.71 the relationship between various aspects of emergency response. A duty holder has specific duties in this regard under the Offshore Installations (Prevention of Fire and Explosion, and Emergency Response) Regulations 1995, which are described in the 2015 Safety Case Regulations as "the internal emergency response duties".[247] The duty holder is now required to carry them out "consistently with the external emergency response plan",[248] namely the national plan prepared by the Secretary of State under the Merchant Shipping Act 1995,[249] and "taking into account the risk assessment undertaken during preparation of the current safety case for the installation".[250]

As regards workforce involvement, the 2015 Regulations further   I-10.72 enhance this dimension of the safety case to the extent that the duty holder must now inform employees, contractors and contractors' employees of the arrangements made by the OSDR in relation to confidential reporting of safety and environmental concerns,[251] and in relation to subsequent investigations respecting the anonymity of individuals making such reports.[252]

The safety case has from the outset, of course, been regarded   I-10.73 as a living document. In other words, it is subject to continuous

---

[245] HSE, The Offshore Installations (Offshore Safety Directive) (Safety Case etc) Regulations 2015, Guidance on Regulations, L154, 2015, p 14.

[246] 2015 Regulations, Reg 29(1) and (2). The duty holder also must notify the competent authority within 24 hours when such measures are taken; see Reg 29(3) and (4).

[247] Reg 30(14).

[248] Reg 30(1)(a).

[249] Section 293(2)(za).

[250] Reg 30(1)(b).

[251] Reg 31(1)(a) and (2). This must also be referred to in training and notices – Reg 30(3). For details of the OSDR's confidential reporting arrangements, see www.hse.gov.uk/osdr/reporting/confidential-reporting.htm (accessed 11 May 2017).

[252] Reg 31(1)(b).

review and updating to take account of changing circumstances and knowledge. Furthermore, it has always been assumed that even if companies are commercial competitors when it comes to exploring for and producing oil and gas, they nevertheless have an incentive to co-operate in relation to safety matters.[253] Perhaps recognising that these assumptions could usefully receive some regulatory under-pinnings, the duty holder is now required to "cooperate with the competent authority to establish and implement a priority plan for the development of standards, guidance and rules which will give effect to best practice in major accident prevention, and limitation of consequences of major accidents should they nonetheless occur".[254] Furthermore, they are obliged to "participate in the preparation and revision of standards and guidance on best practice in relation to the control of major hazards throughout the design and operational lifecycle of offshore oil and gas operations" and in so doing must consider a range of issues with a view to establishing priorities, namely: "effective risk management; management and supervision of major hazard operations; competency of key post holders; reliable decision making; effectively integrating safety and environ-mental management systems between operators and owners and other entities involved in oil and gas operations; key performance indicators; improving well integrity, well control equipment and barriers and monitoring their effectiveness; improving primary containment; improving secondary containment that restricts escalation of an incipient major accident, including well blowouts; and reliability assessment for safety and environmental-critical systems".[255]

I-10.74   Finally, as regards new obligations imposed by the 2015 Regulations, the duty holder is also required to notify the OSDR "without delay" of a "major accident" or "a situation where there is an immediate risk of major accident". That this is again focused on learning lessons so that they may be more widely shared is evident from the further obligation to describe in the notification "the circum-stances, including, where possible, the origin, the potential impacts on the environment and the potential major consequences".[256]

I-10.75   Regarding the new powers in the hands of the regulator, the OSDR has the power to prohibit operations where it is of the opinion that "the measures for preventing or limiting the consequences of a major accident" contained in a safety case, notification of combined opera-

---

[253] An early assertion of the degree of cooperation evident in the industry may be found in the evidence presented by UKOOA to the Burgoyne Committee.

[254] Reg 32(1).

[255] Sch 11.

[256] Reg 33.

tions or notification of well operations "are insufficient to fulfil the requirements set out in the relevant statutory provisions".[257]

Finally, the 2015 Regulations also impose a duty on the regulator to inform the licensing authority immediately where it "determines that an operator no longer has the capacity to meet the requirements of the relevant statutory provisions".[258] This duty is echoed by a provision in the Offshore Petroleum Licensing (Offshore Safety Directive) Regulations 2015, which requires the OSDR to notify the licensing authority in the event that it "determines that an operator no longer has the capacity to meet the 2013 Directive operator requirements for the operations in respect of which that operator was appointed".[259] The consequences of any such notification are severe, because where the licensing authority receives one "it must terminate the appointment of the operator".[260]   I-10.76

SELECTED DEVELOPMENTS IN CRIMINAL LAW

Throughout the foregoing discussion there are references to breaches of regulations constituting an offence. Whilst it may rightly be suggested that the shift in the approach to health and safety at work in the context of the 1974 Act provided regulators with a range of options short of prosecution, there has in more recent years been a growing sense that serious failures in relation to occupational health and safety, even resulting in serious injury or death, have not been punished to the extent that they might be, and perhaps not sufficiently to act as a deterrent. Legislative developments in this regard a decade ago in the shape of the Corporate Manslaughter and Corporate Homicide Act 2007 and the Health and Safety (Offences) Act 2008 were supposed to address this issue, and while it is not clear that they have so far had the impact that they might have had, it is a question whether recent changes to sentencing guidelines suggest that the consequences of future serious breaches of regulations may be much more severely punished. These issues are addressed in the following paragraphs.   I-10.77

The Corporate Manslaughter and Corporate Homicide Act 2007

Even before this Act came into force on 6 April 2008, it was possible for a company to be prosecuted in the event that a person had died as a result of the company's activities. The law was, however, criticised   I-10.78

---

[257] Reg 26.
[258] Reg 6.
[259] Reg 7.
[260] Reg 8(2).

insofar as the requirement that the "directing mind" of the company be identified meant that prosecution of small companies was considerably easier than that of large companies, notwithstanding that the activities of the latter perhaps had greater potential to result in more serious accidents and multiple fatalities.[261]

I-10.79   An organisation[262] is guilty of the new offence (corporate manslaughter in England and Wales or Northern Ireland, corporate homicide in Scotland[263]) "if the way in which its activities are managed or organised (a) causes a person's death, and (b) amounts to a gross breach of a relevant duty of care owed by the organisation to the deceased".[264] The offence is only committed if the way in which the organisation's "activities are managed or organised by its senior management is a substantial element" in the breach of duty.[265]

I-10.80   There are clearly a number of concepts within this definition of the offence which require further clarification. A "relevant duty of care" includes duties owed by the organisation under the law of negligence (or under a supervening statutory provision[266]) "to its employees or to other persons working for the organisation or performing services for it",[267] thus readily covering the situation on an offshore installation where in addition to the employees of the operator there may be many workers present who are employed by a range of contractors and subcontractors. With regard to the potential liability of contractors or subcontractors in the same setting, it is significant that the concept of a "relevant duty of care" is further defined *inter alia* to include "the supply by the organisation of goods or services", "the carrying on by the organisation of any construction or maintenance operations", "the carrying on by the organisation of any other activity on a commercial basis" or "the use or keeping by the organi-

---

[261] The first successful prosecution was precisely of a small company with only one director: *R v Kite and OLL Ltd.* (1994) (unreported). For an accessible discussion of this case, see M G Welham, *Corporate Killing: A Manager's Guide to Legal Compliance* (2002), pp 51–59. See also G Slapper, "Litigation and corporate crime", (1997) *Journal of Personal Injury Litigation* 220–233. For the preceding position in Scotland, see *Transco plc v HMA* 2004 SLT 41. For a discussion, see J Chalmers, "Corporate culpable homicide: *Transco plc v H M Advocate*", 8 (2004) *Edinburgh Law Review* 2, 262–266.

[262] An organisation for the purposes of the Act includes a corporation and a partnership. Corporate Manslaughter and Corporate Homicide Act 2007, s 2. "Corporation" is defined in s 25 to include a "body corporate" and is thus broad enough to encompass a limited liability partnership.

[263] 2007 Act, s 1(5). Note the Act applies in relation to offshore installations by virtue of s 28(3)(e).

[264] *Ibid*, s 1(1).

[265] *Ibid*, s 1(3).

[266] *Ibid*, s 2(4).

[267] *Ibid*, s 2(1)(a).

sation of any plant, vehicle or other thing".[268] A breach of a relevant duty of care is a "gross breach" where the conduct in question "falls far below what can reasonably be expected of the organisation in the circumstances".[269] This raises intriguing possibilities should there ever be a prosecution for a fatality on an offshore installation given that this is an industry where an operator will have set out in detail how it is going to behave in an installation's safety case.[270] The court will thereby be provided with a ready guide as to what the operator itself (and by implication the HSE through the fact of acceptance of the safety case[271]) would have regarded as reasonable in the circumstances. As regards "senior management", this is defined as the persons in an organisation who "play significant roles in (i) the making of decisions about how the whole or a substantial part of its activities are to be managed or organised, or (ii) the actual managing or organising of the whole or a substantial part of those activities".[272] Note that the individuals themselves are not the target of the Act – only an organisation can be prosecuted[273] – but the behaviour of the senior management will be subject to close scrutiny because it is ultimately this that will determine whether there has been the required breach of duty.[274]

Where it is established that a relevant duty of care is owed, it is then a matter for the jury to decide whether or not there has been a gross breach of that duty. In so doing, the Act specifies that there are certain issues which the jury *must* consider and certain that it *may* consider or have regard to. With regard to the former, these are stated to be "whether the evidence shows that the organisation failed to comply with any health and safety legislation that relates to the alleged breach, and if so (a) how serious that failure was; (b) how much of a risk of death it posed".[275] As regards the latter, the jury may "consider the extent to which the evidence shows that there were attitudes, policies, systems or accepted practices within the organisation that were likely to have encouraged any [such failure] or to have produced tolerance of it" (raising questions of what a court would have made of the HSE's findings, for example, of a problematical balancing of financial and safety risk in KP3) and may "have regard to any health and safety guidance that relates to the

I-10.81

---

[268] *Ibid*, s 2(1)(c).

[269] *Ibid*, s 1(4)(b).

[270] See the discussion in this regard at paras I-10.48 to I-10.54.

[271] *Ibid.*

[272] *Ibid*, s 1(4)(c).

[273] See also 2007 Act, s 18 in this regard.

[274] It remains open to the prosecution, of course, to bring separate charges against any individual in connection with the death, including members of the senior management.

[275] 2007 Act, s 8(2).

alleged breach."[276] The jury may also have regard to any other factor which it considers relevant.[277]

I-10.82   Note that an organisation charged under the 2007 Act may simultaneously face charges under health and safety legislation.[278]

I-10.83   As regards the penalty which may be imposed in the event of conviction, this takes the form of an unlimited fine.[279] Since 15 February 2010, sentencing guidelines have been in place for convictions for corporate manslaughter in England and Wales[280] and new guidelines came into force on 1 February 2016 which will apply to those sentenced after that date irrespective of the date of the offence.[281] These guidelines provide courts with criteria with which to judge the seriousness of the offence they are dealing with, including the foreseeability of serious injury, the extent to which the defendant has fallen short of the appropriate standard, whether this is an isolated or more common event, and whether there was more than one death, or a high risk of further deaths or serious personal injury in addition to death.[282] The offence is then categorised as A or B depending upon whether the application of those criteria indicate a high or low level of culpability,[283] with the court then directed to tables to identify the starting point for the penalty and the range within which it will lie.[284] Having identified the starting point, the court must then consider whether there are any factors that increase or reduce the seriousness of the offence.[285] Among the factors included in the former list, one stands out as particularly relevant to the offshore oil and gas industry in light of the findings of the KP3 Asset Integrity Programme discussed previously, namely "cost-cutting at the expense of safety".[286] Interestingly, a factor included in this list in the earlier guidelines which was also relevant in the light of the KP3 findings, namely "failure to heed warnings or advice", especially from HSE inspectors and health and safety representatives, or failure "to respond appropriately to

---

[276] *Ibid*, s 8(3).

[277] *Ibid*, s 8(4). See also s 8(5) for the broad definition of guidance in this context.

[278] See also para I-10.86 below.

[279] 2007 Act, s 1(6).

[280] Sentencing Guidelines Council, *Corporate Manslaughter and Health and Safety Offences Causing Death: Definitive Guideline*, February 2010 (hereinafter "Sentencing Guidelines 2010").

[281] Sentencing Council, *Health and Safety Offences, Corporate Manslaughter and Food Safety and Hygiene Offences: Definitive Guideline*, 2015 (hereinafter "Sentencing Guidelines 2015").

[282] *Ibid*, p 22.

[283] *Ibid*, p 22.

[284] *Ibid*, pp 23–24.

[285] *Ibid*, p 25.

[286] *Ibid*, p 25.

'near misses'", no longer appears.[287] At first sight, this appears like a retrograde step. In fact, however, these matters are now considered at the earlier stage of categorising the offence.[288] Thus, where such factors are in evidence, then it is more likely that the offence will be categorised as more serious, meaning that a higher starting point for the penalty will already be in play. On the other hand, where, for example, a convicted organisation has no previous (or no relevant or recent) convictions, taken steps to remedy the problem, shown a high level of co-operation, has a good health and safety record, or has self-reported, co-operated and accepted responsibility, these factors are likely to have a mitigating effect.[289] As regards the final fixing of the fine, a series of further steps are set out in the Guidelines,[290] but it is stressed that "[t]he fine must be sufficiently substantial to have a real economic impact which will bring home to management and shareholders the need to achieve a safe environment for workers and members of the public affected by their activities".[291] As an indication of what this will mean in practice, the guidelines suggest that even for the smallest organisation with an annual turnover up to £2 million, the minimum fine for a category B offence would be £180,000 with a starting point of £300,000, whereas for a large organisation with an annual turnover in excess of £50 million the fine for a category A offence could reach £20 million with a starting point of £7,500,000.[292] Given that the definition of a large organisation does not come close to the turnover of many oil companies, it is important to note that the guidelines also state that "[w]here an offending organisation's turnover or equivalent very greatly exceeds the threshold for large organisations, it may be necessary to move outside the suggested range to achieve a proportionate sentence".[293] While there is no such guidance in Scotland, it must be assumed that the Scottish courts would take similar issues into account in sentencing following a conviction for corporate homicide under Section 1 of the 2007 Act. Note also that the Scottish Sentencing Council was established in 2015, but that its first Business Plan makes no mention of the establishment of guidelines in this area.[294]

Whether the 2007 Act will ultimately result in fines of the level anticipated for deaths at work to a great extent remains to be   I-10.84

---

[287] Sentencing Guidelines 2010, para 7.
[288] Sentencing Guidelines 2015, p 22.
[289] *Ibid*, p 25.
[290] *Ibid*, pp 25–28.
[291] *Ibid*, p 25, emphasis removed.
[292] *Ibid*, p 24.
[293] *Ibid*, p 24.
[294] Scottish Sentencing Council, Business Plan 2016–2018, Edinburgh, 2016.

seen. Whilst a number of prosecutions have by now been taken forward under the Act, and it has been observed that convictions are increasing,[295] some are sceptical. For example, Field and Jones note that it remains the case that the number of prosecutions is low compared to the number of workplace fatalities, that only small companies are being prosecuted and that fines are below the level that would have been expected under the 2010 Sentencing Guidelines.[296] Where a large oil company (Total) was implicated in the death of a worker at the Lindsey Oil Refinery in North Lancashire in June 2010, the prosecution was brought forward under the COMAH Regulations rather than under the 2007 Act and the fine imposed, whilst large at £1.4 million, was well short of what would have been expected under the guidelines had that approach been taken.

The Health and Safety (Offences) Act 2008

I-10.85   Section 1 of this Act, which came into force on 16 January 2009, amends Section 33 of the Health and Safety at Work, etc. Act 1974. Section 33 specifies the offences which may be committed under the 1974 Act and the amendments to that section give effect to a new schedule[297] which sets out the mode of trial and the maximum sentence which applies in the case of each of the offences listed in Section 33(1). With respect to prosecution for breach of the key duty under Section 2 of the 1974 Act,[298] the penalty on summary conviction rises from a fine not exceeding £20,000 to imprisonment for a period not exceeding six months or a fine not exceeding £20,000 or both, while the penalty for conviction on indictment rises from an unlimited fine to imprisonment for a period not exceeding two years or an unlimited fine or both.[299]

I-10.86   With regard to the prosecution of bodies corporate under the 1974 Act for offences causing death, while imprisonment is of course not a possibility, it is very important to note that the sentencing guidelines discussed in the previous section of this chapter will also apply. Accordingly, even where a body corporate is not charged with corporate manslaughter or corporate homicide under the 2007 Act,

---

[295] "Courts increase manslaughter convictions", *Health and Safety Monitor*, March 2015, 1–2.
[296] S Field and L Jones, "Are directors getting away with manslaughter? Emerging trends in prosecutions for corporate manslaughter", *Business Law Review*, 35(5), 158–163.
[297] Sch 3A.
[298] The employer's duty "to ensure, so far as is reasonably practicable, the health, safety and welfare at work of all his employees".
[299] Sch 3A, para 1.

but only with offences under the 1974 Act causing death,[300] the possibility of very substantial fines must be considered.[301]

Furthermore, insofar as Section 37 of the 1974 Act provides I-10.87
that where an offence committed by a body corporate "is proved to have been committed with the consent or connivance of, or to have been attributable to any neglect on the part of, any director, manager" or other corporate officer, that individual shall also be guilty of the offence and liable to be dealt with accordingly, it is clear that the 2008 Act also has very profound implications for senior management.[302]

Taken together, the 2007 and 2008 Acts have radically altered the I-10.88
health and safety landscape for bodies corporate and their senior managers, with considerably more severe penalties now available to the courts. Quite how these will play out in practice remains to be seen, but the HSE made its intentions clear when the 2008 Act came into force, stating that it would "continue to target those who knowingly cut corners, put lives at risk and who gain commercial advantage over competitors by failing to comply with the law".[303]

CONCLUSIONS

The regulation of health and safety at work offshore has had a long, I-10.89
complex and sometimes troubled history. Four distinct approaches have been successively adopted, ranging from what amounted to self-regulation in the first phase, through detailed prescriptive regulation in the second phase, to the permissioning approach of the third phase, which approach is now enhanced in the current phase by the simultaneous consideration of environmental risks alongside health and safety risks. This chapter has deliberately adopted an historical approach to this subject because the significance of the current approach is more easily appreciated if what has gone before is understood. It is also striking to note how early on in the process the particular problems the current approach seeks to deal with were recognised but not acted upon. The appropriateness of the enhanced permissioning approach becomes more apparent when that point is grasped. Be that as it may, the fact that the character of the UKCS is changing quite profoundly in the context of a maturing province will challenge the permissioning approach. Insofar as that

---

[300] Or, of course, is charged under the 2007 Act and under the 1974 Act in the alternative.
[301] Sentencing Guidelines 2015, para 24.
[302] For further discussion of the 2008 Act generally, see B Barrett, "The Health and Safety (Offences) Act 2008: The Cost of Behaving Dangerously at the Workplace", *Industrial Law Journal* 38, 73–79.
[303] HSE Press Release E011:09, 15 January 2009.

maturity must increasingly be read to include not only the ageing of assets but also the push into deeper water and the engagement with more demanding reservoirs, the challenges become all the clearer. The enhanced permissioning approach accordingly has much to commend it, but nagging questions nevertheless remain about its implementation.

I-10.90   In this regard, recall the problems identified by KP3 in 2007 discussed above. Whereas a subsequent HSE report in 2009 suggested good progress in remedying them, the underlying question of how they could have arisen in the context of an appropriate implementation of the safety case approach (the idea of a living document) remains. Nevertheless, if these problems have indeed been overcome, then there are surely grounds for optimism. It is worth noting, however, that in the aftermath of the Macondo disaster in 2010, there were further indications in the findings of the House of Commons Energy and Climate Change Select Committee that all was not well in the offshore industry in relation to the proper implementation of the safety case approach. While the Select Committee essentially endorsed the current approach to regulating safety and environmental protection offshore, generally accepting the evidence submitted to it by the HSE, the DECC and the MCA,[304] the question nevertheless arises as to whether its finding that "the UK has high offshore regulatory standards, as exemplified by the Safety Case Regime", and that the UK approach is "superior" to that in force in the US at the time of the Macondo disaster, can be squared with its finding that the "offshore oil and gas industry is *responding* to disasters, rather than *anticipating* worst-case scenarios and *planning* for high-consequence, low-probability events".[305] Insofar as the Safety Case as a living document is supposed specifically to ensure the continuous questioning of assumptions, it must be wondered whether the Select Committee should have chosen its words more carefully and taken the opportunity to call for a closer look at what was happening rather than possibly offering inadvertent support for ongoing complacency. It is to be hoped that one lesson from the past has been clearly learned on the UKCS: that it would be best to fix what can be seen to be problematical now, rather than waiting for a disaster and an inquiry to force the issue.

I-10.91   The HSE has been active in this regard. KP3 was followed by a further inquiry focused in particular on the challenges of the maturing province: Key Programme 4 (KP4) Ageing and Life

---

[304] House of Commons Energy and Climate Change Committee, *UK Deepwater Drilling – Implications of the Gulf of Mexico Oil Spill, Second Report of Session 2010–11*, HC 450–451.

[305] *Ibid*, emphasis added.

Extension Programme. Insofar as this reported in 2014, it allows a more recent opportunity to consider whether the safety case is indeed encouraging the continuous questioning of assumptions and a forward-looking approach to the assessment of health and safety in the context of ever more challenging circumstances. Whilst the regulator reports that there is much that is encouraging in this regard, it is nevertheless striking to note findings which point in the opposite direction. Thus, the HSE observes that "[t]here is good evidence that industry leaders have recognised the importance of [ageing and life extension], but there is much to do to ensure arrangements are implemented in practice, and there is no room for complacency".[306] In addition, it noted "[i]nsufficient plans in place to manage the consequences of creeping changes ('normalisation of deviance')"; that "[m]ore focus is needed on forecasting potential long-term future failure mechanisms"; and that "approaches to audits were generally to follow what had gone before, rather than addressing what is needed in the future".[307] In other words, the same sorts of issues that go to the heart of the safety case approach are still being flagged up almost 20 years after its inception.

The question now is whether the changes introduced by the 2015 Regulations contribute in any way to the alleviation of these problems. On the face of it, the indications are relatively encouraging. A number of the innovations discussed above in the foregoing section appear to bolster the learning dimension that is integral to the safety case approach. The challenge will be to ensure that these innovations result in meaningful change at the level of infrastructure and operations as opposed to generating only additional paperwork. If one were to point to potential obstacles in the way of such positive outcomes, one would naturally highlight the very challenging price environment that has been ongoing at the time of writing since the middle of 2014 and which has resulted (as such conditions always do in the industry) in a very considerable focus on cost cutting. Given that the HSE identified the precursor to the problems considered in the KP3 Report as low prices at the start of the first decade of the current century, it is not a radical suggestion to state that there is a pressing need to ensure that the old mistakes are not repeated. If any evidence were needed that this is not a hypothetical problem, one need look no further than the improvement notices served on the operator of the Gryphon FPSO in May 2015. One of these noted that the company had "failed to: (a) Make a suitable and sufficient assessment of the risks to the health and safety of persons that may

I-10.92

---

[306] KP4, p 8.
[307] Ibid, pp 8–9.

arise as a result of the rescheduling of maintenance work orders, as recorded on your system on 21 May 2015, (b) Record the significant findings of the assessment, and (c) Identify necessary control measures".[308] The other improvement notice stated that the company had "no comprehensible written instructions on procedures to be observed for rescheduling maintenance that are safety critical, or may affect the health and safety of persons; no written evidence that technical authorities/system responsible engineers participate in OFD [original finish date] extensions to new dates, or that relevant parts of such instructions have been brought to the attention of every person who is to do anything to which that part relates".[309] That these problems are occurring 20 years after the original introduction of the "living document" safety case is a clear indication of the challenge facing all concerned with the implementation of the 2015 Regulations. KP4 usefully and insightfully quotes a comment made by Judith Hackitt, the HSE Chair, at the Piper 25 conference in 2013, which neatly encapsulates the importance of addressing the remaining issues with the safety case approach: "there are no new accidents. Rather there are old accidents repeated by new people."[310]

---

[308] Notice 306403674 served against Maersk Oil North Sea UK Limited on 5 June 2015, available at www.hse.gov.uk/notices/notices/Notice_details.asp?SF=CN&SV=306403639 (accessed 8 May 2017).

[309] Notice 306403639 served against Maersk Oil North Sea UK Limited on 5 June 2015, available at www.hse.gov.uk/notices/notices/Notice_details.asp?SF=CN&SV=306403639 (accessed 8 May 2017).

[310] KP4, p 10.

CHAPTER I-11

ENVIRONMENTAL LAW AND REGULATION ON THE UKCS

Luke Havemann and Tina Hunter[1]

In the almost 50 years since the commencement of exploration for, **I-11.01** and development of, petroleum on the UK Continental Shelf (UKCS), there have been very few environmental incidents. The oil spill and fluid discharge at the Clair production platform in October 2016[2] and the major oil spill at the Gannet Alpha production platform in August 2011[3] are two notable (and recent) incidents from the last 45 years. The causes of each of these spills provide examples of the two primary sources of environmental pollution associated with petroleum activities, and, therefore, the range of environmental regulation that is required to regulate the environmental impact of offshore petroleum activities.

The first source of environmental harm associated with offshore **I-11.02** petroleum activities are those that arise as a result of the everyday operations of petroleum activities. Such *operational harm* includes spills from pipelines, chemical and oil spills associated with drilling activities, oil contaminated drill cuttings, produced water, unloading of oil, human-generated waste such as sewerage and rubbish, and chemical spills from maintenance activities (such as pipeline cleaning). The spill at the Clair platform, which occurred due to a technical issue with the system designed to separate the mixed production fluids of water, oil and gas (production fluid spill), is an

---

[1] Luke Havemann contributed the text for the second edition of this work. This text has been updated for the third edition by Tina Hunter. The Afterword on Civil Liability was authored by Greg Gordon and James Cowie.

[2] The Clair Platform is located approximately 45km west of the Shetland Islands. In the incident, it was estimated that approximately 700 barrels of oil (bbl) were spilled.

[3] The Gannet Alpha Platform is located approximately 180km offshore of Aberdeen.

example of such operational environmental harm.[4] It can contribute to a variety of environmental impacts, of varying intensity, ranging from a minor oil or chemical spill to a national emergency. Such a national emergency occurred in 2007 when thousands of barrels of oil were spilled from the Statfjord B platform during the loading of an oil tanker from the platform in rough weather.[5] Operational discharges are a continuous source of contaminants to the offshore environment, and can have long-lasting biological effects given the accumulation of polyaromatic hydrocarbons in the food chain.[6]

I-11.03   The second, and thankfully far less frequent, source of environmental harm is those incidents associated with oil spills arising from the loss of control of a well, typically a well blowout. Such environmental harm can be best described as *consequential environmental harm*, as it arises as a consequence of a major event such as a well blowout, rather than directly as a result of day-to-day petroleum operations. The oil spill from the Gannet Alpha platform is an example of such a spill. The worst such incident was the blowout of the Macondo Well in the Gulf of Mexico in April 2010 (known as "Deepwater Horizon" (DWH), named after the drilling platform Deepwater Horizon that exploded and burned to the water line), claiming 11 lives and leading to the uncontrolled release of approximately 134 million gallons of oil until the well was capped 87 days later.[7]

I-11.04   The leak of such a huge volume of volume of oil from DWH may not have been the first time that hydrocarbons have flowed uncontrolled from a well into the sea, but it is certainly the largest.[8] As expected, the DWH catastrophe prompted a review of the regulation of well operations, and petroleum activities in general, which encompassed both the regulation of petroleum operations as well as the regulators themselves. Such review was not confirmed to

[4] BP, "Statement regarding Clair oil release" (3 October 2016), available at www.bp.com/en_gb/united-kingdom/media/press-releases/statement-regarding-clair-oil-release.html (accessed 3 May 2017).
[5] See discussion in www.offshorepost.com/offshore-oil-spill-at-north-sea-statfjord-field (accessed 3 May 2017).
[6] T Bakke, J Klungsøyr and S Sanni, "Environmental impacts of produced water and drilling waste discharges from the Norwegian offshore petroleum industry", 92 (2013) *Marine Environmental Research*, 154.
[7] An estimation of the amount of oil released varies. For this chapter the estimation from the *Smithsonian National Museum of Natural History Ocean Portal* has been selected due to independence of source. See http://ocean.sI-edu/gulf-oil-spill (accessed 22 March 2017).
[8] A spill of a similar size occurred in the Gulf of Mexico as a result of the Ixtoc 1 well blowout and oil spill that occurred on 3 June 1979. An exploratory well (in the Bay of Campeche, Gulf of Mexico) being drilled by the Semi-submersible drilling rig the Sedco 135-F blew out in 50m of water. The well was capped some ten months later on 23 March 1980, with 140 million gallons of oil spilling into the GoM.

the USA where DWH occurred. The huge loss of life, the enormity of the consequential oil spill and the failure to stop the source of the oil spill for almost three months prompted a re-examination of environmental regulation in many petroleum jurisdictions. In the UK, the Energy and Climate Change Select Committee considered the implications of the Gulf of Mexico oil spill, concluding that the existing regulatory arrangements are fit for purpose and the committee was not minded to depart from that assessment.[9] The European Union (EU) did not share such confidence, and set about developing and implementing a new Directive (Directive 2013/20/EU on the Safety of Offshore Oil and Gas Operations (hereinafter "the platform directive") to "establish minimum requirements for preventing major accidents in offshore oil and gas operations and limiting the consequences of such accidents".[10] Such wording of the subject and scope of the platform directive clearly addresses consequential environmental harm – that is environmental harm that arises as a consequence of the failure to control a well or a platform, rather than the constant, day-to-day operational harm.

This chapter examines the regulation of both operational and I-11.05 consequential environmental damage that arises as a result of petroleum activities on the UKCS. Firstly, it addresses the regulation of the operational environmental impacts, examining in detail the regulation of sources of environmental harm, including: drilling, produced water, pipelines, drill fluids and drill cuttings, and atmospheric emissions. Secondly, it examines how environmental regulation in the post-DWH era is not just confined to operational environmental regulation. It examines the regulation of consequential harm associated with oil spills arising from a loss of well control. Such an examination will encompass international law and EU law, which the UK must abide by, as well as UK legislation pertaining to oil spill planning and response.

EARLY YEARS OF ENVIRONMENTAL REGULATION

To say that environmental regulation was a low priority at the I-11.06 commencement of offshore petroleum exploration is something of an understatement. Indeed, as analysed in Chapter I-10, the regulation of safety, surely a much higher priority, was dealt with in one clause of the model clauses which stated that "the Licensee shall comply with any instructions from time to time given by the minister

---

[9] Available at www.parliament.uk/business/committees/committees-a-z/commons-select/energy-and-climate-change-committee/inquiries/uk-deepwater-drilling (accessed 29 April 2017).
[10] Directive 2013/20/EU on the Safety of Offshore Oil and Gas Operations, Art 1.

in writing for security the health, safety and welfare of the persons employed in or about the licensed area".[11] To implement worker protection under the model clauses, the Institute of Petroleum's *Model Code of Safety Practice in the Petroleum Industry* (1964) essentially left the regulation of health and safety to the industry itself. The regulation of the environment was an even lower priority than safety in these early days, as reflected by Robinson:

> "When the North Sea was first opened up for exploration and production, the Government did not consider it necessary to put in place any special legislation to protect the environment ... the environmental impact of the offshore industry was not obvious; exploration and production activity was not seen as a serious source of pollution."[12]

I-11.07 Historically, the regulation of the environment during offshore exploration and production activities on the UKCS has not figured prominently in the public or, indeed, political imagination. This is primarily attributable to the fact that such operations were considered less of a risk to environmental pollution than other sources, particularly tanker operations.[13] This attitude to the regulation of the environment has given way, in due course, to a quite different appreciation of the environmental risks involved, and a veritable raft of regulatory provisions governing the environmental aspects of offshore petroleum activities'.[14] It is these regulatory provisions that we now turn our attention to.

REGULATION OF THE OPERATIONAL IMPACTS OF PETROLEUM OPERATIONS

General Environmental regulation

The Offshore Petroleum Activities (Conservation of Habitats) Regulations 2001 (Conservation Regulations)[15]

I-11.08 The Conservation Regulations are designed to ensure the protection of specific habitats and species from the potentially harmful activities of the offshore industry. To achieve this objective, Regulation 5 of

---

[11] Petroleum (Production) Act 1934, Sch 2, Cl 18.

[12] J Rowan Robinson, "Environmental and Planning Law", in A Hill (ed.), *Daintith, Willoughby and Hill: United Kingdom Oil and Gas Law* (2000) (hereinafter "Rowan Robinson, 'Environmental and Planning Law'"), at 1281.

[13] For details, see J Sheail, *An Environmental History of Twentieth-Century Britain* (2002), pp 221*ff*; and M Regester and J Larkin, *Risk Issues and Crisis Management in Public Relations: A Casebook of Best Practice* (4th edn, 2008), pp 176*ff* and 180*ff*.

[14] J Kearns, "Environmental Management", in J Wils and E C Neilson (eds), *The Technical and Legal Guide to the UK Oil and Gas Industry* (2007), p 537.

[15] SI 2001/1754.

the Conservation Regulations requires an assessment to be made of any activity that is likely to have a significant effect on a "relevant site", prior to the granting of a licence by the Secretary of State in terms of the Petroleum Act.[16] A "relevant site" is defined as one of the following: (1) a special area of conservation;[17] (2) a site of European Community (EC) importance as listed in terms of Article 4(2) of Council Directive 92/43/EEC on the Conservation of Natural Habitats and of Wild Fauna and Flora (hereinafter "the Habitats Directive");[18] (3) a site hosting a priority natural habitat type or priority species;[19] (4) an area classified in terms of Article 4(1) or (2) of Council Directive 79/409/EEC on the Conservation of Wild Birds (hereinafter "the Birds Directive");[20] (5) a site included in the list of sites submitted to the European Commission (the Commission) by the UK in accordance with Article 4 of the Habitats Directive;[21] and (6) a site which the Secretary of State, after consultation with Joint Nature Conservation Committee (JNCC),[22] believes would be likely to be included in the list of sites submitted to the Commission in accordance with Article 4 of the Habitats Directive.[23]

In determining whether or not an activity is likely to have a significant effect on a relevant site, the Secretary of State is obliged to consult with the JNCC.[24] Once the Secretary of State has consulted with the JNCC and any other bodies as required, any assessment that demonstrates that the proposed activity is likely to have an adverse effect on the integrity of a relevant site will prohibit the Secretary of State from authorising the activity.[25] However, authorisation may be granted despite the likelihood of the proposed activity having adverse effects on a relevant site if, in the opinion of the Secretary of State, there is no satisfactory alternative[26] and the Secretary of State certifies that the activity should be carried out for reasons of overriding public interest, including reasons of a social or economic nature.[27] Fortunately, where the Secretary of State gives a certification in support of overriding public interest, he is obliged to consult with the appropriate nature conservation body so that suitable compen-

I-11.09

---

[16] Reg 5(1).
[17] Reg 2(1)(a).
[18] Reg 2(1)(b).
[19] Reg 2(1)(c).
[20] Reg 2(1)(d).
[21] Reg 2(1)(e).
[22] Reg 2(2).
[23] Reg 2(1)(f).
[24] Reg 5(2).
[25] Reg 5(4), read with Reg 5(1), (2) and (3).
[26] Reg 6(1)(a).
[27] Reg 6(1)(b).

satory measures may be taken to ensure that the overall coherence of Natura 2000 is protected. Natura 2000 is the European network of protected sites established under the Habitats and Birds Directives aimed at protecting rare, endangered or vulnerable species.[28] The Habitats Directive requires the establishment of Special Areas of Conservation (SACs) and the Birds Directive requires the creation of Special Protection Areas (SPAs); and, together, SACs and SPAs make up the Natura 2000 series.

I-11.10   Where the Secretary of State is satisfied that something pursuant to a licence, authorisation or approval has, or may have, adverse effects on the integrity of a relevant site, he may give a direction to the person concerned to take certain steps to avoid, reverse, reduce or eliminate such effects, and to submit a plan of the steps to be taken.[29] Notably, before issuing any such direction, the Secretary of State is again obliged to consult with the appropriate nature conservation body.[30]

I-11.11   The Conservation Regulations also specifically provide for the protection of certain creatures and birds. To this end, it is forbidden to carry out oil and gas activities in such a way as, *inter alia*, to harm any creatures (as well as their eggs or breeding sites) that are members of any of the species listed in Annex IV(a) of the Habitats Directive,[31] or of the species listed in Article 1 of the Birds Directive.[32] The Secretary of State may, however, consent to such activities where, in his opinion, there is not only an overriding public interest in the activity being carried out but also a lack of a satisfactory alternative and, having consulted with the JNCC, the activity will not be detrimental to the maintenance of the populations of the species.[33]

I-11.12   Offences under the Conservation Regulations are constituted by failure to comply with a direction by the Secretary of State[34] or by breaches of the regulations protecting certain creatures, birds or plants[35] or of those prohibiting certain methods of killing or capturing specified species.[36] Conviction in relation to a failure to comply with a direction may attract a fine,[37] whereas other convictions in relation to other offences under these regulations may

---

[28] See Natura 2000, available at www.natura.org (accessed 3 May 2017).
[29] Reg 7(1)(a) and (c) and (2)(a).
[30] Reg 7(3).
[31] Regs 10 and 11.
[32] Reg 11(a).
[33] Reg 14.
[34] Reg 7.
[35] Regs 10, 11 and 12.
[36] Regs 16, 17 and 18.
[37] Reg 19(2).

attract a fine or imprisonment.[38] The protection offered by the corporate veil is avoided by means of the fact that where an offence is committed by a body corporate with the consent, connivance or neglect of a director, manager, secretary or other similar officer, or a person purporting to act in such capacity, both the body corporate and the person in question shall be guilty of an offence.[39] In sum, the Conservation Regulations demonstrate that, during the licensing process, the UK has clearly forced potential licensees to see to it that the potential effects of their activities are fairly stringently assessed in relation to certain habitats and species.

The Conservation Regulations dictate that offshore operators   **I-11.13** may not carry out certain geological surveys (including seismic surveys) or shallow drilling, or test any equipment to be used for those purposes, unless the Secretary of State has granted prior written consent thereto.[40] Consent to undertake a survey or shallow drill is independent from any form of permission granted under exploration and Production Licences.[41] Accordingly, an unlicensed area cannot be subjected to a survey or to shallow drilling without both an Exploration Licence and approval to undertake the activity. Similarly, where Production Licences apply, those areas may not be surveyed or shallow drilled unless approval for the activity has been obtained. For internal waters, operators are required to notify the Oil and Gas Authority (OGA) of their intention to undertake landward surveys and/or offshore surveys and shallow drilling.[42]

In addition to the requirement for an EA to be carried out by the   **I-11.14** prospective licensee where a proposed activity will have a "likely significant effect" on the environmental sensitivities of a relevant site, the Department for Business, Energy and Industrial Strategy (BEIS) will also conduct a so-called appropriate assessment (AA).[43] It does not matter whether the relevant activity will in fact occur within

---

[38] Reg 19(3).

[39] Reg 19(4).

[40] Reg 4(1)(a), (b) and (c). See, generally, *Guidance Notes for Oil and Gas Surveys and Shallow Drilling*, available at www.gov.uk/government/uploads/system/uploads/attachment_data/file/50000/3606-PON14a_guide_110906.pdf (accessed 30 April 2017). Note that as at April 2017 these Guidance notes are being reviewed and are likely to change.

[41] *Ibid*, at 6–7.

[42] OGA, *PON 14b*, available at www.ogauthority.co.uk/exploration-production/petroleum-operations-notices/pon-14b (accessed 30 April 2017).

[43] BEIS, *Oil and Gas: offshore environmental legislation, available at* www.gov.uk/guidance/oil-and-gas-offshore-environmental-legislation#the-offshore-petroleum-activities-conservation-of-habitats-regulations-2001-as-amended (accessed 29 April 2017) (hereinafter "BEIS, *Oil and Gas*"); also Reg 5.

the boundary of the relevant site or not. Instead, what is of cardinal importance is simply that the activity will have a significant effect on the relevant site's environmental sensitivities. BEIS stresses that the term "significant effect" must be understood in light of a document produced by the Commission, entitled *Assessment of plans and projects significantly affecting Natura 2000 Sites – Methodological guidance on the provisions of Article 6(3) and 6(4) of the Habitats Directive 92/43/EEC* (hereinafter "the Commission Guidance").[44] The Commission Guidance states that whether or not there will be a "significant effect" may be determined through one of two approaches.[45] The first is through what is essentially a judgement of the possible impacts of a proposed activity, irrespective of whether they may be adverse or beneficial. Such judgements will be made through a process of assessment that is based on numerous factors including, *inter alia*, the magnitude, spatial extent and duration of anticipated change; the resilience of the environment to cope with change; the existence of environmental standards against which the proposal could be assessed; the degree of public interest; and the scope for mitigation, sustainability and reversibility.[46] The second approach for determining whether or not a proposed activity will have a significant effect focuses on the adverse consequences thereof through reference to sets of significance criteria, which relate specifically to various aspects of the environment.[47]

I-11.15   Irrespective of which of these two approaches may be adopted to determine whether a proposed survey or shallow drilling operation may have a likely significant effect on a relevant site, to date, only five fairly small areas of the UKCS have been identified as candidates for the designation and there is no indication that these sites are in areas which might become of interest to the offshore industry.[48] Thus, it would appear that AAs are presently not required for the vast majority of offshore surveys and shallow drilling operations in the UK. Nonetheless, BEIS has adopted a precautionary approach and is prepared to treat areas of the UKCS that may be selected as relevant sites as if they were already selected as such. Accordingly, where the proposed activities of those wishing to conduct offshore surveying or shallow drilling operations in the UK may interact with

---

[44] A copy of the Commission Guidance is available online at http://ec.europa.eu/environment/nature/natura2000/management/docs/art6/natura_2000_assess_en.pdf (accessed 29 April 2017).

[45] *Ibid*, at 62.

[46] *Ibid*.

[47] *Ibid*.

[48] JNCC, "Offshore Marine Protected Areas", available at www.jncc.gov.uk/page-4534 (accessed 25 April 2017).

the environmental sensitivities of what may one day be considered to be a relevant site, AAs may be required. It is noteworthy that, in accordance with PON 9 (Record and Sample Requirements for Seaward Surveys and Wells),[49] upon completion of seismic surveys, what is generally referred to as a "close-out" form must be completed and submitted to BEIS, which will then be used by the UK Government to report seismic survey statistics in furtherance of the objectives of the Agreement on Small Cetaceans of the Baltic and North Seas (ASCOBANS).[50] The information is also be submitted to Digital Energy Atlas and Library (DEAL), which is an initiative sponsored by Oil & Gas UK that provides a web-based service designed to facilitate access to information relevant to the exploration and production of hydrocarbons on the UKCS.[51] Finally, it should be noted that, in addition to a close-out form, each operator must also submit a survey report together with all related marine mammal observations to the JNCC.[52]

Environmental regulation of exploration and production activities

As a member of the EU, the environmental regulation of offshore   I-11.16
petroleum activities in the UK has, to date, been subject to the EU Directive Assessment of the Effects of Certain Public and Private Activities on the Environment (85/337/EEC) as amended by Council Directive 97/11/EC (hereinafter "the Assessment Directive"). The Assessment Directive sets out a mandatory requirement for member states to conduct an environmental assessment for any proposed offshore oil and gas activity.[53] The Offshore Petroleum Production and Pipelines (Assessment of Environmental Effects) Regulations 1999 (as amended) (hereinafter "the Assessment Regulations")[54] implement this Directive. Section 5 of the Assessment Regulations requires offshore operators to assess the environmental impact of offshore activities and to submit an environmental statement when applying for consent to undertake a relevant project (hereinafter

---

[49] OGA PON9, available at www.ogauthority.co.uk/exploration-production/petroleum-operations-notices/pon-9 (accessed 27 April 2017).

[50] *Ibid.* See also ASCOBANS, available at www.ascobans.org (accessed 23 April 2017).

[51] DEAL, available at www.ukoilandgasdata.com (accessed 2 May 2017).

[52] BEIS, *Oil and Gas*, at 18.

[53] Council Directive on the Assessment of the Effects of Certain Public and Private Activities on the Environment (85/337/EEC), as amended by Council Directive 97/11/EC.

[54] The 1999 Assessment Regulations Regulations were amended by the Offshore Petroleum Production and Pipe-lines (Assessment of Environmental Effects) (Amendment) Regulations 2007 (SI 2007/933), to increase public access to environmental information and allowing for greater public participation in the environmental decision-making process.

referred to as "the activity"),[55] prior to submitting an Environmental Statement (ES).

I-11.17   The ES is required to demonstrate whether there will be a significant impact on the environment, and how this will be addressed. Schedule 2 to the Assessment Regulations explains in some detail the required content of an ES, which can essentially be divided into five points. Firstly, an ES must describe the site, design and size of the project,[56] including: the seabed use requirements during the construction and operational phases;[57] a description of the production processes including the nature and quantity of the materials used;[58] and an estimate by type and quantity of the expected residues and emissions resulting from the proposed project.[59] Secondly, an ES must describe measures that will be employed to avoid, reduce and, where possible, remedy significant adverse effects on the environment.[60] Thirdly, an ES, with regard to current knowledge and methods of assessment, must contain the necessary data to assess the effects that the proposed project is likely to have on the environment including, in particular, "fauna, flora, water including the sea and any aquifers under the seabed, air, climatic factors, the landscape or the seascape ... and the interaction between any of the foregoing".[61] Moreover, the data contained in the ES must also include a description of the likely significant effects on the environment arising from the existence of the project, the use of natural resources, the emission of pollutants, the creation of nuisances and the elimination of waste,[62] as well as details of the forecasting methods used to assess such effects.[63] Fourthly, the main alternatives to the proposed activity together with the environmental implications thereof and the reasons for the relevant operator's choice must also be provided.[64] Fifth, an ES must also contain a non-technical summary (NTS) of all the information required by the previous four points[65] so that a non-specialist reader

---

[55] A relevant project is defined in Section 4 of the Assessment Regulations as (1) the drilling of an exploration well; (2) a development; (3) the construction of a pipeline for the conveyance of petroleum other than one which is to form an integral part of any development, or (4) the use of a mobile installation for the extraction of petroleum where the principal purpose of the extraction is the testing of any well.

[56] Sch 2(a).

[57] Sch 2(a)(i).

[58] Sch 2(a)(ii).

[59] Sch 2(a)(iii).

[60] Sch 2(b).

[61] Sch 2(c)(i).

[62] Sch 2(c)(ii).

[63] Ibid.

[64] Sch 2(d).

[65] Sch 2(e).

will be able to understand the principal environmental impacts of a proposed activity without having to refer to the main ES.[66]

This approach to the preparation of the ES has a number of advantages. First, it ensures that an ES not only describes all the foreseeable potential environmental impacts of a proposed offshore operation but also identifies solutions to eliminate or mitigate such impacts.[67] Second, it requires explicit acknowledgement in the ES of any lacunae in environmental information and any other difficulties (including technical difficulties and lack of know-how) encountered by the operator in compiling the required information, and strategies to address such lacunae where acknowledged in the ES. This explicit recognition of uncertainties, coupled with the requirement that alternatives must also be described together with the reasons for the ultimate selection, provides the Secretary of State with a better view of the extent to which environmental considerations are being given due priority. It is also noteworthy that the exact content of every individual ES is to a large extent influenced by formal and informal consultation processes that are mandated and strongly recommended by the Assessment Regulations.

As part of the ES process, operators are required to participate in formal consultations with the relevant environmental agency and they are encouraged to participate in various informal consultations with interested parties, such as special interest and conservation groups as well as users of the sea. All ESs are required to be submitted to various environmental authorities, which are entitled to make representations in relation thereto to BEIS,[68] including: the Joint Nature Consultative Committee (JNCC) and, depending on the location of the proposed activity, either the Scottish Government Department for Environment and Rural Affairs (SDEFRA) and its agency the Scottish Fisheries Research Services (FRS), or to the relevant Department of Environment, Food and Rural Affairs (DEFRA) and its agency the Centre for Environment, Fisheries and Aquaculture Science (CEFAS). In addition, if the activity which is the subject of an ES is within 40km of the coast then that ES must be submitted to other relevant authorities, including, *inter alia*, the Countryside Council for Wales (CCW) for Welsh waters; Natural England for English waters; Scottish Natural Heritage (SNH) and the Scottish Environment Protection Agency (SEPA) for Scottish waters; and one,

I-11.18

I-11.19

---

[66] Rowan Robinson, "Environmental and Planning Law", at p 1287.

[67] DECC (now BEIS), *Guidance Notes for Industry: Guidance Notes on Offshore Petroleum Production and Pipe-lines (Assessment of Environmental Effects) Regulations 1999*, available at www.gov.uk/government/uploads/system/uploads/attachment_data/file/193705/eiaguidancenote.pdf (accessed 3 May 2017).

[68] Reg 9.

or possibly more, of the 12 Sea Fisheries Committees for English and Welsh waters.[69] The importance of such formal consultations is highlighted by the fact that consent to a project will not be granted unless consideration has been given to the representations made by any environmental authority to which a copy of the ES was required to be sent.[70]

I-11.20   The Regulations were amended in 2007 by the Offshore Petroleum Production and Pipe-lines (Assessment of Environmental Effects) (Amendment) Regulations 2007 to implement the requirement for public participation in respect of the drawing up of certain plans and programmes relating to the environment. These amended Regulations came into force on 16 April 2007. Since the Brent Spar case in the mid-1990s, the industry has become much more aware of ensuring proactive dialogue with interested stakeholders.[71]

I-11.21   In sum, it may be said that the foregoing discussion of the Conservation, Assessment, Exploration and Production Regulations has demonstrated that UK law (albeit in many instances implementing European obligations) takes cognisance of the potentially adverse environmental effects of its offshore oil and gas activities even before the licensing process commences. We will now go on to consider the environmental regulations an activity is subject to once an exploration or Production Licence has been granted and activity begins.

Regulation under the Petroleum Licensing (Exploration and Production) (Seaward and Landward Areas) Regulations 2004 (Exploration Regulations)

I-11.22   Petroleum exploration, development and production activities on the UKCS proceed on the basis of licences issued by the Secretary of State under powers granted to him by Section 3(1) of the Petroleum Act 1998. Such licences may be granted on such terms and conditions as the Secretary of State thinks fit.[72] The Exploration Regulations are divided into Schedules according to the type of licence to which they apply.[73] Exploration Licences are governed by the model clauses contained in Schedule 1 of the Exploration Regulations,[74] while Production Licences issued prior to the advent

---

[69] Rowan Robinson, "Environmental and Planning Law", at p 1288.
[70] Reg 5(4)(b)(ii).
[71] See the discussion in Chapter I-10.
[72] Petroleum Act 1998, s 3(1).
[73] See Model Cl 3(1)–(8). In accordance with Model Cl 3(7) and (8), read with the definition of "Petroleum Exploration and Development Licence" contained in Model Cl 2, Schs 6 and 7 apply to landward licences and will therefore not be discussed.
[74] Model Cl 3(2).

of the Production Regulations on 6 April 2008, are governed by the model clauses contained in Schedules 2–4 of the Exploration Regulations.[75] Therefore, a good starting point in any consideration of the environmental obligations of a licensee is the conditions stipulated in Schedule 1 of the model clauses.

Perhaps the first noteworthy model clause of Schedule 1 is Model Clause 1(2), which provides that, whenever the licensee is more than one person, all obligations are joint and several. Thus, although it is not an expressly environmental clause, joint and several liability is a positive enforcement mechanism as it ensures that, should persons renege on the terms of their licence (including any environmental obligations), others who are party thereto can be held accountable. Additionally, obligations may be enforced against whoever is the most suitable of the persons constituting the licensee: for example, where there are two companies constituting the licensee, the election may be made to pursue the wealthier company, which may be far more capable of ensuring adherence to the obligations attached to the relevant Exploration Licence. Concomitantly, where a licensee is more than one person, such persons are unable to allocate the obligations among themselves and thus to choose a potential "scapegoat" should they renege on their obligations. In other words, such persons are prohibited from arranging that the Minister has access to only the "shallowest pocket". This may appear to be a very obvious point, but such joint and several liability is not a feature of the petroleum licences of every jurisdiction, potentially exposing the states in question to the risk that they will be left to deal with the aftermath of any unfulfilled obligations.

I-11.23

Model Cl 7(1) and (2) provide that a licensee may not commence (or recommence) the drilling of a well, or abandon a well, without the written consent of the Minister, the environmental benefit of which is that all drilling and abandonment activities are brought to the attention of the relevant authorities. It is only once BEIS is aware of a licensee's desire to undertake such activities that it is able to take steps to ensure that they are conducted in accordance with the relevant environmental regulations.[76] There are two noteworthy instances where, in relation to abandonment, a lack of written

I-11.24

---

[75] Model Cl 3(3)–(6).

[76] Notably, Model Cl 7(4) provides that where the granting of ministerial consent in accordance with Model Cl 7(1) is conditional upon the position, depth or direction or casing of the well to be drilled, the Minister may direct that the well and all records relating thereto are examined by persons of his choosing. Likewise, Model Cl 7(4) provides that when determining whether or not to grant consent to plug or seal a well in accordance with 7(2), the Minister may direct that the well and all records relating thereto are examined by persons of his choosing.

consent from the Minister is not sufficient cause not to commence with the plugging of wells.[77] First, there is an obligation on a licensee to plug all the wells in the exploration area to which the licence applies not less than one month prior to the expiry (or further determination) of its licence rights, unless the Minister determines otherwise.[78] Second, where a licence is already in force in a particular area, an Exploration Licence for that area may be granted only through agreement between the prospective licensee and the original licence-holder. Where such an agreement results in the original licence-holder's rights ceasing to be exercisable for the time being, or when the agreement between the original and prospective licensee terminates, the licensee must plug its wells within one month after the date on which its rights cease to be exercisable, unless the Minister determines otherwise.[79] Irrespective of whether the abandonment and plugging of wells occurs with or without ministerial consent, Model Cl 7(7) provides that "[t]he plugging of any well shall be done in accordance with a specification approved by the Minister ... and shall be carried out in an efficient and workmanlike manner".

I-11.25   Abandonment and plugging of wells aside, Exploration Licensees are bound to maintain in "good repair and condition" all apparatus, appliances and wells which have not been abandoned or plugged[80] and carry out all operations in a "proper and workmanlike manner in accordance with methods and practice of exploration customarily used in good oilfield practice".[81] However, the phrases "proper and workmanlike manner" and "good oilfield practice" are problematic, having been described as "simplistic and vague".[82] BEIS has, however, stated that good oilfield practice "relates largely to technical matters within the disciplines of geology and ... engineering and to the impact of ... [oil and gas] development[s] on the environment".[83] This statement does not, however, constitute a binding definition and fails to address the issue of how competing interests are balanced.

---

[77] Model Cl 7(2), read with Model Cll 2, 7(5) and (6).

[78] Model Cl 7(6).

[79] Model Cl 7(5).

[80] Model Cl 9(1).

[81] Model Cl 9(2).

[82] Z Gao (ed.), *Environmental Regulation of Oil and Gas* (1998), p 13. See also T Daintith and G Willoughby, *UK Oil and Gas Law* (1996), para 5386, who note, for example, that "[t]he expression 'good oilfield practice' is widely used ... but no indication is given of where such practice may be found". See generally M A G Bunter, *The Promotion and Licensing of Petroleum Prospective Acreage* (2002), pp 309–310; and I L Worika, "Environmental Terms and Concepts in Petroleum Legislation and Contracts", in Gao (ed.), *Environmental Regulation of Oil and Gas*, at pp 393–413.

[83] OGA, *Guidance notes on procedures for regulating offshore oil and gas developments*, available at www.ogauthority.co.uk/media/2423/onshorefdpguidanceaug2016.pdf (accessed 22 April 2017).

For example, from a geological and reservoir engineering perspective it may be highly appropriate to conduct extensive seismic surveys, yet from an environmental impact point of view such conduct may be highly inappropriate.[84]

In addition to carrying out operations in accordance with good oilfield practice (or perhaps in furtherance thereof), licensees are required to take all steps practicable in order, *inter alia*, to prevent the escape of waste or petroleum into the exploration area, its waters or any waters in the vicinity thereof.[85] Failing to do so requires the licensee immediately to give notice to the Minister and the Maritime and Coastguard Agency.[86] Another noteworthy environmental consideration of the Exploration Regulations is the stipulation that licensees "shall not carry out any operations ... in or about the Exploration Area in such manner as to interfere unjustifiably with navigation or fishing in the waters of the Exploration Area or with the conservation of the living resources of the sea".[87] Despite the environmental purport of this stipulation, it is submitted that there are two problems therewith. First, what constitutes unjustifiable interference is not defined; and the problem is compounded by the fact that there has not been any judicial interpretation of this concept. From the perspective of the fishing and shipping industries, the fact that, by implication, there can be justifiable interference, yet there is no guidance provided regarding when such interference becomes unjustifiable, may be a point of concern. It is submitted that this may also be a point of concern for the offshore industry in that it may be in the offshore industry's interests to have its relationships with the fishing and shipping industries based on clearly defined concepts, thereby reducing the possibility of disputes arising.

The second of the aforementioned problems is the fact that the phrase "in the waters of the Exploration Area" appears to be an unnecessary geographical limitation on a licensee's obligation not to have its operations interfere with navigation or fishing. It may be possible to argue that this limitation falls away in relation to fish and other living marine resources by means of the phrase "or with the conservation of the living resources of the sea"; however, this argument

I-11.26

I-11.27

---

[84] The phrase "good oilfield practice" is a vague and ill-defined concept. Notably, s 6 of Australia's Offshore Petroleum and Greenhouse Gas Storage Act 2006, which defines "good oilfield practice", provides the following particularly vague definition that does not address competing issues: "'good oilfield practice' means all those things that are generally accepted as good and safe in (a) carrying on of exploration for petroleum; or (b) petroleum recovery operations".

[85] Model Cl 9(1). Note that Model Cl 1 defines the "exploration area" as "the area for the time being in which the Licensee may exercise the rights granted by this licence".

[86] Model Cl 9(3).

[87] Model Cl 10.

fails to cure the fact that licensees do not appear to be obliged not to interfere unjustifiably with navigation outside their exploration areas. Nevertheless, despite the foregoing concerns, the Exploration Regulations do require licensees to appoint fisheries liaison officers, whose task it is to promote good working relationships between the owners and masters of vessels (including seismic survey vessels) employed by the licensee and organisations representing the local fishing industry.[88] Licensees are furthermore required to consult with local fisheries organisations regarding the sea routes to be used by the licensees' vessels and to ensure that such sea routes are adhered to.[89] What is more, should licensees' activities result in debris, they are obliged, without reasonable delay, to locate and remove such debris.[90] As for the method of removal, they are required to consult with the relevant fishing organisations and to inform the Secretary of State.[91] Importantly, should fishing gear be lost or damaged, or a loss of fishing time result from reported debris, a licensee is obliged to deal with any such loss promptly.[92] What, specifically, a licensee is obliged to do when dealing with loss promptly is not defined.[93] Nevertheless, it is submitted that the abovementioned provisions are, from an enviro-legal perspective, generally praiseworthy as they provide fairly detailed obligations aimed at negating the occurrence of pollution in the form of interference with fishing and shipping activities during the operations under an Exploration Licence.

I-11.28   In addition to the foregoing, the Exploration Regulations contain various record-keeping provisions of environmental significance. First, a licensee must maintain accurate records of the drilling, deepening, plugging or abandonment of wells,[94] and, on or before the 15th day of each month, for the duration of the licence, supply the Minister with a return containing, *inter alia*, "a statement of the areas in which any geological work, including surveys by any physical or chemical means, has been carried out".[95] Second, an

---

[88] Model Cl 23(1).

[89] Model Cl 23(a) and (c). Note that Model Cl 23(c) disallows non-adherence to agreed sea routes unless safety of navigation or security of cargo dictate otherwise. After consultation with local fisheries organisations, licensees are to inform the Secretary of State for Environment and Rural Affairs (as well as the Scottish Government Minister of Environment and Rural Affairs) of the results of the consultations and then to agree on the measures to be employed to minimise interference with fishing activities.

[90] Model Cl 23(2).

[91] *Ibid.*

[92] *Ibid.*

[93] See, however, in this regard the UK Fisheries Offshore Oil and Gas Legacy Trust Fund Limited at www.ukfltc.com and, as an example of one of its initiatives, FishSAFE at www.fishsafe.eu/en/home.aspx (both accessed 7 May 2017).

[94] Model Cl 11(1).

[95] Model Cl 12(1)(a).

annual return must be submitted showing the situation of all wells, as well as all works executed.[96] Third, all records, papers and so forth kept in pursuance of an Exploration Licence may be inspected by duly appointed persons.[97] Logically, the creation and consideration of such records may assist in combating environmental harm caused by, amongst other things, seismic surveys, interference, oil and offshore chemicals. Although the information that licensees submit to the Minister may not be disclosed to any person who is not in the service or employment of the Crown,[98] there are, from an environmental perspective, two noteworthy exceptions. First, the Minister may furnish the Natural Environment Research Council (NERC) with, *inter alia*, any records, returns, samples, and information obtained from a licensee.[99] Second, the Minister and NERC may use the information submitted by the licensee to prepare and publish reports and surveys of a geological, scientific, technical and general nature.[100] It should also be noted that a consequence of the Environmental Information Regulations 2004 may be that such information could become public in some shape or form, albeit subject to protections for, *inter alia*, commercial confidentiality.[101]

Aside from the abovementioned record-keeping provisions, the Exploration Regulations also entitle the Minister to authorise persons to inspect a licensee's installations and equipment and to examine the state of repair and condition thereof.[102] In doing so, such persons may, in certain circumstances, execute any works or provide and install any equipment.[103] For example, where a licensee reneges on its obligation to maintain equipment in good repair and condition so as to prevent the escape of petroleum into the waters of the exploration area, the Minister may, after reasonable notice, execute any works, including the installation of any equipment, which in the Minister's opinion are necessary to secure performance of the obligation.[104] In addition to executing works and installing equipment, the Minister is also entitled to revoke an Exploration Licence where there has been breach or non-observance of any of

I-11.29

---

[96] Model Cl 12(2).

[97] Model Cl 15(a).

[98] Model Cl 14.

[99] Model Cl 14(b). Notably, the Minister may also furnish any other body that conducts substantially similar geological activities to those conducted by NERC with the said information.

[100] Model Cl 14(c) and (d). The right to use such information is also extended to any additional bodies that have been furnished therewith.

[101] SI 2004/3391, Reg 12(5)(e).

[102] Model Cl 16(a).

[103] Model Cl 16(b).

[104] Model Cl 17, read with Reg 9(1)(e).

the terms and conditions thereof.[105] Notably, bankruptcy and liquidation are specifically mentioned as instances when the Minister may revoke a licence,[106] which, it is submitted, may help to ensure that companies that are financially unable to meet the various environmental obligations attaching to Exploration Licences are forbidden from operating offshore.[107]

Regulation under the Petroleum Licensing (Production) (Seaward Areas) Regulations 2008

I-11.30   As with the Exploration Regulations, the first environmentally noteworthy provision of the Production Regulations is that, whenever the licensee is more than one person, all their obligations are jointly and severally applicable.[108] A provision with obvious environmental significance is the stipulation that a licensee may not erect or carry out any relevant works without ministerial consent or approval having been obtained by means of the submission of a programme specifying what works it intends to erect or carry out, the purpose and location thereof, as well as when such works will commence and be completed.[109] Although the Minister maintains the ability to reject a programme if it would be contrary to good oilfield practice or if it would not be in the national interest, the proviso of a programme having to be in the national interest relates only to the maximum and minimum quantities of petroleum that the licensee proposes to acquire[110] and not, for example, whether a proposed operation may detrimentally affect major fishing or mariculture activities which arguably might be in the national interest.[111] Additionally, as previously discussed, the phrase "good oilfield practice" is not defined and the uncertainty surrounding how competing interests will be considered in relation thereto begs the question whether activities that are potentially harmful to the

---

[105] Model Cl 20(2)(a).

[106] See Model Cl 20(2)(c), (e) and (f).

[107] In this regard, see also now Model Cl 20A, inserted by the Energy Act 2008, s 77 and Sch 3, which grants the Minister the power partially to revoke a licence in respect of a person who has become bankrupt etc.

[108] Model Cl 1(2).

[109] Model Cl 17(1)(a) and (b). Notably, Model Cl 17(9) defines "relevant works" as "any structures and any other works whatsoever which are intended by the Licensee to be permanent and are neither designed to be moved from place to place without major dismantling nor intended by the Licensee to be used only for searching for Petroleum". There is no equivalent provision in the Exploration Regulations, as such activity is not in contemplation where only surveying or shallow drilling is permitted.

[110] Model Cl 17(4)(c)(ii), read with Model Cl 17(2)(c) and (1)(b).

[111] Model Cl 18(6) does, however, grant the Minister the discretion to determine what may or may not be within the national interest as regards his interpretation of Model Cl 17 in relation to minimum and maximum quantities of petroleum.

marine environment would ever be sufficient for the Minister to reject a programme on that basis alone when faced with favourable economic considerations.

The Production Regulations contain various provisions pertaining to the drilling, plugging and abandonment of wells that essentially mirror those of the Exploration Regulations.[112] The only noteworthy distinction is that the Minister may direct that upon expiration of production licensees' rights, they need not plug and seal a well; instead they must leave it "in good order and ... fit for further working".[113] There is, however, no explanation as to whether leaving a well in "good order" incorporates environmental concerns. Accordingly, it must be assumed that the concept of "good order" incorporates environmental concerns.[114]

I-11.31

The Production Regulations specifically govern development wells, which are those wells that will be used not merely for the searching for petroleum, but for the actual getting thereof from the licence area.[115] Any work, such as the installation of a casing or equipment, for the purpose of bringing a well into use as a development well is referred to as "completion work".[116] Although licensees may not undertake any completion work except in accordance with a programme approved by the Minister,[117] the Production Regulations do not stipulate that such programmes ought to take note of environmental concerns. They do, however, dictate that all operations be conducted in a "proper and workmanlike manner in accordance with methods and practice of exploration customarily used in good oilfield practice" while taking all steps practicable, *inter alia*, to prevent the escape of petroleum into the waters of the licence area.[118] The concerns that were raised against the same provisions of the Exploration Regulations can again be raised here. There are two unique provisions in the Production Regulations which attract praise rather than concern: first, flaring may not be undertaken without ministerial consent;[119] and, second, licensees are bound to comply with any reasonable instructions from the Minister that are aimed

I-11.32

---

[112] Model Cl 19.

[113] See Model Cl 19(12)(a) and (b).

[114] Note, however, that BEIS (formerly DECC) directs operators, in this regard, to guidance prepared by Oil & Gas UK: *OP006 – Guidance on Suspension and Abandonment of Wells/North Sea Well Abandonment Study* (2009).

[115] Model Cl 21.

[116] Model Cl 21(4).

[117] *Ibid.*

[118] See Model Cl 23.

[119] Model Cl 23(3)(a). There are, however, certain exceptions to this stipulation, such as where there is a risk of injury to persons: see Model Cl 23(7)(a).

at ensuring that funds are available "to discharge any liability for damage attributable to the release or escape of [p]etroleum".[120]

I-11.33   Aside from those provisions discussed above, the noteworthy environmental considerations contained in the Production Regulations, such as those pertaining to fishing and record-keeping, are practically identical to those of the Exploration Regulations and thus will not be reconsidered. In sum, it is submitted that the environmental considerations contained in the Exploration Regulations and Production Regulations are generally praiseworthy as they are fairly detailed and the few criticisms thereof that have been raised pertain mostly to the fact that certain concepts could have been more appropriately defined. That these definitional issues may not be as troubling as might be feared becomes evident, however, when the significant amount of additional legislation and regulation directed specifically at the environmental dimension of offshore hydrocarbon operations is taken into consideration. In short, the licence is but one layer in the complex legal and regulatory arrangements that seek to ensure environmental protection offshore, as the remainder of this chapter reveals.

Produced water

I-11.34   Produced water is water that is extracted from the subsurface along with oil and gas, which may have originated in the reservoir or through injection into the formation so as to maintain pressure and a particular rate of production. Produced water consists of, among other things, a mixture of oil droplets, trace metals, dissolved organic compounds and production chemicals.[121] The key items of legislation governing produced water from the UK's offshore oil and gas operations, are twofold, namely the Offshore Petroleum Activities Regulations and FEPA 1985.

Offshore Petroleum Activities (Oil Pollution Prevention and Control) Regulations 2005 (OPA Regulations)

I-11.35   The OPA Regulations aim to reduce the quantities of hydrocarbons discharged during the course of offshore operations and they do so via, *inter alia*, introducing a permitting system for oil discharges[122] and strengthening the powers of inspection and investigation into

---

[120] Model Cl 23(9).

[121] See IMO Joint Group of Experts on the Scientific Aspects of Marine Environmental Protection, *Impact of oil and related chemicals and wastes on the marine environment* (GESAMP Reports and Studies No 50, 1993), p 170.

[122] Reg 3(1).

such discharges.[123] The definition of "oil" under the OPA Regulations is broad enough to include hydrocarbons found in produced water[124] and thus any discharge of produced water will require a permit under the OPA Regulations.

It is by means of the OPA Regulations that the UK Government I-11.36 implemented its obligations under OSPAR Recommendation 2001/1 and achieved, by the end of 2006, compliance with a 30mg/l monthly average dispersed oil in water discharge, as well as a 15 per cent reduction in oil tonnage discharged in produced water, when compared with discharges from 2000.[125] Interestingly, until 2009, regulation of produced water in the UK was based on a system of allowances regulated by the Dispersed Oil in Produced Water Trading Scheme (DOPWTS) which was, in essence, a cap-and-trade scheme in terms of which DECC (now BEIS) issued discharge allowances out of an overall discharge allowance pot.[126] Although the DOPWTS has been revoked, BEIS is considering various proposals for the future management of produced water, which include maintaining certain aspects of the status quo such as the 30mg/l monthly average of dispersed oil in produced water discharge limit, the 100mg/l maximum dispersed oil concentration limit and the fact that operators' predictions of annual discharge tonnage of dispersed oil in produced water must take into account the principles of Best Available Technique (BAT) and Best Environmental Practice (BEP).[127] Notably, one of the new proposals for the management of produced water is the so-called Risk Based Approach that the OSPAR Commission is in the process of preparing and which will place greater emphasis on risk assessment and furtherance of the objectives of OSPAR Recommendation 2001/1, in particular reaching the ultimate goal of zero harmful discharge.[128]

---

[123] Reg 12(1)(a) and (b).

[124] Reg 2.

[125] Oil & Gas UK, "Produced Water", available at www.ukooaenvironmentallegislation. co.uk/Contents/Topic_Files/Offshore/Produced_water.htm#newdevelopments (accessed 28 April 2017).

[126] The allowance pot for 2006–2007 was approximately 4,888 tonnes of dispersed oil to sea in produced water discharges and was based on the aforementioned OSPAR stipulation of a 15 per cent reduction of oil tonnage discharged in produced water, relative to the discharges in 2000. See European Commission, *Dispersed oil in produced water trading scheme* C(2006)3194 final, available at http://ec.europa.eu/competition/state_aid/cases/203609/203609_589192_27_2.pdf (accessed 30 April 2017).

[127] See the letter from S J Kydd, DECC, Head of Environmental Operations, to the UK offshore industry, entitled "Response to Consultation on the Future of the UK Dispersed Oil in Produced Water Trading Scheme", dated 31 July 2008, available at webarchive. nationalarchives.gov.uk/20111013141145/https://www.og.decc.gov.uk/environment/opaoppcr_letter_310708.pdf (accessed 11 April 2011).

[128] Oil & Gas UK, "Produced Water"; S J Kydd, *Ibid*.

Food and Environment Protection Act 1985

I-11.37 FEPA 1985 regulates produced water where water is transported to another field for reinjection and requires a licence.[129] Thus, the discussion above regarding the need for a FEPA 1985 licence in relation to the export of drill cuttings from one field to another for reinjection is equally applicable to produced water. Nevertheless, it should perhaps be pointed out that applications for a FEPA 1985 licence must include a description of, *inter alia*, how the production water is to be transported and how the reinjection operation will be undertaken (including a technical explanation in support of how containment of the disposed produced water will be undertaken), and an assessment of alternative means of disposal must also be provided so as to demonstrate that offsite reinjection is in line with BAT and BEP.[130]

Pipeline and production chemicals

I-11.38 The cardinal environmentally oriented regulations governing pipeline and production chemicals in the UK are the Offshore Chemical Regulations 2002 (the OC Regulations).[131] This is not to say that the Offshore Installations (Emergency Pollution Control) Regulations 2002[132] (the EPC Regulations) do not play an important role in the regulation of offshore chemicals – they do, but they apply only in so far as they grant the UK Government certain powers of intervention following an accident involving an offshore installation. The thrust of the OC Regulations is that they are only intended to apply to chemicals used and discharged through exploration, exploitation, offshore processing and decommissioning.[133] Thus, non-operational chemicals that might otherwise be used on an offshore installation, such as paints, are not subject to the OC Regulations. When an offshore operator applies for a permit under

---

[129] See s 5(a) and (b) of FEPA 1985 and Regs 14, 15 and 15A of the Deposits in the Sea (Exemptions) Order 1985 (SI 1985/1699); and Oil & Gas UK, "Produced Water".

[130] Oil & Gas UK, "Produced Water".

[131] SI 2002/1355. The OC Regulations implement, in UK law, OSPAR Decision 2000/2 on a Harmonised Mandatory Control System for the Use and Reduction of the Discharge of Offshore Chemicals, which operates in conjunction with two OSPAR Recommendations which are fundamental to the implementation of the Decision, namely, OSPAR Recommendation 2000/4 on a Harmonised Pre-Screening Scheme for Offshore Chemicals and OSPAR Recommendation 2000/5 on a Harmonised Offshore Chemical Notification Format.

[132] SI 2002/1861.

[133] Reg 2; see www.gov.uk/government/uploads/system/uploads/attachment_data/file/448262/OCR_Guidance_Notes__Revised_March_2011_.pdf (accessed 30 April 2017).

the OC Regulations, irrespective of whether it is a production or term permit, not only must the application list the chemicals and the amounts thereof that will be discharged, it must also contain a risk assessment of the effects thereof on the receiving environment. Notably, applications for permits by operators may only include those chemicals that have been assessed by CEFAS and listed on its website. Where such so-called pre-screening has identified a chemical as hazardous, less hazardous substitutes must be sought. Consequently, it is submitted that although the aforementioned pre-screening procedure is praiseworthy in that it possesses the possibility of negating the introduction of hazardous chemicals into the marine environment, it is nonetheless flawed by virtue of the fact that operators are not obliged to choose less hazardous alternatives.

Aside from the abovementioned pre-screening procedure, it is significant to note that specific offshore activities require permits under the OC Regulations. These activities have been divided into three groups, namely: new and existing producing developments; wells, pipelines, and decommissioning of installations; and unforeseen use and spillage.[134]   I-11.39

Wells, pipelines and decommissioning of installations
The use and discharge of offshore chemicals during the drilling of a well requires a term permit to be issued under the OC Regulations. In order to obtain such a permit, a Drilling Operation Permit (DOP) (formally a PON 15B) must be submitted via the UK Oil Portal. Interestingly, the DOP is dual purpose as it includes both a request for a direction that an environmental statement need not be prepared as well as the application for a permit to use and discharge chemicals. Where, however, an environmental statement is deemed necessary, the DOP may again be used to apply for the necessary term permit. In those instances where offshore chemicals are to be used for, among other things, the installation or decommissioning of pipelines, a term permit is required for which a Pipeline Operation Permit (formally a PON 15C) must be submitted. Similarly, when an offshore installation is to be decommissioned and chemicals are to be employed in that regard, the operator must apply for a term permit. In this instance, however, a Decommissioning Operation Permit (formerly a PON 15E) must be submitted.   I-11.40

---

[134] See, generally, DECC (now BEIS), Guidance Notes on the Offshore Chemical Regulations 2002 at 11–12, available at www.gov.uk/government/uploads/system/uploads/attachment_data/file/448262/OCR_Guidance_Notes__Revised_March_2011_.pdf (accessed 2 May 2017).

Unforeseen use and spillage

I-11.41　There are two situations that may involve the unforeseen use of offshore chemicals. First, certain chemicals may need to be used on a contingency basis, such as those that are utilised in the event of drilling problems being encountered. Second, situations may arise on very short notice where unforeseen use is necessary. In such situations, it may be impractical for operators to apply for the necessary permits and so provision is made for them to receive emergency permission telephonically or electronically, provided that the DECC is satisfied that the environmental consequences of such use have been considered. If emergency permission is granted, the operator must, as soon as possible, submit a request for a variation of his existing permit so as to maintain proper records.

I-11.42　　Where there has been an accidental spillage of offshore chemicals, the DECC stresses that the incident must be reported immediately and that the relevant operator must complete a PON 1 within 24 hours of the release.[135] It is in such situations where there has been a chemical spill that the EPC Regulations come into play. As previously mentioned, these Regulations grant the UK Government certain powers of intervention when there is, or may be, a risk of significant pollution, or where the operator has failed to implement proper control and preventative measures.

Drilling fluids and drill cuttings

I-11.43　The UK regulates the environmental threat posed by the adherence of hydrocarbons to drill cuttings by means of the OPA Regulations. In view of the fact that the OPA Regulations have already been discussed above, there is no need for repetition in relation to this issue. Suffice to state that any contamination of drill cuttings by hydrocarbons will require a permit under the OPA Regulations if such cuttings are to be discharged overboard or reinjected into the well.

I-11.44　　As for the various chemicals that adhere to the drill cuttings, the relevant law in the UK is the Offshore Chemical Regulations 2002[136] (the OC Regulations). FEPA 1985 also plays a regulatory role in relation to drill cuttings, in so far as a licence is required in terms thereof for the export of cuttings from one field to another for reinjection.[137] Additionally, as will be discussed in Chapter I-12,

---

[135] Overview of PON 1 is found at BEIS, *Oil and gas: environmental alerts and incident reporting including anonymous reporting*, available at www.gov.uk/guidance/oil-and-gas-environmental-alerts-and-incident-reporting#pon-1 (accessed 7 May 2017).

[136] SI 2002/1355.

[137] Oil & Gas UK, "Reinjection of Mud and Cuttings", available at www.ukooaenviron-

which considers the environmental regulation of decommissioning, certain obligations arising from the provisions of the OSPAR Convention also have a bearing on the environmental threat posed by drill cuttings.[138] For present purposes, however, the specific roles of the OC Regulations and FEPA 1985 will be discussed in more detail below.

Offshore Chemical Regulations 2002

The OC Regulations forbid the use or discharge of offshore chemicals, save in accordance with one of two possible permits granted by the Secretary of State.[139] Significantly, there are currently no exceptions to this stipulation and the term "offshore chemicals" is broadly defined so as to encompass "any chemical, whether comprising a substance or a preparation, intentionally used in connection with offshore activities".[140] The relevant permits are either "production permits", which pertain to installations that use or discharge chemicals, or "term permits" which cover time-limited use and discharge of chemicals during such activities as the drilling of wells and decommissioning activities.[141] Importantly, in determining whether or not to grant a permit, the Secretary of State must have regard to any opinion expressed by CEFAS or the FRS (as appropriate), any State that is party to OSPAR which may be affected by the use or discharge of the chemicals[142] and the general public.[143] Moreover, the Secretary of State may not grant a permit where certain publicity requirements have not been met, including making the permit application available for public inspection and publishing the details thereof in newspapers on occasions when it is likely to

I-11.45

---

mental legislation.co.uk/Contents/Topic_Files/Offshore/Reinjection.htm (accessed 2 May 2017).

[138] See Chapter I-12.

[139] Reg 3(1), read with Reg 4(1).

[140] Reg 2.

[141] See DECC (now BEIS), *Guidance Notes on the Offshore Chemical Regulations 2002* at 15, available at www.gov.uk/government/uploads/system/uploads/attachment_data/file/448262/OCR_Guidance_Notes__Revised_March_2011_.pdf (accessed 2 May 2017).

[142] Reg 4(1)(a), read with Reg 2. Whether or not the Secretary of State consults with CEFAS or FRS depends on the relevant area to which the permit would relate.

[143] Reg 4(1)(b). Note that Reg 7(2) states that the Secretary of State need not consult with the general public if the permit application is made: "(a) in connection with a relevant project for which the Secretary of State gives a direction, pursuant to Reg 6 of the Offshore Petroleum Production and Pipelines (Assessment of Environmental Effects) Regulations 1999, that no environmental statement need be prepared; (b) in connection with a discharge for a pipeline, being a discharge to which the Secretary of State gives a consent pursuant to an authorization issued under Part III of the Petroleum Act 1998; or (c) in connection with activities carried out in accordance with an abandonment programme approved by the Secretary of State under Part IV of the Petroleum Act 1998".

come to the attention of persons interested in, or affected by, the use or discharge of the relevant chemicals.[144]

I-11.46   It is noteworthy that a permit application must include not only a description of the relevant offshore installation and the proposed technology and techniques to be used for the prevention and reduction of the use and discharge of chemicals, as well as measures that will be employed to monitor the use or discharge of such chemicals, but also an assessment of the possible risk to the environment as a consequence of the use or discharge of the chemicals.[145] As to the conditions that may be attached to a permit, these can take almost any form as the Secretary of State is given an overriding discretion to attach such conditions as he thinks fit.[146] More specifically, however, he is entitled to attach conditions relating to, *inter alia*, the quantity of the discharge and measures to prevent both pollution and accidents that may affect the environment.[147] Where a permit is subject to a time limit and an operator wishes to renew it, an application for renewal must be made three months before the expiry date of the permit and the Secretary of State is bound to consult once more with CEFAS or the FRS, as well as any State that is party to OSPAR and which may be affected by the use or discharge of the chemicals.[148] Interestingly, however, the Secretary of State is not bound to take into account representations made by the general public, as he is required to do in relation to the application for a new permit. Nevertheless, not only may renewed permits be subject to such further terms and conditions as the Secretary of State thinks fit,[149] all permits, together with the terms and conditions that attach to them, are to be kept by the Secretary of State[150] in a register which must be open for public inspection.[151]

I-11.47   Once granted, permits are not set in stone and may be varied by means of an application from the operator or by means of a review by the Secretary of State.[152] Applications for substantial variations may be refused,[153] while applications for other variations require the

---

[144] Regs 4(2) and 7(1) and (3).

[145] Reg 6(1)(a)–(d).

[146] Reg 5(1).

[147] Reg 5(2).

[148] Reg 10(1), (2) and (3), read with Reg 2. Surprisingly, the Secretary of State may issue permits that do not have time limits, which from an environmental perspective may not be advisable. For example, where it transpires that particular chemicals have previously unforeseen harmful effects on the marine environment, an operator may nonetheless use these chemicals until such time as the Secretary of State opts to review the relevant permit in the light of this new knowledge and issue new restrictions in relation thereto.

[149] Reg 10(5).

[150] Reg 14(1).

[151] Reg 14(2).

[152] Regs 11 and 12.

[153] Reg 11(4).

Secretary of State to take into account any relevant representations from CEFAS or the FRS, as well as any State that is party to OSPAR and which may be affected by the use or discharge of the offshore chemicals.[154] Again, however, no provision is made for taking the opinion of the general public into account. As regards the reviewing of permits, the Secretary of State is bound to undertake reviews where pollution, or the risk thereof, is of such significance that restrictions on the use or discharge of offshore chemicals ought to be revised or new restrictions ought to be included.[155] Interestingly, it is only in such instances that the Secretary of State has the discretion to take into account any relevant representations from CEFAS or the FRS, as well as any State that is party to OSPAR and which may be affected by the use or discharge of the offshore chemicals.[156]

Bearing in mind the fact that the submission of false or misleading information, or the breach of permit conditions, entitles the Secretary of State to revoke the relevant permit,[157] it is interesting to note that operators are bound to provide details of any incident or accident involving an offshore chemical, including the breach of permit conditions or "where there has been, or may be, any significant effect on the environment".[158] Although the term "effect" is given a broad definition, in that it "includes any direct, indirect, cumulative, short, medium or long-term, permanent or temporary, or positive or negative effect[s]",[159] that which constitutes a "significant effect" remains an unanswered question. Breach of permit conditions constitutes an offence for which certain employees of juristic persons, such as directors, can in certain instances (such as their having consented thereto) be punished.[160] The possibility of detecting breaches of permit conditions is augmented by the fact that the OC Regulations allow for the appointment of inspectors who, when monitoring the use or discharge of offshore chemicals, may board offshore installations and, *inter alia*, take samples found thereon or in the atmosphere, in the water and on the seabed.[161]

I-11.48

Food and Environment Protection Act 1985
As previously mentioned, FEPA 1985 mandates that a licence is required for the deposit of any substances or articles in the sea

I-11.49

---

[154] Reg 11(2), read with Reg 2.
[155] Reg 12(2).
[156] Reg 12(7).
[157] Reg 13(1).
[158] Reg 15(1)(a) and (b).
[159] Reg 15(2).
[160] Reg 18(4). As regards offences and penalties see, generally, Reg 18.
[161] Reg 16.

or under the seabed.[162] In relation to drill fluids and drill cuttings, however, the Deposits in the Sea (Exemptions) Order 1985 exempts the deposit on site or under the seabed of any chemicals, drilling muds or drill cuttings from the FEPA licensing requirement. Nevertheless, the export of cuttings from one field to another for reinjection is not exempt from the FEPA licensing requirement. As such exportation for reinjection is not commonplace, only the most cardinal of FEPA 1985's environmental provisions pertaining thereto will be discussed. First, a licence under FEPA 1985 must, where the licensing authority thinks it necessary or expedient, include provisions to protect, among other things, the marine environment and the living resources which it supports.[163] Second, compliance with such provisions may be achieved through enforcement officers, who may board vessels, aircraft, hovercraft and marine structures if they have reasonable grounds for believing that they contain drill cuttings that are to be deposited in the sea or under the seabed.[164] Third, FEPA licences may be varied or revoked if, among other things, there has been a breach of any of their provisions, or a change in circumstances relating to the marine environment or the living resources it supports, or an increase in scientific knowledge relating thereto.[165] Fourth, persons who act without a licence when one is required, or cause or permit another person to do so, are guilty of an offence.[166] Fifth, and finally, FEPA 1985's provisions governing liability of persons specifically refers to bodies corporate and the like, thus essentially mirroring the relevant provisions of the OC Regulations, save, however, for the fact that FEPA 1985 does not provide that where the commission of an offence is due to the act or default of some other person, that other person may be charged therewith, irrespective of whether or not proceedings are taken against the first-mentioned person.[167]

Atmospheric emissions

I-11.50   The law of the UK governing atmospheric emissions from offshore installations is the subject matter of various sets of regulations and consequently it is outside the scope of this chapter to consider all of them. Those considered here are those regulations that govern

---

[162] Section 5.

[163] Section 8(3).

[164] Section 11.

[165] Sections 8(10) and (11)(b) and 9(11)(a).

[166] Section 9(1)(a) and (b). See also s 21(6). Persons guilty of an offence under this section shall, in accordance with s 21(2A)(a) and (b), be liable: on summary conviction, to a fine of an amount not exceeding £50,000; and, on conviction on indictment, to a fine or to imprisonment for a term not exceeding 2 years or to both.

[167] See Reg 18(6) of the OC Regulations.

the primary sources of atmospheric pollution arising from offshore installations, namely power generation and flaring.

Merchant Shipping (Prevention of Air Pollution from Ships) Regulations 2008 (PAPS Regulations)[168]

Section 128(1)(e) of the Merchant Shipping Act 1995 was specifically   I-11.51
formulated so as "to provide a power to make secondary legislation regarding air pollution from ships ... [that] would implement Annex VI [of MARPOL 73/78]",[169] which, among other things, sets limits on the emission of SO_x and NO_x while also prohibiting the emission of certain ozone-depleting substances such as CFCs. It was not until 8 December 2008, however, that the requisite regulations, namely the Merchant Shipping (Prevention of Air Pollution from Ships) Regulations 2008 (the PAPS Regulations), came into force.

As a starting point, the PAPS Regulations specifically apply to   I-11.52
any vessels "of any type whatsoever including ... a platform, which is operating in the marine environment".[170] The thrust of the PAPS Regulations is that they implement a certification procedure for the regulation of atmospheric emissions from such vessels. In relation to offshore installations, the relevant certificates may take one of two possible forms. The first is an International Air Pollution Prevention Certificate (an IAPP Certificate) which applies to "a platform which is or will be engaged in voyages to waters under the sovereignty of a Contracting Government other than the United Kingdom",[171] a Contracting Government being any government bound by MARPOL 73/78.[172]

The second is a United Kingdom Air Pollution Prevention   I-11.53
Certificate (UKAPP Certificate) which applies to "a platform which is not or will not be engaged in voyages to waters under the sovereignty or jurisdiction of a Contracting Government other than the United Kingdom".[173] A concern in relation to the applicability of these two forms of certificate is that they apply to platforms "engaged in voyages", yet the term "voyage" is not defined and there is not an equivalent regulatory requirement for platforms when stationary. To this end, the *Oxford English Dictionary*'s definition of a "voyage" is that it is a "long journey involving travel by sea or space", which implies that the relevant certificates are not of appli-

---

[168] SI 2008/2924, as amended by the Merchant Shipping (Prevention of Air Pollution from Ships) (Amendment) Regulations 2010 (SI 2010/895).
[169] Lord Davies of Oldham, *Hansard* HL, vol 672, col 1130 (14 June 2005).
[170] Reg 2(1).
[171] *Ibid.*
[172] *Ibid.*
[173] *Ibid.*

cation to stationary platforms. Fortunately, however, a "platform" is described as including "fixed and floating platforms and drilling rigs".[174] Nevertheless, an inconsistency clearly exists between the definition of a platform and the definition of when the relevant certificates apply to platforms. Presumably, the PAPS Regulations apply to platforms only when they are in motion and thus, to some extent, take on the characteristics of a ship. Given that in the usage of the industry "platform" generally applies to a fixed structure, while "rig" is used to describe a mobile installation, the wording of these provisions could usefully be clearer.

I-11.54   The PAPS Regulations dictate that platforms must, on various occasions, be surveyed by a Certifying Authority, which must be satisfied that, *inter alia*, the equipment and systems thereof are such that they will fully comply with particular regulations pertaining to, among other things, the control of emissions of SO_x and NO_x, failing which platforms will not be permitted to proceed to, or remain at, sea.[175] The consequent burden of appropriately maintaining platforms and their equipment falls expressly on the owners and managers thereof and, notably, they are bound to ensure that platforms will not present "an unreasonable threat of harm to the marine environment".[176] What constitutes "an unreasonable threat" is, however, undefined.

I-11.55   Regulations specifically applying to particular forms of atmospheric pollution can be found in Part 3 of the PAPS Regulations. As regards ozone-depleting substances, the deliberate emission thereof is expressly prohibited.[177] Surprisingly, however, the installation of new equipment, systems and so forth, which involve the introduction onto a platform of ozone-depleting substances, other than hydrochlorofluorocarbons, is prohibited, but only in relation to platforms that do not belong to UK-based operators.[178] In relation to NO_x, the emission thereof from diesel engines is, subject to certain exceptions, prohibited where, in line with the procedures set out in the NO_2 Technical Code, such emissions exceed either: 17.0g/kWh when the rated engine speed (crankshaft revolutions per minute) is less than 130 rpm; or, $45.0 \ 3 \ n^{-0.2}$ g/kWh when the rated engine speed is 130 or more but less than 2,000 rpm; or, 9.8g/kWh when the rated

---

[174] *Ibid.*
[175] See Regs 5–15.
[176] Reg 9(1), read with Reg 2(5)(b).
[177] Reg 20(1). The term "deliberate emission", in accordance with Reg 20(2), "includes an emission occurring in the course of maintaining, servicing, repairing or disposing of systems or equipment".
[178] Reg 20(3) and (4). Note that the hydrochlorofluorocarbon exclusion will expire on 1 January 2020.

engine speed is 2,000 rpm or more.[179] Notably, however, the PAPS Regulations do not apply to any emissions from diesel engines that are "solely dedicated to the exploration, exploitation and associated offshore processing of seabed mineral resources".[180] As far as the emission of SO_x is concerned, the relevant stipulation is that the content thereof in any fuel oil used onboard a platform, subject to certain exceptions, may not exceed 4.5 per cent by mass.[181] The most noteworthy exception is the fact that stricter emission limits are placed on platforms within a SO_x emission control area (SECA), with only 1.5 per cent by mass being permitted.[182] Interestingly, the North Sea SOX Emission Control Area (the "North Sea SECA") is such that most of the offshore installations in the UK's waters fall within its borders.[183] Nevertheless, there are numerous installations that fall outside the geographical ambit thereof which would thus be subject to less stringent emissions standards. It is submitted that this disparity is unfortunate and that the PAPS Regulations should be amended so as to apply the stricter emissions of the North Sea SECA to all offshore installations in UK waters.

As regards volatile organic compounds, it appears that the PAPS Regulations do not apply to offshore installations but rather to oil or chemical tankers as well as to terminal operators or harbour authorities operating vapour emission control systems.[184] As regards the regulation of emissions arising from incineration, the PAPS Regulations dictate that, subject to certain exceptions, the incineration of, among other things, polychlorinated biphenyls, garbage containing more than trace amounts of heavy metals and refined petroleum products containing halogen compounds is prohibited.[185] Although "garbage" is defined so as to include "all kinds of ... operational wastes generated during the normal operation of a ship and liable to be disposed of continuously or periodically",[186] the incineration of "substances that are solely and directly the result of exploration, exploitation and associated offshore processing of sea-bed mineral resources"[187] is excluded from the ambit of the PAPS Regulations. Such specifically excluded activities include flaring, as

I-11.56

---

[179] Reg 21.
[180] Reg 3(13)(e).
[181] Reg 22. In accordance with Reg 2(1), "fuel oil" refers to "such substances as may be specified by the Secretary of State in a Merchant Shipping Notice".
[182] Reg 22(3)(a) and(b).
[183] The North Sea SECA lies southwards of latitude 628N and from the north of Scotland it falls eastwards of longitude 48W, while south of England it falls eastwards of 58W.
[184] Reg 23, read with Reg 2(1).
[185] Reg 24(4).
[186] Reg 24(5).
[187] Reg 3(13)(c).

well as the burning of drill cuttings, muds and stimulation fluids during well completion and testing operations.[188] Finally, as regards fuel oil, the PAPS Regulations dictate that such fuel must meet particular specifications, including the fact that it may not exceed the appropriate SO_x content, or cause an engine to exceed stipulated NO_x emission levels, or include any added substance or chemical which causes additional air pollution.[189] Notably, however, hydrocarbon produced and used on platforms as fuel, if that use has been approved by the Secretary of State, is excluded from the relevant fuel oil regulations.[190]

I-11.57   The PAPS Regulations contain various enforcement provisions. Notably, the submission of false information, or the use of equipment that is damaged or defective, entitles the Secretary of State to cancel a certificate.[191] The appointment of inspectors to monitor compliance is also provided for. Enforcement provisions of particular significance are the detention provisions that allow for the fact that a UK-registered platform may be detained until such time as "a surveyor of ships is satisfied that it can proceed to sea without presenting any unreasonable threat of harm to the marine environment".[192]

Offshore Combustion Installations (Prevention and Control of Pollution) Regulations 2001[193]

I-11.58   The OCI Regulations are designed to control atmospheric pollution arising from so-called "qualifying combustion installations" (QCIs) which are permanently installed on a platform and that singularly, or in combination with any other such installation installed on the same site, have rated thermal inputs exceeding 50 megawatts.[194] Notably QCIs are:

> "any technical apparatus in which fuels are oxidised to use the heat thus generated and includes gas turbines and diesel and petrol-fired engines and any equipment on a platform connected to such apparatus which could have an effect on emissions from that apparatus or could otherwise give rise to pollution but does not include any apparatus the main use of which is the disposal of gas by flaring or incineration".[195]

I-11.59   The operation of a QCI requires a permit, and applications for such permits must stipulate, inter alia, the nature and quantities of

---

[188] Reg 3(13)(c)(i).
[189] See, generally, Reg 25.
[190] Reg 25(1)(c).
[191] Reg 18.
[192] Reg 28(1), read with Regs 3(5)(a) and 16(1).
[193] SI 2001/1091.
[194] Reg 2.
[195] Ibid.

foreseeable emissions from the relevant installation, the significant effects they may have on the environment and measures to monitor the emissions.[196] Note that an application for a permit may not need to contain all of the above where an ES regarding the effects of the operation of the QCI in question already exists. In such instances, a copy of the ES must accompany the application for a permit.[197]

When considering an application for a permit, the Secretary of State must not only take into account any relevant ES but also any public representations as well as any representations made by States that are party to the Agreement on the European Economic Area (EEA States) and that are affected by the operation of the combustion installation.[198] Moreover, the Secretary of State must be satisfied that various publicity requirements have been complied with, including the submission of a copy of the application to any EEA State whose environment is likely to be significantly affected.[199] However, what constitutes a significant effect is not defined under the OCI Regulations or under the Pollution Prevention and Control Act 1999 in terms of which the OCI Regulations were created. Nevertheless, the Secretary of State is bound to attach specific environmentally orientated conditions to all permits, which must ensure, *inter alia*, that: appropriate measures are taken to prevent pollution, in particular through the use of BAT;[200] no significant pollution occurs; there are controls on the emission of pollutants;[201] and there are provisions to minimise long-distance or transboundary pollution.[202]

I-11.60

When it comes to enforcement of the OCI Regulations, it is worth noting that not only is the Secretary of State entitled to review, update and revoke permits,[203] he is also empowered to appoint inspectors, serve enforcement and prohibition notices, and take further action

I-11.61

---

[196] Regs 3 and 5.

[197] Reg 5(3), read with Reg 2.

[198] Reg 4(1), read with Reg 2.

[199] Reg 7(3) and (4).

[200] Notably, in relation to the concept of BAT, Reg 2 defines "available techniques" as those "which can be implemented on platforms under economically and technically viable conditions, balancing the costs of their implementation against the benefits to the environment". In relation to such techniques, the terms "best" and "techniques" are respectively defined as "the most effective in achieving a high general level of protection of the environment as a whole" and "the technology used and the way in which the installation is designed, built, maintained, operated and decommissioned".

[201] Such controls may include emission value limits (Reg 4(2)(g)(i)), equivalent parameters or technical measures (Reg 4(2)(g)(ii)), or a combination of emission value limits and equivalent parameters and technical measures (Reg 4(2)(g)(iii)).

[202] Reg 4.

[203] Regs 9 and 10.

in the event that such notices are not complied with.[204] The OCI Regulations create various offences, including: the operation of a combustion installation without, or in breach of a condition attached to, a permit; and failure to comply with an enforcement or prohibition notice.[205] Notably, a body corporate, as well as particular officers thereof, such as directors who through consent, connivance or neglect allow the commission of an offence, may be held liable for any such offence.[206]

Petroleum Licensing (Production) (Seaward Areas) Regulations 2008

I-11.62   In the UK, flaring contributes substantially to the offshore industry's atmospheric emissions, with venting providing a nominal contribution.[207] Flaring is regulated by the previously discussed Production Regulations, and the National Emission Ceilings Regulations 2002[208] (the NEC Regulations). As for venting, the regulation thereof falls under the Energy Act 2008.

I-11.63   The Production Regulations forbid a licensee from flaring any gas unless written consent has been obtained and any conditions attached thereto have been adhered to.[209] Exceptions include instances where there is a risk of injury to persons, or where it is necessary to maintain a flow of petroleum from the well.[210] If flaring without consent has occurred, the licensee is obliged to inform the Minister thereof; and, where it was done so as to maintain the flow of petroleum, the licensee must stop flaring when directed to do so.[211] It is surprising that operators may be permitted to flare gas until such time as they are requested to stop as it would make greater sense if, for example, they were obliged to stop as soon as reasonably possible based on objective scientific criteria related to the flow of petroleum. Similarly, when a licensee wishes to apply for

---

[204] Regs 13–16.

[205] Reg 18.

[206] Reg 18(4).

[207] See The World Bank, "Overview of Onshore and Offshore Gas Flaring and Venting in the United Kingdom", available at http://siteresources.worldbank.org/INTGGFR/Resources/unitedkingdom.pdf (accessed 26 April 2017), where it is noted that in 2001, 20 per cent of the offshore CO_2 emissions arose from flaring while only 0.05 per cent arose from venting.

[208] SI 2002/3118.

[209] Reg 23(3)(a).

[210] Reg 23(7)(a) and (b). This is, however, subject to the proviso in Reg 23(7) that such exceptional flaring should only occur as a consequence of an event that the licensee did not foresee in time to deal therewith in an alternative manner.

[211] Reg 23(7).

consent to flare, a written submission specifying the date of proposed commencement thereof is required,[212] but the licensee is not required to specify a date for the cessation thereof.

National Emission Ceilings Regulations 2002
The NEC Regulations require that the Secretary of State must I-11.64 ensure that, in 2010 and every year thereafter, the emission of certain pollutants, namely sulphur dioxide (SO_2), nitrogen oxides (NO_x), volatile organic compounds (VOC) and ammonia (NH_3), do not exceed the permitted amounts thereof as stipulated in the Schedule to the NEC Regulations.[213] To achieve this objective, the Secretary of State is bound to prepare a programme for the progressive reduction in emissions of the relevant pollutants.[214] Although the NEC Regulations apply to emissions from land, the territorial sea and the continental shelf,[215] Oil & Gas UK notes that the UK Government is of the belief that it will be able to meet the targets of the NEC Regulations without having to apply these targets offshore.[216] Whether or not the targets will be met solely through onshore application remains to be seen. However, as Oil & Gas UK points out, "this does not preclude future targets for offshore operations".[217]

Sewage, garbage and other matter left behind

Entering into force on 1 February 2009, the Merchant Shipping I-11.65 (Prevention of Pollution by Sewage and Garbage from Ships) Regulations 2008[218] (the PPSGS Regulations) are the UK Government's domestic implementation of Annexes IV and V of MARPOL 73/78 which address pollution from ships by sewage and garbage respectively. Importantly, the concept of a "ship" includes fixed or floating platforms operating in the marine environment.[219] Unfortunately, the use of the word "operating" raises the question whether the PPSGS Regulations apply to non-operational

---

[212] Reg 23(4).
[213] Reg 3, read with Reg 2. Presently, the national emission ceilings in the Schedule are: 585 kilotonnes for SO_2; 1,167 kilotonnes for NO_x; 1,200 kilotonnes for VOCs; and 297 kilotonnes for NH_3.
[214] Reg 4(1).
[215] Reg 2(2)(a) and (b).
[216] Oil & Gas UK, "Atmospheric Emissions – Offshore Cargo Loading", available at www.ukooaenvironmentallegislation.co.uk/contents/Topic_Files/Offshore/VOC.htm (accessed 2 May 2017).
[217] *Ibid.*
[218] SI 2008/3257, as amended by SI 2010/897.
[219] Reg 2(1).

offshore installations which, although they may for some reason not be involved in the drilling and production of oil and gas, are nonetheless present in the marine environment with personnel onboard who will certainly be producing sewage and garbage. Leaving this point aside, the definitions of "garbage" and "sewage" under the PPSGS Regulations are particularly broad. What constitutes "garbage" includes:

> "all kinds of victual, domestic and operational wastes generated during the normal operation of a ship and liable to be disposed of continuously or periodically, but does not include fresh fish and parts thereof, sewage, or any other substance the disposal of which is prohibited or otherwise controlled under an Annex to the Convention other than Annex V".[220]

I-11.66   "Sewage" is defined as including, among other things, drainage and other wastes from any form of toilet or urinal, as well as other waste waters that may be mixed with any such drainage.[221]

I-11.67   With regard to sewage in particular, the PPSGS Regulations prohibit the discharge thereof into the sea[222] and provide for surveys to be conducted of offshore installations and their compulsory sewage management systems as well as for the certification thereof.[223] There are some notable exceptions to the prohibition on discharging sewage. First, sewage discharge may take place if it has been through a sewage treatment plant that complies with the requirements of the Merchant Shipping (Marine Equipment) Regulations 1999,[224] the said plant has been tested and the results thereof are reflected in the relevant sewage certificate, and, there will be no visible solids or discolouration of the seawater.[225] Second, treated sewage may be discharged more than 3 nautical miles from the nearest land if it has been through a comminuting and disinfecting system approved by the Maritime and Coastguard Agency (MCA) which meets the standards set out in Merchant Shipping Notice MSN No 1807.[226] Third, untreated sewage may be discharged at a specific rate at a distance of more than 12 nautical miles from land where offshore installations are en route and proceeding at not less than 4 knots.[227] Fourth, the general prohibition on the discharge of sewage does not apply to "old ships", a term which indirectly incorporates offshore

---

[220] *Ibid.*
[221] *Ibid.*
[222] Reg 23(1).
[223] Reg 7.
[224] SI 1999/1957.
[225] Regs 24(a), (b) and (c) and 21(1)(a).
[226] Regs 25(1)(a)(b) and 21(1)(b).
[227] Reg 25(2) and (3), read with Reg 2(1).

installations constructed before 2 October 1983.[228] The owners of such installations are, however, bound to ensure that they are "equipped, so far as is practicable" to discharge sewage through a sewage treatment plant that complies with the requirements of the Merchant Shipping (Marine Equipment) Regulations 1999 or a comminuting and disinfecting system approved by the MCA which satisfies the standards set out in Merchant Shipping Notice MSN No 1807.[229]

As regards the aforementioned sewage certification requirements, I-11.68 offshore installations may not "be put into service, or ... continue in service" unless an "initial survey" has been conducted, the relevant equipment has been deemed to be satisfactory and a Sewage Certificate has been issued.[230] When Sewage Certificates require renewal, the PPSGS Regulations dictate that offshore installations may not "proceed to sea, or ... remain at sea" without a "renewal survey" having been conducted and a new Sewage Certificate having been issued.[231] Notably, the responsibility of appropriately maintaining offshore installations and their sewage equipment falls on the owners and managers thereof, who are specifically bound to ensure that offshore installations "remain fit to proceed to sea without presenting an unreasonable threat of harm to the marine environment".[232] As with the aforementioned PAPS Regulations, however, what constitutes "an unreasonable threat" is undefined.

The relevant sewage management equipment must consist of I-11.69 either: a sewage treatment plant, which, as previously mentioned, must comply with the requirements of the Merchant Shipping (Marine Equipment) Regulations 1999; or a comminuting and disin-fecting system approved by the MCA which meets the standards set out in Merchant Shipping Notice MSN No 1807; or a holding tank that is, among other things, constructed in accordance with standards set out in Merchant Shipping Notice MSN No 1807, and of sufficient capacity relative to, *inter alia*, the number of persons on the installation.[233] Notably, owners of installations constructed before 2 October 1983[234] need only ensure that they are "equipped, so far as is practicable" to discharge sewage through a sewage treatment plant that complies with the requirements of the Merchant

---

[228] Reg 23(3), read with Reg 2(1) and (4).
[229] Reg 21(2), read with Regs 24 and 25.
[230] Reg 7. Reg 2(1) defines a Sewage Certificate as "an International Sewage Pollution Prevention Certificate referred to in Reg 5 of Annex IV".
[231] Reg 8.
[232] Reg 9(1), read with Reg 2(6)(b).
[233] Reg 9(1), read with Reg 2(6)(b).
[234] Reg 2(4), read with Reg 2(1) and 21(1).

Shipping (Marine Equipment) Regulations 1999, or a comminuting and disinfecting system approved by the MCA and meeting the standards set out in Merchant Shipping Notice MSN No 1807.[235]

I-11.70   Regarding garbage, its disposal into the sea from offshore installations is prohibited,[236] with the exception of food wastes that have been ground or comminuted and disposed of from an installation which is 12 nautical miles or more from the nearest land.[237] Offshore installations are also bound to carry garbage management plans and garbage record books that comply with guidelines developed by the IMO and set out, respectively, in Schedules 3 and 4 to Merchant Shipping Notice MSN No 1807.[238]

I-11.71   In relation to the enforcement of the PPSGS Regulations, provision is made for the inspection of offshore installations so as to verify, for example, the existence and validity of a Sewage Certificate.[239] It is noteworthy that provision is made for the detention of offshore installations until, *inter alia*, the relevant surveyor is satisfied that they "can proceed to sea without presenting an unreasonable threat of harm to the marine environment".[240] Unfortunately, guidance is not provided as to what a surveyor's subjective assessment of "an unreasonable threat" may be based on. Aside from the fact that the PPSGS Regulations repeatedly fail to provide an explanation of this particular concept, it is submitted that they go a long way towards satisfying the objectives of Annexes IV and V of MARPOL 73/78.

I-11.72   In addition to the aforementioned Conservation Regulations, FEPA 1985 and Coastal Protection Act 1959 (CPA 1949) contain certain permitting requirements that may need to be met by prospective surveyors and shallow drillers and which, if met, reduce the risk of environmental harm arising. In accordance with FEPA 1985, Sections 5 and 7, a licence is required for all deposits (whether temporary or permanent) that are to be made in the sea or under the seabed, unless they can be specifically exempted by virtue of the Deposits in the Sea (Exemptions) Order 1985.[241] Accordingly, it may be necessary for those wishing to conduct geological surveys or shallow drilling operations to obtain a licence for any deposits that may be made during the course of their activities, such as the abandonment of sacrificial anchors that secured equipment to the seabed. Logically, these require-

---

[235] Reg 21(2), read with Regs 24 and 25.
[236] Reg 29, read with Reg 2(1). Under Reg 26(1) the disposal of plastics is specifically prohibited.
[237] Reg 29(2).
[238] Reg 32.
[239] Reg 36.
[240] Reg 38.
[241] SI 1985/1699.

ments help to reduce the risk that harm in the form of interference, such as the snagging of fishing gear on sacrificial anchors, will arise. Similarly, CPA 1949 dictates that consent is required to: (1) construct, alter or improve any works in or on the seashore below the level of mean high water springs; or (2) deposit any object or materials there; or (3) remove any objects or materials from the seashore below the level of mean low water springs (for example, through the process of dredging).[242] In essence, these requirements are applicable to shallow waters within the limits of the territorial sea where the aforementioned actions could have navigational consequences.[243] However, Section 4(1) of the Continental Shelf Act 1964 has extended the ambit of points (1) and (3) so that they are now applicable in any part of the UKCS where oil and gas exploration and development is designated to take place. Thus, BEIS recommends that potential surveyors and shallow drillers whose actions may result in temporary or permanent deposits on the seabed, contact and obtain authorisation from the relevant authorities in accordance with CPA 1949.[244]

Interference with other users of the sea

What follows is a brief discussion of those laws that address the   I-11.73
possibility of interference with fishing and navigation caused by the presence of offshore installations and pipelines. The cardinal statutes addressing this form of pollution are the CPA 1949 and the Petroleum Act 1998.

Coast Protection Act 1949
In accordance with Section 34 of CPA 1949, the consent of the   I-11.74
Secretary of State is required, not only in relation to the location of offshore installations and pipelines, but also in relation to any construction or alteration of any works on, under or over any part of

---

[242] CPA 1949, s 34.

[243] See, generally, Marine Maritime Organisation (MMO), available at www.gov.uk/topic/planning-development/marine-licences (accessed 2 May 2017).

[244] It is worth pointing out that the Marine Works (Environmental Impact Assessment) Regulations 2007 (SI 2007/1518), which provide a framework for carrying out environmental impact assessments in relation to, *inter alia*, the removal or disposal of articles, and the construction or alteration of certain works within the UK marine environment, do not apply to such articles or works resulting from surveying or shallow drilling. The reason for this is that the two annexes of Council Directive 85/337/EEC, which list the various projects in relation to which environmental impact assessments may be conducted, fail to list any surveying or shallow drilling-related works. This is the situation despite the fact that the marine works to which these regulations apply are defined as those works for which regulatory approval is required, which in turn are defined to include the pertinent provisions of FEPA 1985 and CPA 1949 – see Reg 2(1).

the seashore lying below the level of mean high water springs, or the depositing of any object or any materials thereon, or removing any object or any materials from any part of the seashore lying below the level of mean low water springs, where such operations may result in obstruction or danger to navigation.[245] Notably, Section 4(1) of the Continental Shelf Act 1964 has extended the application of Section 34 of CPA 1949 to all areas of the UKCS.

I-11.75   Obtaining consent requires the submission to the Secretary of State of "such plans and particulars of the proposed operation as he may consider necessary".[246] The plans and particulars that the Secretary of State will consider as necessary will be those that enable him to evaluate the risk to navigation.[247] Every proposed location and each installation will be evaluated in relation to, among other things, shipping movements close to a proposed location and the danger of passing vessels colliding with the installation.[248] The Secretary of State's consent will also depend on the deployment of specified navigation aids such as flashing obstruction lights, foghorns and buoys, which will be subject to regular seaward inspections by the General Lighthouse Authority (GLA).[249] Additionally, obtaining consent in relation to mobile rigs requires notification of the movements thereof being sent, generally 48 hours beforehand, to the consent issuing office and the relevant Coastguard Maritime Rescue Co-ordination Centre as well as the Hydrographer to the Navy.[250]

I-11.76   Where the Secretary of State is of the opinion that the location of a proposed offshore installation or pipeline may obstruct or pose a danger to navigation, he will either refuse to grant his consent thereto or will grant consent subject to such conditions as he thinks fit.[251] Any person who fails to obtain the requisite consent, or fails to comply with any conditions subject to which consent has been granted, shall be guilty of an offence.[252]

I-11.77   Prior to considering the relevant provisions of the Petroleum Act 1998, it should be borne in mind that BEIS considers proposed operations in Deep Water Routes and Traffic Separation Schemes, as well as the approaches thereto, to be such that they will usually obstruct and endanger navigation.[253]

---

[245] Section 34(1)(a), (b) and (c).

[246] Section 34(2).

[247] Oil & Gas UK, available at www.ukooaenvironmentallegislation.co.uk/contents/Topic_Files/Offshore/Fishing_Navigation_Installation.htm (accessed 2 May 2017).

[248] *Ibid.*

[249] Section 36A.

[250] *Ibid.*

[251] Section 34(3).

[252] Section 36(1).

[253] See www.hse.gov.uk/offshore/notices/on_14.htm (accessed 11 May 2017).

The foregoing discussion of the CPA 1949 demonstrates that the **I-11.78**
UK possesses particularly detailed legislation aimed at addressing
pollution in the form of interference with fishing and navigation. As
will be demonstrated below, the Petroleum Act also plays a significant role in relation to offshore pipelines.

Petroleum Act 1998
Under the Petroleum Act, no person may construct or use such **I-11.79**
pipelines without the Secretary of State's prior authorisation.[254]
This prohibition applies to any pipeline in, under or over controlled
waters which are defined as the territorial sea adjacent to the UK and
the sea in any area designated under Section 1(7) of the Continental
Shelf Act 1964.[255] Importantly, authorisations to construct or use
pipelines (commonly referred to as Pipeline Work Authorisations)
may contain conditions pertaining to the route and the design
thereof, as well as steps that must be taken to avoid or reduce interference with fishing or other activities connected with the sea or the
seabed or subsoil.[256]

The Petroleum Act provides that persons who construct or use a **I-11.80**
pipeline without the Secretary of State's authorisation, or fraudulently
obtain such authorisation, shall be guilty of an offence. Notably,
operators may be particularly apprehensive about constructing or
utilising a pipeline without the requisite consent in view of the fact
that, should they do so, the Secretary of State may serve upon them a
notice requiring the removal of any relevant works and, should they
fail to do so, the Secretary of State may attend thereto and recover
any expenses reasonably incurred.[257] The Petroleum Act contains
additional deterrents including provisions relating, where appropriate, to the institution of criminal and civil liability.[258]

REGULATING CONSEQUENTIAL ENVIRONMENTAL HARM

Implementation of international obligations into UK law

The UK legislature implemented regulations to protect the marine **I-11.81**
environment from oil pollution occasioned by the offshore industry
in response to international obligations. These take the form of

---

[254] Section 14(1)(a) and (b).
[255] Section 14(2). Note that s 26(1) defines a "pipeline" as meaning, except where the context otherwise requires, "a pipe or systems of pipes (excluding a drain or sewer) for the conveyance of any thing, together with any apparatus and works associated with such a pipe or system".
[256] Section 15(3)(c)(i), (iii) and (iv).
[257] Section 21(2) and (3).
[258] Sections 22 and 23.

the Merchant Shipping (Oil Pollution Preparedness, Response and Co-operation Convention) Regulations 1998[259] (the "Merchant Shipping (OPRC) Regulations"), the Offshore Installations (Emergency Pollution Control) Regulations 2002[260] (the "EPC Regulations") and the Offshore Petroleum Activities (Oil Pollution Prevention and Control) Regulations 2005[261] (the "OPA Regulations"). The Deepwater Horizon Disaster threw the relevance of these regulations into sharp relief. It is therefore pertinent to ask how these regulations might have been engaged by similar events on the UKCS.

Merchant Shipping (Oil Pollution Preparedness, Response and Co-operation Convention) Regulations 1998

I-11.82   Created under the provisions of the Merchant Shipping Act 1995, the Merchant Shipping (OPRC) Regulations were the means by which UK Government introduced into law the oil spill planning and reporting requirements of the OPRC Convention. Importantly, the Merchant Shipping (OPRC) Regulations define an "oil pollution incident" broadly enough to incorporate discharges of oil from offshore installations. To this end, an oil pollution incident is defined as:

> "an occurrence or series of occurrences having the same origin, which results or may result in a discharge of oil and which poses or may pose a threat to the marine environment, or to the coastline or related interests of the United Kingdom and which requires emergency action or other immediate response".[262]

I-11.83   In order to deal effectively with any such incident, every operator of an offshore installation must have an oil pollution emergency plan.[263] Such plans must be submitted to the Secretary of State at least two months prior to the commencement of drilling or production,[264] and must be reviewed and resubmitted within five years after submission. Additionally, should any major change affecting, or possibly affecting, such plans arise, operators are required, within three months of knowledge thereof, to submit a new or an amended pollution emergency plan.[265] However, what constitutes a "major change" is not defined under the Merchant Shipping (OPRC) Regulations. Nevertheless, where an operator fails to submit, re-submit, maintain or implement an oil pollution

---

[259] SI 1998/1056.
[260] SI 2002/1861.
[261] SI 2005/2055, as amended by SI 2011/983.
[262] Reg 2.
[263] Reg 4(1)(c).
[264] Reg 3(a), read with Reg 4(a)(iii) and (iii)(bb) as well as (7).
[265] Reg 4(5)(b).

emergency plan when required to do so, it shall be guilty of an offence.[266] At the time of writing, it appears that the difficulties encountered by BP in the Gulf of Mexico in controlling the leak and thus shutting off the source of the oil pollution arose, in the main, from the depth at which the operation was being conducted. It is not immediately evident that there are lessons to be drawn for oil pollution emergency plans on the UKCS, given that the deepest developments so far in the UK waters lie at about one-third of the depth in the vicinity of the leak in the Gulf of Mexico,[267] but it is by no means inconceivable that in due course the experience gained from this event may require to be incorporated into emergency plans on this side of the Atlantic.[268]

Notably, although the Merchant Shipping (OPRC) Regulations I-11.84 do not incorporate a provision allowing for circumvention of the "corporate veil", Section 277 of the Merchant Shipping Act 1995 contains the requisite provision, which is applicable "[w] here a body corporate is guilty of an offence under this Act or any instrument made under it". A related environmental consideration is the Secretary of State's right to authorise any person to inspect any offshore installation to which the Merchant Shipping (OPRC) Regulations apply,[269] which is every offshore installation in the UK waters and in any area designated under the Continental Shelf Act 1964.[270]

A final point for consideration is the fact that the Merchant I-11.85 Shipping (OPRC) Regulations contain certain noteworthy reporting obligations, including the requirement that should masters of UK ships become aware of events involving the discharge of oil from offshore installations, they are bound to report such events either to the Maritime and Coastguard Agency, if the event occurred in UK waters, or to the nearest coastal state, if the event occurred outside UK waters.[271] Similarly, where an individual in charge of an offshore installation is aware of any event involving the discharge of oil at sea from another installation, he must report the incident to the Maritime and Coastguard Agency.[272]

---

[266] Reg 7(1).
[267] Existing fields west of Shetland, such as Schiehallion, lie beneath waters of a maximum depth of around 450m, whereas the leak in the Gulf of Mexico in 2010 occurred in water over 1,500m deep.
[268] Note that the seismic survey recently ordered by Chevron to the west of Shetland will take place in depths of up to 1,200m.
[269] Reg 8.
[270] Reg 3(2).
[271] Reg 5(1).
[272] Reg 5(2).

Offshore Petroleum Activities (Oil Pollution Prevention and Control) Regulations 2005

I-11.86   At the heart of the OPA Regulations is a ban on the discharge of oil, except in accordance with a permit.[273] The definition of "oil" is particularly broad and covers:

> "any liquid hydrocarbon or substitute liquid hydrocarbon, including dissolved or dispersed hydrocarbons or substitute hydrocarbons that are not normally found in the liquid phase at standard temperature and pressure, whether obtained from plants or animals, or mineral deposits or by synthesis".[274]

I-11.87   The broad nature of the above definition clearly outlaws the release of any form of oil from an offshore installation without a permit.[275] An operator of an offshore installation who is desirous of obtaining a permit to discharge oil bears the responsibility of furnishing the Secretary of State with information describing, *inter alia*, the installation, its location, the relevant oil, the circumstances in which the oil is to be discharged and the measures that will be employed to monitor the discharge.[276] The submission of false or misleading information may result in any permit that has been granted being revoked.[277]

I-11.88   Upon receipt of the requisite information, the Secretary of State may determine not only whether to grant or refuse the application,[278] but also the time period for which it will be valid[279] and what conditions, if any, ought to be attached thereto.[280] Such conditions may relate to, among other things, the concentration, frequency, quantity, location and duration of the discharge, measures to minimise pollution, and appropriate monitoring techniques and procedures,[281]

---

[273] Reg 3(1).

[274] Reg 2.

[275] There are, however, two exceptions. First, in accordance with Reg 3(2)(a), (b) and (c), discharges that are either regulated by the Offshore Chemical Regulations 2002 or by the Merchant Shipping (Prevention of Oil Pollution) Regulations 1996 and 1998 do not require a permit under the EPC Regulations. Second, in accordance with Reg 3(3)(a), where an existing exemption applies and oil is being discharged, a permit need not be granted until such time as the Secretary of State specifies by notice in writing that such a permit is required.

[276] Regs 3(4), 4(1) and 5(1)(a), (b) and (c). Note that, in accordance with Reg 5(2), the Secretary of State is also empowered to call upon the operator to furnish him with any additional information that he may require.

[277] Reg 9(1)(a).

[278] Reg 4(1).

[279] Reg 4(3).

[280] Reg 4(2) and (4).

[281] Reg 4(2).

as well as any condition that the Secretary of State thinks fit.[282] Importantly, the Secretary of State is entitled to review the conditions attached to any permit and to revoke a permit where there has been a breach of its conditions.[283] Additionally, where the Secretary of State is of the opinion that the operation of an offshore installation involves imminent risk of serious pollution as a consequence of any discharge of oil, a prohibition notice may be served on the permit holder.[284] Significantly, the risks for which prohibition notices may be served need not relate to the contravention of a permit issued under the OPA Regulations, but to any aspects of the operation of an offshore installation.[285]

With a view to securing compliance with the OPA Regulations, the Secretary of State may appoint inspectors to investigate compliance therewith including the monitoring of any discharge of oil.[286] Such inspectors may, *inter alia*, board offshore installations and inspect relevant records, as well as make such examinations or investigations as they consider necessary, including the taking of samples.[287] Failure by, *inter alia*, corporations or their directorial and managerial staff to comply with the requirements of the OPA Regulations constitutes an offence.[288] Defences that can be raised are either that the contravention arose as a consequence of something that could not reasonably have been prevented, or that the contravention was due to something done as a matter of urgency for the purpose of securing the safety of a person.[289] Significantly, where a defendant wishes to rely on the latter of these two defences, he is prohibited from doing so if the necessity to do the thing in question was due to the fault of the defendant.[290]

I-11.89

Offshore Installations (Emergency Pollution Control) Regulations 2002

Crafted in accordance with Section 3 of the Pollution Prevention and Control Act 1999, the EPC Regulations grant the UK Government certain powers of intervention following an accident involving an offshore installation so as to prevent and reduce possible pollution.

I-11.90

---

[282] Reg 4(a).
[283] Regs 7(1) and (2) and 9(1)(b). Note that, in accordance with Reg 8, should a permit holder wish to assign his permit to another person, the permit holder and the proposed assignee must both make application to the Secretary of State.
[284] Reg 14(1).
[285] Reg 14(2).
[286] Reg 12(1)(a) and (b).
[287] See, generally, Reg 12.
[288] See, generally, Reg 16.
[289] Reg 16(2)(a) and (b).
[290] Reg 16(3)(b).

Although an "accident" is defined as "any occurrence causing material damage or a threat of material damage to an offshore installation",[291] a definition of what constitutes "material damage" is not provided. Nevertheless, a definition of "pollution" is provided which specifically refers to oil that may, *inter alia*, harm living resources and marine life, or interfere with other legitimate uses of the sea.[292]

I-11.91   In order to achieve the objective of the EPC Regulations, the Secretary of State is granted the broadly phrased authority to give directions to the operator or manager of an offshore installation (as well as any servant or agent of the operator)[293] that "may require the person to whom they are given to take, or refrain from taking, any action of any kind whatsoever".[294] Somewhat more specifically, the Secretary of State is also entitled to direct the relevant person to, *inter alia*: relocate the installation, or any part thereof; not discharge any oil or other substance; and take remedial measures.[295] If the Secretary of State is of the opinion that any such directions are inadequate, he may take "any action of any kind whatsoever"[296] including taking over control of the relevant offshore installation,[297] or sinking or destroying the offshore installation or part thereof.[298] Importantly, however, it must be borne in mind that the Secretary of State may only exercise such powers where an accident has occurred and, in the opinion of the Secretary of State, the use of such powers is urgently needed as the accident will cause "significant pollution".[299] Neither the EPC Regulations nor the Pollution Prevention and Control Act 1999 provide an explanation of what constitutes significant pollution.

I-11.92   The Minister has extensive powers to direct how an operator or manager should respond to an accident that may give rise to pollution and indeed to intervene himself. While these powers are undoubtedly necessary to allow steps to be taken in the case of some egregious failure on the part of the operator or manager, it might be suggested that the events in the Gulf of Mexico in 2010 would give any Minister pause for thought before intervening in a way that may, in due course, raise questions as to his competence to make such decisions.

---

[291] Reg 2.
[292] Reg 2.
[293] Reg 3(2).
[294] Reg 3(3).
[295] Reg 3(3)(a), (b) and (c).
[296] Reg 3(4).
[297] Reg 3(4)(c).
[298] Reg 3(4)(b).
[299] Reg 3(1)(a), (b) and (c).

Lastly, it should be noted that it is an offence to fail to comply **I-11.93** with a direction,[300] which is, among other things, punishable on summary conviction to a fine not exceeding £50,000.[301] It might be suggested that this fine may not be substantial enough relative to the harm that may be occasioned by a corporation that has unambiguously opted not to comply with a direction.

THE EU OFFSHORE SAFETY DIRECTIVE

As a response to the vast amount of oil spilled and consequential **I-11.94** environmental damage caused by the DWH catastrophe in 2010, the European Union (EU) undertook to address the safety of humans and the environment from harm arising from offshore oil and gas activities. Initially, the EU recommended that offshore platform safety be implemented as a regulation, with the EU indicating that self regulation and discretion, hallmarks of the safety regulation system in the UK,[302] were inadequate means of ensuring safety and environmental protection.[303] Instead the EU proposed clear, robust and ambitious rules, sparking fears amongst the industry and academics alike that the EU was seeking to return to the prescriptive approach to regulation that had been abandoned after the Piper Alpha Disaster.[304]

After much consultation and discussion,[305] the EU eventually imple- **I-11.95** mented offshore safety and environment protection as a Directive: Directive 2013/20/EU on the Safety of Offshore Oil and Gas Operations ("Platform Directive"). The Platform Directive seeks to mitigate the risk of oil spills and injuries, arising from both well and platform accidents, by requiring member states to ensure that participating companies are well financed and have the necessary technical expertise to undertake the petroleum operations.[306] Before an activity commences, companies are required to prepare a Major Hazard Report for their offshore installation, which contains a risk assessment and an emergency response plan for oil spills.[307] Similarly, prior to the operation of the installation, technical solutions that are critical for the safety of operators' installations must be independently verified.[308]

---

[300] Reg 5(2).

[301] Reg 5(4).

[302] Refer to Chapter I-10 for a discussion of the current regulatory framework.

[303] G Gordon, "Offshore safety: the European Commission's legislative initiatives", in M Roggenkamp and H Bjornebye (eds), *European Energy & Law Report X* (2014), p 143.

[304] *Ibid*, p 143.

[305] *Ibid*, pp 144–151 for an analysis of the process.

[306] Directive 2013/20/EU on the Safety of Offshore Oil and Gas Operations, Art 3.

[307] *Ibid*, Art 6

[308] *Ibid*, Art 11.

Perhaps most importantly, independent, competent national authorities are required to verify safety provisions, environmental protection measures and the emergency preparedness of all facilities, and if companies do not respect the minimum standards, there is the right to impose sanctions, including halting production.[309] To minimise the possibility of loss of well control and well blowouts, well operations can only be undertaken by an installation that is technically capable of controlling the foreseeable hazards, has an accepted major hazards report and where the well design and construction must have been verified by an independent expert examination.[310]

I-11.96   Should an oil spill occur, the Platform Directive sets out the requirements for emergency preparedness and response.[311] According to Article 28, member states are required to have an internal emergency oil spill response plan, and to ensure that the operator and owner of an installation are required to have on hand the necessary resources to address an oil spill and the capacity to respond quickly were an accident to occur. In addition, under the Platform Directive, companies are fully liable for the prevention and remediation of environmental damage in EU marine waters, including exclusive economic zones (EEZs) and continental shelves.[312]

I-11.97   Given its status as secondary law, all EU countries are required to implement the principles and objectives set out in the Platform Directive into their national offshore petroleum regulatory framework to prevent accidents that may contribute to oil spills. Such implication in the UK occurred under the Offshore Installations (Offshore Safety Directive, Safety Case etc) Regulations 2015 (Platform Directive Regulations), to which guidance has been provided by the UK Health and Safety Executive (HSE).[313] The Offshore Safety Directive Regulator (OSDR) has been implemented as the Competent Authority required under the Platform Directive. The OSDR is a partnership between the HSE and BEIS Offshore Oil and Gas Environment and Decommissioning Team.

I-11.98   Effectively, the creation of the OSDR as a partnership between safety and environmental regulators indicates the integral relationship between safety and environment in consequential environmental regulation. This compares to operational environmental regulation, which focuses on the actual operational activity itself. Such integration between health, safety and environment is undertaken in several juris-

---

[309] *Ibid*, Art 8(3).

[310] *Ibid*, Art 6(6).

[311] *Ibid*, Art 28.

[312] *Ibid*, Art 7.

[313] HSE Guidance on Regulations, available at www.hse.gov.uk/osdr/guidance/guidance-regulations.htm (accessed 30 April 2017).

dictions, including Norway (with the Norwegian Petroleum Safety Authority – PTIL) and the Australian (National Offshore Petroleum Safety and Environmental Management Authority – NOPSEMA).

CONCLUSION

Environmental harm does occur as a result of offshore petroleum activities. The regulatory framework has developed in the UK from humble beginnings, where such harm was furthest from the regulator's mind, to a world-class framework that has, to date, avoided *severe* environmental harm that has characterised other jurisdictions, notably the USA as a result of the DWH disaster. I-11.99

In relation to operational environmental harm, where harm arises as a consequence of day-to-day operation (including chemical spills, produced water and leaks from valves), a strong regulatory framework has been developed and enhanced by the regulator and the relevant department. This framework is robust and continues to be refined, particularly since the creation of the Oil and Gas Authority as a consequence of the Wood Review in 2014. I-11.100

On the other hand, there is consequential environmental harm, which arises as a consequence of the loss of well control, particularly a well blowout. Such an event can cause environmental catastrophe, as was the case in the DWH accident. Furthermore, where a blowout does occur, the response required does not include just the clean-up of the spill; there is also a need to address the root cause of the spill. In several recent cases, including the Montara Oil Spill in Australia and the DWH event, oil continued to leak for several months.[314] It is this consequential environmental harm that continues to undergo changes in the regulatory regime as recent events such as DWH have exposed regulatory gaps. Such reform is bringing together the regulation of safety and environment in the UK, with the establishment of the OSDR as a partnership between the heath/safety and environmental regulators. Such joining of health/safety and environment regulation emulates the regulation that occurs in other jurisdictions. I-11.101

Afterword: Civil liability for environmental harm caused by oil released from offshore installations

This chapter is primarily focused upon environmental regulation, not civil liability for environmental harm caused by pollution I-11.102

---

[314] The Montara oil spill took 74 days to contain, with the Deepwater Horizon spill taking 87.

emanating from an offshore installation. However, given the practical importance of liability, we will briefly outline the relevant law.

I-11.103   There is no specific piece of UK legislation that addresses civil liability for harm caused by pollution incidents originating on the UKCS. This has a positive and negative aspect. The positive aspect is that where such legislation exists, it frequently imposes a cap on what might otherwise quickly amount to indeterminate liability; or in the case of the relevant Norwegian legislation, provides less favourable treatment for those outside Norway than for those within.[315] The negative dimension is that, *absent* a statutory system imposing strict liability or a reverse evidentiary burden and clearly stipulating what constitutes a compensable harm, it can be extremely difficult for a claimant to make a successful claim.

I-11.104   Harm caused by pollution from tankers or caused by a loss of bunker oil is compensated via funds established by a series of International Conventions.[316] These Conventions do not, however, apply to pollution emanating from an offshore installation.[317] Instead, there are three broad areas which require consideration: state claims, the law of tort/delict and the Offshore Pollution Liability Agreement (OPOL).

I-11.105   Firstly, the state – which might incur very significant loss as a result of a major pollution incident, and will have required the operator to provide financial security as a condition of issuing a licence[318] – will, by virtue of the contractual nature of the licence, be able to claim in damages, if it can be shown that the pollution incident occurred as a result of the breach of any of the terms of the licence (for instance, model clause 23, which requires licensees to operate according to good oilfield practice). Alternatively, the Environmental Liability Directive[319] (as amended by the Offshore

---

[315] K Svendsen, "Compensation of Harm to the Marine Natural Environment Caused by Petroleum Spills in the Barents Sea: An Analysis of Norwegian and Russian Law", 30 *Ocean Yearbook* (2016), ss 304–344.

[316] See eg the International Convention on Civil Liability for Bunker Oil Pollution Damage 2001; International Convention on Civil Liability for Oil Pollution Damage 1969; the International Convention on the Establishment of an International Fund for Compensation for Oil Pollution Damage 1971 with subsequent Protocols.

[317] Attempts in the 1970s to negotiate such a treaty for the North Sea area were unsuccessful, with the Treaty being agreed but never entering into force. The text of the treaty is available to download from www.gov.uk/government/uploads/system/uploads/attachment_data/file/604867/Convention_Oil_Pollution_Damage.pdf (accessed 8 May 2017).

[318] Usually, this will be by membership of OPOL, discussed further below; however if the OGA does not consider OPOL membership to be sufficient, it can insist upon further security.

[319] Directive 2004/35/CE of the European Parliament and of the Council of 21 April 2004

Safety Directive)[320] could potentially allow the state to receive compensation for a range of different forms of accident response and remediation measures. The Waste Directive would also provide a remedy for clean-up costs, but only if fault can be demonstrated on the part of the licensees.[321]

I-11.106 Neither the licence nor the Environmental Liability Directive would provide any legal footing for claims by non-state actors who have suffered a loss as a result of a pollution incident. Private individuals might be able to make a claim under tort/delict, but this will generally involve the need to prove fault and some heads of claim – for instance, pure economic loss – would be irrecoverable at common law. Thus the fisherman whose nets have suffered oil damage and whose catch is tainted with oil, rendering it unmarketable, would be likely be able to maintain a claim while the fisherman stuck in port due to oil in the fishing grounds would not.[322]

I-11.107 Thirdly, and perhaps most promisingly, is the potential for liability under OPOL.[323] Through OPOL, the operating oil companies commit to make good direct losses suffered as a result of an offshore pollution incident up to a limit of £250 million per incident. Liability is primarily undertaken by the operator of the installation from which the pollution emanated, but the scheme provides that in the event of non-payment by that operator, others will make good the loss. OPOL has never yet been claimed upon and consequently, there is lack of clarity on issues such as the precise scope of coverage (what, for instance, is a "direct loss" for the purpose of the scheme?) and the mechanism by which a claim on the guarantee would be made in the event that the polluting operator cannot or will not pay. That said, the offer of strict liability and the prospect of the claim being handled expeditiously and without the need for recourse to court means that for many claimants, OPOL will be the most attractive compensation option open to them.

I-11.108 Finally, it should be noted that following Deepwater Horizon the EU took an interest not just in offshore safety, but also in civil

---

on environmental liability with regard to the prevention and remedying of environmental damage.

[320] Directive 2013/30/EU of the European Parliament and of the Council of 12 June 2013 on the safety of offshore oil and gas operations.

[321] *Commune de Mesquer v Total France SA and Total International Ltd*, European Court Reports 2008 I-04501.

[322] For a more detailed consideration of issues of tort/delict in this context, see G Gordon, "Oil, Water and Law Don't Mix" 25(1) 2013 *Environmental Law and Management*, 1–11.

[323] OPOL, available at www.opol.org.uk (accessed 7 May 2017).

liability for pollution from offshore installations. It commissioned two reports with a view towards assessing whether this was an area where the laws of member states should be harmonised.[324] Ultimately, it decided not to proceed with this highly complex and contentious project at this point in time.[325]

---

[324] University of Maastricht, *A study on Civil Liability, financial security and compensation claims for offshore oil and gas activities* (2013), available for download from https://euoag.jrc.ec.europa.eu/files/attachments/liability-study-offshore-final-report-22-oct-2013.pdf; bio by Deloitte, *Civil Liability, Financial Security and Compensation Claims for Offshore Oil and Gas Activities in the European Economic Area* (2014), available for download from https://euoag.jrc.ec.europa.eu/files/attachments/201408_offshore_oil_and_gas_activities_liabilitystudy_final_report.pdf (both reports accessed 3 September 2017).

[325] European Commission, *Official Report on Liability, financial security and compensation claims for offshore oil and gas activities* (2015), available for download from https://euoag.jrc.ec.europa.eu/files/attachments/celex_52015dc0422_en_txt.pdf (accessed 3 September 2017).

CHAPTER I-12

DECOMMISSIONING OF OFFSHORE OIL AND GAS INSTALLATIONS

John Paterson

It might have been assumed that the low oil price that has been such **I-12.01** a feature of industry thinking since the last edition of this work was published would mean that decommissioning[1] would by now have become the defining characteristic of the United Kingdom Continental Shelf (UKCS). All indications at the time of writing, however, are that the picture is much more nuanced. While it is true that a considerable number of decommissioning plans are in development – Oil & Gas UK report in *Decommissioning Insight 2016* that 52 have been added on the UKCS since the previous year, bringing the total to 153 – it is also the case that successful efforts to extend the life of assets have resulted in a not inconsiderable number of such plans being postponed – Oil & Gas UK report 17 in this category in the same

---

[1] Note that the terms "abandonment" and "decommissioning" can be used interchangeably. While the Petroleum Act 1998 as amended (discussed below at para I-12.42) uses "abandonment", the competent authority (formerly the Department of Trade and Industry, then the Department for Business, Enterprise and Regulatory Reform, followed by the Department of Energy and Climate Change and now the Department for Business Energy and Industrial Strategy, hereinafter referred to as "the Department"), the regulator, the Oil and Gas Authority, and the industry generally avoid this term. The Department notes simply in its Guidance Notes for Industry (discussed below at para I-12.57) that "decommissioning" is the "preferred and generally accepted term" (para 2.1). Industry sources suggest that there is concern that "abandonment" conveys the wrong image of what is involved. This sensitivity to public perceptions of the terminology is explained in large part by the Brent Spar case (discussed below at para I-12.20). For a discussion, see A D M Forte, "Legal Aspects of Decommissioning Offshore Structures", in D G Gorman and J Neilson (eds), *Decommissioning Offshore Structures* (1997) (hereinafter "Gorman and Neilson"), pp 125–140, 126–127.

period.[2] All of that said, however, the dramatic pictures of the Brent Delta topsides being removed from the gravity base structures that support it in April 2017[3] send a powerful set of messages. Firstly, as one of the giant first-generation fields, the decommissioning of Brent perfectly illustrates the reality of the mature hydrocarbon province. Secondly, as the heaviest ever offshore lift, the Brent Delta operation also demonstrates the scale of challenge ahead and surely dispels any idea that this stage in the lifecycle of the province is a mere tidying-up exercise lacking the glamour and pioneering spirit of the early days of exploration and first oil. Thirdly, that such a complex operation has attracted only positive (or at worst neutral) media reporting could be taken as indicating that the problems that beset one of the earliest decommissioning operations (the Brent Spar in the mid-1990s, discussed below at para I-12.20) have been overcome and that a robust regulatory regime enjoying broad support is now in place. It is suggested, however, that whilst the first two of these messages is relatively uncontroversial, there are indications that the third cannot yet be unequivocally supported. Whilst the removal of the Brent Delta topsides has not given rise to any controversy, the same cannot be said of the operator Shell's plans in relation to the remainder of the platform: the gravity base structures. The overall decommissioning programme for the Brent field,[4] which at the time of writing is being considered by the regulator, has been rejected by a number of NGOs, including Greenpeace UK, on the basis that it "could breach international law".[5] The response from the operator has been to welcome the feedback from stakeholders, but there can be no doubt that there will be considerable concern that despite all the efforts since the mid-1990s to avoid disagreements in relation to decommissioning, Shell could be on the verge of another serious problem, with Greenpeace UK refusing to rule out direct action of the sort that was such a dramatic feature of their campaign in relation to the Brent Spar.[6] It is important, therefore, to gain an understanding of the scale of what lies ahead on the UKCS in relation to decommissioning as well as of the current state of legal and regulatory arrangements.

I-12.02   There are some 552 installations on the UKCS. Of these, 21 are floating, 278 are subsea, 244 are fixed steel structures and eight

---

[2] Oil & Gas UK, *Decommissioning Insight 2016*, p 6, available at http://oilandgasuk. co.uk/decommissioninginsight.cfm (accessed 29 August 2017).

[3] Available at www.shell.co.uk/sustainability/decommissioning/brent-field-decommissioning/brent-delta-topside-lift.html (accessed 4 May 2017).

[4] For details, see www.shell.co.uk/sustainability/decommissioning/brent-field-decommissioning/brent-field-decommissioning-programme.html (accessed 4 May 2017).

[5] See www.bbc.co.uk/news/uk-scotland-scotland-business-39528090 (accessed 4 May 2017).

[6] *Ibid.*

are gravity-based concrete structures. In addition, there are some 15,000km of pipelines.[7]

Perhaps more informative than the numbers of installations and lengths of pipeline are the costs involved. Estimates vary, but have risen over the years. The previous edition of this work reported Government estimates of around £20 billion and an industry figure of £26 billion.[8] More recently, in early 2014, the Wood Review suggested a figure of £35 billion while noting that this could "escalate significantly and easily exceed £50 billion".[9] The importance of the legal treatment of decommissioning is thus clear: precisely what the law requires to be removed, for example, will have an impact on precisely what the final costs will be.

I-12.03

It is worth noting at the outset that part of the reason for these substantial sums is the fact that when the first-generation structures were designed and built, the practicalities of decommissioning were not considered. Notwithstanding that relatively detailed projections of costs and returns were calculated over the lifetime production profile of a reserve, decommissioning costs were notable by their absence.[10] As time went on, however, this factor began to loom larger in industry thinking,[11] though even then it is important to bear in mind that structural engineers were often more focused on responding to emergent problems as ongoing experience in the North Sea threw up new challenges.[12] The deep water of the Northern North Sea is another reason that is often cited as an explanation for

I-12.04

---

[7]   Presentation by K Mayo, Head of the Offshore Decommissioning Unit, Kuala Lumpur, 1–2 October 2009. For full details of the installations on the UKCS see OSPAR Commission, *2009 Biennial Update of the Inventory of Oil and Gas Offshore Installations in the OSPAR Maritime Area* (OSPAR Publication No. 334/2009). This lists all the installations located within the sea area covered by the OSPAR Convention of 1992 (discussed below at para I-12.13).

[8]   Oil & Gas UK, *2010 Oil & Gas UK Activity Survey*, p 17, available at http://oiland-gasuk.co.uk/wp-content/uploads/2015/05/EC020.pdf (accessed 29 August 2017).

[9]   Sir I Wood, *UKCS Maximising Recovery Review: Final Report*, available at www.gov.uk/government/uploads/system/uploads/attachment_data/file/471452/UKCS_Maximising_Recovery_Review_FINAL_72pp_locked.pdf (accessed 4 May 2017), p 50.

[10]   For a view of field project calculations at this time, see F E Banks, *The Political Economy of Oil* (1980), p 52.

[11]   See eg P H Prasthofer, "Decommissioning Technology Challenges", III (1998) *Offshore Technology Conference* 379.

[12]   The industry entering the North Sea had assumed that experience gained in the Gulf of Mexico could simply be transferred. The first 10 to 15 years, however, revealed that instal-lation design assumptions based on maximum wave height were not sufficient and that fatigue failure was more of a problem in the North Sea. No sooner had this been factored in than dynamic response emerged as an issue. For an overview of these problems, see R J Howe, "Evolution of Offshore Drilling and Production Technology", IV (1986) *Offshore Technology Conference* 593.

the high costs,[13] though some studies suggest that a variety of factors, including, paradoxically, the very flexibility of the UK's approach to decommissioning, contribute to the problem.[14]

I-12.05   The legal treatment of decommissioning therefore needs to be sensitive to these cost[15] and technical[16] issues. Equally, as the Brent Spar case of the mid-1990s demonstrated, it needs to take account of the potential for possibly considerable public interest in decommissioning decisions.[17]

I-12.06   This chapter is accordingly structured as follows. Following this Introduction, the evolution of the international legal regime is considered, both at the global and regional levels prior to the Brent Spar case. Thus, there is consideration of the Convention on the Continental Shelf 1958, the London Dumping Convention 1972, the United Nations Convention on the Law of the Sea 1982 and the International Maritime Organization's Guidelines and Standards 1989, and the OSPAR Convention 1992. Thereafter, the extraordinary events of the Brent Spar case and its implications are examined. This involves consideration of the initial regulatory approach, the Greenpeace protest, the Stakeholder Dialogue initiated by Shell and the impact of the case on OSPAR and the UK Government. This is followed by a review of international and domestic legal developments post-Brent Spar, with particular attention paid to the 1996 Protocol to the London Convention, OSPAR Decision 98/3 and the Petroleum Act 1998 as amended by the Energy Acts 2008 and 2016. Noting that the Department has not utilised its regulatory powers under the 1998 Act but rather has preferred to operate on the basis of guidance, the chapter then examines its updated Guidance Notes for Industry in some detail, specifically the treatment of Section 29 notices and the decommissioning programme process. The interaction of the Department's responsibilities in relation to decommissioning with the Oil and Gas Authority's responsibilities in relation to the Maximising Economic Recovery Strategy are then

---

[13] Oil & Gas UK suggest that 44 per cent of the total expected bill will be attributed to the Northern North Sea, with the Southern and Central areas accounting for 24 per cent and 15 per cent respectively. Oil & Gas UK, *2010 Oil & Gas UK Activity Survey*, p 17.

[14] P E O'Connor, B R Corr, S Palmer and R C Byrd, "Comparative Assessment of Decommissioning Applications of Typical North Sea and Gulf of Mexico Approaches to Several Categories of Offshore Platforms in the Middle East" (2004) *Proceedings of The Fourteenth International Offshore and Polar Engineering Conference* 460.

[15] See A G Kemp and L Stephen, "Economic and Fiscal Aspects of Decommissioning Offshore Structures", in Gorman and Neilson, pp 80–123.

[16] See P A Meenan, "Technical Aspects of Decommissioning Offshore Structures", in Gorman and Neilson, pp 23–56.

[17] See A G Jordan and L G Bennie, "Political Aspects of Decommissioning", in Gorman and Neilson, pp 141–162.

considered. A continuing area of uncertainty related to residual liabilities is briefly discussed before concluding remarks are made.[18]

THE EVOLUTION OF INTERNATIONAL LAW 1958–1992

United Nations Convention on the Continental Shelf 1958

The starting point for any discussion of decommissioning outside   I-12.07 of territorial waters, involving as it does operations on the continental shelf, must be international law. The foundation document in this regard is the United Nations Convention on the Continental Shelf 1958. At the same time as this Convention granted states "sovereign rights for the purpose of exploring [the continental shelf] and exploiting its natural resources"[19] and entitled them "to construct and maintain or operate ... installations" to that end,[20] it also provided in blunt terms that "[a]ny installations which are abandoned or disused must be entirely removed".[21] While this may have seemed to be an entirely reasonable proposition in 1958, bearing in mind that offshore operations were then only in their infancy and very much confined to shallow waters, it became apparent with the passage of time that this requirement might not always be realistic. This realisation posed no problems for states that were not party to the Convention, but for a country such as the United Kingdom, the situation was different. It undoubtedly faced some of the most difficult challenges in complying with Article 5(5) given the fact that substantial structures were located in deep water on the UKCS. The Government argued in 1987, however, that the 1958 Convention needed to be interpreted purposively. This purpose could be discovered in Article 5(1) which stipulated that "[t]he exploration of the continental shelf and the exploitation of its natural resources must not result in any unjustifiable interference with navigation, fishing or the conservation of the living resources of the sea". Given that the circumstances of the industry had changed considerably since 1958 and that it was, by 1987, difficult – if not indeed impossible – to meet the complete removal requirement, the UK would remove installations to the extent necessary to ensure that there was no unjustifiable interference with navigation, fishing or the conservation of the living resources of the sea.[22]

---

[18] Note that the tax treatment of decommissioning is discussed in Chapter I-7, the health and safety dimension in Chapter I-10 and decommissioning security in Chapter I-13.

[19] 1958 Convention, Art 2(1).

[20] 1958 Convention, Art 5(2).

[21] 1958 Convention, Art 5(5).

[22] T Daintith, G Willoughby and A Hill, *United Kingdom Oil and Gas Law* (3rd edn, looseleaf, 2000–date) (hereinafter "Daintith, Willoughby and Hill"), para 1-1304. See also

UNCLOS 1982 and the IMO Guidelines 1989

I-12.08   Considerations such as these had in any event prompted the United Nations itself to revisit the issue of redundant installations. The 1982 Convention on the Law of the Sea (UNCLOS), the preamble to which explicitly recognises "that developments since ... 1958 ... have accentuated the need for a new and generally acceptable Convention on the law of the sea", thus stipulates a less draconian position:

> "Any installations or structures which are abandoned or disused shall be removed to ensure safety of navigation, taking into account any generally accepted international standards established in this regard by the competent international organization. Such removal shall also have due regard to fishing, the protection of the marine environment and the rights and duties of other States. Appropriate publicity shall be given to the depth, position and dimensions of any installations or structures not entirely removed."[23]

I-12.09   It may be wondered, as a consequence, why the UK Government needed to argue for a purposive interpretation of the 1958 Convention as late as 1987 in view of this more favourable provision in the 1982 Convention. This is explained by the fact that the UK did not accede to the latter convention until 1997.[24] Nevertheless, with that accession, it is now UNCLOS 1982 that sets out the UK's international obligations with regard to decommissioning.

I-12.10   Given the wording of Article 60(3), it is important to know the identity of "the competent international organization" and whether it has established "any generally accepted international standards". In this regard, the body in question is the International Maritime Organization,[25] and in particular its Maritime Safety Committee.[26]

A D M Forte, "Legal Aspects of Decommissioning Offshore Structures", in Gorman and Neilson, pp 125–140, at p 129.

[23] 1958 Convention, Art 60(3) as applied to the continental shelf by Art 80.

[24] See D H Anderson, "British Accession to the UN Convention on the Law of the Sea", 46(4) (1997) *ICLQ* 761.

[25] The IMO is an agency of the United Nations, established by the Inter-Governmental Maritime Consultative Organisation Convention of 1948. Art 1 of the Convention (as amended) lists the purposes of the IMO, including "(a) To provide machinery for co-operation among Governments in the field of governmental regulation and practices relating to technical matters of all kinds affecting shipping engaged in international trade, and to encourage the general adoption of the highest practicable standards in matters concerning maritime safety, efficiency of navigation and prevention and control of marine pollution from ships; and to deal with administrative and legal matters related to the purposes set out in this Article".

[26] Its functions include consideration of "aids to navigation, construction and equipment of vessels, manning from a safety standpoint, rules for the prevention of collisions, handling of dangerous cargoes, maritime safety procedures and requirements, hydro-

This body produced Guidelines[27] which were adopted by the IMO's Assembly in 1989[28] and which state that "[a]bandoned or disused offshore installations or structures on any continental shelf or in any exclusive economic zone are required to be removed, except where non-removal or partial removal is consistent" with the guidelines and standards it goes on to set out.[29] These Standards and Guidelines are not legally binding: the IMO Assembly resolution adopting them simply "recommends" that they be taken into account by "Member Governments ... when making decisions regarding the removal of abandoned or disused installations or structures".[30] Such whole or partial removal is to be carried out "as soon as reasonably practicable after abandonment or permanent disuse" of an installation or structure.[31] The Guidelines provide that the treatment of installations or structures should be on a case-by-case basis by the relevant coastal state. Where it is proposed to allow the whole or part of an installation or structure to remain in place, account must be taken of a range of factors including: potential effects on navigation; environmental effects; the costs, technical feasibility and risks to personnel involved in removal; and any new use or other justification for allowing all or part of the installation or structure to remain.[32] Notwithstanding the fact that the Guidelines allow for the possibility of only partial removal, the Standards provide that where an installation or structure stands in less than 75m of water and weighs less than 4,000 tonnes in air (excluding topsides) it should be entirely removed.[33] For installations and structures put in place after 1 January 1998, the water depth is increased to 100m.[34] The only exception permitted in the case of installations or structures falling within these parameters is where complete removal is "not technically feasible or would involve extreme cost, or an unacceptable risk to personnel or the marine environment".[35] It should be noted, however, that the Guidelines further provide that installations or structures located in

---

graphic information, log-books and navigational records, marine casualty investigation, salvage and rescue and any other matters directly affecting maritime safety". 1958 Convention, Art 29.

[27] Guidelines and Standards for the Removal of Offshore Installations and Structures on the Continental Shelf and in the Exclusive Economic Zone (hereinafter "IMO Guidelines").

[28] Resolution A.672(16), adopted 19 October 1989.

[29] IMO Guidelines, para 1.1.

[30] Resolution A.672(16).

[31] IMO Guidelines, para 1.2.

[32] *Ibid*, para 2.1.

[33] *Ibid*, para 3.1.

[34] *Ibid*, para 3.2.

[35] *Ibid*, para 3.5.

certain defined areas important for navigation "should be entirely removed and should not be subject to any exceptions".[36] It should also be noted that para 3.13 of the Guidelines provides that no installation or structure should be emplaced on or after 1 January 1998 unless its "design and construction ... is such that entire removal ... would be feasible". It could thus be suggested that the increase in the water depth criterion to 100m for installations or structures of 4,000 tonnes or less emplaced on or after 1 January 1998 in para 3.2 is superfluous, insofar as para 3.13 appears to impose a more stringent standard for *all* installations or structures emplaced after that date. In any case where there is partial removal with no part of the installation or structure projecting above the surface, this must leave an unobstructed water column of at least 55m.[37] It should be noted finally that as regards the situation where there is a new use or other justification for allowing all or part of the installation or structure to remain, the Guidelines specifically envisage their reuse as artificial reefs where they can serve to enhance fisheries.[38]

London Convention 1972

I-12.11   It is one thing to remove an installation or structure, whether wholly or partly; what happens to it thereafter is quite another. Reuse as an artificial reef is clearly a possibility in some situations, but where there is any proposal to dispose of an installation or structure in the sea where no new use is intended, then this must be considered in terms of the various dumping Conventions. In this regard, there are both global and regional instruments to be considered. The first global instrument is the London Convention on the Prevention of Marine Pollution by Dumping of Wastes and Other Matter of 1972. This divides wastes into three categories, in each case specifying what action may be taken in relation to them: dumping of wastes listed in Annex I to the Convention is prohibited;[39] dumping of wastes listed in Annex II requires a prior special permit;[40] and

---

[36] *Ibid*, para 3.7.

[37] *Ibid*, para 3.6.

[38] See *Ibid*, paras 3.4.1 and 3.12. This approach has particularly been adopted in the Gulf of Mexico where it has been overseen by the Minerals Management Service of the US Department of the Interior under Title II of the National Fishing Enhancement Act of 1984 (P.L. 98-623) and the National Artificial Reef Plan (NOAA Technical Memorandum NMFS OF-6, November 1985, as amended) developed by the National Marine Fisheries Service. At least 128 installations have been reused in this way. For further details see: www.bsee.gov/what-we-do/environmental-focuses/rigs-to-reefs (accessed 11 May 2017).

[39] The list in Annex I to the 1972 Convention includes, *inter alia*, certain heavy metals, persistent synthetics, heavy oils and high-level radioactive substances.

[40] The list in Annex II to the 1972 Convention includes, *inter alia*, certain metals, fluorides,

the dumping of all other wastes requires a prior general permit.[41] No permit is to be issued prior to careful consideration of factors mentioned in Annex III.[42]

As regards offshore installations, these are specifically covered **I-12.12** by Article III(1)(a)(ii) which defines "dumping" as including "any deliberate disposal at sea of vessels, aircraft, platforms or other man-made structures at sea". It would accordingly appear that as far as the London Convention is concerned, provided a structure or installation did not fall foul of the Annex I prohibition, its disposal at sea would be possible subject to a prior special permit.

OSPAR 1992

The regional instrument on dumping affecting the United Kingdom is **I-12.13** the Convention for the Protection of the Marine Environment of the North-East Atlantic of 1992. This is more commonly known as the OSPAR Convention, reflecting the fact that it combined and updated the pre-existing Oslo Convention for the Prevention of Marine Pollution by Dumping from Ships and Aircraft 1972 and Paris Convention for the Prevention of Marine Pollution from Land-based Sources 1974. The extent of the Convention's coverage is precisely defined in Article 1(a), but for present purposes it is sufficient to note that the whole of the UKCS is included. Dumping is defined so as to include specifically "any deliberate disposal in the maritime area of ... (2) offshore installations and offshore pipelines",[43] but it does not include "the leaving wholly or partly in place of a disused offshore installation or disused offshore pipeline, provided that any such operation takes place in accordance with any relevant provision of the Convention and with other relevant international law".[44]

The general obligations imposed on contracting parties are set out **I-12.14** in Article 2. The principal obligation is:

> "to prevent and eliminate pollution and ... take the necessary measures to protect the maritime area against the adverse effects of human activities so as to safeguard human health and to conserve marine ecosystems and, when practicable, restore marine areas which have been adversely affected".[45]

---

pesticides and "[c]ontainers, scrap metal and other bulky wastes liable to sink to the sea bottom which may present a serious obstacle to fishing or navigation".

[41] 1972 Convention, Art IV(1).

[42] These factors include the characteristics and composition of the matter, the characteristics of the dumping site and the method of deposit, all as further defined.

[43] OSPAR 1992, Art 1(f).

[44] *Ibid*, Art 1(g).

[45] *Ibid*, Art 2(1)(a).

I-12.15 In adopting "programmes and measures" to this end, the contracting parties are required to apply the precautionary principle[46] and the polluter pays principle.[47] They must also "ensure the application of best available techniques and best environmental practice", both of which terms they must define taking account of criteria set out in Appendix 1 to the Convention.[48]

I-12.16 The key provisions dealing with offshore installations are contained in Annex III to the Convention. Article 5(1) of that Annex provides:

> "[n]o disused offshore installation or disused offshore pipeline shall be dumped and no disused offshore installation shall be left wholly or partly in place in the maritime area without a permit issued by the competent authority of the relevant Contracting Party on a case-by-case basis".

I-12.17 Article 5(3) further provides that where a Contracting Party intends to issue such a permit for dumping after 1 January 1998 it "shall, through the medium of the [OSPAR] Commission, inform the other Contracting Parties of its reasons for accepting such dumping, in order to make consultation possible". This echoed pre-existing notification and consultation arrangements contained in Guidelines for the Disposal of Offshore Installations issued in June 1991 under the Oslo Convention.[49]

I-12.18 It should be noted that the Convention does admit exceptions. Firstly, the requirements relating to the disposal of offshore installations do not apply "in case of *force majeure,* due to stress of weather or any other cause, when the safety of human life or of an offshore installation is threatened".[50] Secondly, Articles 8 and 10 of Annex III envisage the possibility of leaving installations in place or emplacing them for purposes other than those for which they were originally intended – in other words, rigs-to-reefs[51] – albeit that this will only be where specifically authorised and in accordance with guidelines to be drawn up by the OSPAR Commission.

I-12.19 It might be suggested, then, that with the OSPAR Convention, a fairly sophisticated and robust international regime was in place, not least in the area including the UKCS, to deal with the removal and ultimate disposal of offshore installations and pipelines. And yet even before the OSPAR Convention entered into force on 25 March

---

[46] *Ibid*, Art 2(2)(a).

[47] *Ibid*, Art 2(2)(b).

[48] *Ibid*, Art 2(3)(b).

[49] See Parliamentary Office of Science and Technology, "Oil Rig Disposal", POST Note 65, July 1995.

[50] OSPAR 1992, Annex III, Art 6.

[51] See note 38 above.

1998, the adequacy of this approach had been profoundly called into question in a remarkably public fashion.

BRENT SPAR

No account of the legal treatment of the decommissioning of **I-12.19** offshore installations would be complete without a discussion of the Brent Spar case, not only because it enjoyed such a high profile, but especially because it led directly to changes in the approach both of OSPAR and the UK regulator at the time, the DTI.

The original disposal plan

The Brent Spar was in many respects a unique structure in that it was **I-12.20** neither a rig nor a platform, but rather a floating oil storage buoy. It was intended as a temporary storage and tanker loading facility for the Brent field in the Northern North Sea – operated jointly by Shell and Esso – until such time as a pipeline could be built. It weighed 14,500 tonnes, was 140m tall and was composed of six huge storage tanks with a capacity of some 50,000 tonnes of oil and a displacement of some 66,500 tonnes. It was finally declared redundant in 1991.

Disposal options for the Brent Spar could be divided into two **I-12.21** broad groups. The first would involve its removal and dumping in deep water; the second, its removal and dismantling ashore. While the structure had been assembled practically in its entirety onshore then floated out to its location in the North Sea, a number of factors made the simple reversal of this process difficult. While the structure was stable where it stood, degradation of the tanks over time coupled with damage to two during operations meant that any attempt to refloat it or to rotate it to a towing position risked buckling or even rupturing the tanks. This was problematical because although the tanks had been drained of oil and filled with seawater, residual sludge that could not be pumped out remained.

The provisions of the relevant international instruments in **I-12.22** force at the time, namely the London Convention and the Oslo Convention, were enacted in the UK by the Food and Environmental Protection Act 1985 and the Petroleum Act 1987. Under the 1987 Act, Shell, as the operator, was required to obtain a licence from the DTI for the disposal of the Brent Spar.[52] Grant of the licence was dependent on acceptance of the Abandonment Plan prepared by Shell. This Plan had to be proportionate, cost-effective and

---

[52] Where the operator proposed to dispose of an installation at sea away from the original site, there was also a requirement for licences to be obtained under the Food and Environmental Protection Act 1985.

consistent with both international obligations and the precautionary principle. In addition, the Plan had to constitute the Best Practicable Environmental Option (BPEO), a concept proposed by the UK Royal Commission on Environmental Pollution in 1988.[53] Demonstrating that an option constituted the BPEO involved a number of factors, including: ensuring that a full range of alternatives had been considered; specifying the origins of data used and their reliability; presenting scientific evidence objectively in order to assist the taking of decisions with social or political significance; and not regarding financial considerations as overriding.

I-12.23   Discussions between Shell and the DTI began in 1992, with some 13 disposal options initially being considered. Of these, six were regarded as viable, with two finally being considered in detail: deepwater disposal and horizontal dismantling. Between 1992 and 1994, different aspects of these two options were examined (including some 30 separate studies) until documentation was submitted to the DTI proposing deepwater disposal at one of three sites identified by SOAEFD[54] as the BPEO – a conclusion that had itself been subject to three independent evaluations. During the same period, surveys of those sites had been commissioned by Shell and SOAEFD. The DTI approved the Abandonment Plan on 20 December 1994, and in May 1995 granted Shell a licence to dispose of the Brent Spar at the North Feni Ridge in the North Atlantic. The deepwater disposal option was chosen over the horizontal dismantling on the basis that it involved significantly lower risks to personnel (by a factor of six), was cheaper (by a factor of four) and would have only a minimal environmental impact.

I-12.24   In view of what has been said above about the international regime for the removal and disposal of offshore installations, nothing in the Brent Spar case to this point should come as a surprise. Indeed, it is significant in this regard that the other Contracting Parties under the Oslo Convention had been notified on 16 February 1995 of the DTI's approval of the disposal plan in accordance with the Guidelines mentioned above[55] and had not raised any objections or concerns. But the story was far from over.

The Greenpeace protest

I-12.25 The environmental NGO, Greenpeace, took the view that the dumping of such a vast structure was unacceptable in all circum-

---

[53] Twelfth Report of the Royal Commission on Environmental Pollution, *Best Practicable Environmental Option* (Cm 310, 1988).

[54] The North Feni Ridge, the Rockall Trough and the Maury Channel, all located in the North Atlantic.

[55] See para I-12.16.

stances and set a dangerous precedent – albeit that the DTI was at pains to stress that it adopted a case-by-case approach in line with the IMO Guidelines.[56] In addition, it believed, following a suggestion from an ex-oil worker, that the inventory of toxic substances on the installation could be much more significant than had been admitted by Shell. It accordingly occupied the Brent Spar and took samples from the storage tanks. From this sample it concluded that perhaps as much as 5,000 tonnes of oil remained on board, in contrast to the nominal amount claimed by Shell. This claim, coupled with the dramatic video footage of the Greenpeace occupation of the installation transformed the Brent Spar case from a peripheral issue of technical interest only to regulators and industry into a major international issue touching the whole question of the attitudes of government and industry to ocean dumping specifically and environmental protection in general. The case became headline news across Europe, and Shell became the target of public protest ranging from a boycott of its products to the firebombing of its petrol stations in Germany. Meanwhile, the UK Government came under sustained pressure from its European partners, including previously quiescent Oslo Convention partners, to reverse the decision to allow deepwater disposal.[57]

The UK Government's response was extremely robust, defending   I-12.26
the regulator's decision on the basis that a rigorous process had identified deepwater disposal as the BPEO.[58] This tough stance was continued even as the issue came to dominate relations with its European partners, most notably at the G7 summit in June 1995. Shell, on the other hand, wavered in the face of the dramatic effects on its business across Europe and finally announced, even as the UK Prime Minister reiterated the Government's stance, that it was abandoning the deepwater disposal plan. Shortly afterwards, Norway granted Shell permission to moor the installation in the Erfjord while it was decided what should happen next. And what happened next was nothing short of extraordinary.

---

[56] See para I-12.10.

[57] For a discussion of the extent to which the Brent Spar case involved continental European environmental norms overriding those applying in the UK, see S C Zyglidopoulos, "The Social and Environmental Responsibilities of Multinationals: Evidence from the Brent Spar Case", 36 (2002) *Journal of Business Ethics* 141.

[58] For a discussion of the relative unimportance of scientific rationality in this context as compared to the "symbolic capital" of the key players, see H Tsoukas, "David and Goliath in the Risk Society: Making Sense of the Conflict between Shell and Greenpeace in the North Sea", 6 (1999) *Organization* 499. See also A D M Forte, "Legal Aspects of Decommissioning Offshore Structures", in Gorman and Neilson, p 127.

Shell's Stakeholder Dialogue[59]

I-12.27   Flying in the face of the Government's insistence that it carry out the approved Abandonment Plan, Shell reopened the whole question of how it would dispose of the Brent Spar, calling for proposals from contractors and setting up what was essentially an entirely new regulatory process. It first commissioned Det Norske Veritas (DNV), an independent, not-for-profit foundation to carry out an audit of the contents of the installation with the aim of resolving the dispute with Greenpeace. Before the publication of DNV's report, which supported Shell's assessment, Greenpeace admitted errors in its sampling.[60]

I-12.28   Shell also announced what it called the "Way Forward". It placed a notice in the Official Journal seeking expressions of interest from contractors regarding the disposal of the Brent Spar. These submissions together with some 200 other proposals that Shell had received were to be developed into a "long list" and the organisations involved were then invited to meet pre-qualification criteria. A list of 21 contractors was eventually published, with those involved required to develop an outline of their disposal plans.

I-12.29   At this point, Shell also announced that there would be a Stakeholder Dialogue Process, which would play a role in identifying the ultimate solution. This process grew out of an earlier approach to the Environment Council,[61] where it had begun to discuss options for the way forward as regards reaching a new disposal decision. The Environment Council first proposed a process by which a Europe-wide panel of 50 to 60 stakeholders would be established, with a view to it being consulted throughout the technical process of developing a new disposal plan as a means of testing ideas and keeping in touch with the various interested constituencies. While Shell was agreeable to this proposal, the response from the UK Government was negative. The latter did, however, accept a modified plan – albeit stressing that whatever disposal option was eventually chosen had to be at least as good as deepwater disposal which the regulator had

---

[59] This section draws on work carried out in the context of the EU Framework Programme 5-funded RISKGOV project which involved interviews with individuals involved in the Brent Spar case, including representatives of Shell, Greenpeace, the DTI and the Environment Council. The Brent Spar Case Study is available in G Brownless and J Paterson, "Complex and contentious risk based decisionmaking in the field of health, safety and the environment: Comparative analysis of two UK examples", Health and Safety Executive, Research Report 448 (2006).

[60] See also M Saunders, "Environmental Protection: abandonment – study vindicates Shell over Brent Spar", 13(12) (1995) Oil and Gas Law and Taxation Review 145.

[61] The Environment Council is an independent charitable organisation that brings together stakeholders from all sectors to develop solutions to environmental problems.

identified as the BPEO. This modified plan was not vastly different from the first proposal and it is probably not insignificant that in the interim between the UK Government's initial rejection and ultimate acceptance, the Energy Minister had left the Government. Equally, a report commissioned by the DTI from the National Environmental Research Council had concluded that, "some means should be sought to take public acceptability into account in evaluating future marine environment impact assessments".[62]

The Environment Council next set about contacting parties who were likely to be interested in being involved in the process. This produced some 200 responses and it was then a matter of arriving at a balanced group of stakeholders who would be prepared to meet periodically, prior to the points at which key decisions about the disposal would have to be taken. These meetings were facilitated by the Environment Council and it was made clear from the outset, firstly, that the deepwater disposal option had to be considered, as this was what the regulators had decided was the BPEO, and, secondly, that the aim of the exercise was not to reach a consensus but rather to ensure that whatever decision was eventually reached emerged from an open and transparent process. The process itself began with the facilitator attempting to draw out from the participants what their concerns were with the various options on the table, with a view to informing the engineering process.

I-12.30

The first meeting in London on 1 November 1996 discussed a range of some 30 disposal options produced by the 21 pre-qualified contractors on the so-called long list. As a result of that initial Stakeholder Dialogue, the list was reduced to 11 disposal options from six contractors by mid-January 1997. Those contractors were then given four months to develop detailed commercial projects. DNV was once again retained to provide an independent evaluation of the projects on technical, environmental and safety grounds.

I-12.31

Over the next few months, Stakeholder Dialogue meetings were held in Denmark and the Netherlands and in June the six short-listed contractors presented nine detailed proposals. Then, in the autumn of 1997, the contractors' prices together with the findings of DNV's evaluations were published and further Dialogue meetings were held in the UK, Denmark, the Netherlands and Germany. On the basis of these interactions, the choice was narrowed down to the original deepwater disposal option and a plan to reuse the Brent Spar in the construction of a quay extension at Mekjarvik near Stavanger in Norway. A final BPEO assessment was conducted and Shell announced in January 1998 that it had chosen the reuse option.

I-12.32

<hr />

[62] NERC, *Scientific Group on Decommissioning Offshore Structures: First Report* (1996).

This choice then had to be approved by the DTI, whose approval was forthcoming in August 1998. The project was completed in July 1999.

I-12.33   The Stakeholder Dialogue process, dovetailing with the technical process, operated according to the BPEO approach. This meant that once the most technically feasible options had been identified, their environmental aspects were addressed first, then their safety considerations and finally their cost. The environmental evaluation covered such areas as energy balance, emissions to air, consumption of resources, waste disposal, containment, ecological effects, aesthetic impacts, local societal effects and the environmental management systems put in place by the contractors.

I-12.34   It is interesting to note that the stakeholders concluded that none of the options put forward would have a significant environmental impact, not even the original deepwater disposal plan which had caused so much controversy.[63] The issue was, therefore, how to choose among the options, given that there were only very small differences between them in terms of environmental impact. Here, the stakeholders in the dialogue process were able to agree criteria to be used in such circumstances. Firstly, projects with a *positive energy balance* were to be favoured, that is, those in which more energy was saved than consumed. Secondly, projects coming higher up the *waste hierarchy* would be favoured over those lower down. This hierarchy, which aims at the minimisation of waste, ranks options as follows: first, reuse; second, recycling; third, disposal. Applying these criteria, the stakeholders favoured the quay extension proposal above others on the short list inasmuch as it had the best energy balance figure and would allow 80 per cent reuse. Turning next to safety, the quay extension and deepwater disposal options had the lowest potential for loss of life or a major accident. Finally, on cost, deepwater disposal was the lowest, with the quay extension coming next.

I-12.35   It is worth noting, however, that although the calculations of risk and cost allowed projects to be ranked according to the different criteria, the actual outcome for the quay extension fell short of expectations: it cost nearly twice as much as expected and failed to achieve a positive energy balance.

The impact of Brent Spar on OSPAR and the UK Government

I-12.36 The Brent Spar case was a wake-up call to those concerned with the legal treatment of the decommissioning of offshore installa-

---

[63] In this regard, see further P A Tyler, "Disposal in the deep sea: analogue of nature or *faux ami?*", 30 (2003) *Environmental Conservation* 26.

tions both at the international and at the domestic levels. Despite the apparently robust and comprehensive regime that was in place in 1995, the public reaction to the proposed dumping of the Brent Spar indicated that it suffered from two major shortcomings. First of all, whatever the scientific thinking about the possibility of deepwater disposal, this was evidently not an approach that was socially acceptable. Secondly, despite the UK Government's firm belief in the regulatory approach under the Petroleum Act 1987, the public perception was rather of a closed conversation between industry and regulator, with neither of these parties enjoying much in the way of public support. And while OSPAR did not have a high public profile in the case, it was acutely aware of the weaknesses in its position that Brent Spar had exposed. None of the contracting parties had been concerned initially by the plan to dump the installation in the North Atlantic – this was, after all, something that the Oslo Convention explicitly counte-nanced provided the relevant assessment had been carried out, which the DTI's notification manifestly demonstrated it had been. Crucially for OSPAR, nothing in the 1992 Convention, which was then shortly due to come into force, would have altered the position. The same dumping plan could have been approved and the same notification would have been made, presumably eliciting a similarly quiet agreement. For the DTI, the main issue was the fact that its robust and considered regulatory approach had been completely rejected by the public, despite its own assurance about the sound scientific basis for the decisions made. By contrast, Shell's Stakeholder Dialogue, initiated in the face of strong opposition from both the DTI and the Government, had taken the considerable heat out of the situation and was widely regarded as a success, even by those who had initially been most critical of the company, such as Greenpeace. Both OSPAR and the DTI drew lessons from this experience and, as will be seen below, the current approach to decommissioning, at both the international and the domestic levels, reflects this fact.[64]

INTERNATIONAL AND DOMESTIC LAW POST-BRENT SPAR

The 1996 Protocol to the London Convention

The situation with regard to dumping is modified somewhat by   I-12.37
the coming into force on 24 March 2006 of the 1996 Protocol to

---

[64] For a discussion of the impact of the Brent Spar case on the integration of environ-mental values into national and international policymaking, see L G Bennie, "Brent Spar, Atlantic Oil and Greenpeace", 51 (1998) *Parliamentary Affairs* 397.

the London Convention 1972.[65] While this adopts a different, and indeed stricter, approach inasmuch as it prohibits *all* dumping *with the exception of* wastes and other matters mentioned in Annex I to the Protocol (the so-called "reverse list"), the fact that this includes "vessels and platforms or other man-made structures at sea" means that the position is broadly the same for the question of decommissioning offshore installations as under the 1972 Convention. A special permit will still be required, and the grant of such a permit must be in accordance with the provisions of Annex II, which is headed Assessment of Wastes or Other Matter that May Be Considered for Dumping. It is also provided that "[p]articular attention shall be paid to opportunities to avoid dumping in favour of environmentally preferable alternatives".[66] This, together with the Protocol's adoption of the precautionary principle[67] as well as the polluter pays principle,[68] means that anyone proposing to dump an installation or structure would have to demonstrate that there were no environmentally preferable options, and that appropriate preventive measures were being taken "when there is reason to believe that wastes or other matter introduced into the marine environment are likely to cause harm even when there is no conclusive evidence to prove a causal relation between inputs and their effects".[69] Note that there is no mention in the London Convention or the 1996 Protocol of pipelines, although this omission is less important in the case of the UK, covered as it is by a regional instrument that specifically mentions pipelines.

OSPAR Decision 98/3

I-12.38   The preamble to OSPAR Decision 98/3 on the Disposal of Disused Offshore Installations,[70] adopted unanimously at the Ministerial Meeting in July 1998, contains an immediate indication of the impact of the Brent Spar case insofar as it recognises that "re-use,

---

[65] For more details see E J Molenaar, "The 1996 Protocol to the 1972 London Convention", 12 (1997) *International Journal of Marine and Coastal Law* 396; E A Kirk, "The 1996 Protocol to the London Dumping Convention and the Brent Spar", 46(4) (1997) *ICLQ* 957; L de La Fayette, "The London Convention 1972: Preparing for the Future", 13 (1998) *International Journal of Marine and Coastal Law* 515. For a discussion of the issue of waste generally in the context of decommissioning, see J Rowan Robinson and L Cowie, "Decommissioning and the regulation of waste", 1 (2003) *International Energy Law and Taxation Review* 1.

[66] 1996 Protocol, Art 4(1.2)

[67] *Ibid*, Art 3(1).

[68] *Ibid*, Art 3(2).

[69] *Ibid*, Art 3(1).

[70] Note that this Decision does not apply to pipelines.

recycling or final disposal on land will generally be the preferred option for the decommissioning of disused offshore installations in the maritime area". The precise way in which this recognition is given effect is through a general prohibition on the "dumping, and the leaving wholly or partly in place, of disused offshore installations within the maritime area"[71] followed by the opening up of the possibility of a derogation from that general prohibition in certain defined circumstances.[72] Derogations, involving leaving all or part of an installation in place or dumping as appropriate, may be permitted by the competent authority of a contracting party in the following cases:

(a) all or part of the footings of a steel installation weighing more than 10,000 tonnes in air emplaced before 9 February 1999;[73]

(b) a concrete installation (including a gravity-based concrete installation, a floating installation and any concrete anchor base which results, or is likely to result, in interference with other legitimate uses of the sea);

(c) any other disused offshore installation when exceptional or unforeseen circumstances resulting from structural damage or deterioration, or from some other cause presenting equivalent difficulties, can be demonstrated.[74]

Any such permission may only be issued following, firstly, an assessment in accordance with Annex 2 to the Decision that satisfies the competent authority that there are "significant reasons why an alternative disposal ... is preferable to re-use, recycling or final disposal on land"[75] and, secondly, consultation with the other Contracting Parties in accordance with Annex 3 to the Decision.[76] Any permit must be in the form specified in Annex 4 to the Decision[77] and must be reported to the OSPAR Commission,[78] with a

I-12.39

---

[71] OSPAR Decision 98/3, para 2.

[72] *Ibid*, para 3.

[73] This would appear to be at odds with the IMO Guidelines which, recall, provide at para 3.13 that no installation or structure should be emplaced on or after *1 January 1998* unless its "design and construction ... is such that entire removal ... would be feasible" (emphasis added). See para I-12.10. This raises the possibility of an installation having been placed on the UKCS after 1 January 1998 but before 9 February 1999 and being allowed under OSPAR to have its footings left in place at decommissioning despite the wording of the IMO Guidelines. The crucial issue here, of course, is that while Decision 98/3 is binding, the IMO Guidelines are not.

[74] OSPAR Decision 98/3, para 3 and Annex 1.

[75] *Ibid*, para 3.

[76] *Ibid*, para 4.

[77] *Ibid*, para 5.

[78] *Ibid*, para 9.

further report following completion of the disposal.[79] In view of the important role played by Det Norske Veritas in the Brent Spar case, it is interesting to note that one of the requirements of Annex 4 is that a permit shall "require independent verification that the condition of the installation before the disposal operation starts is consistent both with the terms of the permit and with the information upon which the assessment of the proposed disposal was based".[80]

I-12.40   The general prohibition on dumping in Decision 98/3 is certainly significant, although it might be contended that the envisaged derogations mean that, in many respects, there has been little change from UNCLOS 1982 where the practical difficulties involved in removal in many cases were first recognised. Two things indicate that such criticism would be wide of the mark. Firstly, it is practically impossible to imagine an operator in a similar position to Shell pushing for dumping if another option were available, even if more difficult or more expensive. Secondly, OSPAR has clearly signalled that it does not want to be caught on the back foot again as it was in 1995. Paragraph 7 of the Decision provides that the Commission will seek to reduce the scope of derogations in the light of ongoing experience with decommissioning. While it was specifically envisaged that the Commission would consider possible amendments initially in 2003, it was noted at that time that "decommissioning activity has not developed as quickly as expected in 1998" and that, consequently, there was no substantive evidence either to suggest that any of the derogation categories was no longer needed or on which revised criteria for these categories could be based.[81] The Commission undertook to review this issue again at its meeting in 2008.[82] Once again, however, the conclusion was that "the number of projects involving concrete structures and substantial steel footings has been very low and there have been no significant developments in the technical capabilities of the industry which would support a reduction in the categories eligible for derogation".[83] The matter was to be reviewed again in 2013, with OSPAR making it clear that the aim was to see whether the categories that may be considered

---

[79] *Ibid*, para 10.

[80] *Ibid*, Annex 4, para 2.b.

[81] *Annual Report of the OSPAR Commission, 2002–2003*, Vol 1, Chapter 5, para 138.

[82] *Ibid*, para 139. For further discussion, see L de La Fayette, "New Developments in the Disposal of Offshore Installations", 14 (1999) *International Journal of Marine and Coastal Law* 523; E A Kirk, "OSPAR Decision 98/3 and the dumping of offshore installations", 48(2) (1999) *ICLQ* 458; J Woodcliffe, "Decommissioning of offshore oil and gas installations in European waters: the end of a decade of indecision?", 14(1) (1999) *International Journal of Marine and Coastal Law* 101.

[83] OSPAR Commission, *Assessment of impacts of offshore oil and gas activities in the North-East Atlantic*, (2009), p 26.

for derogation could be reduced such that "derogations from the dumping ban remain exceptional".[84] However, once again in 2013, the conclusion was that "no new evidence had come to light to require a change to current derogations categories, so these would continue in their existing form".[85] The matter will be reviewed again in 2018. It is interesting to note that whereas only a few years ago it would have been more or less unthinkable in the OSPAR region to consider attempting to widen the derogation cases, there are recent indications that this is no longer the case. One reason for this is undoubtedly a concern on the part of industry (and indeed governments and taxpayers) to minimise the costs involved in removing infrastructure. Perhaps more surprising are indications that concerns are beginning to emerge that current decommissioning requirements may actually result in environmental damage. This is because where infrastructure has been *in situ* for decades, it essentially becomes part of the local ecosystem and can actually play a role in promoting its sustainability.[86] One could well imagine countries with large amounts of offshore infrastructure such as the UK and Norway being sympathetic to a new approach which widened the scope of derogations based on a more open assessment of the environmental effects associated with different approaches to decommissioning, including leaving in place, but whether other contracting parties would have a similar view is very much an open question.[87]

The Petroleum Act 1998 as amended by the Energy Acts 2008 and 2016

The impact of Brent Spar is not immediately obvious in the legislation **I-12.41** that now governs decommissioning on the UKCS, notwithstanding that this received the Royal Assent in 1998. This is because Part IV of the Petroleum Act 1998 largely serves to consolidate the pre-existing provisions to be found in the Petroleum Act 1987 Parts I and II as well as other enactments. But the echoes of the case are certainly to be detected, as will be seen below, in the Guidance Notes that the Department has issued to supplement the legislation.

---

[84] *Ibid*, p 27.
[85] *Annual Report of the OSPAR Commission*, 2012–2013, p 3.
[86] For an excellent review of the literature in this field, see O Langhamer, "Artificial Reef Effect in relation to Offshore Renewable Energy Conversion: State of the Art", *The Scientific World Journal* (2012), doi:10.1100/2012/386713.
[87] For a more extensive discussion of this issue, see J Paterson, "Decommissioning Offshore Installations: international, regional and domestic legal regimes in the light of emergent commercial, political, environmental and fiscal concerns", *AMPLA Yearbook* (2015).

I-12.42   The UK's approach to decommissioning is based on the proposition that no one may commence or continue decommissioning without an approved abandonment plan, failing which they will be guilty of an offence.[88] Precisely who may be liable to effect abandonment is established by Section 29(1) of the 1998 Act, which empowers the Secretary of State to serve a notice upon a variety of parties requiring them to submit "a programme setting out the measures proposed to be taken in connection with the abandonment of an offshore installation[89] or submarine pipeline".[90] The service of such a Section 29 notice may be initiated by the Secretary of State or at the request of a party subject to Section 29(1).[91] The notice will either specify the date by which the abandonment programme is to be submitted to the Secretary of State or, as is more usual in practice, provide for it to be submitted on or before a date to be specified in future.[92] The recipient of the notice must consult with the OGA prior to submitting their plan to the Secretary of State and must so frame their plan as to minimise the cost in so far as is reasonably practicable.[93] Section 29(3) allows the Secretary of State to require, in the notice, that the person upon whom it is served must also carry out such other consultations as he or she may specify. The required contents of the abandonment programme are briefly listed by Section 29(4), but, as will be seen below, this is now substantially supplemented by the Department's Guidance Notes.[94]

I-12.43   The parties upon whom such a notice may be served are listed in Section 30(1) and include:

   (a) the person having the management of the installation or of its main structure [that is, the operator];
   (b) the licensee;[95]
   (ba) a person who has transferred his interest under a licence without the consent of the Secretary of State;[96]

---

[88] Section 28A.

[89] As defined by s 44.

[90] As defined by ss 26 and 45.

[91] Section 29(1A).

[92] Section 29(2).

[93] Section 29(2A). For further discussion of the interaction of the decommissioning regime with the MER regime, see Chapter I-5. Note in general, however, that the OGA is required to "consider and advise on ... alternatives to abandonment or decommissioning" and how to ensure that costs are minimised. See s 29 (2B).

[94] See para I-12.57.

[95] As defined by s 30(5) and (6).

[96] Inserted by the Energy Act 2008, s72(2)(a) to deal with a situation which the Secretary of State had become aware of in the context of increased transfers of assets in the mature province. See also Sch 5, para 10 for the concomitant duty to carry out an approved programme.

(c) a person falling outside of the above categories who is party to a joint operating agreement or similar agreement;

(d) a person falling outside the above categories who owns any interest in the installation otherwise than as security for a loan;

(e) a body corporate[97] outside the above categories but which is associated with a body corporate within any of those categories.[98]

As regards pipelines,[99] the parties upon whom a Section 29 notice **I-12.44** may be served are listed in Section 30(2) and include:

(a) a person designated as the owner by an order made by the Secretary of State under Section 27;[100]

(b) a person falling outside the above category who owns an interest in the whole or substantially the whole of the pipeline, otherwise than as security for a loan; and

(c) a body corporate outside the above categories but which is associated with a body corporate within any of those categories.[101]

An amendment introduced by the Energy Act 2008 also extends **I-12.45** the scope of those on whom a Section 29 notice may be served to those who fall into one of the above categories and who are not yet actually engaged in relevant activities on an offshore installation but who intend to become so involved.[102] The same Act has also extended the list of parties on whom a Section 29 notice may be served so as to ensure that installations used for gas storage and importation (which is itself the subject of Chapter 2 of Part 1 of the 2008 Act)

---

[97] The term "body corporate" is substituted for "company" here and in s30(2)(c) by the Energy Act 2008, s72(2)(b) and (3) to ensure that limited liability partnerships are covered as well as companies. This is another example of the way in which the law has had to adapt to keep pace with the changing profile of the actors on the UKCS as a mature province. The test for establishing whether one body corporate is associated with another is contained in s 30(8)–(8D) substituted by s 72(5) of the 2008 Act, once again with the intention of ensuring that limited liability partnerships are brought within the scope of those on whom a s 29 notice may be served.

[98] As defined by s 30(8) and (9).

[99] Note that the definition of "submarine pipeline" now includes a pipeline which is intended to be established, allowing earlier service of a s 29 notice in respect of pipelines than had heretofore been the case. See Energy Act 2008, Sch 5, para 11 amending s45 of the Petroleum Act 1998.

[100] Section 27(1) provides that "'owner' in relation to a pipeline ... mean[s] the person for the time being designated as the owner of the pipeline ... by an order made by the Secretary of State".

[101] As defined by s 30(8) and (9).

[102] Section 30(5)(b) as amended by s 72(4) of the 2008 Act.

are also subject to the same decommissioning regime.[103] Installations used for Carbon Capture and Storage (the subject of Chapter 3 of Part 1 of the 2008 Act) will also be covered by the decommissioning provisions of the 1998 Act.[104] The comprehensive reach of the Section 29 notice is enhanced by the ability of the Secretary of State to require any party appearing to fall within the Section 30 categories to furnish him with the name and address of every other person whom that party believes to fall within those categories,[105] on pain of a criminal penalty.[106] The aim of this approach, bluntly stated, is to ensure that whoever else ends up having to foot the bill for decommissioning installations (which, recall, is likely to be in the region of £35 billion to £50 billion) it will not be the British taxpayer (at least beyond the liability that already falls on the taxpayer as a consequence of decommissioning allowances).[107] This consideration explains why under Section 31(1) persons falling into categories (d) and (e) will not have a Section 29 notice served on them where the Secretary of State is satisfied that parties under categories (a) to (c) have made "adequate arrangements, including financial arrangements ... to ensure that a satisfactory abandonment programme" is made. This provision will not apply, however, where a notice has not been complied with or where the Secretary of State has rejected a programme.[108] It is noteworthy, however, that a Section 29 notice once served may be withdrawn.[109] This recognises the fact that as the UKCS matures as a hydrocarbon province, there is increasing activity with regard to the transfer of assets. The Government is keen to encourage this so as to maximise recovery. It is accordingly willing to relieve a party divesting itself of assets of ongoing liabilities for decommissioning costs through withdrawal of a Section 29 notice. That said, however, this will only happen where the Secretary of State is satisfied that the party acquiring the assets has the ability, technical and financial, to meet the decommissioning responsibilities. Where this is not the case, the parties will be required to enter into a financial security agreement.[110] Furthermore, it is important to realise that the fact that a Section 29 notice is withdrawn does not mean that a further notice cannot be served at some future date.[111]

---

[103] Energy Act 2008, Sch 1, paras 10 and 11.
[104] Energy 2008 Act, s 30.
[105] Section 30(3).
[106] Section 30(4).
[107] See Chapter I-7.
[108] Section 31(3).
[109] Section 31(5).
[110] See Chapter I-13.
[111] Section 31(5).

Nor does it mean that a party who has had their Section 29 notice withdrawn cannot be recalled, as will be seen below.[112]

The Secretary of State may approve or reject an abandonment **I-12.46** plan,[113] or approve it with modifications or conditions.[114] Such modifications or conditions may relate to the requirements of the MER Strategy, for example, collaboration with others or cost reductions. Importantly, there is a specific provision ensuring that any such MER-related modification would not result in increased costs for any Section 29 holder under the plan or indeed any other abandonment plan.[115] They may also, however, require that the duty holder carry out and publish a review of the programme and its implementation, including recommendations for the future (or at least make this information available to the Secretary of State and the OGA).[116] The Act, as amended in 2016, also imposes new correlative obligations on the Secretary of State who must now consult with the OGA before reaching a decision on the programme, specifically taking "into account the cost of carrying out the programme ... and whether it is possible to reduce that cost by modifying the programme or making it subject to conditions".[117] Similarly, when so consulted, the OGA is obliged to consider and advise on alternatives and whether costs could be reduced, and if so how.[118] In the case of a rejection or of a failure to comply with a Section 29 notice, the Secretary of State may prepare an abandonment plan.[119] In so doing, he may call on the recipients of a Section 29 notice to provide records, drawings or other information,[120] on pain of a criminal penalty,[121] and to reimburse the costs of preparing the programme.[122] Where the Secretary of State does prepare a programme under this section, he must consult the OGA as outlined above[123] and the OGA must similarly consider and advise the Secretary of State.[124]

In view of the fact that the circumstances surrounding an instal- **I-12.47** lation may change, whether from a technical perspective or from that of the parties having an interest in it, Section 34 provides for

---

[112] See para I-12.48.
[113] Section 32(1).
[114] Section 32(2).
[115] Section 32(2A)(a).
[116] Section 32(2A)(b).
[117] Section 32(6).
[118] Section 32(7).
[119] Section 33(1).
[120] Section 33(2).
[121] Section 33(3).
[122] Section 33(4).
[123] Section 33(3A).
[124] Section 33(3B).

the possibility that an approved programme may be revised. Thus, either the Secretary of State or the persons who submitted the abandonment programme may propose an alteration to it or to any condition attached to it.[125] With the advent of the MER Strategy, this section is also modified by the 2016 Act to ensure that, firstly, where such alterations are proposed by a Section 29 notice holder, they must ensure that costs are kept to the minimum reasonably practicable[126] and, secondly, where they are proposed by the Secretary of State with a view to reducing costs, they must not increase the costs borne by any individual with obligations under the programme or any other programme.[127] Furthermore, either the Secretary of State or the persons who submitted the abandonment programme may propose that any person who has a duty to carry out the programme may be relieved of that duty or that another person may have the duty imposed upon them.[128] It is this seemingly innocuous provision that raises the possibility (not so far acted upon) that the Secretary of State may recall a party who has had a Section 29 notice withdrawn and who has no current interest in the licence or the installation. Note that persons falling within the categories listed in Section 30(1)(d) and (e) and 30(2)(b) and (c) will not have such a duty imposed upon them unless, in the Secretary of State's view, another person already with that duty has failed or may fail to discharge it.[129] Where the Secretary of State proposes a change to the programme, then the persons having the duty to carry out the programme have an opportunity to make representations.[130] Where a proposal has been made to remove the duty from a person or to give it to another, then all the affected parties have a similar opportunity to make written representations.[131] The decision on any change to an approved abandonment programme lies with the Secretary of State who must give reasons for it.[132] Before making any such determination, however, where any proposal under subsection (1) appears to the Secretary of State to be likely to "have an effect on the cost of carrying out the programme", he must consult with the OGA and take the effect into account,[133] and the OGA must "consider and advise on ... alternatives" and on whether Subsection (4A) applies and, if so, whether it has been

---

[125] Section 34(1)(a).
[126] Section 34(4).
[127] Section 34(4B)
[128] Section 34(1)(b) and (2).
[129] Section 34(3).
[130] Section 34(5).
[131] Section 34(6).
[132] Section 34(7).
[133] Section 34(7A).

complied with – that is, that costs have been kept to the minimum level practicable.[134]

It is also possible that a programme that has been approved by   I-12.48
the Secretary of State may include a provision "by virtue of which the programme may be amended".[135] Where this is the case and a proposed amendment is likely to have an effect on the cost of carrying out the programme, then it must be framed so as to minimise the cost to the lowest reasonably practicable level.[136] In such circumstances, the OGA must be consulted[137] and that body must "consider and advise on … alternatives" and on whether Subsection (2) applies and, if so, whether it has been complied with – that is, that costs have been kept to the minimum level practicable.[138] Any person responsible for approving amendments under this section must take account of their effect on the cost of carrying out the programme.[139]

It is also possible for one or more of those who submitted a   I-12.49
programme to ask that the approval be withdrawn.[140] Where this occurs and not all of the persons initially involved in the submission are now involved, the Secretary of State will notify the others and give them an opportunity to make written representations.[141] The Secretary of State's determination of any such application will be notified to all of those who made the initial submission.[142]

The duty to secure the carrying out of an approved abandonment   I-12.50
programme (as well as compliance with any conditions) is imposed by Section 36 on *each* of the persons who submitted it, meaning that the duty is joint and several. Section 36A, inserted by the Energy Act 2016, allows the Secretary of State to serve a notice on any person who submitted an abandonment programme with a view to reducing the cost of carrying it out whether by requiring or prohibiting action,[143] for example in relation to the timing of action or collaboration with others.[144] This modification to the original duties contained in the 1998 Act is subject to the same protection seen above regarding increases in costs, whether in relation to the programme in question or any other programme.[145] Where the

---

[134] Section 34(7B).
[135] Section 34A(1).
[136] Section 34A(2).
[137] Section 34A(3).
[138] Section 34A(4).
[139] Section 34A(5).
[140] Section 35(1).
[141] Section 35(2).
[142] Section 35(3).
[143] Section 36A(2).
[144] Section 36A(3).
[145] Section 36A(5).

Secretary of State proposes to issue a notice under this section, he must not do so without giving the person concerned "an opportunity to make written representation as to whether the notice should be given".[146] Failure to comply with such a notice is an offence[147] and, in such circumstances, the Secretary of State may take action to carry out the programme and recover costs[148] subject to interest.[149]

I-12.51   Bearing in mind the joint and several liability of the holders of Section 29 notices, if there is any default in relation to the carrying out of the programme, then the Secretary of State may require any of those who submitted the programme to take such remedial action within such time as he may specify in a written notice.[150] In practice, therefore, one party – namely the one with the greatest assets – may be left to bear the full burden of the decommissioning costs and then to attempt to recover the shares due by the others. Failure to comply with such a notice is an offence, unless the party on whom the notice was served can demonstrate that they exercised due diligence to avoid it.[151] Equally, where there is non-compliance, the Secretary of State may carry out the required remedial action and recover the cost from the person on whom the notice was served,[152] with interest running at commercial rates from the date of notification of the sum due until payment.[153] Note, however, that the commercial realities of the UKCS, which see licences divided into separate sub-areas, mean that a party to a licence or to a joint operating agreement (JOA) may have no commercial interest in an installation located in one or another sub-area of the licensed area. This is now recognised in Section 31(A1)–(D1) which prevents the Secretary of State from serving a Section 29 notice on a party who has never derived a commercial benefit from the installation in question.[154]

I-12.52   The Government's concern to ensure that the taxpayer does not end up footing the bill for decommissioning (beyond the amount that is borne indirectly through tax allowances)[155] is reflected in the

---

[146] Section 36A(6).
[147] Section 36A(7).
[148] Section 36A(8).
[149] Section 36A(9) and (10).
[150] Section 37(1). The obligation to consult with the OGA where such remedial action appears likely to have an effect on the cost of carrying out the programme is also inserted into this section by the 2016 Act, as is the duty of the OGA to consider and advise in such circumstances. See Section 37A.
[151] Section 37(2).
[152] Section 37(3).
[153] Section 37(4) and (5).
[154] Inserted by s 72(7) of the Energy Act 2008. There is a concomitant adjustment to s 34 by virtue of s 72(8) of the 2008 Act.
[155] See Chapter I-7.

provisions of Section 38. First of all, after a Section 29 notice has been served and before an abandonment programme has been submitted, the Secretary of State may require a person to provide information relating to their financial affairs together with supporting documentation.[156] Secondly, in order to allow the Secretary of State to confirm that a person is actually capable of fulfilling their obligations under an abandonment programme, he may at any time require such information and documentation as may be specified.[157] Failure to comply with such requests is an offence, as is the knowing or reckless supply of false information.[158] Note also that given the sensitive nature of the information involved, it is also an offence for anyone to disclose such information as has been provided to the Secretary of State in these regards.[159]

The Secretary of State has very widely drawn powers where he I-12.53
is not satisfied that a person will be capable of fulfilling their duties under Section 36 (that is, to carry out an approved abandonment plan and any conditions attaching thereto). Section 38(4) provides that he may require such a person to "take such action as may be specified ... within such time as may be specified". This includes the provision of financial security. Such a notice cannot be served without the person concerned having the opportunity to make written representations,[160] but, once served, failure to comply is an offence.[161] It was previously the case that the Secretary of State could only take such action after an abandonment programme had been approved, but this can now be done as soon as a Section 29 notice has been served.[162]

As may be imagined, the extent of the Secretary of State's powers I-12.54
in these regards has been the subject of some concern on the part of the industry. Given the increasing number of transfers of assets that naturally accompany the maturing of a hydrocarbon province, the possibility that a party who has, with full consent of the Secretary of State, divested itself of its interests in a licence and in an installation may be recalled to implement and bear the costs of an abandonment plan is seen as a potential obstacle to such transfers. The consultation period prior to the passing of the Energy Act saw lobbying on the

---

[156] Section 38(1)–(1B). The scope of this power was extended by s 73(1) and (2) of the 2008 Act which substituted these subsections for the original s 38(1).

[157] Section 38(2). The range of information that may be requested was extended by s 73(3) of the 2008 Act which amplified the wording of the original s 38(2).

[158] Section 38(3).

[159] Section 38(6).

[160] Section 38(5).

[161] Section 38(6).

[162] Section 38(4A) read with s 38(2A) as inserted respectively by s 73(5) and s 73 (4) of the 2008 Act.

part of the industry for the repeal of this aspect of Section 34, but to no avail. The Department reminded the industry about the Ardmore case where the parties developing the field went into liquidation and although "[t]he decommissioning costs were very low at around £5 million … considerable effort has been required to ensure the costs did not fall to the taxpayer".[163] The Government's view has been that the uncertainties associated with decommissioning costs at what is still a relatively early stage of the process are such that it requires to have this protection in place so that the interests of taxpayers may be ensured.[164] In this regard, one further change effected by the Energy Act 2008 requires consideration at this stage, namely the protection from creditors of funds set aside for decommissioning costs. As will be discussed further below, the industry had taken an initiative to establish a model decommissioning security agreement that would facilitate transfers of assets, but a remaining concern was whether such an arrangement would be proof against the claims of creditors in the event of the insolvency of one of the parties to the arrangement.[165] Any lingering uncertainty is removed by the disapplication of any provision of insolvency law which could interfere with the arrangement being used for its intended purpose of meeting decommissioning costs.[166]

I-12.55   The provisions of Part IV of the Petroleum Act considered so far concentrate especially on ensuring that there is clarity about who will be responsible for the preparation and implementation of an abandonment programme and that the financial burden will not fall on the tax payer. As regards the substantive content of an abandonment programme, Section 39 empowers the Secretary of State to make regulations. It is envisaged that these may prescribe standards for dismantling, removal and disposal; standards and safety requirements where there is only partial removal; make provision for the prevention of pollution; and make provision for inspection. The Secretary of State is also empowered to charge fees[167] and to make it an offence to contravene any regulations.[168] The usual requirements are imposed regarding consultation before the making

---

[163] DTI, "Decommissioning Offshore Energy Installations: A Consultation Document", June 2007, p 33.

[164] *Ibid*, p 34.

[165] The joint opinion for UKOOA on this point by Gabriel Moss QC and Mark Arnold concluded that "a trust which functions as a security mechanism may be challenged by a liquidator. However, a carefully drawn trust mechanism should survive such a challenge."

[166] Section 38A inserted by s 74 of the Energy Act 2008. Section 38B ensures that information is provided so that creditors and potential creditors are aware that such funds are protected in the case of insolvency.

[167] Section 39(2)(d) and (e).

[168] Section 39(3) and (4).

of any such regulations.[169] It is noteworthy, however, that no regulations have ever been made. This is undoubtedly because the regulator is of the view that detailed prescriptive regulations would interfere with the case-by-case approach that it was at pains to stress it would take during the events surrounding the decommissioning of the Brent Spar.[170] It is also worth noting that by the time of the 1998 Act, the problems caused by the attempt to produce detailed prescriptive regulations for health and safety at work offshore under the Mineral Workings (Offshore Installations) Act 1971 were fully recognised following the criticism in the Report of the Cullen Inquiry into the Piper Alpha Disaster.[171] Instead of regulations, the Department has produced extensive Guidance Notes, which in their current incarnation extend to over 130 pages.

THE DEPARTMENT'S GUIDANCE NOTES

The Department's Guidance Notes on Decommissioning were first   I-12.56
produced in August 2000 and have been updated since, with the most recent version, the sixth, being issued in March 2011.[172] It is noteworthy that the introduction to the Guidance Notes states explicitly that they have been prepared taking account of "views expressed by operating companies and other interested parties" and that they "provide a framework and are not intended to be prescriptive".[173] The echoes of Brent Spar can also be heard in this introduction where it is stated that the approach adopted will allow "adequate time for full and considered consultation". But it is equally clear that the regulator has the interests of industry in mind, with stress being laid on flexibility (within the constraints of law and policy) and on the avoidance of unnecessary delay.[174]

The influence of the Brent Spar case is evident too in the   I-12.57
Government's overall policy on decommissioning. This seeks "to achieve effective and balanced decommissioning solutions, which

---

[169] Section 39(5) and (6).
[170] See para I-12.23.
[171] See Chapter I-10.
[172] Offshore Decommissioning Unit, DECC, *Guidance Notes: Decommissioning of Offshore Installations and Pipelines under the Petroleum Act 1998*, March 2011, available at www.ogauthority.co.uk/decommissioning/programmes-guidance (accessed 11 May 2017) (hereinafter "Department's Guidance Notes"). Where reference is made below to earlier versions of the Guidance Notes, this is specifically mentioned; otherwise, the 2011 version should be assumed. Note that at the time of writing new Guidance Notes are expected imminently. These are likely to provide further clarity on the interaction of the Decommissioning and MER regimes.
[173] Department's Guidance Notes, p 1.
[174] *Ibid*, p 1.

are consistent with international obligations and have a proper regard for safety, the environment, other legitimate uses of the sea, economic considerations and social considerations".[175] The overall policy is further specified in a list of matters that the regulator will seek to ensure, including that "decommissioning decisions are consistent with waste hierarchy principles[176] and are taken in the light of full and open consultations".[177] It is worth noting, however, that whereas earlier versions of the Department's Guidance Notes stated at the outset that the "Government will act in line with the principles of sustainable development", this is replaced in the current version with a recognition of the need to "maximise energy production as a contribution to UK energy security" on one hand, while taking account of "impacts on climate change"[178] on the other – a subtle, but perhaps not insignificant, shift in emphasis given the concerns discussed in Chapter I-3 above. This change is amplified by the addition of the suggestion that the DECC (now BEIS) will ensure that "decommissioning will be regarded as the last option after re-use of the facilities for energy or other projects has been ruled out" but equally that "comparative assessments of decommissioning options take account of impacts on climate change".[179]

I-12.58   The Guidance Notes then go on to flesh out the implementation of the provisions of Part IV of the 1998 Act (as now amended by the Energy Acts 2008 and 2016) and other relevant legislation[180] in accordance with the UK's international obligations, with particular reference to OSPAR Decision 98/3[181] and the IMO Guidelines and Standards 1989.[182] The following aspects of the Guidance Notes will be discussed in turn below: the treatment of Section 29 notices; and

---

[175] *Ibid*, para 1.1.

[176] The waste hierarchy is discussed further in para 6.2 of the Department's Guidance Notes. It is a framework that prefers waste reduction ahead of any other option, failing which re-use and recycling in that order. Only where none of these options is possible will disposal be considered. Note that this approach was adopted in the Brent Spar Stakeholder Dialogue instituted by Shell. See para I-12.25.

[177] Department's Guidance Notes, para 1.2.

[178] *Ibid*, para 1.1.

[179] *Ibid*, para 1.2.

[180] This includes the Coast Protection Act 1949, the Offshore Installations (Safety Case) Regulations 2005, the Pipeline Safety Regulations 1996, the Food and Environment Protection Act 1985, the Environment Protection Act 1985, the Environment Protection Act 1990 and the Radioactive Substances Act 1993. Note that the oil and gas operations already covered by the Petroleum Act 1998 and the Energy Act 2008 are exempt from the provisions of the Marine and Coastal Access Act 2009 and the Marine (Scotland) Act 2010.

[181] Discussed in detail in s 7 of the Department's Guidance Notes.

[182] Discussed in detail in s 8 of the Department's Guidance Notes.

the decommissioning programme process, including discussions with the Department and other stakeholders.

Treatment of Section 29 notices

The process for the service of Section 29 notices commences when **I-12.59** a field development is approved, with the Department sending the operator a Facility Information Request. This asks the operator to confirm the accuracy of the information held by the Department with regard *inter alia* to the "companies"[183] involved in the field.[184] Thereafter, a so-called "warning letter" will be issued to the operator, the owner and "relevant licensees and JOA partners" advising them that the Secretary of State is considering serving a Section 29 notice and giving each 30 days to make written representations as to why this should not happen.[185] Service of the notice will usually then follow. Prior to the amendments introduced by the Energy Act 2008, the involvement of parties other than the operator and the owner would only have been possible after production had started. As regards pipelines, a similar procedure is followed, normally only in respect of the owner.[186] Whereas it was previously the case that the procedure would only commence once the pipeline is emplaced or production has started, it is now the case that this will happen "when the pipeline works authorisation is given and construction has commenced".[187] A considerable period of time may then elapse before the Secretary of State calls for the submission of an abandonment programme.[188] While the Section 29 notice includes advice on the need to consult interested bodies in preparing an abandonment programme, nearer the point of decommissioning, any party who received such a notice will also receive a list of the organisations they should consult in respect of their programme.[189]

As regards the possibility that a person who has had a Section 29 **I-12.60** notice withdrawn under Section 31(5) may be "called back" under

---

[183] Although this should now be read as "corporate bodies".

[184] Department's Guidance Notes, para 3.3.

[185] *Ibid*, para 3.4. The Department also makes clear that where it has concerns about financial resources or other issues affecting the satisfactory achievement of decommissioning service may also be contemplated in respect of parents or associates in accordance with s 30 of the 1998 Act as amended, as discussed above at para I-12.40.

[186] Although again parents or associates may be included where there are concerns about financial resources or other issues that may have an impact on satisfactory decommissioning.

[187] Department's Guidance Notes, para 3.6. This change is a consequence of the amended definition of "submarine pipeline", discussed at para I-11.45.

[188] Department's Guidance Notes, para 3.8.

[189] *Ibid*, para 3.7. See further para I-12.64.

Section 34, the Department stresses in the guidance that it regards this as a "measure of last resort ... which we endeavour to avoid by the use of prudent security arrangements".[190] While this situation has not so far arisen, the Department gives some indication of how it would act. In particular, it suggests that where more than one company is involved it would "aim to agree a fair and reasonable distribution of the liabilities in discussion with the companies concerned".[191] There is an interesting change in the wording at this point between the latest version of the Department's Guidance Notes and earlier versions. Whereas it used to be provided that a person who *has* had a notice withdrawn under Section 31(5) would not normally be liable for any new installation emplaced in a field after assignment of their interest (although they may be liable for new equipment fitted to an installation in respect of which they had previously held a Section 29 notice),[192] this is now expressed as applying in a situation where a Section 29 notice is *not* withdrawn.[193] At first sight this seems contradictory until it is realised that in each case what is contemplated is the treatment of a party who has assigned their interest in an installation and who has no current financial involvement. The change in wording would thus appear to indicate that withdrawal of a Section 29 notice is less likely than was previously the case. One might want to argue that the change in wording opens up the question of what happens when a Section 29 notice *is* now withdrawn and then the recipient is called back under Section 34, but there appears little doubt that they would be treated in the same way in this respect as if the notice had never been withdrawn. In any case, the Department's Guidance Notes also state that "[i]f a company [for which we should now presumably read "body corporate"] has concerns relating to a specific Section 29 case, they should contact DECC's [now BEIS's] Offshore Decommissioning Unit for further clarification".[194] All of this may give some comfort to the industry as regards the implementation of a provision that sometimes provokes strong criticism.

I-12.61   Some comfort may also be derived from the clarification that the latest Guidance Notes offer in terms of whether a range of parties might find themselves inadvertently and unexpectedly liable for decommissioning costs as a result of the complex contracting and infrastructure arrangements that characterise the UKCS. Thus, the industry has been worried about the position both of contractors offering management services to operators, and of the operator of a

---

[190] *Ibid*, para 3.11.

[191] *Ibid*, para 3.11.

[192] Department's Guidance Notes (September 2006 version), Annex F, para 11.

[193] Department's Guidance Notes, para 3.13.

[194] *Ibid*, para 3.14.

host installation in respect of a tieback. The DECC (now BEIS) has been at pains to offer reassurance in each case,[195] although in respect of the latter it has seen fit to set out some parameters that it considers in "determining when it is reasonable and proportionate to treat tiebacks as separate installations".[196]

Decommissioning programme process

As mentioned above, the regulator characterises the process of approving a decommissioning programme as "flexible, transparent and subject to public consultation"[197] and envisages that it will typically have five main stages: preliminary discussions with BEIS; detailed discussions and submission of a consultation draft programme to BEIS, other interested parties and the public for consideration; formal submission of a programme and approval under the Petroleum Act 1998; commence main works and undertake site surveys; and monitoring of the site.[198] I-12.62

Preliminary discussions with the Department
The regulator stresses the importance of entering into discussions well in advance of the cessation of production, perhaps as much as three years in the case of multi-installation fields and even longer where a derogation under OSPAR Decision 98/3 is a possibility. The wording of Section 29(2) of the 1998 Act, until the changes introduced by the Energy Act 2016, may have given the impression that an operator should wait until called upon by the Secretary of State to submit a decommissioning programme before taking any action, but the guidance makes clear that the "onus rests with the Operator to initiate these discussions".[199] The regulator's expectation at this initial stage is that the operator will outline the "likely timetable of future events to form a basis of agreement on when more detailed discussions should commence and what documentation should be prepared in advance".[200] The Department is keen to ensure that the burden on operators is as light as it can reasonably be and it accordingly will act as far as possible as a "one-stop shop" for the I-12.63

---

[195] *Ibid*, paras 3.15 and 3.16 respectively.
[196] *Ibid*, paras 3.17 to 3.21.
[197] *Ibid*, para 5.2.
[198] *Ibid*, para 5.3. A useful overview of the process, both in straightforward cases and in those where a derogation from the general rule in OSPAR Decision 98/3 is sought, is provided in the form of a flowchart in Annex J to the Department's Guidance Notes. Earlier versions of the Department's Guidance Notes divided the process into six main stages, with what is now Stage two being divided into two separate stages.
[199] Department's Guidance Notes, para 5.8.
[200] *Ibid*, para 5.10.

operator's contacts with government. While it will, thus, liaise with many of the other interested departments, there will be occasions when the operator must do this itself – perhaps most notably the OGA in relation to the MER Strategy[201] and the Health and Safety Executive in respect of the modification of the safety case to take account of decommissioning.[202]

I-12.64   The expected content of the decommissioning programme is outlined in Section 6 and Annex C to the Guidance Notes. Templates for both non-derogation and derogation cases are now available on the OGA website, meaning that the process of developing the programme is as streamlined and standardised as possible, whilst recognising that the case-by-case approach is still adhered to.[203]

Detailed discussions and submission of consultation draft programme to BEIS, other interested parties and the public for consideration

I-12.65   In straightforward cases, it is envisaged that there will be little distinction between the first and second stages of the Decommissioning Programme process. Where a derogation from the general rule in OSPAR Decision 98/3 is sought, however, then the Department will be required at this stage to determine whether a case has been made out in accordance with the procedure specified in Annex 2 to the Decision. In this regard, the regulator encourages operators to use the criteria and methodology it has itself developed and which are contained in Annex A to the Guidance Notes. This essentially involves the completion of a matrix, which allows comparison of the safety and technical risks and the environmental, societal and economic impacts associated with different decommissioning options. While the matrix approach is not compulsory, the Department notes that its use "will help to provide a clear overall indication of the acceptability of the derogation case".[204]

I-12.66   While under previous versions of the Department's Guidance Notes the Decommissioning Programme Process was divided into six rather than five main stages – with detailed discussion with the Department being placed prior to consultation with other interested parties and the public – it was always the case that such consultation could actually be required in parallel with discussions with the Department. Thus, even when the two stages were separated, stakeholder engagement was actually required earlier in a derogation

---

[201] See Chapter I-5.
[202] See Chapter I-10.
[203] The templates can be found at www.ogauthority.co.uk/decommissioning/programmes-guidance (accessed 17 May 2017).
[204] Department's Guidance Notes, Annex A, para 6.

case: the Notes accompanying Annex A in the February 2008 version of the Department's Guidance Notes already indicated that the completion of the matrix in respect of societal impacts would depend upon stakeholder engagement.[205] The merging of these two stages in the current version of the Guidance Notes means, however, that consultation will also be required in parallel with discussions with the Department in straightforward cases. At one level, these will be the statutory consultations with parties notified to the recipient of a Section 29 notice, as provided for by Section 29(3). The Guidance Notes contain more information in this regard insofar as Annex H lists those parties who are normally included in a notification.[206] Beyond that, the guidance indicates that operators will also be asked to announce proposals publicly and to make copies of the draft programme available for consultation. The results of such consultation should be discussed in the next draft of the programme.[207] In derogation cases, "operators will need to develop and manage a wide-ranging public consultation process", the "form and timing" of which is to be discussed with the regulator.[208] Interestingly, the wording of the guidance suggests that only where such a consultation indicates that a derogation should be sought will the Department then consult with the other OSPAR contracting parties.[209] This may be another indication that a derogation will not be sought where there is strong public feeling against the idea, unless, of course, there is some overriding consideration – safety of personnel, for example – which renders it necessary. Note also that the guidance places a responsibility on the operator to prepare documentation supporting the derogation case, even though it is the Department's responsibility to carry out the consultation with the other OSPAR contracting parties.[210]

It will also be necessary to carry out an Environmental Impact   **I-12.67** Assessment (EIA) and prepare an Environmental Statement (ES) at this stage, especially in a derogation case, in order to support the completion of the environmental impact aspects of the matrix in Annex A to the Department's Guidance Notes, but also in straightforward cases where the EIA will be restricted to addressing "the impacts of the proposed decommissioning activity on the

---

[205] In this regard, the Department directs operators to the Oil & Gas UK Guidelines on Stakeholder Engagement for Decommissioning Activities.

[206] Namely, the various fishing organisations as well as Global Marine Systems, which provides submarine cable laying and maintenance.

[207] Department's Guidance Notes, paras 6.25 and 6.26.

[208] *Ibid*, para 6.27.

[209] *Ibid*, para 6.28.

[210] *Ibid*, para 6.29.

environment".[211] The EIA and ES are creatures of European law,[212] and are required in the case of certain public and private projects. Certain oil and gas projects are now subject to the mandatory preparation of an ES,[213] but this is *not* the case with the decommissioning of installations and pipelines. Notwithstanding the absence of any legal requirement, the Guidance Notes make clear that "a decommissioning programme will nevertheless need to be supported by an EIA".[214] An Environmental Impact Assessment will thus be required in respect of *all* decommissioning programmes.[215]

I-12.68   Finally, note that in derogation cases, while reputational issues are not to be included in the matrix in Annex A, the Guidance Notes make clear that these may well influence the final decision on whether a derogation will be permitted. Implicit in the wording here is the fact that the Department is thinking not only about the reputation of the operator but also about that of the Government.[216] It may thus be suggested that were there ever to be public antipathy to a decommissioning programme on the scale witnessed in the Brent Spar case, the Department might not be minded to argue on the basis of scientific rationality in the way that it did then, absent some overriding consideration such as safety.

Formal submission of a programme and approval under the Petroleum Act 1998

I-12.69   The extent to which the decommissioning process goes forward on the basis of guidance rather than statutory or regulatory requirements is evident from the fact that it is only at this third stage, when the operator and the Department are in a position to agree a final version of the decommissioning programme, that the Secretary of State will call for its submission in terms of Section 29(1) of the 1998 Act.

Commencement of main works and conduct of site surveys

I-12.70   Stage four of the decommissioning process covers the implementation of the physical removal work together with the notification to

---

[211] *Ibid*, para 12.1.

[212] Council Directive 85/337/EEC on the Assessment of the Effects of Certain Public and Private Projects on the Environment as amended by Council Directive 97/11/EC (see Chapter I-4 and Chapter I-11).

[213] The directives are transposed in the UK in respect of these projects by the Offshore Petroleum Production and Pipe-lines (Assessment of Environmental Effects) Regulations 1999 (SI 1999/360) as amended by the Offshore Petroleum Production and Pipe-lines (Assessment of Environmental Effects) (Amendment) Regulations 2007 (SI 2007/933).

[214] Department's Guidance Notes, para 12.1.

[215] *Ibid*, Annex C, para 10.

[216] "Companies and government will also wish to take account of reputational issues from their own perspective".

the Department of the progress of that work. The means by which this notification will take place as well as the milestones for review of progress will be specified in the decommissioning programme.[217] Any variations of the programme require the Secretary of State's approval under Section 34.[218] When the work is completed (including the clearing of any debris and the carrying out of any required surveys of the site), a Close-out Report must be submitted to the Department within four months. The content of the Report is detailed in Section 13 of the Guidance Notes and includes: the outcome of the process and how it was achieved; details of variations and consequent permits required; results of clearance operations and monitoring; results of post-decommissioning sampling; details of any future monitoring; measures taken to manage risks from any legacies; details of actual costs and explanations of differences from forecasts.[219] Note that the operator will be asked to place a copy of the Close-out Report on its website.

Monitoring of the site
Where a derogation from the general position specified by OSPAR   I-12.71
Decision 98/3 has been agreed and all or part of an installation remains in place, then the operator will be required to undertake ongoing monitoring of the site in order that the condition of the remains may be kept under review.[220] The precise monitoring regime will be agreed with the Department and the schedule will be included in the decommissioning programme.[221] Reports on the monitoring are to be submitted to the Department along with any proposals for remedial work. The operator must publish these monitoring reports (for example, on the internet), but there is ambiguity in the Guidance as to whether this requirement extends to proposals for remedial work. It would be safest to assume that it does.[222]

As discussed above,[223] Annex 4 of OSPAR Decision 98/3 requires   I-12.72
that there be independent verification of the condition of the installation before work on the disposal begins. The Guidance indicates that "it will be for the Operator to propose a suitable organisation to carry out the independent verification".[224] The unspoken element here is that the Department will need to approve this proposal.

---

[217] Department's Guidance Notes, para 5.15.
[218] As discussed above at para I-12.42.
[219] Department's Guidance Notes, para 13.1.
[220] *Ibid*, para 5.17.
[221] *Ibid*, para 14.1.
[222] *Ibid*, para 14.2.
[223] See para I-12.36 above.
[224] Department's Guidance Notes, para 14.3.

I-12.73   As regards the post-decommissioning report required by para 10 of Decision 98/3, the Guidance indicates that the Department will prepare this on the basis of the Close-out Report and will give the operator the opportunity to review it before submission to the OSPAR Commission.[225]

I-12.74   Where all or part of an installation is to remain in place, it is the operator's responsibility to ensure that the UK Hydrographic Office is informed so that navigational charts may be updated accordingly.[226] Where there is projection above the surface, the operator will also be responsible for the installation and maintenance of the appropriate navigational aids, in consultation with the relevant interested parties.[227]

THE ROLE OF THE OIL AND GAS AUTHORITY

I-12.75   As was mentioned above, and as is evident from the references to consultation with the OGA inserted in the Petroleum Act 1998 by the Energy Act 2016, the Guidance Notes just considered will now need to be read in conjunction with that body's responsibilities under the MER Strategy, in general, and in relation to its Decommissioning Strategy, in particular. At the time of writing, new Departmental Guidance is imminently expected, which should provide clear indications of how Section 29 notice holders should manage their interactions with both BEIS and the OGA. In the meantime, it is sufficient to note the key features of the latter's Decommissioning Strategy.[228] This document outlines three priorities: cost certainty and reduction; decommissioning delivery capability; and decommissioning scope, guidance and stakeholder engagement.[229] Of these, the first and third are most likely to impact the issues considered earlier in this chapter. In relation to cost certainty and reduction, the fact that this is further defined as "[d]riving targeted cost efficiency programmes including innovative and regional approaches with extensive and effective knowledge sharing and best practice adoption"[230] indicates what the references to "collaboration" in the 2016 amendments to the 1998 Act might mean in practice: decommissioning in some cases will no longer be a matter for an individual operator to determine in a one-to-one conversation with BEIS, but via the intervention of the OGA, may well require timings and

---

[225] *Ibid*, para 14.4.
[226] *Ibid*, para 15.1.
[227] *Ibid*, para 15.4.
[228] Oil and Gas Authority, Decommissioning Strategy, 2016.
[229] *Ibid*, p 6.
[230] *Ibid*, p 6.

methodologies to be discussed and determined with other operators in the vicinity, whether directly connected to the infrastructure in question or not. In relation to decommissioning scope, guidance and stakeholder engagement, the OGA recognises that:

> "[t]he decommissioning stakeholders are very diverse, reflecting numerous groups and interests. Managing interfaces between all stakeholders will be complex and demanding, particularly in the Non-Governmental Organisation (NGO) area. The OGA will aim to simplify and align positions, and will work very closely with [BEIS]".[231]

Given the issues identified in the introduction to this chapter **I-12.76** regarding the current controversy surrounding Shell's Brent decommissioning programme, specifically in relation to the derogations sought under OSPAR Decision 98/3, it looks as if the OGA will have an early opportunity to test its abilities in this regard.

RESIDUAL LIABILITIES

One of the issues exercising the minds of those concerned with **I-12.77** decommissioning is the question of residual liability for an installation left wholly or partly in place under a derogation from the general position described by OSPAR Decision 98/3. In case anyone should be under any illusions in this regard, the Department's Guidance Notes discussed in the foregoing Part of this chapter inform readers that "[a]ny residual liability remains with the owners in perpetuity"[232] and that "[a]ny claims for compensation from third parties arising from damage caused by any remains will be a matter for the owners and the affected parties and will be covered by the general law".[233] In this regard, the position under both English and Scots law is that the owner of such an installation would be liable in damages for loss arising from his negligence in circumstances where a duty of care is owed to the other party. The test for the existence of a duty of care is laid down in *Caparo Industries Ltd. v Dickman*.[234] There Lord Bridge of Harwich stated that the test is threefold: is there sufficient proximity between the parties? Was the

---

[231] *Ibid*, p 6.
[232] Department's Guidance Notes, para 16.1.
[233] *Ibid*, para 16.3. Note that in terms of a UK seaward area petroleum licence, well casings and fixtures relating to wells that are not plugged and abandoned at the expiry or determination of a licensee's interest in a licence "shall be left in left in good order and fit for further working" and "shall be the property of the Minister". See The Petroleum Licensing (Production) (Seaward Areas) Regulations 2008 (SI 2008/225), Model Clauses 12(b) and 14.
[234] [1990] 2 AC 605.

loss foreseeable? Is it fair, just and reasonable in the circumstances to impose the duty of care?[235] It may be said that in all of the most likely scenarios involving loss arising from the remains of a decommissioned installation, this test would be passed. For example, imagine a fishing boat snagging its nets on the remains and sinking. There would certainly be sufficient proximity between the owners of the remains and a legitimate user of the seas above them. The scenario suggested is entirely foreseeable. It does not appear unfair, unjust or unreasonable to impose a duty of care on the owners of remains, which by their very nature constitute a hazard to other users.

I-12.78   Insofar as this is accepted to be the case, the question which then arises is how the owner may discharge his duty of care and thus avoid liability in the event that damage occurs in such circumstances. It may be suggested that insofar as the owner has fulfilled all of his obligations under Part IV of the Petroleum Act 1998, had his Close-out Report accepted by the Department, notified the UK Hydrographic Office of the situation and conformed to the schedule of ongoing monitoring agreed in the decommissioning programme, then it is unlikely that any liability would arise. It is also the case that in derogation situations where there continues to be a projection above the surface, the 500m safety zone will continue to exist around the remains.[236] Insofar as anyone entering that area would be guilty of an offence,[237] then should they incur losses there would undoubtedly be a finding of contributory negligence.

I-12.79   The prudent owner, accordingly, probably has little to fear with regard to residual liability – at least as far as the Scottish and English courts are concerned. It has been pointed out, however, that a problem may arise where a claim is raised in a foreign court "claiming extra-territorial jurisdiction and applying principles of strict liability".[238] This issue, together with the sheer uncertainty involved in retaining residual liability for decades or longer, persuade some in the industry that an approach similar to that adopted in the Norwegian sector of the North Sea would be preferable. Section 5-4 of the Norwegian Petroleum Activities Act of 1996 as amended states that:

> "In the event of decisions for abandonment, it may be agreed between the licensees and the owners on one side and the State on the other

---

[235] At 617–618.

[236] See ss 21 and 22 of the Petroleum Act 1987. See also Department's Guidance Notes, paras 15.5 and 15.6.

[237] Petroleum Act 1987, s 23.

[238] Daintith, Willoughby and Hill, para 1-1326.

side that future maintenance, responsibility and liability shall be taken over by the State based on an agreed financial compensation."[239]

There are presently no indications that an exactly similar approach is on the cards in the UK, and it is in any case perhaps not insignificant that the wording of the Norwegian provision reads "*may* be agreed" rather than "*shall* be agreed".   I-12.80

This is not only an issue that concerns those who bear the residual liability, however. Other users of the waters above the UKCS face potential difficulties if they suffer harm as a result of negligence (however remote a possibility this may be) but the owners of the installation in question are then defunct. This is one of the drivers behind work undertaken within the PILOT Brownfields initiative to develop a Fisheries Legacy Trust Company.[240]   I-12.81

The foregoing discussion proceeds, however, without any specification of who the owners of the infrastructure in question actually are. In this regard, the Guidance Notes state simply that "[t]he persons who own an installation or pipeline at the time of its decommissioning will remain the owner of any residues".[241] The unspoken assumption is that the owners will be the operator and its co-venturers at the time of decommissioning. Insofar, however, as we are discussing infrastructure that is impossible to remove from the seabed, it is surely necessary to ask who the owner of the seabed is. Whether one were to apply Scots or English property law, the residues would be treated as fixtures that would be owned by the owner of the seabed. The first consideration that presents itself, of course, is whether property law is relevant on the UKCS in the way that it would be onshore or even in the territorial sea. Insofar as a distinction is drawn in UNCLOS 1982 between the rights of the coastal state in the territorial sea (sovereignty) and the continental shelf (sovereign rights) it might be suggested that it is not appropriate to speak in terms of anyone, including the coastal state, "owning" the seabed on the latter. It is certainly the case that all of the hydrocarbon operations on the UKCS have proceeded on the basis of a licence (which, as is shown elsewhere in this work, is a hybrid contractual/regulatory instrument) rather than a lease (which would imply the exercise of a property right). But even if that is true, it is interesting to note that other operations on the continental shelf, specifically the siting of renewable energy infrastructure and carbon sequestration activities, require not only a licence from BEIS but also   I-12.82

---

[239] Available in English on the Norwegian Petroleum Directorate's website at www.npd.no/en/Regulations/Acts/Petroleum-activities-act/#Section 5-4 (accessed 11 May 2017).
[240] Department's Guidance Notes, para 16.4. See also www.ukfltc.com
[241] Department's Guidance Notes, para 16.1.

a lease from the Crown Estate.[242] Insofar as this latter requirement does indeed indicate that a property right in the continental shelf is being exercised, the question surely arises as to how the Crown Estate could enjoy the benefits of ownership in relation to renewable energy and carbon sequestration operations whilst simultaneously avoiding the liabilities associated with ownership in relation to hydrocarbon infrastructure. The arrangements allowing transfer of the Crown Estate's Scottish functions to Scottish Ministers under the Scotland Act 2016[243] only adds to the interest of this intriguing question.

CONCLUSION

I-12.83 The approach to decommissioning on the part of the Department and of the industry has changed considerably in the two decades since the Brent Spar case. Previous editions of this work have commented upon the low public profile of this issue in the succeeding years and suggested that this may be said to be a measure of the success of the current way of doing things. At the time of writing, the questions surrounding the Brent decommissioning programme raised by an influential group of NGOs and the forthcoming five-yearly review of the scope of the derogations permitted under OSPAR Decision 98/3 raise the possibility that the profile of decommissioning may be about to be raised and that a degree of contention that has been absent may be about to return. In many respects, the legal and regulatory arrangements discussed above, which have been developed in such a way as to enhance the transparency of decision-making, should be robust enough to cope with any forthcoming challenges. What may prove to be a sterner test for current arrangements, however, may be a realisation that what has so far not really been addressed in any meaningful way is what environmental protection actually means in the context of decommissioning. Is it a return to the status quo ante irrespective of all other considerations? Or is it a more open-minded consideration of what the environmental consequences of any decision will be, even if that means that one possible outcome is that infrastructure is left in place whether or not it fits a current derogation case? That this question can even be asked indicates how far things have changed in the recent past. The coming years will be of considerable interest for all with a stake in decommissioning.

---

[242] Under the Energy Act 2004, s 84, and the Energy Act 2008, s 1, respectively.
[243] Section 36, amending the Scotland Act 1998.

CHAPTER I-13

DECOMMISSIONING SECURITY

Judith Aldersey-Williams

INTRODUCTION

The question of the scale of the decommissioning challenge facing **I-13.01** the United Kingdom Continental Shelf (UKCS) has been increasingly in the headlines in recent years, whether in the context of the cost to the national purse in tax relief, the opportunities for a new business stream for the supply chain or simply to marvel at the engineering undertaking it involves.[1] Although estimates vary, it is likely that over the next decade alone, well in excess of £17 billion will be spent on decommissioning,[2] while the total bill may exceed £47 billion.[3] The cost of plugging and abandoning a single development well can exceed £10 million, so individual decommissioning projects can have costs running into hundreds of millions of pounds.[4] Given that, by definition, this expenditure occurs when the owner of the installation is no longer generating revenue from it, the question of potential default inevitably arises.

Owners of oil and gas assets may make provision in their accounts **I-13.02** for the estimated costs of decommissioning, but there is currently no regulatory requirement in the United Kingdom to set aside cash

---

[1] See eg *Financial Times*, "North Sea oil: the £30bn break-up", 8 June 2016; T Baxter, "Myths about North Sea decommissioning", available at www.scotsman.com/news/opinion/tom-baxter-myths-about-north-sea-decommissioning-1-4384732 (accessed 30 March 2017).

[2] Oil & Gas UK Decommissioning Insight 2016, available for download at http://oiland-gasuk.co.uk/decommissioninginsight.cfm (accessed 30 March 2017).

[3] Oil & Gas Authority Decommissioning Strategy, Chapter 4, available for download at www.ogauthority.co.uk/news-publications/publications/2016/decommissioning-strategy (accessed 30 March 2017).

[4] See note 2 above.

for decommissioning. Moreover, the tax regime actively discourages parties from setting aside funds for decommissioning, by requiring physical decommissioning to have occurred before tax relief on expenditure is granted.[5] The assumption is that owners will pay the costs of decommissioning out of cash flow from other producing assets, but the history of insolvency of oil and gas companies on the UKCS, particularly following the oil price collapse in 2014, demonstrates the inherent risk in this assumption.[6]

I-13.03   These events have made parties significantly more cautious regarding decommissioning costs, such that it is now standard procedure for parties at risk of becoming liable for decommissioning costs, on the default of another party, to ask for decommissioning security to be provided under a decommissioning security agreement (DSA).

WHICH PARTIES MAY SEEK SECURITY?

I-13.04   Chapter I-12 describes the wide powers of the Secretary of State under the Petroleum Act 1998 (hereinafter "the Act") to serve and withdraw Section 29 notices in relation to the decommissioning of installations and pipelines, and to impose liability for decommissioning on parties who were previously or could have been served. In practice, those notices are served primarily upon the operator and its co-venturers with a beneficial interest in the installation or pipeline. Those parties are then jointly and severally liable to carry out the approved decommissioning programme. Given such joint and several liability, it is not surprising that the parties to a joint operating agreement (JOA) are among those most interested in obtaining decommissioning security.

I-13.05   Parties to new JOAs in the UKCS now routinely agree to enter into a DSA if they develop a field on the relevant licence area.[7] However, while having a DSA in place is taken into consideration by BEIS[8] when considering matters such as the withdrawal of

---

[5] See ss 162–165 of the Capital Allowances Act 2001 (for RFCT/SCT), and s 3(1)(i)(hh)–(j) Oil Taxation Act 1975 (for PRT).

[6] For example, Acorn Oil & Gas and Tuscan Energy became insolvent following the failed development of the Ardmore Field in 2005 and Oilexco North Sea Ltd went into administration in 2009. More recently Iona Energy Company (UK) Plc, Iona UK Huntington Limited and First Oil plc have gone into administration while Noreco Oil UK is understood to be subject to a forfeiture procedure in relation to its interest in the Huntington Field.

[7] For instance the Oil & Gas UK Model Form JOA (January 2009) assumes that the parties will enter into a DSA in the form of the Oil & Gas UK Model Form DSA (March 2009).

[8] The Department for Business, Energy and Industrial Strategy.

Section 29 notices,[9] there is currently no regulatory requirement to have a DSA. This is therefore entirely up to the parties to the JOA to agree, and many older JOAs contain minimal or no provision for decommissioning security. If parties are required to provide security, the relevant clauses in older JOAs may amount to no more than an agreement to agree, without the detailed assumptions and operational provisions which reflect current best practice. Moreover, even where such provisions exist, parties have been known simply to ignore contractual obligations to put in place decommissioning security arrangements by a certain point, for instance before seeking consent to a development, and if all parties are keen to avoid the cost of providing security, no-one will challenge this failure. Similarly, where the JOA contains no obligation to enter into a DSA, a single party who is reluctant to incur the costs of security can often effectively veto any change.

The other persons with a significant interest in protection against decommissioning liability are sellers of assets. As noted at para I-12.48, the Secretary of State may decline to release a seller from Section 29 notices based on concerns regarding the financial capability of the incoming purchaser;[10] or, following such a release, may under Section 34 of the Act, recall the seller (or an associated company) to prepare a decommissioning programme and carry it out, if the Secretary of State is concerned about the ability of those with Section 29 notices to do so.[11] Faced with this ongoing liability, either actual under Section 29 or contingent under Section 34, sellers now routinely seek security from buyers against the risk that they may default in carrying out their decommissioning obligations, in addition to indemnities against such liability in the sale and purchase agreement.[12] I-13.06

The wording of Section 30 is wide enough to impose liability for the decommissioning of an installation on any licensee of the licence area within which it falls. However, Section 31A of the Act provides that a licensee may not be served with a notice under Section 29 in respect of an installation, purely on the basis of being a licensee or a JOA party if it has never been entitled to derive any "financial or other benefit" from the use of that installation. While the question of what constitutes a financial or other benefit is not specified in the Act, BEIS in its Guidelines states that "the intention is to capture benefits which are the substantive equivalent of an ownership or equity interest in the field and installation e.g. by I-13.07

---

[9] See para I-12.46 *et seq.*
[10] *Ibid.*
[11] See para I-10.48.
[12] See para II-9.72.

receiving production or payments, royalties or bonuses in lieu of production".[13] Consequently, in practice, security is not currently sought by other licensees who are not and never have been beneficially interested in the field in question, even if, for example, such licensees receive benefits in the form of transportation or processing services or fuel gas from the installation in question, or conversely receive tariffs for the provision of such services to the installation in question.

I-13.08   Section 30(1)(a) of the Act provides that Section 29 notices may be served on "the person having the management of the installation or of its main structure". The BEIS Guidelines[14] state that "the wording of Section 30(1)(a) indicates only one person can manage the installation and our interpretation is that the operator approved by the Secretary of State under the Petroleum Act licence would be the manager. We do not treat contractors providing a service to the operator as a manager within Section 30(1) (a) of the Act." Consequently, while "total facilities contractors" or providers of operating services from another "host" installation may seek an indemnity in their management contract from the relevant field owners in respect of any liability for the decommissioning of that infrastructure, it is not current practice in the UKCS for such persons to request security for such liability. However, it should be noted that the BEIS Guidance is subject to amendment at any time or to being overruled by a court decision on the interpretation of the legislation. Therefore, a risk remains that such a person could still be found liable for decommissioning in future.[15]

I-13.09   Section 29 notices are a relatively blunt instrument in that they can relate either to an installation or a pipeline, but not to a specific part of it. For instance, those who "own an interest in the installation" may be served with a Section 29 notice under Section 30(1)(d) of the Act. The owners of drilling rigs which are used for production purposes or Floating Production, Storage and Offloading units (FPSOs) are routinely served with Section 29 notices which do not distinguish between responsibility for disconnecting and sailing away the relevant vessel and responsibility for decommissioning the wells, subsea manifolds, risers and flowlines to which it is connected.

---

[13] BEIS Guidance Notes on Decommissioning of Offshore Oil and Gas Installations and Pipelines under the Petroleum Act 1998, Version 6 (March 2011) para 3.23, available at www.gov.uk/guidance/oil-and-gas-decommissioning-of-offshore-installations-and-pipelines (accessed 20 March 2017). Note that revised guidelines are expected in early 2017.
[14] Ibid, para 3.15.
[15] No such case has occurred to date in the UKCS but if there were to be a major failure of licensees, it is to be expected that the authorities would look at all possible means to insulate the taxpayer from further liability.

The contracts for the lease of such vessels should therefore address liability for decommissioning in detail and, where appropriate, require each party to give security for its share of such liabilities. In some cases, a field group may own equipment on a host platform, perhaps because this is needed solely to process the field group's production and the host wishes the field group to take responsibility for decommissioning such equipment. BEIS will not normally serve the field group with a Section 29 notice in relation to the host installation on the basis of such ownership, and in these cases practice is for the field group and the host group to rely on an indemnity from the other in relation to their respective decommissioning liabilities under the contract.

It is perhaps surprising that the party least likely to require security for decommissioning is the state. The UK Government has obligations under international law[16] to ensure that installations are decommissioned, and, therefore, might seem to have a strong interest in ensuring that the costs of such decommissioning are borne by the appropriate parties and do not end up with the taxpayer. However, although Section 38 of the Petroleum Act 1998 gives BEIS the right to seek security in appropriate cases,[17] this power is not widely used. There are understood to be around a dozen such securities in existence. It appears that BEIS has, to date, tended to rely on the security of having at least one substantial oil and gas company subject to a Section 29 notice in respect of a field, with others subject to Section 34 contingent liability. If all of the licensees on a field are less substantial entities, BEIS may choose to serve a Section 29 notice on a parent with greater resources rather than look to security. However, as noted above, the fact that contractual security is in place within a joint venture is something taken into account when assessing the need for further action.

I-13.10

It is apparent from the above analysis that a party acquiring an interest in an existing field may find itself subject to requests for security from the seller, the other participants in the field and potentially BEIS. In theory, it could find itself giving security to each of them for the same risk. The Oil & Gas UK Model Form DSA provides the option for both seller and BEIS to be a party to avoid

I-13.11

---

[16] See Chapter I-12.

[17] If the Secretary is not satisfied that a person will be capable of discharging their duty to carry out a decommissioning programme, he may by notice require that person to take specified action – eg providing security. It is notable that until amendments made by the Energy Act 2008, this right arose only after approval of a decommissioning programme which might be very late in the life of a field and gave rise to problems where a field failed to perform as expected and failed shortly after development (viz., the Ardmore development). Now BEIS can ask for security at any point in the life of a development.

the need for double (or even treble) security. However, BEIS will only be a party in the rare instance that it expressly requires security. In such a case it will be necessary to select from among the options given in the model those which meet BEIS's requirements in respect of security as set out in Annex G of its Guidance.[18] In particular, BEIS does not accept parent company guarantees as a form of security,[19] so any party from which BEIS requires security must provide this by means of cash, letter of credit or on-demand bond, although it may be possible, with suitable drafting, to allow other parties the option to provide PCGs.

DECOMMISSIONING SECURITY AGREEMENTS

I-13.12 From the analysis set out above, it can be seen that DSAs may be entered into between co-venturers pursuant to, or in connection with, a JOA; between seller and purchaser as part of, or annexed to, a sale and purchase agreement; or, occasionally, between a purchaser and BEIS.

I-13.13 Whatever the context, the basic form of a DSA is broadly similar. Security is given based on an annual estimate of decommissioning costs prepared by the operator under a field DSA or by the provider of security in other cases. The estimated decommissioning costs are generally multiplied by a contingency and then discounted to a net present value. In the case of field DSAs, it is usual then to deduct the net present value of the remaining reserves from the cost estimate, on the basis that if any licensee defaults, its co-venturers will be able to claim its production, or ultimately forfeit its interest, and use the proceeds to pay for its share of decommissioning. The reserves value is not always deducted in the case of M & A transactions, since the seller is not able to benefit from any remaining reserves in a case of default, but the seller may be persuaded to do so depending on the bargaining power of the parties. Security is then given for the security giver's participating interest share of the resulting figure. (In the case of a field-wide DSA, even though each co-venturer could potentially be liable for 100 per cent of the costs, it is sufficient for them each to provide for their individual share of the costs. In M & A transactions, it is impracticable to expect a purchaser to provide security for more than this share, even though the seller could potentially be liable for 100 per cent of the cost.)

I-13.14 The contingency or risk factor applied to the decommissioning cost estimate reflects the fact that such estimates are notoriously

---

[18] See note 13 above.
[19] See paras 15–18 of Annex G of Guidance (see note 13 above).

inexact, and that the overall calculation contains a number of assumptions which may or may not be correct, such as assumptions as to the value reserves will generate, the costs of extracting those reserves and the appropriate discount rate. This contingency can vary but has historically been up to 50 per cent on top of the estimated costs. However, this element is one of the key commercial negotiating points in a DSA and in a buyer's market will tend to be reduced.

The effect of deducting reserves in the calculation is that security I-13.15 will not be due until the point in the field's life at which the value of remaining reserves falls below the estimated decommissioning costs inflated by the risk factor. This point is usually referred to as the trigger date. If reserves are not deducted then security may be required immediately and the trigger date is effectively the date of the transaction.

In a field DSA, the estimate prepared by the operator must be I-13.16 approved by the co-venturers who may refer the matter to an expert if agreement cannot be reached. Similar expert resolution may be included in M & A transactions, while BEIS may simply require each estimate to be automatically referred to an expert for independent review. Expert reviews at least three yearly, and possibly annually, are a requirement if BEIS is a party to the DSA. Regular independent expert reviews may also be included in field DSAs as a method of keeping the operator honest without an individual party having to make a reference and bear the risk of costs.

The calculation is reviewed annually, and the security renewed in I-13.17 the case of LOCs or PCGs with a fixed value, to take account of: changes in the value of the reserves, as they are produced and as commodity prices change, and in the estimate of decommissioning costs, as plans for decommissioning are firmed up and as contractor rates change. There may also be a requirement for a review and replacement during the year if there is a significant shift in the estimate of costs or, more likely, reserves.

Security is most often given in the form of a parent or affiliate I-13.18 guarantee, or on-demand bond from a parent or affiliate, or a letter of credit (LOC) from a bank. Cash is sometimes included as an option, but rarely used in practice; although, occasionally, a company may agree to make payments into trust from its cash flows to build up a decommissioning fund, where for some reason the other options are unavailable or unattractive – this may be an option used, for instance, by infrastructure funds buying oil and gas assets. Bonds from insurance companies are increasingly being used as they appear to be competitive with letters of credit. The bank or insurance company providing a LOC/bond, or any parent or affiliate providing a guarantee or bond, must meet certain standards stipulated in the agreement, including as to credit rating. As fewer banks have the required ratings, insurance companies have stepped in to the market.

I-13.19   Banks and insurance companies will charge fees for providing security instruments and may also ask for collateral security from the person required to give security, depending on its financial strength. Where the person required to give security is unable to provide collateral, the amount of security may be deducted from that company's borrowing base. The effect is that the provision of decommissioning security may significantly reduce the ability of purchasers, particularly the smaller new investors the Government seeks to encourage in the UKCS, to invest in their newly acquired assets in order to maximise late-life production, and this, in turn, may render the transaction uneconomic. Decommissioning security has therefore been viewed as one of the major barriers to successful asset transfers in the UKCS in recent years.

I-13.20   From the trigger date, security may be provided for a number of years while the field continues to produce. Each year, as the reserves fall, the amount of the security will tend to increase until, by the time of cessation of production, it should equal the anticipated cost of decommissioning plus any agreed contingency. The structure of the DSA ensures that each year the new security is provided in advance of the old security expiring, so that, if it fails to be provided in due time, there is still time for the old security to be enforced. Provided the company which is giving the security remains solvent and carries out its decommissioning obligation, then the security is never enforced. When decommissioning is complete, the obligation to provide security will cease and the final LOC, bond or guarantee will simply expire and not be replaced.

I-13.21   However, if a company which has given security is in default then its security may be enforced. A default could include insolvency; a failure to meet its obligations under the DSA, such as the obligation to replace an expiring security; a failure to pay for decommissioning; or the issuer of a security ceasing to meet the requirements of the DSA in terms of credit rating and no replacement security being provided within the stipulated grace period. If security is called, the bank or affiliate in question will be required to pay the secured amount, usually to a trustee on behalf of the beneficiaries of the security. In this situation, the funds are held by the trustee separate from the assets of the defaulting licensee and invested in a low-risk manner (usually either in cash or gilt-edged securities) until the time of decommissioning. The discount rate used in the calculation is intended to ensure that in such a case the funds held by the trustee grow over the remaining life of the field, such that they are sufficient to meet the estimated costs of decommissioning at the time it is expected to occur. The monies could be held for a considerable period if the default occurs early in the life of a field.

A fully termed DSA may provide for the remedy of a default, so I-13.22 that, for instance, if a party which has failed to renew its security provides an appropriate security instrument, the cash proceeds of the earlier call on its security may be returned to it. If this does not occur, then the party in default may be permitted to use any cash in its trust fund to meet its share of decommissioning obligations when the time comes, or may be required to meet those obligations out of cash flow but then be reimbursed out of the trust fund. Once decommissioning is complete, any funds left in the trust will be returned to it. If the defaulting party does not recover sufficiently to meet its decommissioning obligations, then the DSA will provide a process by which any beneficiary of the security, which is obliged by the Government to carry out decommissioning in place of the defaulting party, will be able to claim reimbursement of the costs of that decommissioning activity from the trustee.

If a party is not in default, then it will pay its share of decommis- I-13.23 sioning costs as they arise but may be allowed to reduce its security over time to reflect those payments. If a party sells all or part of its interest in the field, then a field DSA will usually provide for the return or reduction of its security once its assignee has provided substitute security in the appropriate amount.

Security is usually provided until decommissioning is complete I-13.24 and then for a further 12 months, often at a much reduced level in case BEIS imposes additional work requirements at a late stage, for instance in relation to sea-bed clearance. After decommissioning is complete, any remaining monies in the trust or any uncalled security should be returned to the party which provided the security. Residual obligations in relation to decommissioned assets may remain, particularly if assets have not been fully removed from the seabed, but these obligations are felt to be too remote and liability too unquantifiable to justify security. The industry has established a scheme, the UK Oil & Gas Offshore Fisheries Legacy Trust Fund, to provide ongoing seabed monitoring and communication of information about those installations and pipelines that are permitted to remain on the seabed after decommissioning.[20]

Since one of the major risks in relation to decommissioning is the I-13.25 insolvency of a licensee before decommissioning is complete, it is important that any fund or instrument provided by way of security should be secure against the other creditors of the licensee in the event of insolvency. It is for this reason that a trust is very often

---

[20] See www.ukfltc.com (accessed 11 May 2017).

established in which the security is held. This on its own would probably be sufficient to alienate the security from the assets of the licensee so as to protect it from other creditors, but Section 38A of the Petroleum Act 1998 (as amended by Section 74 of the Energy Act 2008) puts the matter beyond doubt. It provides that funds set aside as financial security for decommissioning are protected in the event of the insolvency of the party required to provide the security where the arrangements establishing the security were put in place after 1 December 2007.

THE OIL & GAS UK MODEL FORM DSA

I-13.26   As DSAs are a relatively recent development in UKCS practice, there was a good deal of divergence in style and form until the publication in 2009 of the Oil & Gas UK Model Form DSA which is now widely used between co-venturers. While it is not designed for use in the context of M & A transactions, its provisions are often adapted and incorporated for use in this context. One aim of the drafters was to remove the risk of the purchaser having to provide security twice in respect of the same liability, to its co-venturers and to the seller of the asset, by developing an agreement which would allow all those potentially affected by default to benefit from the security provided to the co-venturers.

I-13.27   The Oil & Gas UK Model Form DSA permits sellers of oil and gas interests to remain party to the DSA as "Second Tier Participants" after they have sold their interest – they are no longer required to provide security, but are given certain rights of approval of the decommissioning estimate and the ability to enforce the security provided by the co-venturers. BEIS can also be a party to the DSA where required, although it is not known if this option has ever been used. Due to the breadth of the Secretary of State's powers under Part IV of the Petroleum Act 1998, there are a range of other parties who are potentially liable to carry out decommissioning, such as the affiliates of co-venturers or former co-venturers. It would not be practicable to include all of these as parties since they are a large, uncertain and changing group, but there is an option in the Oil & Gas UK Model Form DSA to give them rights as third parties (known as "Third Tier Participants") (using the Contracts (Rights of Third Parties) Act 1999) to enforce the security if they were ever to be made liable for decommissioning.

INTERACTION OF FIELD DSAS AND SPA-LINKED DSAS

I-13.28   The Oil & Gas UK Model Form JOA assumes that the parties will negotiate and execute a DSA at the same time as agreeing their JOA.

However, parties are sometimes reluctant to spend time negotiating a DSA at the outset of a venture when there may never be a development requiring decommissioning. Parties may opt instead to include a requirement to negotiate and agree a DSA, perhaps based on the Oil & Gas UK Model Form DSA, before any application is made for Field Development Plan consent. However, the risk with this approach is that at the key time this requirement is simply overlooked or waived. The issue tends to then be postponed until such time as one co-venturer is looking to sell their interest to a third party. At this stage, the remaining field parties may have considerable leverage. They may insist on any field DSA containing "grandfathering" provisions under which only the new entrant and any subsequent entrants have to provide security. Negotiating a DSA at the outset of a venture, when everyone is aligned, is far more likely to produce a balanced outcome than leaving it until later in the life of a field when some parties are seeking to exit.

If there is a field DSA, the seller may feel comfortable in seeking I-13.29 no separate security for the decommissioning liabilities linked to the assets being transferred. In the past, sellers were not always prepared to accept the deduction of the value of reserves when calculating decommissioning security, on the basis that they had no ability to forfeit reserves under the JOA. However, if they are benefiting from security under a field DSA, it is reasonable to assume that the co-venturers will exercise this right and that reserves should be taken into account. Some sellers in such circumstances would in the past have required a "top-up security" in respect of the difference between the security provided under the field DSA and the security it would normally have expected under a Sale and Purchase Agreement (SPA) DSA, or an element of security provided before the normal "trigger date", to ensure that the parties in fact met their obligations to provide substantive security at the appropriate time. However, as noted below, sellers are increasingly finding it difficult to achieve the level of decommissioning security that they may once have expected and such provisions are now uncommon. If no field DSA exists, and either the co-venturers are uninterested in putting one in place or the seller is concerned that this will take too long, the purchaser may still find itself having to provide security to the co-venturers and separate security to the purchaser. In these cases, however, the seller may agree to release its own security if and when a field DSA is executed in a form satisfactory to it and to which it is party.

As noted below, the need to provide decommissioning security I-13.30 is a barrier to the sale of assets and therefore sellers have looked for more innovative solutions where, particularly for some late life assets, the cost of providing security in traditional form may render the asset unsaleable. Some sellers have therefore chosen to retain some or all of the decommissioning liability, although this is not straightforward.[21]

KEY COMMERCIAL ISSUES

I-13.31 While DSAs can be long and complex documents, the vast majority of their provisions are mechanical and administrative, designed to ensure that the correct security is provided, to deal with challenges to the operator's estimates and to provide for the circumstances in which the security may be enforced and funds held by the trustee released. There are relatively few contentious issues to be negotiated. The main ones are usually the questions of the ratings required of the banks, insurance companies or affiliates providing LOCs and guarantees, and the risk factor to be applied to decommissioning costs. The Oil & Gas UK Model Form DSA suggests a rating of AA-/Aa3, while DECC's Guidance suggests a rating of AA/Aa2 will be required. However, following the financial crisis far fewer banks and corporates meet these criteria and there is considerable pressure to adopt lower thresholds, particularly where the choice will determine whether a party is able to provide a parent company guarantee or faces the cost of providing an LOC or, even more significantly, the requirement to provide collateral. The risk factor chosen has a significant impact on both the timing of the trigger date and the level of security to be provided. Reaching agreement on these issues has become harder following the financial crisis – fewer banks meet existing rating requirements and the cost of LOCs has risen (this will depend on the financial strength of the group seeking the LOC but may be several per cent per annum). At the same time, levels of security have increased as discount rates have fallen.

I-13.32 The other area which may be contentious is the rights to be accorded to "Second Tier Participants", former owners who may remain party to the DSA in order to have the benefit of the security provided under it. Second Tier Participants do not provide security and have little or no interest in reducing the amount of security provided by others. The Oil & Gas UK Model Form DSA provides options for such Second Tier Participants to be given a right to object to the estimate provided by the operator and refer the matter to an Expert or simply to make representations.[22]

I-13.33 The Oil & Gas UK Model Form DSA contains a relatively detailed set of assumptions regarding the calculation of costs and revenues, designed to minimise the operator's discretion and thus

---

[21] See paras II-9.74 and II-9.75.

[22] See further B Holland, *Decommissioning in the United Kingdom Continental Shelf: Decommissioning Security Disputes*, 28 (2016) *Denning LJ* 19.

reduce the scope for disagreement over the calculation of the amount to be secured. However, these provisions should always be reviewed carefully for their suitability for a particular installation.

TAX ISSUES

Decommissioning costs are subject to significant tax relief as set out in Chapter I-7. However, for the purposes of decommissioning security, decommissioning costs were historically calculated "gross", without any allowance for such tax relief. This reflected a number of factors including the lack of confidence that the reliefs would still be available at the time of decommissioning,[23] concern that the company carrying out decommissioning would not qualify for relief,[24] or would have insufficient profits in the United Kingdom against which to offset them.[25] Finally, there was concern about the application of so-called "subsidy rules" which disallowed expenditure where the taxpayer has been reimbursed by a third party for expenditure. If applied to decommissioning, these rules could have prevented a party, carrying out decommissioning after the insolvency of the person primarily liable, from claiming tax relief if that party had been able to call on security to reimburse him for his expenditure.

I-13.34

The result was that decommissioning security was calculated on the basis of gross decommissioning costs rather than the much lower net cost which was ultimately likely to be incurred.[26] The effect of this was to increase the amount of capital required by way of security. This in turn made it harder for assets to change hands, by rendering some transactions uneconomic as it restricted the funds available for new owners to develop their acquired assets. There was also an impact on investment by historic owners who, when contemplating new investment, had to factor in the risk that tax relief would no longer be available when the time came for decommissioning.

I-13.35

In order to address this issue, the industry and HM Treasury worked together to develop a means of giving certainty as to the

I-13.36

---

[23] See paras I-7.45 to I-7.54.

[24] Because reliefs were available only to licensees or former licensees and the party doing the decommissioning might be an affiliate or parent.

[25] The Government had introduced legislative changes to address this latter concern to some extent. See paras I-7.45 to I-7.48.

[26] In the case of petroleum revenue tax (PRT) paying fields, where the effective tax relief rate has been up to 75 per cent, the effect was that the security required could be four times the ultimate net cost of decommissioning after tax relief (or even higher where a contingency is built in to the formula). In non-PRT paying fields, where the effective tax relief rate was, until recently, 50 per cent, the effect was that the security required could be twice the ultimate net cost after tax relief.

availability of decommissioning tax relief. The aim was to allow parties to accept security net of tax relief, reducing the barrier to asset transactions and releasing capital for investment. A legislative solution was not considered workable, as the doctrine of parliamentary sovereignty would enable a later Parliament to amend or repeal any statute attempting to entrench tax relief. The solution which was chosen was therefore contractual – Decommissioning Relief Deeds (DRDs), contracts in which the Government effectively guarantees a specified level of decommissioning tax relief.[27] Although Parliament could in principle pass legislation to invalidate such contracts this would have legal, commercial and political consequences which were thought by the industry to render it unlikely (however, there might be difficulties in relation to the transfer of these contractual liabilities if Scottish independence were to occur). The introduction of DRDs required certain legislative changes, to permit payments under DRDs and to address some technical aspects of the tax rules, including disapplication of the subsidy rules in certain circumstances – these changes were included in the Finance Act 2013 and the first DRDs were issued in September of that year.

I-13.37   The DRD is a standard form contract between HM Treasury and an oil or gas company which is or has been within the oil taxation regime (or an associate of such a company). The guarantee extends to the signatory company and to its associated companies (so that each corporate group needs only one DRD), and covers any assets with which they are, or have been, associated. The guarantee has two parts – one in relation to the counterparty (or associated company) decommissioning its own assets in the normal course, and the other in relation to decommissioning for which the counterparty (or an associated company) is liable as a result of the default of another party. In relation to the decommissioning of its own assets in the ordinary course, the Government effectively guarantees the existing tax capacity of the counterparty and its associates – if the tax relief regime for decommissioning were to change, the Government would have a contractual obligation to make a compensating payment to the affected companies. However, if the tax relief regime remains unchanged, the Exchequer would pay no more in tax relief than it is expecting to pay today. The Government is free to increase or reduce tax rates which will affect the amount of tax capacity built up in the future but the companies will still have the benefit of any tax capacity built up in earlier periods. If decommissioning is carried

---

[27] "Qualifying companies" are defined in the Finance Act 2013, s 80(3). The pro-forma DRD and guidance on how to apply for it are available at www.gov.uk/government/consultations/decommissioning-relief-deeds-increasing-tax-certainty-for-oil-and-gas-investment-in-the-uk-continental-shelf (accessed 15 May 2017).

out by one party as the result of the default of an unrelated party, the Government had to address the risk that the party carrying out the decommissioning has no tax capacity – if it had not done so, then a party would continue to ask for decommissioning security on a gross basis as it could not be guaranteed any tax relief. Therefore, the Government guarantees that the party carrying out decommissioning as a result of a default by an unrelated party will receive tax relief against Ring Fence Corporation Tax and Supplementary Charge of 50 per cent, regardless of its tax capacity. In addition, in the case of Petroleum Revenue Tax (PRT) paying fields, the party carrying out the decommissioning will be entitled to the benefit of the PRT capacity of the defaulting party. Parties seeking security can now, therefore, be confident in allowing the party providing security to provide it net of that same level of tax relief, provided they, or a company in their group, hold a DRD.

The introduction of the DRD resulted in changes to the Oil & Gas UK Model Form DSA to provide for security to be given net of tax relief. The necessary changes include amending the calculation of the amount of decommissioning security, and the provisions dealing with payment out of the trust (since payments may now be net of tax relief). The most contentious changes to the standard were those to introduce protective provisions dealing with the residual risk that the guarantee provided by the DRD is in future challenged or removed. Many existing DSAs were also amended in a similar manner although dealing with historical DSAs can be complex and lengthy. Nonetheless, the Treasury reported in December 2016 that it had entered into 76 Decommissioning Relief Deeds and that Oil & Gas UK estimates that these deeds have so far unlocked more than £5.9 billion of capital, which can now be invested elsewhere.

I-13.38

APPENDIX I-A

MATURE PROVINCE INITIATIVES

Greg Gordon and John Paterson

Editors' note: This Appendix contains a reprint of the Chapter on Mature Province Initiatives originally contained in the second edition of this work. Although the Mature Province Initiatives have been overtaken by the implementation of the Wood Review, we have included this chapter for the readers' convenience as, in view of the Initiatives' influence on policy, the third edition makes extensive reference to them.

Cross-references to other parts of this work have been included as appropriate. The chapter has not been otherwise updated since the second edition and as such states the law as at January 2011.

I-A.1   As a hydrocarbon province such as the UKCS matures, industry and government must play close attention to the changing economics of both existing operations and new developments.

I-A.2   From the point of view of government, it is important to extend the life of existing fields for as long as possible, to encourage exploration for new fields and to encourage the development of known but perhaps marginal discoveries. There are two principal reasons for this. First, there is the question of energy security. While some observers believed that this issue had essentially disappeared from the political agenda at the end of the last century, with the liberalisation of energy markets,[1] there is no question but that it is now once again a major concern for governments.[2] Second, there is the impact on the

---

[1] See eg J Mitchell, "Energy Security", in Mitchell et al. (eds), *The New Economy of Oil: Impacts on Business, Geopolitics and Society* (2001), pp 176–208.

[2] For example, the Government's 2050 Pathways Analysis notes that the "pathways show an ongoing need for fossil fuels in our energy mix, although the precise long term role of oil, coal and gas will depend on a range of issues, such as development of CCS":

national economy: whether one is a net importer or exporter of oil and gas may make a significant difference to the balance of payments.

From an industry standpoint, the maturing of a province alters its relative attractiveness compared with others. Thus, for a major company, well used to operating on the global stage, the smaller return on investment available, whether from the smaller discoveries likely to be made or during the downward slope of an existing field's production profile,[3] may mean that bigger rewards are possible in the context of frontier provinces elsewhere in the world where larger discoveries may still be made.[4] Such companies may have little interest either in exploring for new fields, in developing marginal discoveries or in taking steps to extend the life of existing fields. There may still be money to be made from such ventures, but the same investment in a frontier province that results in a commercial discovery is likely to produce a bigger yield. On the other hand, smaller and often newer companies may perceive rich pickings in the same circumstances, perhaps due to variations in risk appetite, or perhaps making use of innovative technology. I-A.3

A range of government policies may, therefore, have an impact on decisions taken by the industry in such a context. Government must walk a fine line if it is to encourage exploration, development and the extension of the life of existing fields while at the same time ensuring that it satisfies legitimate societal expectations regarding the nation's take from such activity. I-A.4

These issues were recognised by government and industry as early as 1998, not least because they became particularly pressing in a high-cost province such as the UKCS in the context of the low oil prices then prevailing. The Oil and Gas Industry Task Force (OGITF) was established at that time with the "overall objective ... to create a climate for the UKCS to retain its position as a pre-eminent active centre of oil [and] gas exploration, development and production and to keep the UK contracting and supplies industry at the leading edge in terms of overall competitiveness".[5] This group established the so-called "Vision for 2010": "The UK Oil and Gas Industry and Government working together in partnership to deliver quicker, I-A.5

Department of Energy and Climate Change, *2050 Pathways Analysis* (URN 10D/764, July 2010), p 35. See also Chapter I-3.

[3] "Average discovery size since 2000 has been 267 million boe per field, with two thirds of all discoveries less than 15 million boe (these should be compared with early UKCS fields each containing hundreds of million boe and, in some cases, 1 to 2 billion boe)": Oil & Gas UK, *Economic Report 2010*, p 7.

[4] "The UKCS has, therefore, to compete with less mature and less costly provinces, where the size of discoveries is much larger": Oil & Gas UK, *Economic Report 2010*, p 42.

[5] PILOT, *About PILOT: What is PILOT?*, available at www.pilottaskforce.co.uk/data/aboutpilot.cfm (accessed 16 December 2010).

smarter, sustainable energy solutions for the new century. A vital UK Continental Shelf is maintained as the UK is universally recognised as a world centre for global business."[6] In order to realise this vision, a list of deliverables was established, including the following: a production level of 3 million boe/d beyond 2010; £3 billion of industry investment per annum; and prolonged self-sufficiency in oil and gas for the UK.[7] The OGITF continued in existence until 1999 and was succeeded in 2000 by the PILOT initiative, which was established both to monitor progress towards the vision for 2010 and to identify what else might need to be done to ensure that it is realised.

I-A.6   As will be seen in this chapter, the work of PILOT has led to significant initiatives aimed at responding to the challenges thrown up by a mature province. The first of these is the Fallow Areas Initiative, which aims to deal with the problem both of allocated but unexplored acreage and of undeveloped discoveries; the second is the so-called Brownfields Initiative, which aims to maximise the economic recovery of hydrocarbons from existing developments and which includes most notably, for present purposes, the Stewardship Initiative. As will be seen, these initiatives are noteworthy not only because of their innovative approach to the complex of technical, economic and political challenges posed by the mature province, but also because (at least in the case of Fallow Discoveries and Stewardship) they reveal a willingness on the part of the industry to acquiesce in schemes for which there is at best a relatively flimsy legal basis.

THE FALLOW AREAS INITIATIVE

Background and introduction

I-A.7   As has already been seen, in the early days of exploration for oil and gas in the UKCS, Production Licences were granted which were of a lengthy duration and which contained very little in the way of provisions designed to control or incentivise exploration or production activity within the acreage let.[8] As Daintith has observed, "the idea that licensees might make significant discoveries but then not develop them does not appear to have occurred to those who first drafted

---

[6] PILOT, *About PILOT: What is PILOT?: PILOT Vision* (hereinafter "PILOT, *Vision*"), available at www.pilottaskforce.co.uk/data/pvision.cfm (accessed 16 December 2010).

[7] PILOT, *Vision*.

[8] Early licences had production periods of up to 40 years and were based upon "the very limited landward experience gained to 1964": T Daintith, G Willoughby and A Hill, *United Kingdom Oil and Gas Law* (looseleaf, Release 53, February 2007) (hereinafter "Daintith, Willoughby and Hill"), para 1-107. See also Chapter I-4.

the offshore licensing arrangements in 1964–1965".[9] Although one should not rush to criticise draftsmen who were operating under significant pressure of time and with little experience of the industry they sought to regulate, the fact remains that the offshore licensing system put in place in 1965 left the state powerless to control the pace either of exploration or production. Major legislative amendments made in 1975[10] and further measures adopted in 1988[11] and 1996[12] went some way towards ameliorating the problem. However, at least in practice, there remained in the areas of exploration and production something of a regulatory black hole into which licences were wont to disappear once they had completed their initial term.[13] By 2002, it was estimated by the PILOT Progressing Partnership Work Group (PILOT PPWG) that there were 247 blocks in the UKCS where no significant exploration activity had taken place for an extended period, and 250 discoveries where no significant progress had been made to produce hydrocarbons.[14] While it was believed

[9] T Daintith, *Discretion in the Administration of Offshore Oil and Gas* (2006) (hereinafter "Daintith, *Discretion*"), at para 4104.

[10] The Petroleum and Submarine Pipelines Act 1975 made substantial unilateral changes to all existing licences with retroactive effect. See Daintith, Willoughby and Hill at paras 1-330 to 1-331. The 1975 Act introduced the Minister's power to serve an additional or supplementary work programme at any time: this power continues to be incorporated into current licences by Model Cl 16; and to exercise controls over development and production, which controls are now incorporated into current licences by Model Cll 17 and 18. The legislation was passed in the face of considerable industry opposition: Daintith, *Discretion*, para 5107. The retroactivity of the Bill, and the fact that it offered no compensation to the industry in respect of any diminution in value between the licences as granted and as they would be after the legislation entered into force, drew particular criticism in Parliament: see eg *Hansard* HC (series 5), vol 891, cols 486 and 503 (30 April 1975) and Standing Committee D, Official Report (1974–1975), vol V, cols 1106 and 1146–1172.

[11] The Petroleum (Production) (Seaward Areas) Regulations 1988 (SI 1988/1213) replaced the two-term licence which had previously been in use with a three-term licence by splitting what had previously been a 30-year second term into a 12-year second term and an 18-year third term and providing that the licence would be permitted to progress into a third term if a field development plan has been approved or consented to.

[12] A 1996 Fallow Field Initiative (on which see Daintith, Willoughby and Hill (looseleaf, Release 62, Feb 2010), para 1-605) led to the Petroleum (Production) (Seaward Areas) (Amendment) Regulations 1996 (SI 1996/2946), which provided for a short initial term which could be extended in the event that the licensee undertook additional exploration activity and linked the amount of acreage which could be retained beyond the initial term to the number of wells drilled. A number of licences issued on these terms continue in existence but the amendments were not adjudged a success and, as shall be seen, were abandoned for future licences in 2002.

[13] The initial term of the licence is less problematic because of the need for the initial work programme to be carried out if the licence is to continue.

[14] PILOT, *The Work of the Progressing Partnership Work Group* (2002) (hereinafter "PILOT PPWG, *Work*"), para 3.1.2, available at www.pilottaskforce.co.uk/files/workgroup/422.doc (accessed 16 December 2010). *Prima facie*, the position seems to have

that in around half of these cases there was good cause for this lack of progress, that still left a large number of blocks and discoveries lying fallow for reasons which had nothing to do with the blocks or discoveries themselves but on other grounds, chiefly those connected with the choices and priorities of the licensees.[15] This would be unfortunate at any time for a state such as the UK, which obtains the vast majority of its oil and gas revenue by taxing the profits made on produced hydrocarbons.[16] But the criticality of this situation increased as the UKCS matured as an oil and gas province and the production from existing fields started to decline,[17] less and less virgin acreage became available to let outside of frontier areas, and the average size of new discoveries became progressively smaller.[18]

I-A.8   PILOT PPWG was established in 2002 to identify, report upon and make recommendations concerning "all commercial [and] behavioural barriers to development"[19] which existed in the maturing UKCS. The deficiencies within the licensing regime discussed above were one of a number of issues[20] PILOT PPWG identified when reporting upon its work to date:

> "The standard UKCS approach of long licence terms and low annual rentals combined with both limited relinquishment and limited

---

been worse than estimated: in a little over two years of operation, the initiative identified 660 fallow blocks or parts of blocks and 240 fallow discoveries: PILOT, *Annual Report 2004–2005* (2005), p 6, available at ww.pilottaskforce.co.uk/templates/relay/communicationrelay.cfm/1653 (accessed 16 December 2010). However, the large disparity between the position reported in 2005 and the 2002 estimate may at least in part be explained by the fact that the 2005 figures include not just blocks, but parts of blocks.

[15] Many oil and gas companies have a portfolio containing assets in diverse locations throughout the world. They will not necessarily wish or be able to press on with the development of all those assets at an even pace but may decide, perhaps for strategic reasons or because of a lack of financial or other resources, to prioritise an asset or group of assets in one province over those in another. They may, for example, wish to retain assets in a relatively stable but high-cost province, such as the UK, as "insurance" while they undertake higher-risk, but potentially more profitable, activity elsewhere.

[16] See Chapters I-4 and I-7.

[17] See DECC, *Digest of United Kingdom Energy Statistics (DUKES) 2010: long-term trends*, Table 3.1.1: *Crude oil and petroleum products: production, imports and exports 1970 to 2009*. Note that total annual production has steadily dropped from a peak of 137,099 tonnes in 1999 (of which 132,814 tonnes were offshore production) to 68,199 tonnes in 2009 (of which 67,018 tonnes were offshore production). Also see Figure I-**3.1** above.

[18] P Carter, *The Regulator's Dilemma: how to regulate yet promote investment in the same asset base – the UK's experience* [2007] IELTR 62, at 62.

[19] PILOT PPWG, *Work*, para 1.2.

[20] Among the other problem areas identified by this influential report were the need to place controls upon rights of pre-emption (PILOT PPWG, *Work*, Chapter 6, discussed further at para II-15.17), and the commercial and supply chain Codes of Practice (PILOT PPWG, *Work*, Chapters 4 and 11 respectively).

activity obligations provides an environment where there is too little pressure on licensees to deliver value from their licences. Under these conditions, misalignments between co-licensees, decisions on marginal or high-risk economic activities, or divestment can be repeatedly deferred and remain unresolved, potentially indefinitely."[21]

The work of PILOT PPWG led to significant changes to the **I-A.9** Production Licence Model Clauses which were implemented in the 2002 Licensing Round and which continue to form the basis of the current standard Production Licence's term and relinquishment structure.[22] As has already been seen, however, in the absence of primary legislation having retroactive effect, amendments to the model clauses affect only new licences incorporating those model clauses.[23] In consequence, Pilot PPWG also had to consider how to remedy the same weaknesses in the context of existing licences. The Work Group decided against recommending retrospective changes to existing licences on the basis that such a measure would be "disproportionate given the willingness of the exploration community to address the issue".[24] However, the Work Group did not have sufficient faith in the community's professed willingness to allow it to recommend a purely voluntary approach. Instead it recognised that "a voluntary scheme was already effectively in place and had failed to fully galvanise activity".[25] Given what has already been said about the number of fallow blocks and discoveries in existence at the time when PILOT PPWG was reporting, this might be thought to be a considerable understatement.[26] Be that as it may, the Work Group was satisfied that the most satisfactory result would be achieved by a scheme which was essentially voluntary in nature, but which was "underpinned by licence powers and the discretion of the Department".[27]

In formulating the detail of the scheme, the Work Group adopted **I-A.10** two guiding principles: "that a group of licensees doing all that a fully resourced and skilled group could reasonably be expected to do should not be disadvantaged and that no group should have reasonable grounds to feel that they had not been given full opportunity to create value from their licence".[28] While the Work Group chose to express its findings in a positive manner, the principles could

---

[21] PILOT PPWG, *Work*, at para 3.1.2.
[22] See Chapter I-4.
[23] See Chapter I-4.
[24] PILOT PPWG, *Work*, para 3.2.3.2.
[25] *Ibid*, para 3.2.3.2.
[26] See para I-A.07 above.
[27] PILOT PPWG, *Work*, para 3.2.3.3.
[28] *Ibid*, para 3.2.3.3.

also be couched in a more negative way: the initiative will not protect a group of licensees who are not doing all that could reasonably be expected; and a group of licensees who have had the opportunity to exploit their acreage but who have not availed themselves of it cannot expect to retain it indefinitely.

I-A.11   The Government adopted PILOT PPWG's recommendations. The Fallow Areas Initiative is made up of two separate but related schemes: the Fallow Blocks Process and the Fallow Discovery Process. Although conceptually similar, the two programmes differ somewhat in detail. They will be discussed in turn.

The Fallow Blocks Process

I-A.12   A fallow block is defined as a block where the initial term of the licence has expired and there has been no drilling, dedicated seismic or other significant activity for a period of three years.[29] The Revised Fallow Blocks Guidance does not define the term "block" but an earlier set of Clarification Notes issued by the DTI states that a block "will be assigned an area agreed between the Department and Licensees" and that determinations made for field or Petroleum Revenue Tax purposes are not relevant to this issue.[30] Information provided by the Department on the progress and operation of the process discloses that, in practice, the Department commonly sub-divides blocks into two or more constituent areas. Areas on the block that are producing hydrocarbons are by definition excluded from the ambit of the Fallow Blocks Initiative, as are areas in which a discovery has been made.[31] However, if there is a fallow area within the block it will be isolated from the non-fallow areas and subjected to the Fallow Blocks Process.[32] As the way in which "block" is to be understood has a considerable bearing upon the scheme's operation and extent,[33] it is perhaps unfortunate that the main items

---

[29] DTI, *Revised Guidance for the Fallow Blocks Process (July 2005)* (hereinafter "Revised Fallow Blocks Guidance"), B. Definitions, available at www.og.decc.gov.uk/UKpromote/fallow/FallowBlocksGuidance.doc (accessed 16 December 2010). When the scheme was first introduced the triggering period was defined differently: four years plus two of no other significant activity: see PILOT PPWG, *Work*, Appendix 1(a), at 42, n 2 to para 1.

[30] DTI, *Fallow Blocks Process – Clarification Notes* (2004) (hereinafter "*Fallow Block Clarification Notes*"), p 3, n 1.

[31] However such areas may respectively be subjected to the Stewardship Process (discussed at paras I-A.35ff) and the Fallow Discovery Process (discussed at paras I-A.23ff).

[32] See DECC, *UK Fallow Assets and Process – 11 January 2010*: www.og.decc.gov.uk/UKpromote/fallow/fallow_assets.htm (accessed 16 December 2010). See the discussion under the heading "2010 Fallow Blocks": "Many of [the fallow blocks published today] have been producing fields or discoveries on the block, but it is the area outside the field or discovery area that now needs significant activity."

[33] As the *Revised Fallow Blocks Guidance* (and all antecedent guidance on the topic) is

of published guidance are less than explicit on how the scheme is implemented in practice.

The Department assesses the status of blocks annually and, having   I-A.13
made a provisional determination of all blocks' status, issues a list to the licence-holders of all the blocks it considers fallow. Licensees are provided with an opportunity to meet with, and make representations to, the Department before the formal classifications are issued.[34]

Blocks which have been classified as fallow are then allocated one   I-A.14
of two classes: A and B.[35] A block will be designated Class A Fallow if "the current licensees are doing all that a technically competent group with full access to funding could reasonably be expected to do" to progress towards activity.[36] In such cases the Department accepts that the lack of activity is not attributable to the licensee. The block will be noted as Fallow A on the Department website and in the instances where a "technical barrier to progress" has been identified, the Department may request that details of that barrier be communicated to LOGIC and/or ITF for the purposes of guiding those organisations' programmes of technical research and development. Beyond this, the Department requires no action from the licensee. The block's status will be reviewed annually.[37]

In addition to the standard route to Fallow A status, described   I-A.15
above, a newly fallow block which has not already been formally classified and which would otherwise be classified Fallow B will be deemed to be Fallow A if:

- a field has commenced production within the previous year; or
- there has within the previous year been a change of operator, and the Department has approved that change; or
- there has been a substantial change in ownership within the previous three months, and the Department has approved that change.[38]

---

expressed in terms of blocks, not parts of blocks, one might reasonably have expected the fact that there was a producing area somewhere within the block to save the block as a whole from the ambit of the regime.

[34] *Fallow Block Clarification Notes*, at p 3, n 2.

[35] In practice the Department further divides these classes into several numbered sub-classes. These sub-classes are of no legal consequence: a block classified A1 is not subjected to a different legal regime to an A5, but they do provide some insight into the particular issues which can arise to frustrate activity. For a summary of some of the sub-categories, see the *Fallow Block Clarification Notes*, at p 4*ff*.

[36] *Revised Fallow Blocks Guidance*, "B – Definitions".

[37] *Ibid*, "C – Fallow Blocks Process: Fallow A Blocks", superseding the biennial assessment programme which subsisted when the scheme was first introduced: see PILOT PPWG, *Work*, Appendix 1a, para 2.

[38] *Revised Fallow Blocks Guidance*, "D – Change of Interest".

I-A.16   A block is designated Class B Fallow if "the current licensees are unable to progress towards activity due to a misalignment within the partnership, a failure to meet economic criteria, [and/or] other commercial barriers".[39] Although the scheme as initially proposed by PILOT PPWG and the Guidance subsequently issued by the DTI scrupulously avoids the language of blame, it is implicit in this definition and from the consequences which flow from Fallow B classification (discussed in the paragraphs that immediately follow) that this designation is applied where the Department believes that the current licensees have, for reasons which cannot be justified, made a less than acceptable level of progress.

I-A.17   After a block has been classified as Fallow B, unless divestment of the asset is under serious consideration, in which case certain relaxations to the process are applied,[40] it will be entered into a process which, in broad terms, is designed to prompt the licensee either to start making use of the asset, sell it to (or otherwise involve the enterprise of) someone who will, or relinquish it back to the state. After formal designation of a block as Class B Fallow, the licensees are given an initial three-month period in which to report to the Department on what activity could feasibly be carried out on the block to alter its Fallow B status. Licensees are specifically asked to consider whether a re-allocation of interests would assist in this process.[41] If no adequate proposals are received by the end of that period, the block will be publicly listed as Fallow B on the Department's website[42] for a period of one year.[43]

I-A.18   During that period the licensee is at liberty to market the block itself,[44] and/or may submit a significant activity plan to the Department. If such a plan is agreed by the Department, the block will be re-classified as "rescued" and the licensee will be given a period of one year, which may be extended at the Minister's discretion, to implement the works specified.[45] Once the work has

---

[39] Ibid, "B – Definitions".

[40] In such cases the block will be temporarily classed Fallow B Hold (if Fallow B status has not yet been formally assigned) or Rescued (if Fallow B status has been assigned and the block released on the Department's website) to permit the transfer to take place: Fallow Block Clarification Notes, p 1f, per the note to numbered point 4.

[41] Fallow Block Clarification Notes, at p 1: see para 3.

[42] Initially the scheme stipulated that the listing would be on the LIFT website. This has been replaced in the most recent Guidance by a reference to the Department's own website, although LIFT (now known as UKLIFT) is still recommended as "a good shop window" for fallow blocks: Fallow Block Clarification Notes, at p 2: see the note to para 6.

[43] Revised Fallow Blocks Guidance, "C – Fallow Blocks Process: Fallow B Blocks", para 4.

[44] Ibid, para 4.

[45] Ibid, para 6.

been undertaken, the block will revert to non-fallow status and, therefore, cannot be brought back into the Fallow Blocks Process for at least three years.[46]

An intermediate checkpoint is encountered after nine months of the Fallow B period have elapsed. At this point, licensees and any interested third parties[47] should report to the Department any plans for significant activity, and any licensee not having a firm plan at that time is expected to assign its interest to any co-licensees or third parties having such a plan if requested by them to do so.[48]   **I-A.19**

If no satisfactory significant activity plan is agreed by the end of the 12-month Fallow B period, the published guidance provides that the licensee "will relinquish the block".[49] The language would suggest that the licensee is under a firm obligation so to do. However, this is rather misleading. As has already been seen, the scheme is essentially a voluntary one, underpinned only by existing licence powers.[50] Thus the Department has no free-standing legal right to demand the relinquishment of the licence at the process's end, but may do so only if a power within the licence permits. The licence term that is of greatest relevance in this regard is Model Cl 16(2),[51] which was discussed at para I-4.51. As has already been seen, the clause provides the Minister with a right to serve notice upon a   **I-A.20**

---

[46] *Ibid*, para 7.

[47] Usually, parties which propose to acquire an interest in the asset.

[48] *Revised Fallow Blocks Guidance*, "C – Fallow Blocks Process: Fallow B Blocks", para 5.

[49] *Ibid*, para 8.

[50] See paras I-A.09 to I-A.11 above. See also *Revised Fallow Blocks Guidance*, "F – DTI Regulatory Position", which states: "If it appears to DTI at any stage in the process that there is a firm activity plan that is not being progressed on a commercial basis, the DTI will consider using its powers under the PSPA to require the licensees to drill the block or forfeit the licence (or part thereof)." This piece of guidance appears to be a cryptic reference to the Minister's willingness to use licence powers, but is poorly drafted in two respects. Instead of referring to ministerial powers under the licence it refers to the "PSPA" (presumably the Petroleum and Submarine Pipelines Act 1975), which, while it effected important changes to the terms of existing licences, does not of itself contain the relevant powers. Second, its description of the Minister's powers under the licence is rather imprecise, and makes them sound rather stronger and less equivocal than they in reality are.

[51] Standard Production Licence, current numbering, as specified by the Petroleum Licensing (Production) (Seaward Areas) Regulations 2008 (SI 2008/225). This provision being one of the ones inserted with retroactive effect by the Petroleum and Submarine Pipelines Act 1975 a like term can be found in all previous iterations of the model clauses. However, the numbering of the sets of clauses varies. For instance, it was Model Cl 12 in the 2004, and Model Cl 16 in the 1988, Seaward Production Licence Model Clauses: see the Petroleum Licensing (Exploration and Production) (Seaward and Landward Areas) Regulations 2004 (SI 2004/352), Sch 4, and Petroleum (Production) (Seaward Areas) Regulations 1988, Sch 4, respectively.

licensee requiring the licensee to prepare and submit an appropriate programme for exploring the licensed area and provides a rather elaborate procedure, which includes the possibility of arbitration, in the event that the licensee does not comply. While it is true that this process may ultimately result in revocation of all or part of the licensed area, such a result is by no means the inevitable outcome of the procedure,[52] and even if that were to be the outcome in a given case, the process could take considerable time to complete, particularly if the licensee chose to arbitrate. Model Cl 16(2) gives the Minister significant powers but is by no means a straightforward right to effect an immediate revocation upon the end of the Fallow Blocks Process.

I-A.21   The Department has stated that fallow blocks adjacent to median lines will be brought into the Fallow Blocks Process but has acknowledged that median line issues may impact upon a licensee's ability to comply with the initiative by stating that it "fully recognises" the potential implications of such issues.[53] Licensees of median line blocks can therefore expect a degree of latitude in the scheme's operation if a cross-border issue can be shown to have hindered progress within the block.

I-A.22   The Department has publicly confirmed that the Fallow Blocks Process is a voluntary one and has committed to applying its rules "in a fair and reasonable manner".[54] On its own, this might be read as no more than an undertaking to follow procedural fairness in applying the process's rules. However, it appears that more is intended. No formal dispute resolution mechanism or right of appeal is provided for within the process, but the Revised Guidance states that licensees should raise their concerns with senior Department officials if they believe that their case has not been fully understood or if the rules are not being applied properly "or are producing an unreasonable outcome".[55] The last-mentioned ground for referral is perhaps the most significant as it strongly suggests that the rules of the scheme are in the final analysis intended to be flexible and that there may be scope for relaxations in their application in any given case.

The Fallow Discovery Process

I-A.23 The Fallow Discovery Initiative is conceptually similar to the Fallow Blocks Scheme but is directed not towards encouraging exploration of

---

[52] For instance, the Minister may be taken to arbitration, and may lose.
[53] *Revised Fallow Blocks Guidance*, "E – Median Line Blocks".
[54] *Ibid*, "G. Right of Appeal".
[55] *Ibid*, "G. Right of Appeal".

blocks but towards promoting the exploitation of individual discoveries[56] which have been made during exploration activity but which have been left lying dormant. The central philosophy of the process is again to require, or at least encourage, oil companies to surrender unused acreage or to engage in significant activity or transfer their interests or otherwise involve those who will. And, again, the Fallow Discovery Process is essentially a voluntary system.[57]

The Fallow Discovery Process follows broadly the same pattern **I-A.24** as that for fallow blocks. As with blocks, discoveries are allocated into two categories, A and B, with Fallow A representing a fallow discovery where the current licensees are doing all that may be reasonably expected and Fallow B representing the situation where the present licensees could reasonably have done more to develop the discovery.[58] In addition, the provisions on appeals and dispute resolution, median line discoveries and the Department's regulatory position[59] are all *mutatis mutandis* the same as those discussed above in the context of the Fallow Blocks Process. There are, however, some material differences between the two schemes. These are noted in the paragraphs which follow.

Material differences between the two schemes
When the Fallow Areas Initiative was introduced, one of the more **I-A.25** significant differences between the two constituent schemes was the absence in the Fallow Discovery Process of a provision equivalent to the Fallow Block Process's stipulation that the licence would be relinquished if at the end of the process the licensee had failed to commit to significant activity. As with the extended timeframe, this was said to be justified on the basis of the significant degree of investment which would have been involved in progressing activities to the point where a discovery had been made.[60] However, the Clarification Notes issued in 2004 incorporated a provision confirming that the Department expected the relinquishment of discoveries which continued to be

---

[56] Defined as "any well where hydrocarbons were encountered": DTI, *Revised Guidance for the Fallow Discovery Process (July 2005)* (hereinafter "Fallow Discovery Revised Guidance") "B – Definitions", available at www.og.decc.gov.uk/UKpromote/fallow/FallowDiscoveriesGuidance.doc (accessed 16 December 2010). Initially the scheme applied only to discoveries having no or only one appraisal well: PILOT PPWG, *Work*, *Appendix* 1b, p 46.

[57] See paras I-A.09 to I-A.11 above; also *Fallow Discovery Revised Guidance*, "H – Right of Appeal".

[58] For the full definition of each of the categories, see *Fallow Discovery Revised Guidance*, "B – Definitions".

[59] That is, the matters which were respectively discussed above in the fallow blocks context at paras I-A.21 to I-A.22 above.

[60] PILOT PPWG, *Work*, para 3.2.5.2.

fallow at the end of the process. This requirement was initially expressed in a rather weak way when first introduced: whereas the published Clarification Notes stated that at the end of the process a fallow block "will" be relinquished,[61] the equivalent guidance for the discovery process stated only that a fallow discovery "should" be relinquished[62] – phrasing which would tend to suggest that there was scope for dialogue about the matter. The Revised Guidance of 2005 is firmer, and uses the word "will" in respect of both categories of fallow asset. The extent to which this apparently firm obligation can be enforced is discussed at paras I-A.28 to I-A.32 below.

I-A.26   The current Discoveries Guidance divides the Fallow A class into three sub-classes: "linked", "stranded" and "active". The guidance recognises a variation upon the standard process for "linked" Fallow A discoveries.[63] However, nothing of legal consequence appears to turn on the "stranded" and "active" designations; these sub-classes are defined by the Guidance but not mentioned again, and seem to be used for administrative purposes only.

I-A.27   The Fallow Discovery Process, although noticeably similar in structure to the Fallow Blocks Process, takes place over the significantly longer timescale of 27 months.[64] This difference has arisen because licensees are likely to have invested considerably more time, energy and resources in a fallow discovery than a fallow block.[65] The authors of the scheme recognised that in these circumstances fairness required the licensee of a fallow discovery to be afforded a greater opportunity to extract value from his licensed interests than the licensee of a fallow block.

---

[61] *Fallow Block Clarification Notes*, 2, numbered para 8.

[62] *Ibid*, 3, numbered para 8.

[63] "Linked" Fallow A discoveries are those which are "explicitly included" in an investment plan pertaining to a neighbouring development: *Fallow Discovery Revised Guidance*, "B – Definitions". Certain relaxations in the fallow process are offered in these circumstances: *Fallow Discovery Revised Guidance*, "C – Fallow Discoveries Process: Fallow A Discoveries", para 2.

[64] As opposed to 15 in the case of fallow blocks: see para I-A.17 above. Like the Fallow Blocks Process, an initial warning that the Department intends to class an asset as fallow is followed by an opportunity to make representations. This in turn is followed by formal intimation of the status of the asset; a three-month period during which a plan can be presented, and then a formal notification that the asset is fallow. There then follows a two-year period in which the discovery is advertised on the Department's website and during which the licensees can present a significant activity plan and/or market the asset: *Fallow Discovery Revised Guidance*, "C – Fallow Discoveries Process".

[65] See PILOT PPWG, *Work*, para 3.2.5.2: "Fallow discoveries differ from fallow blocks significantly in that licensees will generally have invested heavily in the exploration activity and the resulting discoveries are likely to have substantial emotional or even 'book value'. In addition, the transition from discovery to development involves, in most cases, a far greater financial commitment and exposure."

Because, as noted at para **I-A.23** above, the Fallow Discovery **I-A.28** Process is directed towards a different phase of operations than is the Fallow Blocks Process, the licence terms that can be used by the Minister to enforce the processes differ. Model Cl 16 is not relevant to the Fallow Discoveries Initiative; the licence provisions of greatest consequence are the Minister's power under Model Cl 17 (as supplemented by Model Cl 18) to exercise control over the production phase of operations by approving, declining to approve, or varying the licensee's development and production programme. One may, however, query the extent to which these clauses are well suited to underpin the Fallow Discovery Process. Daintith raises a fundamental issue by observing: "the drafting of [Model Cl 17] assumes that the licensee, not the Minister, initiates the development process, and is not well designed to compel development".[66] Certainly, Model Cl 17 contains no direct equivalent to Model Cl 16(2), which, as has been seen, specifically provides that the Minister may "by notice in writing" require the licensee to submit a work programme "at any time".[67] Does the Minister have the power to initiate the development process? Model Cl 17(6) does provide the Minister with the power to prepare a production programme and serve it upon a licensee, but only *after* the licensee has had a proposed programme rejected by the Minister and has failed satisfactorily to avail itself of the opportunity to submit a modified programme.[68] Model Cl 17(1) provides the Minister with no assistance in this regard: it contains not a forcing obligation to produce a development and a production plan within a certain period of time, but a prohibition upon carrying out relevant work except in accordance with an approved programme or ministerial consent.[69] Moreover, Model Cl 17(2) opens with words that strongly suggest that the licensee initiates the process.[70] It does, however, go on to state that the programme must be submitted "in such form and by such time and in respect of such period during the term of this licence as the Minister may direct". Although those words are capable of alternative construction,[71] and while it would certainly have been preferable for wording equivalent to that used in Model Cl 16 to be used, it is suggested that this provides sufficient authority to enable the Minister to request, through a direction

---

[66] Daintith, *Discretion*, at para 4311.

[67] See para I-A.14.

[68] Model Cl 17(6) read together with Model Cll 17(4) and 17(5).

[69] Model Cl 17(1).

[70] That is, "The Licensee shall prepare and submit to the Minister ...".

[71] One could seek to argue, for example, that they do no more than provide the Minister with the power to make stipulations concerning matters of form and timescale in the event that a licensee chooses to submit a production and development programme.

issued to a licensee, that a programme be submitted. Certainly, it does appear that those who promoted the Petroleum and Submarine Pipelines Bill, which introduced what is now Model Cl 17, intended the Minister to have such a power.[72]

I-A.29   Even if the specific example given by Daintith is not made out, he is certainly correct to observe that Model Cl 17 is not a sound foundation for the Fallow Discovery Process. Like Model Cl 16, Model Cl 17 dates from legislation passed in 1975,[73] a time when the UKCS was in the early stages of its development as an oil and gas province. Enough lessons had been learned to allow the Government to appreciate that the early licences had been granted in terms which were deficient, at least from the standpoint of the State;[74] however, petroleum policy was dominated by considerations substantially different from those the UKCS now faces as an increasingly mature province. This is perhaps best demonstrated by the fact that Model Cl 18, entitled "Provisions supplementary to clause 17", is primarily concerned not with *maximising* recovery but with the circumstances in which it may be appropriate for the state to impose a *limit* upon it;[75] something that seems almost unthinkable in the present production-hungry climate. Moreover, the Act was politically controversial and extensively amended in its passage through Parliament. Several amendments were made in order to meet the concerns of the international oil industry who were reluctant to cede control over their licensed interests to the state.[76] In these circum-

---

[72] See eg the speech of Under-Secretary of State for Energy (John Smith) on 8 July 1975: Standing Committee D, Official Report (1974–1975) vol V, col 1278: "We believe it is a necessary part of depletion controls that the Secretary of State should be able to require a licensee to submit a programme or different programmes for different parts of the area. Otherwise, the licensee could sit on commercially exploitable discoveries and the nation would lose the benefits of these reserves ... we think it important that the Secretary of State should have this power to require, through the programme, the licensee to exploit the find."

[73] That is, the Petroleum and Submarine Pipelines Act 1975, on which see para I-A.07 above. The numbering of this model clause has changed from time to time. For instance, it was Model Cl 13 in the 2004, and Model Cl 17 in the 1988 Seaward Production Licence Model Clauses: see the Petroleum Licensing (Exploration and Production) (Seaward and Landward Areas) Regulations 2004 (SI 2004/352), Sch 4 and the Petroleum (Production) (Seaward Areas) Regulations 1988, Sch 4.

[74] See Chapter I-4.

[75] See also the so-called Varley assurances, discussed by Daintith, Willoughby and Hill (looseleaf, Release 59, March 2009) at para 1-704.

[76] The debate in the Select Committee on what are now Model Cll 16–18 spans more than 370 columns: 1106–1383. None of these model clauses escaped amendment and Model Cl 18 was, in the face of stark opposition from the oil industry and its financiers, withdrawn in its entirety and replaced with a clause which sought to strike a more careful balance than had the original between "effective depletion control and the legitimate commercial interests of the industry": per Under-Secretary Smith at col 1307.

stances, it is hardly surprising either that Model Cl 17 is not as explicitly focused upon maximising recovery as one would expect of a like clause drafted today, or that it does not contain broad statements of ministerial power but a more modest and detailed system of checks and balances.

Bearing this background in mind, when considering the utility of Model Cl 17 as an underpin to the Fallow Discovery Initiative, at least two further issues seem to require discussion. Does the scheme of the clause envisage that the Minister is entitled to demand the preparation of one development programme, or a succession of them? And on what basis is the Minister entitled to refuse to accept a programme submitted to him? These issues will be considered in turn. **I-A.30**

One programme, or a succession? As has already been seen,[77] Model Cl 17(2) provides that "[t]he Licensee shall prepare and submit to the Minister, in such form and by such time and in respect of such period during the term of this licence as the Minister may direct, a [production] programme". Model Cl 17(3) provides that the Minister may direct the licensee to prepare separate programmes in respect of separate parts of the licensed area[78] or, where the submitted programme relates only to a particular period within the term of the licence, to prepare a further programme in respect of additional periods.[79] Powers to vary, generally by *limiting*, the amount of petroleum to be produced, are contained in Model Cl 18.[80] Beyond this, however, the Minister has no obvious power to demand a succession of production programmes. Applying the above to the present context, it is suggested[81] that the Minister will be in a strong position against a licensee who, having been directed to do so, has not prepared a development and production programme for the licensed area at all. However, in the absence of a clear and unequivocal right to serve notice "at any time",[82] what is more doubtful is the Minister's position in a scenario where he has been provided with a programme, approved it without making any Model Cl 17(3) qualifications and then subsequently come to the view, perhaps years later, that the programme is inadequate. At best, **I-A.31**

---

[77] See para I-A.28 above.

[78] Model Cl 17(3)(a).

[79] Model Cl 17(3)(b).

[80] Limitation is permissible in the national interest provided that the Minister has served an appropriate notice and further notice; the Minister may require an increase in quantity of petroleum which the licensee is required to get from the licensed area only by reason of a national emergency: Model Cl 18(4).

[81] *Quaere* Daintith, *Discretion*, para 4311; see para I-A.28.

[82] Present in Model Cl 16(2) for work programmes: see para I-A.20 and Chapter I-4.

the open-textured nature of the drafting provides ample opportunity for an obdurate licensee to delay matters. At worst, Model Cl 17 provides no meaningful regulatory underpin to the Fallow Discovery Process in such cases.

I-A.32 **The grounds for rejecting a programme.** Given the amount of discretion that is generally afforded to the Minister by the licensing system, the grounds on which the Minister may reject a programme submitted in accordance with Model Cl 17(2) may seem surprisingly narrow.[83] There are only two grounds: that the proposals are contrary to good oilfield practice,[84] or that they are, in the opinion of the Minister, not in the public interest.[85] No other factors will suffice. "Good oilfield practice" is not defined within the model clauses but is generally taken to relate "largely to technical matters within the disciplines of geology and reservoir, petroleum and facilities engineering and to the impact of the development on the environment".[86] Thus the term is generally considered to carry technical, rather than economic or policy-related connotations. The national interest criterion would seem to provide the Minister with surer grounds for rejecting a programme. However, while it is clear that at the macro level the national interest is served by a timely and thorough recovery of oil and gas reserves, it is by no means clear to the present authors that an individual programme which does not offer to develop a fallow discovery of small to medium potentiality can seriously be stated to be contrary to the national interest.

Conclusion on the Fallow Assets Initiative

I-A.33 There can be no doubting the impact of the Fallow Assets Initiative. In the first three years of its operation "over 660 blocks and 240 discoveries [were] identified as inactive or 'fallow'".[87] Statistics released by the Department in early 2010 show that the scheme continues to be effective in stimulating exploration, drilling and other significant activity, promoting the transfer of assets or causing licensees to relinquish blocks, either in whole or in part. Perhaps most

---

[83] The extent of ministerial discretion in licensing matters is discussed throughout Chapter I-4 and forms one of the major themes in Daintith, *Discretion*.

[84] Model Cl 17(4)(c)(i).

[85] Model Cl 17(4)(c)(ii).

[86] DECC, *Offshore Field Development Guidelines*, (hereinafter "Field Development Guidance"), available at www.og.decc.gov.uk/regulation/guidance/reg_offshore/reg_offshore_guide.doc (accessed 16 December 2010).

[87] *PILOT Annual Report 2004–2005*, at 6. See also further information to similar effect in the PILOT Report relative to 2005–2006, available at www.pilottaskforce.co.uk/templates/relay/communicationrelay.cfm/1839, at 11.

eye-catching of all is the large number of blocks and part-blocks that have been re-licensed following relinquishment. Of those blocks that were designated as Fallow B, 85 were re-licensed by the end of the 23rd licensing round offered in 2005. The next licensing round witnessed the re-licensing of an additional 52 blocks relinquished under the Fallow Process.[88] In the Department's latest release in 11 January 2010, 17 new Fallow B blocks that "must have significant activity or be relinquished by 31 December 2010"[89] are added to the list, as well as six new Fallow B discoveries that "must have significant activity or … be relinquished by 31 December 2011".[90]

The scheme has therefore been, by any practical measure of achievement, a remarkable success. But this should not blind us to the fact that the initiative is, from a regulatory perspective, a very curious animal. The scheme seeks to compel parties to engage in expensive and risky commercial activities, and/or to divest themselves of assets for which they have paid substantial sums of money and which could potentially realise millions of pounds of revenue, under pain of losing those assets without receiving a penny in compensation. Moreover, it does so on the flimsiest of legal foundations. One might have imagined that the industry would resent such a scheme and resist it with great force; instead, it has gone along with it voluntarily. The practical consequences of the disconnections between what the licence empowers the Minister to do and what the scheme purports to require have gone untested. Nor is there evidence that the delaying tactics that could have been used to frustrate the scheme without openly challenging it have been utilised. The substantial tightening up which has taken place, particularly in the Fallow Discovery Process, since the scheme was brought into effect has not been seriously opposed by the industry. Why? The fact that the process is underpinned by the provisions of the licence is frequently referred to.[91] The fact that it is also a process which takes place in the

I-A.34

---

[88] See the summary of activity in formerly fallow blocks contained at DECC, *UK Fallow Assets and Process –11th January 2010*, at "Relicensing of Fallow B Blocks", available at www.og.decc.gov.uk/UKpromote/fallow/fallow_assets.htm (accessed 17 December 2010).

[89] DECC, *Fallow Blocks 11th Release – 11 January 2010*, note, available at www.og.decc.gov.uk/UKpromote/fallow/FallowBBlocks_2010.doc (accessed 17 December 2010).

[90] DECC, *Fallow Discoveries 11th Release – 11 January 2010*, available at www.og.decc.gov.uk/UKpromote/fallow/ReleaseBDiscoveries_2010.doc (accessed 17 December 2010). Note that these six Fallow B Discoveries are added to the already existing 27 Fallow B Discoveries from the immediately preceding release, the licensees of which ought to carry out significant activity before 31 December 2010 in order to avoid the issue of relinquishment.

[91] PILOT PPWG, *Work*, para 3.2.3.3; *Revised Fallow Blocks Guidance*, "F – DTI Regulatory Position; *Revised Fallow Discoveries Guidance*", "G – DTI Regulatory Position".

shadow of the licensee's desire to remain in good standing with the Department, which is in a position quite legitimately to discriminate against the licensee when allotting new licences,[92] is less frequently discussed but must surely provide at least as great an incentive to the licensee to co-operate in the scheme's operation.[93] But the desire to remain in good standing can only be relied upon to moderate the behaviour of individual licensees for so long as they obtain a benefit for so doing. As the UKCS becomes progressively more mature and the number of new or recycled fields to be let dwindles, and their relative size and value decrease, certain licensees may take the view that they are prepared to retain their existing acreage, but that they have no interest in bidding for new lets. When this point is reached, the threat of loss of good standing will lose its sting. It becomes much more likely that at least some licensees will start to act in a less co-operative manner and test the boundaries of the scheme.[94] The Fallow Areas Initiative has been a notable and welcome success, but its success is by no means guaranteed to continue.

STEWARDSHIP

I-A.35   While PILOT's Progressing Partnership Work Group was developing the ideas that would lead to the Fallow Areas Initiatives, another had the task of considering issues related to existing fields, also known as brownfields. In 2002, this Brownfields Work Group launched a benchmarking exercise in order to establish the contribution that could be made to the attainment of the Vision for 2010 by mature fields.[95] This study considered 23 mature fields, which together accounted for half of the production and reserves contributed by such fields. Reporting in 2003, it concluded that the development of brownfields could help in "closing the gap to the PILOT targets". It also noted, however, that this would "require a concentrated effort on moving brownfield projects forward".[96]

---

[92] See Chapter I-4.

[93] See Daintith, *Discretion*, para 3421 and the discussion in Chapter I-4.

[94] The industry's tendency to conduct operations jointly, discussed at Chapter II-2, may mitigate this problem. It is possible that unco-operative participants in a JOA may find themselves outvoted. However, there are arguments that for some in the industry, specifically among the independents, this point of willingness to challenge the status quo may have already been reached and that it is only a matter of time before the UKCS witnesses similar action to the challenges mounted to the industry's regulators in the United States. See E Üşenmez, "*Increase in Influence: How the Independents are Challenging the Authority of the US Government and Why the UK Government Should Take Note*", 5 (2009) IELR 171–181.

[95] See para I-A.03.

[96] Aupec, *2002 Brownfield Benchmark Study*, June 2003, p 1.

Building on this foundation, PILOT launched the so-called I-A.36
Brownfield Studies in 2004, which reported their findings in March
2005.[97] This report has resulted in some important new industry and
regulatory initiatives. This part of the chapter considers some of the
key findings of the March 2005 Report before looking in some more
detail at the most significant regulatory development to emerge from
it: the Stewardship Process.

THE BROWNFIELD STUDIES REPORT

The Report identifies three "levers" to be used in realising the Vision I-A.37
for 2010: "(1) Increase the size of the resource base. (2) Ensure that oil
and gas is extracted as efficiently as possible while the infrastructure
exists. (3) Extend the life of the infrastructure."[98] As regards the first
"lever", and in line with the observations above regarding the nature of
a mature province, the Report's focus is particularly on maximising the
recovery from existing fields and developing discoveries around them
rather than on looking for new discoveries. Whereas the preceding
benchmarking study mentioned above had noted that a "concentrated
effort" would be required to move brownfield projects forward, the
Report is more direct, suggesting that progress towards the Vision for
2010 will require "a change of mindset across the industry".[99]

The received wisdom of the industry has been that in so far as I-A.38
maximising the recovery from a field is principally about ensuring
that the right assets are in the right hands, the focus must be upon
facilitating asset transfer if it is always to be possible to increase
investment when required. There is certainly a great deal of truth in
this belief, and the Government is clearly playing a role in this regard
with the Fallow Blocks and Fallow Discovery Initiatives discussed
previously in this chapter.[100] But the Brownfield Studies reveal that
a focus on asset transfer with regard to producing fields could be
obscuring the important insight that similar increases in investment
(and thus in recovery) as were achieved with the involvement of a
new operator or owner were also possible where only the existing
joint venture (JV) partners were involved.[101]

That said, however, the research conducted in the course of I-A.39
the Brownfield Studies reveals a wide variation between the best

---

[97] PILOT, *Maximising Economic Recovery of the UK's Oil and Gas Reserves: Context for the Brownfields Challenge* (Report of the PILOT 2004 Brownfields Studies), March 2005 (hereinafter *"Brownfield Studies Report"*).
[98] *Brownfield Studies Report*, p 4.
[99] *Ibid*, p 6.
[100] See in particular para I-A.17.
[101] *Brownfield Studies Report*, p 11.

and worst performance in this regard. In view of this finding, the Report indicates that the key issue for existing fields is not transfer of assets, but rather ensuring that existing JV partners are applying the standards of *stewardship* displayed by those operating the best-performing fields. This concept of stewardship is understood, *inter alia*, to be about the "asset owners consistently doing the right things to identify and exploit opportunities".[102] While at first sight it may thus appear that stewardship is primarily concerned with the voluntary activities of the JV partners in implementing good oilfield practice, the Report makes clear – indeed from the outset – that intervention by the regulator is envisaged where problems exist in this regard. The first of its four recommendations highlighted in the Introduction is "Improving stewardship – screening all fields to focus more detailed annual reviews on those fields where stewardship could be a concern".[103]

I-A.40   The apparent willingness of the industry to agree to enhanced regulatory scrutiny of, and even intervention in, its affairs is thus a very striking feature of this report. A Brownfields Work Group dedicated to stewardship set about developing a model for the screening process, the basic detail of which is outlined in the Report.[104] This gives a first indication of why the industry has behaved in what may appear to be a counterintuitive way. The model aims to reduce the administrative load on the vast majority of JVs, first, by introducing a simplified approach to the annual reporting of field data and, second, by focusing the detailed attention of the regulator only on those fields where, in its view, "further conversation" is necessary.[105] In short, a similar attitude on the part of the industry as was apparent in respect of fallow blocks and discoveries[106] is evident here: there is no sympathy for JV partners who are not doing all that they could to maximise the recovery from the assets they have been entrusted with. It remains to consider, however, how this basic model has been developed and implemented by the Department.

The Stewardship Process

I-A.41   The Stewardship Process is not a creature of statute or even of regulation. In common with the treatment of fallow blocks and fallow discoveries, stewardship is dealt with on the basis of the Secretary of State's powers under the licence. One will hunt in vain,

---

[102] *Ibid*, p 11.
[103] *Ibid*, p 5.
[104] *Ibid*, p 24.
[105] *Ibid*, p 24.
[106] See para **I-A.16**.

however, among even the most recent iteration of model clauses for details of the process. Instead the details are elaborated in the Department's *Offshore Field Development Guidelines*. In so far as this guidance indicates that the regulator's *"overall aim* is to maximise the economic benefit to the UK of its oil and gas resources" and that its first policy objective is "ensuring the recovery of all economic hydrocarbon reserves",[107] it may readily be seen how stewardship, as a process of checking to see that asset owners are "consistently doing the right things to identify and exploit opportunities", fits into this picture. And indeed these sentiments are repeated practically verbatim in the section of the guidance dedicated to stewardship.[108] The Department is at pains to emphasise that there will usually be alignment between its objectives in terms of maximising economic recovery for the nation and the commercial interests of the JV partners.[109] It notes, however, that there are instances where there could be divergence, including the following:

- "Where a field covers more than one block, with different owners. Attempts to gain higher shares in total output (ie capturing other companies' reserves) could damage reservoirs and result in needless expenditure.
- Where production is via a floating production system [and] high operating costs [produce] ... an incentive to cream off high early production and move to the next location, rather than produce all the economic oil.
- Where company capital constraints points them towards a lower cost, but less economic, development option which could leave potentially economic reserves unproduced.
- Where severe cash constraints lead Licensees to prefer options which emphasise the need for early cash at the expense of additional recovery, or result in additional gas flaring.
- Where partners disagree amongst themselves."[110]

The first four of these are clear cases of a lack of alignment between   I-A.42 the commercial interests of the JV and the state's interests as represented by the regulator. Some, however, as well as producing a situation which takes the JV out of alignment with the regulator may also be examples of a divergence among the interests of the partners in an individual JV. This indicates perhaps the most compelling reason for the industry's willingness to countenance the enhanced regulatory scrutiny and intervention that stewardship

---

[107] *Field Development Guidance*, s 2.1 (emphasis in original).
[108] *Ibid*, s 6.1.
[109] *Ibid*.
[110] *Ibid*, Appendix 3.

seems to involve: the very threat might be enough to persuade recal-citrant partners that they must either agree to further investment or divest. And indeed, in addition to the idea that stewardship is about asset owners consistently doing the right thing, the guidance notes mention an additional "key factor": that "[a]ssets are in the hands of those with the *collective will, behaviours and resources* to achieve this".[111] The Department's discussion on licence extension policies[112] mentions this threat more directly:

> "Where a field is not being operated to DECC's satisfaction, [the Department] will press the existing licensee to raise his game to the level expected. If it continues to fall below the required standard and reaches the end of the licence's lifespan, the Secretary of State reserves the right to refuse an extension, *and instead to invite competitive bids for a new licence over the field.*"[113]

I-A.43   Thus, the philosophy of asset transfer discussed at para I-A.38 has not wholly been left behind in this initiative.

I-A.44   The Stewardship Process is described as involving two stages. In the first, the operator, acting on behalf of the licensee(s), prepares an annual return which summarises "key aspects of the field's performance".[114] This return is to be completed during February and covers the previous calendar year, together with some forecasting for the year ahead.[115] Running in parallel with this exercise is a requirement for a further return relating to Production Efficiency, the results of which may be fed into the stewardship process. This is also the responsibility of the operator with the same deadline for submission of February.[116] The general idea is that the reporting burden is reduced for most fields.[117] Once the data is submitted, the guidance states that it will be screened quickly by the Department to identify "those fields where

---

[111] *Field Development Guidance*, s 6.1 (emphasis added).

[112] In September 2010, the offshore licences awarded in the 1st licensing round expired. The licences from the 2nd round will expire on 24 November 2011.

[113] DECC, *Licence Extensions*, at "Poorly-stewarded Fields" (emphasis added), available at www.og.decc.gov.uk/upstream/licensing/licextent.htm (accessed 17 December 2010).

[114] *Field Development Guidance*, s 6.1.

[115] *Ibid*, Appendix 11.

[116] *Ibid*, Appendix 12. Note that the Production Efficiency Review Process is also the product of a joint government-industry initiative, this time involving the DECC–Oil & Gas UK Production Efficiency Work Group.

[117] Note, however, that there is some confusion in this regard in the published information. Whereas the *Brownfield Studies Report*, p 24, suggested that the "Annual Field Report is replaced by a simplified and more focused data request", the Department's website notes that while there has been since 2005 a waiver until further notice of "the requirement for periodic reports on licensed activities ... [t]his waiver has no effect on any other reporting requirement, such as ... Annual Field Reports" DECC, *Upstream: Reporting*, at "Other Reports", available at www.og.decc.gov.uk/upstream/field_reporting/index.htm (accessed 17 December 2010).

more detailed discussions are required".[118] As regards the screening process, the Department will focus upon "simple, objective performance indicators based on reserves replacement, production decline, facilities performance, investment levels and well utilisation".[119] If further clarification of the data submitted is required, then the regulator will discuss this informally with the operator. It is anticipated that feedback on the submission will be made in May. This brings the first stage to a close – and indeed brings the whole Stewardship Process to a close for most operators.

I-A.45   The second stage thus involves only those fields where issues have been identified during the first. It begins with formal written notification to the operators concerned and will involve discussions between the Department and the JV partners as to how these issues may be resolved. It may also involve audit of specific aspects of field management or indeed a full audit of the field as a whole. Third-party experts may be involved "to help resolve technical issues".[120] The Department expects that in so far as good stewardship equates to attractive economic investment, then in most cases "improvement can be secured by normal commercial means by the JV; perhaps by realignment, the introduction of 3rd party investment or, possibly, divestment".[121]

I-A.46   If, however, the discussions (and perhaps audits and the involvement of technical experts) do not lead to alignment between the Department and the JV, then the guidance indicates that the regulator will use its powers under the licence "to require the JV to improve its Stewardship of the field".[122] The text of the guidance mentions two possibilities, namely the specification of a development and production programme requiring the JV to carry out economic investment, and the replacement of the operator in cases where this party is identified as being the cause of the problem. The diagram illustrating the process mentions additionally, somewhat menacingly, "other sanctions".[123] Presumably, in the absence of any further specification, this is a reference to the possibility that in the ultimate the Department may revoke the licence.

I-A.47   It is one thing to say that Stewardship fits into the regulator's policy aims and objectives. It is quite another to say that there is a legal basis for the process. In so far as it is the forerunner to Model Cl 17[124] that is extracted in the Appendix to the guidance where

---

[118] *Field Development Guidance*, s 6.1.
[119] *Field Development Guidance*, Appendix 11, at "Stewardship Process Timing".
[120] *Ibid*, s 6.1.
[121] *Ibid*, s 6.1 and Appendix 11.
[122] *Ibid*, s 6.1.
[123] *Ibid*, s 6.1 and the associated diagram.
[124] That is, Cl 15. Only sub-cll (1) and (2) are included in the Appendix.

the Department lists the "most relevant" model clauses,[125] it would appear that the regulator assumes that this clause provides it with the necessary power. The guidance certainly notes that in reviewing Field Development Plans the Department "will need to be satisfied ... that the proposals address all the recoverable reserves of a field and do so over a long enough time period" and that "Licensees take into account implications for other developments in the area".[126] In other words, in describing its approach to Model Cl 17, the regulator clearly has in view precisely the sorts of issues covered by Stewardship. But while this clause may be read as giving the regulator the power to consider these issues *at the outset*, it is a question whether it allows him to *revisit* these issues in later years as the Stewardship Process assumes. As has been noted above in relation to the discussion of fallow blocks and discoveries, it is by no means clear that Model Cl 17 may be read in this way.[127] There is nothing in that clause that explicitly allows the regulator to serve a development notice (as envisaged by the guidance) in the absence of a new programme. It is then a question of whether the regulator could get round this problem by requiring a new development programme. Once again as noted above, the fact that Model Cl 17(2) refers only to "a programme" singular raises doubts in this regard. It might be argued that the reference in Model Cl 17(3) to the Minister directing the licensee "to prepare a programme or programmes ... in respect of a further period or further periods" during the term of the licence could be taken to imply that the regulator may act in this way, but this would appear to be excluded in cases where the original programme related to the entire term of the licence. It might thus be a matter of arguing that the regulator would have to rely on a purposive interpretation of the model clauses in order to prevail in the service of a development notice where a JV raises objections.

I-A.48　　On the other hand, if one assumes, as the regulator clearly does, that it *does* possess the required powers under the licence to impose the "sanctions" listed in the guidance at the end of the second stage of the Stewardship Process, it could be argued that the process itself is actually a sign of the Department's moderation and forbearance. Inasmuch as the licence requires no such discussion before the service of notices, the regulator could be said to be going further to accommodate problem JVs than strictly it is required to. The fact that one can reach two such divergent conclusions, however, is surely a confirmation of the legal uncertainty surrounding this issue.

---

[125] *Field Development Guidance*, Appendix 2.
[126] *Ibid*, s 2.2.
[127] See in particular para I-A.31.

Conclusion on Stewardship

It may be said, then, that Stewardship is in many respects a logical   I-A.49
progression from the treatment of fallow blocks and discoveries.
Whereas the latter aims to ensure that allocated acreage is actually
explored and assessed and that discoveries are actually developed,
the focus of Stewardship is on ensuring that producing assets fulfil
their potential. The apparent willingness of the industry to countenance the enhanced regulatory scrutiny and intervention involved
can at first sight appear counterintuitive. Closer inspection, however,
reveals that this willingness is essentially motivated by self-interest.
At the most basic level, the simplified reporting required by the
Stewardship Process means that the vast majority of fields experience
a reduced administrative burden. More importantly, where there
are misaligned JVs, the more dynamic partners can use the threat
of intervention by the Department as a lever to encourage the more
recalcitrant either to agree to investment or to divest themselves of
their stake. Finally, where either this threat does not work or the JV
as a whole is simply underperforming, then the regulator will indeed
take steps to enforce the changes necessary to maximise recovery.

It is important to note, however, that this apparent win–win   I-A.50
scenario may mask problems. In common with the situation
pertaining to fallow blocks and discoveries, the precise legal basis
for the Stewardship Process is far from clear. This is not a problem
at the level of the industry overall where there is broad agreement
with the Department on the need for the process. But it is a question
what would happen should a recalcitrant JV partner on the receiving
end of the Department's attentions challenge the legitimacy of any
demands made by the regulator. Vague references to licence powers
would not then be of any avail and a court may struggle to see
how Model Cl 17 justifies the sort of action that the Department's
guidance suggests it may be minded to take when "further conversa-
tions" do not result in an alignment of views between it and the JV.
The analysis above suggests that the Department may have to rely on
a purposive interpretation of the model clauses in order to prevail in
such a situation. The fact that courts have not so far been required
to consider such clauses only serves to mask the holes that appear to
exist.

This absence of court action up to this point can of course be   I-A.51
explained by the fact that companies that might have been minded
to litigate are "repeat players". In other words, knowing that they
have to deal with the Department on an ongoing basis means that
they are unlikely to take any action that would be calculated to incur
the regulator's wrath. Whether this calculus will continue to apply as
the province matures further must be open to question.

INDEX

Location references are to paragraph numbers; authors are indexed if they are mentioned in the main text.

AAPL Model Form 610, II–2.03, II–2.43, II–2.44
AAPL Model Form 810, II–2.27
abandonment *see* decommissioning
abuse of dominance
    EU law, II–11.10–14, II–11.18–20
    ICoP and, II–11.52–3
    infrastructure access and, I–6.64,
        I–6.70–2, II–11.46, II–11.51–4
    prohibition, II–11.08, II–11.18–20
    tying, II–11.20, II–11.56
    vertical agreements, II–11.58
ACAS, II–15.47
account bank agreements, II–10.49,
    II–10.82–4
acreage
    competing applications, I–4.35–7,
        I–4–50
    decline, II–3.15
    Fallows Area Initiative, I–A.6
    frontier areas, I–A.7
    public announcement, I–4.19
    recycling discarded acreage, I–4–54
    rental payments, I–4–45–6
    selection and environmental issues,
        I–4.24–31
    unused acreage, I–A.23
adjudication
    arbitration *see* arbitration
    enforcement, II–15.78–81
    litigation *see* litigation
    statutory adjudication, II–15.62–4
administration
    contractual rights and, II–10.99
    decommissioning liabilities, II–10.105
    environmental considerations,
        II–10.106–9
    expenses, II–10.100–5
    health and safety and, II–10.106–9
    moratorium over legal proceedings,
        II–10.98–9

objectives, II–10.97
overview, II–10.95–109
salvage principle, II–10.104–5
ADR *see* alternative dispute resolution
affiliates
    affiliate guarantees, I–13.18, I–13.31
    confidential information to, II–12.57
    contractor groups, II–4.64, II–6.57
    decommissioning, I–13.27, II–9.77
    definition, II–5.58–9, II–5.67, II–9.49,
        II–9.50
    IMHH Agreement, II–6.66
    market share, II–11.14
    operator groups, II–6.52
    pre-emption rights and, II–9.48–54
    transfer of assets to, II–9.48–54
agency law, operators, II–2.37–8
agency workers
    contracts, II–14.10
    definition, II–14.04, II–14.10–18
    employees or, II–14.12–18
    end-users, II–14.10
    implied contracts of service, II–14.12–18
    length of employment relationship,
        II–14.17
    working time rights, II–14.46
Agreement on Small Ceteceans of the
    Baltic and North Seas (ASCOBANS),
    I–11.15
AIM listing, II–10.11
AIPN *see* Association of International
    Petroleum Negotiators
air pollution
    combustion installations, I–11.58–61
    merchant shipping, I–11.50–7
    national emission ceilings, I–11.64
    offshore regulation, I–11.50–64
    petroleum licensing and, I–11.62–4
air transport, IMHH Agreement and,
    II–6.71
ALARP standard, I–9.14, I–10.50
Aldersley-Williams, J, II–2.70
Algeria, I–3.55
alternative dispute resolution (ADR)
    civil procedure and, II–15.43–4

concerns, II–15.49–54
confidential information, II–15.37,
   II–15.53–4
consensual nature, II–15.35
early neutral evaluation, II–15.55,
   II–15.64
enforcement, II–15.79
EU law, II–15.46
expert determination *see* **expert**
   determination
foreign jurisdictions, II–15.45–6
litigation and, II–15.21
med-arb, II–15.57
mediation *see* **mediation**
mini-trials, II–15.56
non-binding nature, II–15.52, II–15.56
option, II–15.04, II–15.35–64
power asymmetry, II–15.49–51
specialist tribunals, II–15.47
terminology, II–15.35
tiered escalation to, II–15.31, II–15.35
United States, II–15.25, II–15.45
American Arbitration Association/
   International Court of Dispute
   Resolution (AAA/ICDR), II–15.66
AMIs (area of mutual interest)
   agreements, II–2.05, II–11.34
Anardako Petroleum, II–2.33
Angola, I–3.56, II–15.81
anti-competitive agreements
   block exemptions, II–11.17, II–11.48,
      II–11.64
   technology transfer, II–11.64,
      II–12.27
   de minimis exclusions, II–11.14–16
   examples, II–11.11
   exemptions, II–11.17, II–11.30, II–11.32
   forms, II–11.12
   horizontal agreements, II–11.28
   infrastructure access and, I–6.64–9
   issues in upstream agreements,
      II–11.30–67
   joint operating agreements, II–11.28,
      II–11.30–3
   joint procurement contracts, II–11.28,
      II–11.35
      infrastructure agreements, II–11.46
   joint sales contracts, II–11.28,
      II–11.36–45
      Britannia decision, II–11.40
      Corrib decision, II–11.42, II–11.45
      EU decisions, II–11.39–45
      infrastructure agreements,
         II–11.46–50

UK-Belgium gas interconnector,
   II–11.41, II–11.44
non-compete clauses, II–11.58
prohibition, II–11.08–17
specialisation agreements, II–11.17,
   II–11.48
territorial restrictions, II–11.15,
   II–11.58
transport agreements, joint supply/
   purchase of capacity, II–11.46–50
UUOAs, II–11.34
vertical agreements, II–11.29,
   II–11.56–67
   exchange of information, II–11.60–2
   market share, II–11.58
   MER UK and, II–11.58–60
   non-compete clauses, II–11.58
   problem areas, II–11.58
   void agreements, II–11.13
apprenticeships, II–14.05, II–14.22
arbitration
   binding nature, II–15.52, II–15.65
   clauses, II–15.65
   costs, II–15.06
   Energy Charter, I–3.22
   enforceability, II–15.66
   enforcement, II–15.78–81
   gas/LNG price reviews, II–8.45
   governing law, II–15.67
   industry preference for, II–15.01,
      II–15.66
   international enforcement, II–15.81
   international regimes, II–15.66
   investor-state arbitration, II–15.48
   jurisdiction, II–15.65
   litigation alternative, II–15.70–1
   med-arb, II–15.57
   New York Convention, II–15.81
   option, II–15.04
   overview, II–15.65–71
   power imbalances, II–15.50
   privacy, II–15.71
   process, II–15.66–8
   reach of clauses, II–15.69
   speed, II–15.70
   Yukos arbitration, I–3.23
Ardmore field, I–12.54, II–2.56
area rental payments, I–4.45–7
Argyll field, I–4.74n
ASCOBANS, I–11.15
Asfari, Ayman, I–10.58n
asset acquisitions
   advantages, II–9.10–18
   consents, II–9.14

corporate acquisitions or, II–9.10–18, II–9.43
due diligence, II–9.16, II–9.19–29
earn-in, II–9.08
farm-in, II–9.07, II–9.81–3
insolvency cases, II–10.117–18
investigation of title, II–9.24
  insolvency cases, II–10.117–18
liabilities, II–9.17, II–9.75–80
monetary consideration, II–9.06, II–9.59–65
no pre-emption, II–9.11–12
operatorship transfer, II–9.15
overview, II–9.04–18
simplicity, II–9.13
swaps, II–9.09, II–9.43, II–9.46
taxation, II–9.18
types, II–9.05–9
asset sale and purchase agreements
assets, II–9.58
completion, II–9.84–8
conditions precedent, II–9.66–7
consents, II–9.66
consideration, II–9.59–65, II–10.118
decommissioning liabilities, II–9.76–80
indemnities, II–9.75–7
interim periods, II–9.68–9, II–9.88
overview, II–9.57–80
waiver of pre-emption rights, II–9.66
warranties, II–9.70–4
asset sales
assignment clauses, II–9.21
auctions, II–9.22
consents, II–9.30–6
corporate capacity, II–9.29
due diligence, II–9.19
farm-out agreements, II–9.07, II–9.81–3
restrictions, II–9.37–56
asset stewardship see stewardship
asset trading
acquisitions see asset acquisitions
agreements see asset sale and purchase agreements
assignment see assignation/assignment
completion, II–9.82
consents, II–9.27, II–9.30–6, II–9.66
consents to transfer, II–9.89, II–9.90
decommissioning agreements and, I–13.06, II–9.28
due diligence, II–9.19–29
execution deeds, II–9.89
farm-in/out, II–9.07, II–9.81–3, II–10.15
increasing activity, I–12.45
JOA consent, II–9.55–6

joint ventures, II–2.12, II–2.22
Master Deed, II–2.13, II–9.32, II–9.43–5, II–9.87, II–9.89–93
MER Strategy and, I–12.45, II–9.03
New Transfer Arrangements, II–9.89–93
notices of transfer, II–9.89, II–9.92
OGA role, II–9.30–6
overview, II–9.01–93
portfolio management, II–9.02
pre-emption rights see pre-emption rights
rationale, II–9.02
restrictions, II–9.37–56
sales see asset sales
stewardship and, I–A.38
Wood Review, I–7.58
assignation/assignment
completion, II–9.84–8
due diligence and, II–9.21, II–9.27, II–10.118
gas/LNG sales agreements, II–8.51
insolvency cases, II–10.118
intellectual property
  collaborations, II–12.43
  future inventions, II–12.36
  requirement, II–12.35
  vertical agreements, II–11.58
JOAs, II–2.13, II–9.55–6
leases, II–13.13, II–13.24
licences, II–9.24
  documentation, II–9.84
  OGA consent, II–9.32
Master Deed, II–2.13, II–9.32, II–9.43–5, II–9.87, II–9.89–93
mortgages, II–10.77
patents, II–12.14
pre-emption rights and, II–9.21, II–9.27
  affiliate route, II–9.48–54
  overview, II–9.37–56
  package deals, II–9.47
  unmatchable deals, II–9.46
restrictions, II–9.21, II–9.37–56
securities, II–10.77, II–10.79
warranties, II–9.71
Association of International Petroleum Negotiators (AIPN)
AIPN GSA, II–8.25, II–8.29, II–8.30, II–8.43, II–8.49, II–8.53
AIPN SPA, II–8.26, II–8.47
standard contracts, II–5.75–6
training programmes, II–15.24
Atiyah, Patrick, II–2.58
atmospheric emissions, I–11.50–61
auctions, II–9.22

Australia
  forfeiture clauses and insolvency,
    II–2.91
  Montara Oil Spill, I–11.101
  production decline, I–5.07
  SEAM, II–15.56
automated platforms, II–12.03
automatic referral notices (ARNs), I–6.86,
  I–6.94–5
aviation, II–6.04

back-to-back indemnity clauses, II–6.21–3,
  II–6.57, II–7.107
banking crisis (2007–8), II–10.94
BAT, I–11.36, I–11.37, I–11.60
BATNA, II–15.23
BDO Stoy Hayward Commercial Disputes
  Survey (2003), II–15.19
BEACH, II–8.30
BEACH 2015, II–8.22
Beck, M, I–10.47
Belgium
  North Sea continental shelf boundaries,
    I–8.09
  UK-Belgium gas interconnector,
    II–11.41, II–11.44
benchmarking, II–11.55, II–11.60
bespoke licences, I–4.03, I–4.74
Best Available Technique (BAT), I–11.36,
  I–11.36, I–11.37, I–11.60
Best Environment Practice (BEP), I–11.36,
  I–11.37
Best Environment Practice (BES), I–11.36
bidding agreements, II–2.05–6, II–2.17,
  II–9.84, II–11.34
block exemptions see anti-competitive
  agreements
blocks
  average size, II–3.15
  contiguous blocks, I–4.22, I–4.23
  fallow blocks see fallow blocks
  grid system, I–4.18–19
  limitation on licensed numbers, I–4.66
  meaning, II–03.02
  part blocks, I–4.18
  PEDLs, II–03.02
  size and shape, I–5.17, I–9.09
bona vacantia, II–10.113
boycotts, I–12.25, II–11.11
BP, I–7.58, I–8.06, II–2.33, II–2.55,
  II–3.15, II–08.06, II–8.53
  see also Deepwater Horizon disaster
Brazil, II–4.41, II–12.03
Brent field, I–10.16n, I–12.01

Brent Spar decommissioning
  consensus building, II–15.17
  controversy, I–12.01, I–12.06, II–2.93
  effect on regulation, I–11.20
  Greenpeace protest, I–12.25–6
  Guidance Notes and, I–12.56–7
  legal impact, I–12.36, I–12.38–40,
    I–12.55
  original disposal plan, I–12.20–4
  OSPAR and, I–12.19, I–12.36, I–12.76
  overview, I–12.19–36
  Stakeholder Dialogue, I–12.06,
    I–12.27–35, I–12.36, II–15.07n
Brexit, I–1.04, I–9.115n, II–11.01n,
  II–14.82
bribery, II–5.65, II–5.67
British Chamber of Commerce, I–9.86
British Gas, I–5.05, II–8.21
British Geological Survey, I–9.13
British National Oil Corporation
  (BNOC), I–5.05, II–2.03, II–5.13
brownfields
  Brownfields Initiative, I–5.06, I–12.81,
    I–A.35–51
  Brownfields Studies Report, I–A.37–40
  Brownfields Work Group, I–A.35,
    I–A.40
  meaning, I–A.35
building contracts, II–2.67, II–2.85,
  II–5.03, II–6.04, II–15.62–4,
  II–15.77
bundling, I–6.72, I–6.89
Burgoyne Report, I–10.23–30, I–10.31,
  I–10.33–5, I–10.38–9

call-off contracts, II–5.57, II–5.59
Canada
  CAPL Operating Procedure JOA,
    II–2.43
  JOAs, fiduciary duties, II–2.43
  servitudes, II–13.23
  standard contracts, II–5.74
Canadian Association of Oilwell Drilling
  Contractors (CAODC), I–5.74
capital allowances, I–3.44, I–7.28–33,
  I–7.62
CAR insurance, II–2.36
carbon budgets, I–3.60
carbon capture and storage, I–3.12,
  I–5.22, I–12.45
carry financing, II–10.26
cartels, II–11.02, II–11.03, II–11.23,
  II–11.36
cash flow finance, II–10–04

casual workers, II–14.03, II–14.08, II–14.29, II–14.30
Center for Strategic & International Studies (CSIS), I–3.06
Centre for Environment, Fisheries and Aquaculture Science (CEFAS), I–11.19, I–11.38, I–11.45, I–11.46, I–11.47
Centrica, I–3.58
CERM, I–3.20–1, I–3.67
CFCs, I–11.51
CFR contracts, II–8.14
Channel Islands, I–8.12
charges, II–10.65, II–10.87, II–10.89
chemicals, I–9.26, I–11.38–42, I–11.44–8
Chevron, II–3.15
China, I–1.08
Chrysaor, I–7.58
Churchill, Winston, I–9.01
CIF contracts, II–08.06, II–8.15, II–8.17, II–8.20
circular indemnities, II–6.06–7
civil procedure
  ADR and, II–15.43–4
  commercial court, II–15.75
  costs, II–15.43
  England, II–15.43, II–15.77
  pre-action protocols, II–15.43–4
  Scotland, II–15.44, II–15.75–6
  summary trials, II–15.76
  Technology and Construction Court, II–15.77
Clair field, II–3.15
Clair Platform, I–11.01, I–11.02
Clark, Greg, I–3.02
Claymore platform, I–10.32
climate change
  Committee on Climate Change, I–3.60
  emission reduction and energy security, I–3.37–42, I–3.60–2
  EU legislation, I–3.37, I–3.40, I–3.69
  National Emissions Reduction Plan, I–3.37
  renewable energy, I–3.30
closed list principle, II–13.03
cluster area allowances (CAAs), I–7.40
coal-fired power stations, I–3.41, I–3.62, I–3.69
coast protection, I–4.30, I–4.81, I–11.74–8
collaboration
  agreements, II–12.40–7
  competition and, I–2.54, I–5.50–1, II–1.01, II–11.30–3, II–11.55, II–11.62–4

deal-building, II–15.20
decommissioning, 1–12.46
dispute resolution and
  consensus building, II–15.17–19
  option, II–15.04, II–15.17–20
  intellectual property and, II–12.40
  indemnities, II–12.46–7
  joint ownership, II–12.43–5
  legal obligation, I–5.33, I–5.46–57
  meaning, II–12.38
  MER Strategy, I–5.46–57, II–11.05, II–15.20
  stakeholders, II–15.17–18
  Wood Review and, I–2.54, I–5.23, I–5.46, II–04.01
Columbus field, II–3.15
combustion installations, I–11.58–61
Commercial Code of Practice (CCoP), I–5.53, I–6.79
commercial court, II–15.75
company groups, definition, II–5.59, II–5.62, II–6.52
company voluntary arrangements, II–10.119–20
Competition and Market Authority (CMA)
  collaboration and competition, I–5.50, II–1.01
  de minimis agreements and, II–11.16
  ICoP and, II–11.53
  investigation powers, II–11.02
  on OGA and competition, II–11.62
  regulatory authority, I–6.64
competition law
  abuse of dominance see abuse of dominance
  administrative powers, II–11.04
  anti-competitive agreements see anti-competitive agreements
  basic prohibitions, II–11.08–20
  collaboration and, I–2.54, I–5.50–1, II–1.01, II–11.30–3, II–11.55, II–11.62–4
  comfort letters, II–11.41
  common oil and gas issues, II–11.30–67
  dawn raids, II–11.03
  de minimis exceptions, II–11.14–16, II–11.23
  EU law, I–6.64, I–6.68
    decisions on joint sales contracts, II–11.39–45
    joint procurement, II–11.35
    sanctions, II–11.02
  exchange of information and, II–11.54–5

importance, II–11.01–7
infrastructure access and, I–6.07,
    I–6.64–72, I–6.98, II–7.65
    ICoP, II–11.52–3
    joint supply/purchase, II–11.46–54
investigation powers, II–11.03
joint services, I–6.68
MER Strategy and, II–11.05,
    II–11.58–60, II–11.63–7
overview, II–11.01–67
price-fixing, I–6.68, II–11.11, II–11.15,
    II–11.58
relevant market, I–6.66–7, II–11.26–7
rule of capture and, II–3.04–5
sanctions, II–11.01–2
transport agreements and, II–7.65
UK or EU jurisdiction, II–11.21–5
whistle-blowing, II–11.03
compulsory purchase, II–13.03, II–13.09
computer programs see software
concerted practices, II–11.12, II–11.55
confidential information
    ADR, II–15.37, II–15.53–4
    arbitration, II–15.66
    competition law and, II–11.54, II–11.55
        vertical agreements, II–11.60–2
    confidentiality agreements, II–5.14,
        II–7.09, II–7.29–32, II–11.56
    due diligence and, II–9.20
    employees and, II–12.33
    intellectual property, II–12.26–32
    joint operation agreements, II–2.42
    LOGIC contracts, II–12.54–7
    mediation, II–15.37
    risk matrix, II–11.55
    standard contracts, II–5.14
    transport agreements, II–7.29–32
Conocophillips, II–3.15
consensus building, II–3.35, II–15.07,
    II–15.17–19, II–15.37
consequential losses
    assessment of exclusion, II–6.80–2
    defining, II–6.81
    drafting issues, II–6.80–1
    exclusion, II–6.16, II–6.77–82
    IMHH, II–6.61, II–6.81
    JOAs, II–2.34
    London Bridge, II–6.79, II–6.80
    meaning, II–6.78–9
consideration
    asset purchase, II–9.06, II–9.59–65
        insolvency cases, II–10.118
    economic date, II–9.60–1
    licences, I–4.32–4

notional interest, II–9.63
    share sales, II–9.65
    working capital, II–9.62
consortium financing, II–10.30, II–10.33
construction and tie-in agreements
    entry into force, II–7.53
    indemnities, II–7.51–2
    transport, II–7.48–53
continental shelf
    boundaries see continental shelf
        boundaries
    UK see UKCS
continental shelf boundaries
    1969 North Sea cases, I–8.07,
        I–8.09–12
    boundary agreements, I–8.06, I–8.08
        unitisation and, II–3.52–65
    early delimitation, I–8.07–8
    North Sea boundaries, I–8.03–12
    Rockall, I–8.13–35
    territorial extent, I–4.61
    UNCLOS, I–8.15–33
    unresolved issues, I–8.13–33
contra proferentem interpretation,
    II–6.34–5, II–6.37, II–6.38–45
contractor groups, definition, II–4.62–4
contractors
    agency workers, II–14.04
    contractor-operator relationship
        dismissal issues, II–14.83–93
        drilling contracts, II–4.04–24
    employment status, II–14.04
    independence, II–14.25
contracts
    agency, II–2.37–8
    categories, II–5.04
    competition and see anti-competitive
        agreements
    construction and tie-in agreements,
        II–7.48–53
    contract manuals, II–5.10
    crude oil sales see crude oil sales
        agreements
    dayrate drilling see drilling contracts
    dispute resolution clauses, II–15.29–30
    freedom of contract, II–8.02
    gas/LNG sales see gas/LNG sales
        agreements
    IMHH Agreement see Industry Mutual
        Hold Harmless Deed
    indemnities see indemnities
    JOAs see joint operating agreements
    joint procurement contracts, II–11.28,
        II–11.35

joint sales contracts, II–11.28,
 II–11.36–45
licences as, I–4.12–14, I–4.82, I–10.05
penalty clauses, II–8.40
primary and secondary obligations,
 II–8.41
risk allocation *see* indemnities
risks, II–13.04
securities over contractual rights,
 II–10.76–80
service contracts, II–5.05–8
SPAs *see* petroleum sales agreements
standard form *see* standard contracts
structures, II–5.04–11, II–6.55,
 II–12.07–8
transport *see* transport agreements
UUOAs *see* unitisation and unit
 operating agreements
convertible bonds, II–10.09, II–10.12
co-operation, international law, II–3.51
copyright
 creation, II–12.17
 exploitation options, II–12.19
 licensing, II–12.19
 overview, II–12.17–25
 requirements, II–12.18, II–12.21
 scope, II–12.22
 software, II–12.23–5
 training materials, II–12.20
corporate debt facilities, II–10.27–8,
 II–10.37, II–10.67
Corporate Major Accident Prevention
 Policy (CMAPP), I–10.65, I–10.66
corporate manslaughter, I–10.78–84
corporate veil, I–11.12, I–11.84
corporation tax
 capital allowances, I–7.31
 new field developments, I–2.30–2
 oil/gas taxation and, I–7.03
 rates, I–2.22, I–2.24
 recent reduction, effect, I–2.52
 RFCT *see* ring fence corporation tax
corruption, II–5.65, II–5.67, II–15.25
cost switch, II–7.77–9
cost-sharing agreements
 cost overruns, II–7.39
 transport agreements, II–7.38–40
Côte d'Ivoire, I–3.56
country risks, II–10.08
Countryside Council for Wales (CCW),
 I–11.19
Court of Session procedure, II–15.44,
 II–15.76
Courtney, W, II–6.03

credit insurance, II–10.25
credit rating, II–2.96, II–8.52
CREST, II–10.86
CRINE, II–5.15, II–11.35, II–12.51
Crommelin, M, II–2.20
cross guarantees, II–10.40, II–10.65,
 II–10.67
cross indemnities *see* mutual indemnities
crossing agreements, II–7.42–7
cross-licence agreements, II–3.09, II–11.45
cross-user liability agreements (CULAs),
 II–7.97, II–7.106, II–7.107–9
Crown
 bona vacantia, II–10.113
 exclusive exploration rights, I–4.08
 leases, I–12.82
 ownership of hydrocarbon deposits,
  I–4.04, I–4.59, II–3.10
 ownership of UKCS, I–4.09, II–3.10
 property rights, I–9.08
 renewable energy and, I–12.82
crude oil sales agreements
 back-to-back contracts, II–08.06
 CFR contracts, II–8.14
 CIF contracts, II–8.06, II–8.15, II–8.17
 collateral support, II–8.52–3
 currency, II–8.19
 delivery, II–08.12–17
 derivatives, II–08.05
 DES contracts, II–8.16
 documents, II–08.05
 FOB contracts, II–8.06, II–8.13, II–8.17
 form and structure, II–8.04–7
 general terms, II–08.05, II–08.06,
  II–08.07
 grade and quality, II–08.08–10
 overview, II–8.04–19
 payment, II–8.19
 price, II–8.18
 quantity, II–08.11
 risk and title, II–8.12–17
 special provisions, II–08.05, II–08.06,
  II–08.07, II–08.08–19
Cuadrilla, I–9.01–2, I–9.91, I–9.100,
 I–9.128
Cullen Report
 causes of disaster, I–10.32–3
 goal setting approach, I–10.36–7,
  I–10.50
 overview, I–10.31–9
 prescriptive approach and, I–9.16,
  I–10.35, I–10.50, I–12.55
 recommendations, I–10.34–8, I–10.61,
  II–3.56

safety case, I–10.41
trade unions and, I–10.39
Culzean field, I–7.40
currency movements, II–10.08

D'Ancona Report, I–6.73
Daintith, T, I–4.09, I–4.13, I–4.74, I–A.7,
    I–A.29
damages
    see also indemnities
    consequential losses see consequential
      losses
    decommissioning liabilities,
      I–12.77–82
    Deepwater Horizon, II–2.33
    torts, II–6.30
data room, II–9.22
Davey, Ed, I–5.07, I–10.63
DEA UK, II–9.35
DEAL, I–6.80, I–6.84, I–11.15
deal-building, II–15.20
debentures, II–10.65, II–10.69, II–10.72,
    II–10.95
debt capital options, II–10.09–14
decommissioning
    1998 Petroleum Act, I–12.41–55
    administration and, II–10.105
    asset trading and, II–9.28
    Brent field, I–12.01
    Brent Spar see Brent Spar
      decommissioning
    chemicals, I–11.40
    close-out reports, I–12.70, I–12.73,
      I–12.78
    collaboration, 1–12.46
    costs, I–12.03–5, I–12.40, I–12.54,
      I–13.01–3
    current situation, I–12.01
    definition, I–7.46
    delaying, II–9.80
    finance, II–10–04
    financial information, I–12.52
    financial security, I–12.53–4
    Guidance Notes, I–12.41, I–12.56–74
      residual liabilities, I–12.77, I–12.82
    health and safety, I–10.44
    IMO Guidelines, I–12.10, I–12.25,
      I–12.38n, I–12.58, II–2.93
    international law
      1996 Protocol, I–12.37
      Continental Shelf Convention,
        I–12.07
      London Dumping Convention,
        I–12.11–12, I–12.37

      OSPAR, I–12.13–18, I–12.19,
        I–12.36, II–93–4
      OSPAR Decision 98/3, I–12.37–40,
        I–12.58, I–12.65, I–12.71–3, I–12.77,
        I–12.83, II–2.93
      overview, I–12.07–18, II–2.93–4
      UNCLOS, I–12.08–9, I–12.40,
        I–12.82
    joint ventures, I–13.04–5, I–13.12,
      I–13.28, II–2.92–6
      Model JOA, II–2.92, II–2.95
    last option, I–12.57
    liabilities, I–12.51, I–12.60–1
      see also decommissioning security
      agreements
      asset sale and purchase agreements,
        II–9.76–80
      foreign courts, I–12.79
      Norway, I–12.79–80
      residual liabilities, I–12.77–82
    licences, Brent Spar, I–12.22–3
    MER Strategy and, I–5.42, I–12.47,
      I–12.57, I–12.63, I–12.75, II–7.113
    ministerial powers, I–12.53–5
    monitoring sites, I–12.71–4
    offshore installations, I–12.01–83
    OGA role, I–12.42, I–12.46, I–12.47,
      I–12.48, I–12.63, I–12.75–6
    strategy, II–5.78
    plans see decommissioning plans
    principles, I–12.57
    remedial action, I–12.51
    section 29 notices see section 29 notices
    security see decommissioning security
      agreements (DSAs)
    site surveys, I–12.70
    tax regime, I–13.02, I–13.34–8
      relief, I–7.45–8, I–13.36–8, II–9.78
      uncertainties, I–7.49–54
    transport agreements and, II–7.110–13
    Wood Review recommendations, I–5.23
decommissioning plans
    approval, I–12.46
    Brent Spar, I–12.22
    conditions, I–12.46
    consultation, 1–12.46, I–12.42, I–12.48,
      I–12.59, I–12.66
    contents, I–12.42, I–12.64
    environmental impact assessments,
      I–12.67
    environmental statements, I–12.67
    financial information, I–12.52
    financial security, I–12.53–4
    implementation, I–12.50–1, I–12.70

liabilities, I–12.51, I–12.60–1
MER Strategy and, 1–5.42, 1–12.46,
    I–12.47
ministerial powers, I–12.53–5
preliminary discussions, I–12.63–8
preparation by Secretary of State,
    1–12.46
process, I–12.62–74
requirement, I–12.42, II–9.77
revision, 1–12.47–8
s 29 notices see section 29 notices
submission, I–12.69
withdrawal of approval, I–12.49
Decommissioning Relief Deed (DRD),
    I–7.51–2
decommissioning security agreements
    (DSAs)
asset trading and, I–13.06, II–9.28
collateral securities, I–13.19
commercial issues, I–13.31–3
end date, II–2.95
enforcing, I–13.21–2
fees, I–13.19
field and SPA-linked agreements,
    I–13.28–30
JOA parties, I–13.04–5, I–13.12,
    I–13.28
joint ventures, II–2.92, II–2.95
Model Form, I–13.11, I–13.26–7,
    I–13.31–3, I–13.38, II–2.03, II–2.92,
    II–2.95, II–2.96, II–5.14
    alternative form, II–2.96
overview, I–13.12–25
parties, I–13.04–11
risk factor, I–13.14, I–13.15, I–13.31,
    II–2.95
run-down period, II–2.95
third tier participants, I–13.27
trigger date, I–13.15, I–13.29, I–13.31,
    II–2.95
Deepwater Horizon disaster
catastrophic scale, I–11.101, II–4.138
criminal liabilities, II–2.33
damage, I–11.03–4, II–2.32
design and construction issues, I–10.68
dispute resolution and, II–15.18
fallout, II–1.01
liabilities, II–2.27, II–2.33
ministerial powers and, I–11.92
qualified indemnities and, II–6.25
regulatory impact, I–10.59, I–10.70,
    I–11.81, I–11.94, I–11.99, I–11.108,
    II–2.27
safety case approach and, I–10.90

source of problems, I–11.83
wilful misconduct, II–2.32–3
delimitation agreements see continental
    shelf boundaries
Delivered Ex Ship (DES) contracts,
    II–8.16, II–8.27
Denmark
Brent Spar controversy and, I–12.32
cross-border fields, II–3.49
dispute resolution, II–15.25
DONG, II–11.43
DUC, II–11.43
joint sales agreements, II–11.43
North Sea continental shelf boundaries,
    I–8.07, I–8.08, I–8.09–12
Rockall dispute, I–8.14, I–8.19,
    I–8.20–2, I–8.25, I–8.26–7, I–8.32,
    I–8.35
deposits in the sea, I–11.49, I–11.72
derivatives, II–08.05, II–8.22
Derman, P and A, II–3.27, II–3.31
Det Norske Veritas (DNV), I–12.27,
    I–12.31, I–12.39
development control, I–9.36–42
development wells, I–4.60, I–11.32,
    I–13.01
Digital Energy Atlas and Library (DEAL),
    I–6.80, I–6.84, I–11.15
directors
breach of permit, I–11.48
corporate manslaughter, I–10.78–84
disqualification, II–11.02
offences, I–10.87, I–11.12, I–11.61
wrongful trading, II–10.116
discoveries
decline, I–2.02–3
early oil discoveries, I–3.08
maturity and undeveloped discoveries,
    I–2.21
Dispersed Oil in Produced Water Trading
    Scheme (DOPWTS), I–11.36
dispute resolution
access to infrastructure, I–6.37
adjudication see adjudication
ADR see alternative dispute resolution
arbitration see arbitration
choices, II–15.01–6
clauses, II–15.29
collaboration option, II–15.04,
    II–15.17–20
costs, II–15.06
dispute anticipation and management,
    II–15.07–8
employment disputes, II–15.47

Energy Charter, II–15.48
exit points, II–15.02
flexibility, II–15.01
gas/LNG price reviews, II–8.45
gas/LNG sales agreements, II–8.51
guided-owner approach, II–3.38
industry culture, II–15.01–8
investor-state dispute resolution,
    II–15.48
lessons learned analyses, II–15.08
litigation *see* litigation
MER Strategy and, II–15.11
negotiation
    clauses, II–15.29–34
    lawyers and, II–15.26–7
    LOGIC standard contracts, II–15.31
    option, II–15.04, II–15.21–34
    practices, II–15.21–3
    teams, II–15.28
    tiered escalation clauses, II–15.31–4
    training, II–15.24–5
OGA recommendations, II–15.10,
    II–15.11
ombudsmen, II–15.09, II–15.11
overview, II–15.01–82
pendulum procedure, II–3.37
speed, II–15.01
stakeholders and, II–15.17–18
state parties, II–15.48
stress, II–15.19
unilateral action
    avoidance, II–15.12–13
    complaints, II–15.09–11
    dominant parties, II–15.14
    health and safety and, II–15.15
    missed opportunities, II–15.15–16
    option, II–15.04
    overview, II–15.09–16
unitisation agreements
    cross-border agreements, II–3.52,
        II–3.55
    expert determination, II–3.35–40
    WTO disputes, II–15.48
diving operations, I–10.45, II–14.44
drainage pipes, II–13.22
drill or drop, I–4.49
drilling
    contractor/ operator relationship
        commercial, II–04.05–12
        operational, II–4.13–24
        overview, II–04.04–24
        supervision, II–4.49
    dayrate drilling contracts *see* drilling
        contracts

fluids and cuttings, I–11.43–9
industry cycles, II–4.05, II–4.07
onshore drilling consent, I–9.12–28
standard contracts, II–5.22
units *see* drilling units
drilling contracts
    appendices, II–4.58
    clauses, II–4.52
    commencement, II–4.77–89
        commencement date, II–4.77,
            II–4.83–9
        effective date, II–4.77, II–4.80–2
    conventional model, II–4.15, II–4.22–3
    dayrates, II–4.06, II–4.121–36
        currency, II–4.124
        escalation clauses, II–4.69, II–4.81
        force majeure rates, II–4.131–2
        operating rates, II–4.123, II–4.125–6
        redrill rates, II–4.134–6
        repair rates, II–4.85, II–4.130
        standby rates, II–4.127–9
        weather rates, II–4.133
        zero rates, II–4.85, II–4.92, II–4.130
    definitions, II–4.61–76
        contractor groups, II–4.62–4
        drilling units, II–4.65–7
        operating areas, II–4.68–70
        wells, II–4.71–6, II–4.107
        worksites, II–4.68–70
    drilling units, II–4.25–48
    duration, II–4.71, II–4.103–10
        delivery deadlines, II–4.89
        options, II–4.109–10
        sidetracks, II–4.107
        term-based contracts, II–4.104–5
        well-based contracts, II–4.106
        wells in progress, II–4.108
    elements, II–4.49–144
    end, II–4.90–103
        completion date, II–4.90–3
        contractors' default, II–4.96
        convenience, II–4.96, II–4.99–102
        force majeure, II–4.89, II–4.96,
            II–4.98
        insolvency, II–4.96, II–4.103
        natural expiry, II–4.90–3
        termination, II–4.55, II–4.94–103,
            II–5.23
    exhibits, II–4.58
    follow-on contracts, II–4.09, II–4.72,
        II–4.93
    form of agreements, II–4.57–60
    general conditions, II–4.58
    indemnities, II–4.55, II–4.137–44

catastrophic loss, II–5.55
downhole equipment, II–5.53
pollution, II–5.54
standard contracts, II–5.52–5
invitations to tender, II–4.49
liabilities, II–4.137–44
models, II–4.16–24
operator/contractor relationship, II–4.04–24
overview, II–04.01–144
project management model, II–4.21
remuneration, II–4.55
site access, II–4.54, II–4.56, II–4.111–20
standard contracts, II–5.22
indemnities, II–5.52–5
LOGIC MODU 97, II–4.52–74,
II–4.86–91, II–4.97, II–4.114,
II–4.116–18, II–5.22–4
structure, II–4.51–6
turnkey model, II–4.17–20
drilling units
categories, II–4.25
definition, II–4.65–7
drillships, II–4.25, II–4.41–8
global numbers, II–4.26
inspection, II–4.28, II–4.82
jack-ups, II–4.25, II–4.26, II–4.27,
II–4.29–34, II–4.112–13, II–4.118
nature, II–4.25–48
operational history, II–4.85
semi-sunmersibles, II–4.25, II–4.26,
II–4.35–40, II–4.112
drillships, II–4.25, II–4.41–8
drinking water, I–10.42
due diligence
asset purchasers, II–9.16, II–9.23
asset sellers, II–9.19
asset trading, II–9.19–29
assignment clauses and, II–9.21, II–9.27
charges, II–9.25
confidentiality and, II–9.20
consents, II–9.27
corporate capacity, II–9.29
data room, II–9.22
decommissioning agreements, II–9.28
encumbrances, II–9.25
existing agreements, II–9.26
insolvency cases, II–10.117–18
investigation of title, II–9.24
royalties, II–9.25
share purchase, II–9.16
third-party rights, II–9.25
duty of care
confidential information and, II–11.56

corporate manslaughter, I–10.79–81
decommissioning and, I–12.77–8
operators, II–2.27–31
principle, II–6.30

early neutral evaluation, II–15.55,
II–15.64, II–15.79
earn-in, II–9.08
easements, II–13.06, II–13.21, II–13.27,
II–13.31
economic date, II–9.60–1
economic modelling
cost reductions, I–2.18–19, I–2.46
exploration, I–2.12, I–2.41–51
maturity and undeveloped discoveries,
I–2.21
Monte Carlo technique, I–2.11, I–2.13,
I–2.41, I–2.45
pre-oil price collapse, I–2.16–20
procedure, I–2.11–15
tax incentives and new field
developments, I–2.22–40
EEA
competition law, II–11.23, II–11.27
control of air pollution, I–11.60
members, II–11.23
Egypt, I–3.55
Ekofisk Bravo accident (1977), I–10.23
Elswick field, I–9.01
emergency preparation, I–10.43
emissions
atmospheric emissions, I–11.50–61
emissions trading, II–7.72, II–7.73
employee protection insurance, II–2.36
employees
agency workers, II–14.04
contracts of service, II–14.05
implied contracts, II–14.12–18
control criterion, II–14.06, II–14.12
definition, II–6.53, II–14.05–9
intellectual property and, II–12.33–7
intention of parties, II–14.07
multi-factor approach, II–14.06
mutual obligations, II–14.08, II–14.11
personal service, II–14.09, II–14.24
posted workers, II–14.36, II–14.42
rights, II–14.02
substitutes, II–14.09, II–14.24
territorial jurisdiction, II–14.33–43
unfair dismissal see unfair dismissal
worker status, II–14.21
working time rights, II–14.46
employment
ADR, II–15.47

agency workers *see* **agency workers**
British regulation of offshore workers
  meaning of offshore work, II–14.44–5
  territorial jurisdiction, II–14.31–43
  working time *see* **working time**
categories, II–14.02–4
dismissal at third party requests,
  II–14.83–93
dispute resolution, II–15.47
employees *see* **employees**
family rights, II–14.02
independent contractors *see* **contractors**
industry practice, II–14.32
legal overview, II–14.01–93
mobile workforce, II–14.32
non-discrimination, II–14.22
post-Brexit regulation, II–14.82
status of personnel, II–14.02–30
workers *see* **workers**
employment agencies
  see also **agency workers**
  indemnities, II–6.12
  industry practice, II–14.10, II–14.18
  status of workers, II–14.10–18
employment tribunals
  indemnities against claims, II–6.12
  personal jurisdiction, II–14.02
  pre-action protocol, II–15.47
  territorial jurisdiction, II–14.33–43
  working time jurisdiction, II–14.68,
  II–14.71
encumbrances, II–8.12, II–9.25
Energy Charter, I–3.22–5, I–3.68, II–15.48
Energy Charter Secretariat, I–3.06, I–3.07
energy consumption
  emission reduction, I–3.60–2
  energy efficiency, I–3.60–2
  EU reduction, I–3.28–9
  UK, I–3.54–63
  UK imports, I–3.54–6, I–3.64
energy security
  Energy Charter, I–3.22–5
  EU legislation, I–3.26–36, I–3.66
  reduction in energy consumption,
  I–3.28–9
  renewable energy, I–3.30
  stock-holding, I–3.26–7, I–3.31–5
  international dimension, I–3.16–25
  International Energy Program,
  I–3.16–21, I–3.32
  National Security Strategy, I–3.03,
  I–3.74
  oil crises, I–3.16, I–3.20, I–3.22, I–3.28
  overview, I–3.01–75

Scottish policy, I–9.119–22
UK, I–3.08–15
  diversification, I–3.63
  emission reduction, I–3.37–42,
  I–3.60–2
  energy consumption, I–3.54–63
  energy efficiency, I–3.14, I–3.60–2
  energy production, I–3.43–53
  exports, I–3.10, I–3.15, I–3.65
  gas storage, I–3.57–9
  imports, I–3.11–13, I–3.54–6, I–3.64
  renewable energy, I–3.63, I–3.73
  Wick Report, I–3.05
Energy Security Strategy (2012), I–3.51
English, W, II–3.09, II–3.19, II–3.26,
  II–3.46
Enoch and Blane, II–3.65
Enquest, I–7.58
environment
  administration and, II–10.106–9
  air pollution, I–11.50–61
  blowouts, I–11.03, I–11.101
  chemicals, I–11.38–42, I–11.44–8
  coastal areas, I–4.30, I–4.81, I–11.74–8
  consequential harm regulation,
  I–11.81–93
  decommissioning, I–12.67
  Deepwater Horizon, I–11.03–4
  drilling fluids and cuttings, I–11.43–9
  early UKCS regulation, I–11.06–7
  EIAs, I–4.26, I–11.14, I–12.67
  emergencies, I–11.02
  environmental statements
    decommissioning plans, I–12.67
    production regulation, I–11.16–19
  exploration regulation, I–11.22–9
    model clauses, I–11.22
  FEPA licences, I–11.37
  habitats, I–4.25, I–11.08–15
  health and safety regulation, I–10.66–7,
  I–11.98
  incidents, I–11.01
  interference with other sea users,
  I–11.73–80, I–12.07
  jurisdictions, I–10.61
  leases and, II–13.13
  liabilities, asset trading and, II–9.76
  marine spatial planning, I–4.28–31
  oil pollution *see* **oil pollution**
  onshore wells, I–9.13, I–9.25–8
  operational harm, I–11.02
  operational impacts of petroleum
  operations, I–11.08–80
  petroleum licensing and, I–4.24–31

produced water, I–11.34–7
production regulation, I–11.16–21,
  I–11.30–3
  environmental statements, I–11.16–19
  risks, II–10.08
  sewage and garbage, I–11.65–72
  spills, I–11.01–2
  strategic environmental assessments,
    I–4.27
  UKCS regulation, I–11.01–108
Environment Agency, I–9.25–6, I–10.61
Environment Council, I–12.29–30
environmental impact assessments, I–4.26,
  I–11.14, I–12.67
environmental statements (ES)
  decommissioning plans, I–12.67
  production regulation, I–11.16–19
E.ON, II–11.03
equality regulation, II–14.04, II–14.22,
  II–14.43
equity finance, II–10.09–14
Esso, I–8.06, I–12.20
EU
  see also Brexit
  Energy Charter, I–3.22–5
  energy imports, I–3.15
  law see EU law
  members, II–11.23
  UK accession, I–3.27
EU law
  ADR, II–15.46
  civil liability for oil pollution,
    I–11.105–6, I–11.108
  climate change legislation, I–3.37,
    I–3.40, I–3.69
  competition, I–6.64, I–6.68
    abuse of dominance, II–11.10–14,
      II–11.18–20
    decisions on joint sales contracts,
      II–11.39–45
    joint procurement, II–11.35
    sanctions, II–11.02
    technology transfer, II–11.64,
      II–12.27
    UK or EU jurisdiction, II–11.21–5
  Emissions Trading Scheme, II–7.72,
    II–7.73
  energy security legislation, I–3.26–36,
    I–3.66
  environmental statements, I–12.67
  health and safety at work, I–10.46
  insolvency, II–10.96
  licensing, I–4.11
    competing applications, I–4.36

offshore safety, I–10.59–76, I–11.94–8
oil pollution, I–11.94–8
patents, II–12.15
pipelines, I–6.66
REACH, I–9.26
trade secrets, II–12.31–2
working time, II–14.44–5
  amending, II–14.81–2
European Federation of Energy Traders,
  MASTER DES LNG SPA, II–8.27,
  II–8.29
European Patent Office, II–11.15
Ewing, Fergus, I–9.108, I–9.119,
  I–9.121–2
exclusive economic zones, I–4.01, I–4.39,
  I–8.23, I–11.96, II–2.93
exclusive purchasing/supply agreements,
  II–11.11
expert determination
  court jurisdiction and, II–3.39
  enforcement, II–15.79
  gas/LNG price reviews, II–8.45
  nature of process, II–15.58–60
  option, II–15.04
  overview, II–15.58–61
  scope, II–15.61
  unitisation, II–3.35–40
exploration
  competition law, relevant market,
    II–11.27
  Crown rights, I–4.08
  decline, I–2.01, I–2.51, I–5.07
  early activity, I–3.08
  environmental regulation, I–11.22–9
    model clauses, I–11.22
  finance, II–10–04, II–10.10
  government funding, I–5.17
  licences see exploration licences
  MER Strategy, I–5.42
  modelling, I–2.12, I–2.41–51
  regional exploration plans, I–5.16
  shale gas, I–9.02–5
  size of exploration blocks, I–5.17
  tax issues, I–7.41–2
  Wood Review recommendations,
    I–5.15–17
Exploration Expenditure Supplement,
  I–7.31–3
exploration licences
  geological surveys and, I–11.13
  landward licences, I–4.79
  Markham field, II–3.53
  meaning, I–4.39
  model clauses, I–4.17, I–4.40

non-exclusivity, I–4.05, I–4.39
onshore petroleum, I–9.08
rights, I–4.39
seaward licences, I–4.03, I–4.39–40
shale gas, I–9.127
export control, II–4.69
export finance, II–10.25

facility information requests, I–12.59
Falkirk Against Unconventional Gas,
 I–9.103
Fallow Areas Initiative
assessment, I–A.33–4
Fallow Blocks Process, I–A.11–22
Fallow Discovery Process, I–A.11,
 I–A.23–32
overview, I–A.7–34
pre-Wood state control, I–5.06
schemes, I–A.11
fallow blocks
classification, I–A.14–16
definition, I–A.12
industry attitude, I–A.40
list, I–A.13
model clauses and, I–A.47
numbers, I–A.7, I–A.9, I–A.33
process, I–A.11–22, I–A.25
 dispute resolution, I–A.22
 timescale, I–A.27
 voluntary system, I–A.22
public policy, I–A.49–50
reasons, I–A.7
Fallow Discovery Process
aims, I–A.23
classification, I–A.26
grounds for rejecting programmes,
 I–A.32
model clauses and, I–A.28–32
overview, I–A.11, I–A.23–32
tightened scheme, I–A.34
timescale, I–A.27
voluntary system, I–A.23
family rights, II–14.02
farm-in/out
agreements, II–9.81–3
finance option, II–9.07, II–10.15
Faroe Islands, I–8.14, I–8.19, I–8.20,
 I–8.21, I–8.27, I–8.30
fee letters, II–10.51
FEPA licences, I–11.49
Ferrers, Lord, I–10.13
FIDIC contracts, II–5.03
fiduciary duties
definition, II–2.39–42

operators, II–2.39–43
Field, S, I–10.84
field agreements
contractual structure, II–5.05
function, II–5.04
parties, II–5.11
standard contracts, II–5.14
field allowances, I–3.47, I–7.12, I–7.34,
 I–7.37
finance
account bank agreements, II–10.49
account controls, II–10.64
on balance sheet debt, II–10.27–8
borrowing base mechanism,
 II–10.54–60
carry financing, II–10.26
consortium financing, II–10.30, II–10.33
corporate debt facilities, II–10.27–8,
 II–10.37
debt capital options, II–10.09–14
documents, II–10.40–53
equity, II–10.09–14
export finance, II–10.25
farm-out, II–9.81, II–10.15
fee letters, II–10.51
forward purchasing, II–10.21–2
hedging documents, II–10.52–3
insolvency see insolvency
intercreditor agreements, II–10.47–8
land rights and, II–13.08
lifecycle, II–10–06
loan agreements, II–10.40–6
net profit interest, II–10.16–18
prepayment facilities, II–10.20
project finance, II–10.23–4
purposes, II–10–04–5
RBL facilities, II–10.35–6
repayment, II–10.61–3
risk factors, II–10.07–8
royalties, II–10.19
securities see securities
sources, II–10.06, II–10.08–39
step-in agreements, II–10.24
streaming, II–10.21–2
syndication, II–10.37–8
trends, II–10.121
vendor finance, II–10.29–32
Fiscal Review, I–1.03
Fisheries Legacy Trust Company, I–12.81
Fisheries Research Services (FRS), I–11.19,
 I–11.45, I–11.46, I–11.47
fixed charges, II–10.69, II–10.72, II–10.77,
 II–10.79, II–10.82–4
fixed-interest agreements, II–3.09

floaters, II–4.26, II–4.27, II–4.35–40
floating charges, II–10.65, II–10.67,
  II–10.69–70, II–10.91, II–10.95,
  II–10.108, II–10.116
flotels, II–14.77
FOB contracts, II–08.06, II–8.13, II–8.17,
  II–8.20, II–8.28, II–8.29
force majeure
  dayrate drilling contracts, II–4.89,
    II–4.96, II–4.98, II–4.131–2
  dispute resolution and, II–15.14
  gas/LNG sales agreements, II–8.37,
    II–8.49–50
  OSPAR Convention, I–12.18
  Rough platform, I–3.58
  volcanic ash, II–4.132
forfeiture clauses
  asymmetrical terms, II–2.75
  building contracts, II–2.67
  decommmisioning and, II–2.96
  enforceability, II–2.57, II–2.70,
    II–2.74–82
  insolvency law and, II–2.55–7,
    II–2.83–91
  joint ventures, II–2.54–91
  penalty clauses, II–2.57, II–2.58–82
  time periods, II–2.82
  unfair preferences, II–2.57, II–2.83–91
  unitisation agreements, II–3.45–6
  withering interest clauses, II–2.63,
    II–2.65
Forties field, I–10.16n
forward purchasing, II–10.21–2
Foster, J, I–10.47
FPSO units
  see also transport agreements
  arrangements, II–7.21
  costs, II–7.18, II–7.21
  industry practice, II–7.20
  meaning, II–7.18
  storage capacity, II–7.19
  transport option, II–7.18–21
  Wood Review on, II–7.20
fracking see shale gas/fracking
France, North Sea boundaries, I–8.12
fraud, I–11.80, II–6.43, II–6.44
free market economics, I–5.05
Fridman, Mikhail, II–9.35
Friends of the Earth, I–9.92, I–9.99
Frigg Agreement, II–3.59–61, II–3.62,
  II–3.63
frontier licences
  26th round, I–4.64
  assessment, I–4.68

flexibility, I–4.86
  introduction, I–3.46, I–4.63, I–4.74
  MER Strategy, I–4.67
  nine-year licences, I–4.65
  numbers, I–4.68
  overview, I–4.61–8
  terms, I–4.64
  withdrawal, I–4.63

G7, I–12.26
Gabon, I–3.56
Gannet Alpha Platform, I–11.01, I–11.03
garbage
  definition, I–11.65
  offshore regulation, I–11.65, I–11.70–2
gas
  LNG, I–3.13, I–3.55, II–11.27
  prices, II–8.42–8
  relevant market, II–11.27
  sales agreements see gas/LNG sales
    agreements
  storage, I–3.57–9
gas/LNG sales agreements
  AIPN GSA, II–8.25, II–8.29, II–8.30,
    II–8.43, II–8.49, II–8.53
  AIPN SPA, II–8.26, II–8.47
  assignment clauses, II–8.51
  BEACH 2015, II–8.22
  CIF contracts, II–8.20
  collateral support, II–8.52–3
  conditions precedents, II–8.51
  delivery, II–8.30–5
  dispute resolution, II–8.51
  excess/shortfall, II–8.36–7
  FOB contracts, II–8.20, II–8.28, II–8.29
  force majeure, II–8.37, II–8.49–50
  form and structure, II–8.20–8
  governing law, II–8.51
  key terms, II–8.29–51
  liabilities, II–8.51
  Master DES LNG SPA, II–8.27, II–8.29
  Master Ex-Ship LNG Sales Agreement,
    II–8.28
  Master FOB LNG Sales Agreement,
    II–8.28
  Model Master Agreement, II–8.22
  National Balancing Point (NBP),
    II–8.22, II–8.30, II–8.42
  NBP 2015, II–8.22
  overview, II–8.20–51
  passing of risk/title, II–8.30
  payment, II–8.51
  price diversion, II–8.47–8
  price review clauses, II–8.44–6

pricing, II–8.42–3
quantity, II–8.29, II–8.51
specification, II–8.51
take or pay clauses, II–8.38–41
taxation, II–8.51
verification clauses, II–8.51
warranties, II–8.51
General Lighthouse Authority (GLA),
I–11.75
geological surveys, I–11.11–15
Germany
Brent Spar controversy and, I–12.32
Brent Spar protest, I–12.25
cross-border fields, II–3.49
North Sea cases, I–8.07–12
North Sea continental shelf boundaries,
I–8.10–12
UK oil exports to, I–3.15
Ghana, I–3.56
Gill Report (2009), II–15.44
good oilfield practice, II–2.27, II–2.29
Greenpeace, I–12.01, I–12.06, I–12.25–7
grid system, I–4.18–19
gross negligence, II–2.28, II–2.33, II–6.24
Gryphon FPSO, I–10.92
guarantees
affiliates, I–13.18, I–13.31
asset transfer and, II–9.55, II–9.66
cross guarantees, II–10.40, II–10.65,
II–10.67
due diligence, II–9.26, II–9.28
leases, II–13.24
loan guarantees, II–10.25
parent company guarantees, II–8.52,
II–8.53, II–9.77
Gulf War (1991), I–3.20

habitats, I–4.25, I–11.08–15
Hackett, Jim, II–2.33n
Hackitt, Judith, I–10.92
Harvard Negotiation Program on
Negotiation (PON), II–15.23
health and safety at work
1974 Act, I–9.15, I–10.18–22
administration and, II–10.106–9
maintenance trend, I–2.10
offshore *see* health and safety offshore
onshore *see* health and safety onshore
Robens Report (1972), I–10.18–22,
I–10.24, I–10.30, I–10.33
working time and, II–14.74, II–14.80
zero tolerance approach, II–15.15
Health and Safety Commission, I–10.10,
I–10.40

Health and Safety Executive
centralisation, I–10.10
cross-border agreements and, UK/
Netherlands, II–3.56
decommissioning and, I–12.63
environmental jurisdiction, I–10.61
guidance, I–10.51, I–10.53
Key Programme 3, I–10.55–8, I–10.90
Key Programme 4, I–10.91, I–10.92
on maintenance trend, I–2.10
methods, I–10.38, I–10.47, I–10.48
onshore safety, I–9.20–8
regulatory authority, I–9.16
safety cases, I–10.41–7
working time enforcement, II–14.68–70
health and safety offshore
1st phase, I–10.03, I–10.04–11
2nd phase, I–10.03, I–10.12–30
3rd phase, I–10.03, I–10.31–58
4th phase, I–10.03, I–10.59–76
1974 Act, I–10.18–22
asset integrity, I–10.55–8, I–10.83,
I–10.90–1
Burgoyne Report, I–10.23–30, I–10.31,
I–10.33–5, I–10.38–9
criminal liability, I–10.41, I–10.77–88
2008 Act, I–10.85–8
corporate manslaughter, I–10.78–84
penalties, I–10.83–4, I–10.85–7
cross-border agreements, UK/
Netherlands, II–3.56
Cullen Inquiry, I–10.31–9, I–10.39,
I–10.41, I–10.50, I–10.60, I–12.55
emergency response, I–10.43, I–10.71
environment and, I–10.66–7, I–11.98
EU law, I–10.46, I–10.59–76
Offshore Safety Directive,
I–10.60–76, I–11.94–8
evolution, I–10.02–3
goal-setting regulations, I–10.36–7,
I–10.42–7
Institute of Petroleum Code, I–10.05,
I–10.07–9
Key Programme 3, I–10.55–8, I–10.81,
I–10.83, I–10.90, I–10.92
Key Programme 4, I–10.91, I–10.92
leadership, I–10.58
licensing approach, I–10.03, I–10.04–11
Macondo disaster *see* Deepwater
Horizon disaster
Major Hazard Report, I–10.62
no blame culture, II–15.15
occupational health, I–10.04
overview, I–10.01–92

permissioning approach, I–10.03,
  I–10.31–58
Piper Alpha *see* **Piper Alpha disaster
  (1988)**
prescriptive approach, I–10.03,
  I–10.12–30, I–11.94, I–12.55
quantified risk assessments (QRAs),
  I–10.35, I–10.38, I–10.53
regulators, I–10.17, I–10.61–3,
  I–10.75–6, I–11.97–8
Safety Case, I–10.34, I–10.37, I–10.62
  1992 legislation, I–10.41–7
  2005 Regulations, I–10.48–54
  2015 Regulations, I–10.64–76
  assessment, I–10.90
  major accidents, I–10.71–4
Sea Gem Inquiry, I–10.06–11, I–10.12,
  I–10.13, I–10.16, I–10.19, I–10.23–5,
  I–10.32, I–10.35–6, I–10.38
workforce involvement, I–10.39,
  I–10.50
health and safety onshore
consent to drill, I–9.12–28
guidance, I–9.21–3
legislation, I–9.15–16
well integrity, I–9.19–24
hedge funds, II–10.09, II–10.52–3
hedging agreements, II–10.76
Heffron, R, I–9.119
helicopter operations, I–10.42, I–10.43
Hewitt, G, II–2.17, II–2.18, II–2.70
Hewitt, T, II–6.28
high-yield bonds, II–10.09, II–10.12
Hill, A, I–4.04, I–4.09, I–4.74
Hinkley Point C power station, I–3.51
Historic Scotland, I–9.53
holidays
annual holiday rights, II–14.62–4
holiday pay, II–14.23, II–14.63–4
home rule *see* **Scottish devolution**
Hong Kong, II–15.56
horizontal agreements, II–11.28
  see also **anti-competitive agreements**
hub development, II–3.11
human rights, I–4.13
Hunt, Lord, I–3.48
hydraulic fracturing *see* **shale gas/fracking**

Iceland, Rockall dispute, I–8.14, I–8.19,
  I–8.20–1, I–8.25, I–8.27–8, I–8.32,
  I–8.35
ICSID, II–15.48
IMHH *see* **Industry Mutual Hold
  Harmless Deed**

indemnities
asset trading agreements, II–9.75
building contracts, II–6.04
circular indemnities, II–6.06–7
collaboration, IP licensing, II–12.46–7
concept, II–6.03–4
consequential loss *see* **consequential
  losses**
construction and tie-in agreements,
  II–7.51–2
contractual indemnities, II–6.03–76
corporate acquisitions, II–6.04
dayrate drilling contracts, II–4.55,
  II–4.137–44
definitions, II–6.03
  company groups, II–6.52
  drafting, II–6.52–4
  employees, II–6.53
  personnel, II–6.53, II–6.61
  property, II–6.54, II–6.61
drafting issues
  definitions, II–6.03, II–6.52–4
  delimiting circumstances, II–6.46–7
  interpretation issues, II–6.37–54
  multi-party issues, II–6.48–9
  negligence issue, II–6.38–45
  stray indemnities, II–6.51
employment agencies, II–6.12
fraud and, II–6.44
full and primary indemnities, II–6.50,
  II–6.60, II–6.61
gas/LNG sales agreements, II–8.51
heads, II–4.142
ICoP, I–6.92–3
indemnity and hold harmless clauses
  back-to-back provisions, II–6.21–3,
  II–6.57, II–7.107
  carve-outs, II–6.23–5
  case law, II–6.33–54
  concept, II–6.05
  drafting, II–6.37–54
  interpreting, II–6.33–54
  mapping, II–6.27
  multi-party issues, II–6.48–9,
  II–6.55–76
  negligence issue, II–6.38–45
  oil and gas contracts, II–6.09–12,
  II–6.21–2
  presumptions, II–6.30
  qualifying provisions, II–6.23–5
  rejection, II–6.26
  statutory control, II–6.28–9
  third parties, II–6.31–2
  unfair contract terms, II–6.28–9

infrastructure access, I–6.92–3,
    I–6.117–18, II–6.26, II–6.85
interpretation
    contextualism, II–6.36
    contra proferentem, II–6.34–5,
        II–6.37, II–6.38–45, II–6.46
    drafting implications, II–6.37–54
    general rules, II–6.33
    traditional approach, II–6.33–5
issues, II–6.28–32
judicial suspicion, II–6.04, II–6.11
liability caps, II–6.83–6
    access to infrastructure, II–6.85
    importance, II–6.86
    third parties, II–6.84
    transport agreements, II–7.43
    unfair terms, II–6.84
LOGIC contracts, II–6.05
    intellectual property, II–12.52,
        II–12.60–1
multiple parties, II–6.48–9, II–6.55–76
    see also Industry Mutual Hold
        Harmless Deed
mutual hold harmless *see* mutual hold
    harmless indemnities
mutual indemnities
    meaning, II–6.04
    mutual hold harmless clauses and,
        II–6.06–12
overall limitations of liability, II–6.83–6
parties, II–4.140
pipeline servitudes, II–13.14
qualified indemnities, II–6.23–5
risk matrix table, II–4.144
servitudes, pipelines, II–13.14, II–13.24
simple indemnity clauses
    meaning, II–6.04
    oil and gas contracts, II–6.09
stray indemnities, II–6.51
strict liability, II–4.144
subrogation rights, II–6.50, II–6.61,
    II–6.66–7
timing, II–4.143
transport agreements, II–7.94, II–7.97–9
    crossing and proximity agreements,
        II–7.43–5
    CULAs, II–7.97, II–7.106, II–7.107–9
    new entrants, II–7.106
    triggers, II–4.141
independent contractors *see* contractors
indirect losses *see* consequential losses
Industry Mutual Hold Harmless Deed
    (IMHH)
    2012 Deed, II–6.58

affiliates, II–6.66
complexity, II–6.76
consequential losses, II–6.61, II–6.81
contractual, II–6.75–6
core provisions, II–6.59–61
dangers of scheme, II–6.76
Deed of Adherence, II–6.63, II–6.75
definitions, groups, II–6.66
effect, II–6.69
entry into force, II–6.63
exceptions, II–6.70–3
    air transport, II–6.71
    landward areas, II–6.72
    operators, II–6.73
full and primary indemnities, II–6.50,
    II–6.60, II–6.61
gaps, II–6.76
general acceptance, II–6.76
geographical extent, II–6.64
group benefits, II–6.66
initial Deed, II–6.58
introduction, II–6.58
new parties, II–6.63
order of precedence, II–6.65
overview, II–6.58–76
parties, II–6.58
personnel, II–6.61, II–6.66
purpose, II–6.59
right to defend, II–6.68
risk allocation, II–5.63
scope, II–6.61
terminology, II–6.05
third parties, II–6.66, II–6.76
waiver of subrogation, II–6.61,
    II–6.66–7
withdrawal from, II–6.69
information *see* confidential information
infrastructure
    access *see* infrastructure access
    categories, I–6.22–7
    collaboration, I–5.33, I–5.61
    competition and, I–6.07, I–6.64–72,
        I–6.98, II–7.65
    integrity, I–10.57
    lack, I–3.46
    LNG, I–3.55
    Norway, I–3.13
    OGA role, II–7.76
    pipelines *see* pipelines
    Piper Alpha disaster and, I–10.57
    specification, II–7.92–6
    taxation, I–7.55–7
    terminology, I–6.22–7
    transport *see* transport

UK-Norway Framework Treaty (2005), II–3.63
West of Shetland, I–3.48
Wood Review, I–2.54, I–5.21

infrastructure access
applications, I–6.94, I–6.99
capacity, I–6.38, I–6.90–1, I–6.102–3
  collective capacity, II–7.100–3
  new entrants, II–7.104–6
competition and, I–6.07, I–6.64–72, I–6.98, II–7.65
  ICoP, II–11.52–3
  joint supply/purchase, II–11.46–54
complex regime, I–6.14
Energy Act 2011, I–6.21–44
factors, I–6.29–30
Guidance, I–6.14, I–6.52, I–6.63, I–6.71
  applications, I–6.99
  capacity, I–6.102–3
  conflicting contracts, I–6.104–5
  goals, I–6.96
  liabilities, I–6.117–18
  non-discrimination, I–6.106
  overview, I–6.96–118
  principles, I–6.106
  statutory factors, I–6.101
  tariffs, I–6.107–16
ICoP *see* **Infrastructure Code of Practice**
indemnities, I–6.92–3, I–6.117–18, II–6.26, II–7.97
  liability caps, II–6.85
Infrastructure Act 2015, I–6.45–8
issues, I–6.01–19
legislative framework, I–6.21–48
MER Strategy, I–6.01, I–6.49–63, I–6.105, I–6.115, I–6.120
overview, I–6.01–123
principles, I–6.12, I–6.62–3
small companies, I–3.50, I–6.05
tariffs, I–6.90, I–6.107–16, I–7.55, II–7.71–6
technical difficulties, I–6.09
ullage, I–7.55
Wood Review, I–2.54, I–5.21

Infrastructure Code of Practice (ICoP)
adhering to, I–5.53, II–7.65
applications, I–6.94
ARNs, I–6.94
automatic referral notices, I–6.86, I–6.94–5
capacity, I–6.90–1
competition issues, II–11.52–3
complex regime, I–6.14, I–6.52
compliance, II–11.52

conflicts of interest, I–6.87
indemnities, I–6.92–3
information sharing, I–6.80–4
introduction, I–6.73–4
liabilities, I–6.92–3
mutual hold harmless provisions, I–6.93
non-discrimination, I–6.72, I–6.88
overview, I–6.73–95
parties, I–6.87
pre-MER document, I–6.15
principles, I–6.79
reasonableness, I–6.86, I–6.90
scope, I–6.75–8
separation of services, I–6.89
tariffs, I–6.90, II–11.52
timelines, I–6.86
transparency, I–6.79, I–6.88
unbundling, I–6.89
voluntary code, I–6.75
Wood Review and, I–5.19

innovate licences
29th round, I–4.03, I–4.41
flexibility, I–4.63, I–4.67, I–4.77, I–4.86
frontier licences and, I–4.63
guidance, I–4.78
meaning, I–4.03
model clauses, I–4.03
overview, I–4.75–8
purpose, I–4.76
success, I–4.79
terms, I–4.78
work programmes, I–4.47

innovation
see also **intellectual property**
collaborations *see* **collaboration**
commercial challenges, II–12.05–10
first-mover advantage, II–12.06
importance, II–12.01–4

insolvency
acquisition of assets in, II–10.117
administration, II–10.95–109
anti-deprivation principle, II–2.83, II–2.90
company voluntary arrangements, II–10.119–20
dayrate drilling contracts and, II–4.96, II–4.103
decommissioning securities and, I–13.21, I–13.25
EU law, II–10.96
liquidation, II–10.110–18
numbers, II–2.56, II–10.93, II–10.122
overview, II–10.93–122
small companies, II–2.55

specialist regimes, II–10.94
unfair preferences, II–2.57, II–2.83–91
wrongful trading, II–10.116
Institute of Petroleum, Model Code of
    Safe Practice, I–10.05, I–10.07–9,
    I–11.06
insurance
    CAR insurance, II–2.36
    credit insurance, II–10.25
    employee protection, II–2.36
    National Insurance, II–14.32, II–14.37
    operators, II–2.36
    overseas investment, II–10.25
    securities over, II–10.76, II–10.81
intellectual property
    assignation
        collaborations, II–12.43
        future inventions, II–12.36
        requirement, II–12.35
        vertical agreements, II–11.58
    collaboration agreements, II–12.40
    confidential information, II–12.26–32
    copyright, II–12.17–25
    know-how, II–12.26–32
    licensing, II–12.14, II–12.19,
        II–12.48–50
    LOGIC contracts, II–12.51–61
        background IP, II–12.53–7
        confidential information, II–12.54–7
        foreground IP, II–12.58–9
        indemnities, II–12.52, II–12.60–1
        title, II–12.52–9
    no-challenge clauses, II–11.64
    overview, II–12.11–37
    ownership
        employee v contractor, II–12.33–7
        joint ownership in collaborations,
        II–12.43–5
    patents, II–12.12–16, II–12.48
intercreditor agreements, II–10.47–8
interests clauses, JOAs, II–2.10–11,
    II–2.33
International Air Pollution Prevention
    Certificates (IAPPs), I–11.52
International Association of Drilling
    Contractors (IADC), standard
    contracts, II–5.16, II–5.73, II–5.76
International Chamber of Commerce
    (ICC), II–15.66
International Court of Arbitration,
    II–15.66
International Court of Justice
    jurisdiction, I–8.24, I–8.29
    North Sea cases, I–8.07, I–8.09–12

International Energy Agency (IEA),
    I–3.16–21, I–3.32, I–3.59, I–3.67
International Energy Program (IEP
    Agreement), I–3.16–21, I–3.32
International Finance Corporation,
    II–15.09
International Group of Liquefied Natural
    Gas Importers
    Master Ex-Ship LNG Sales Agreement,
    II–8.28
    Master FOB LNG Sales Agreement,
    II–8.28
International Maritime Organization
    (IMO), decommissioning guidelines,
    I–12.06, I–12.10, I–12.25, I–12.38n,
    I–12.58, II–2.93
International Swaps and Derivatives
    Association, II–08.05, II–8.22
interpretation rules
    contextualism, II–6.36
    contra proferentem, II–6.34–5, II–6.37,
        II–6.46
    negligence issue, II–6.38–45
    general rules, II–6.33
investment allowances
    assessment, I–7.62
    introduction, I–7.34
    mechanism, I–7.34–40
    new field developments, I–2.22, I–2.28,
        I–2.34
    value allowance, I–7.39
investment exit risk, II–10.08
investor-state dispute resolution, II–15.48
Iran, I–1.08, I–3.20
Iraq, invasion of Kuwait, I–3.20
Ireland, Rockall dispute, I–8.14, I–8.19,
    I–8.20–2, I–8.25, I–8.30, I–8.32,
    I–8.35
irritancy, II–10.98
Israel
    Six-Day War (1967), I–3.24
    Yom Kippur War (1973), I–3.16, I–3.28
ITLOS, I–8.24, I–8.29

jack-up drilling units, II–4.25, II–4.26,
    II–4.27, II–4.29–34, II–4.112–13,
    II–4.118
Japan, I–3.11
JCT contracts, II–5.03
Jennings, A, II–6.81
joint bidding agreements, II–2.05–6,
    II–2.17, II–9.84, II–11.34
joint operating agreements (JOAs)
    see also operators

assessment, II–2.97
cash calls, II–2.54, II–10–04
classes of members, II–2.23
common interests, II–6.26
competition law issues, II–11.28,
    II–11.30–3
confidential information, II–2.42
consent to transfers, II–9.55–6
consequential losses, II–2.34
contractual nature, II–2.08–9
contractual structure, II–5.04,
    II–5.06–7, II–5.09, II–5.11
co-operation and competition,
    II–11.30–3
decommissioning, II–2.92–6
    security, I–13.04–5, I–13.12, I–13.28,
    II–2.95–6
default, II–2.54–7
description, II–2.75
express written terms, II–2.43
fiduciary duties, II–2.42–3
forfeiture clauses, II–2.54–91, II–10.109
    decommmisioning and, II–2.96
    enforceability, II–2.57, II–2.70,
    II–2.74–82
    insolvency law and, II–2.57,
    II–2.83–91
    penalty clauses, II–2.57, II–2.58–82
    unfair preferences, II–2.57,
    II–2.83–91
functional-related functions, II–2.09,
    II–2.14–22
functions, II–2.08–9
good oilfield practice, II–2.27, II–2.29
horizontal agreements, II–11.28
interests clauses, II–2.10–11, II–2.33
interests transfer, II–2.12, II–2.22
model forms, II–2.03–4
no mutual liability, II–2.02
non-consent clauses, II–2.50–3
non-operators, II–2.23, II–2.24, II–2.42,
    II–2.44
novation, II–9.84, II–9.89
OGA Model JOA, II–2.03–4, II–5.13
    agency, II–2.38
    consequential losses, II–2.34
    decommissioning, II–2.92
    interests clause, II–2.10
    Opcom, II–2.44
    wilful misconduct, II–2.29
opcoms see joint operating committees
operators' obligations, II–2.23–43
partnerships, II–2.02, II–2.07,
    II–2.15–22, II–7.12–15

pass marks, II–2.47
pre-emption rights, II–2.13, II–9.11,
    II–9.37, II–9.43–54
proprietorial functions, II–2.09–13
securities over rights, II–10.79–80
sole risk clauses, II–2.50–3
standard contracts, II–5.13
tenancy in common, II–2.11
tendering process, II–5.36
UUOAs and, II–3.17–18, II–3.46
wilful misconduct, II–2.28–9
work programmes, II–5.06
joint operating committees
    agency and, II–2.38
    authorisations for expenditure, II–2.46,
    II–2.49
    budgets, II–2.49
    consents, II–2.43
        contracts, II–2.35
    importance, II–2.44
    information to, II–2.48
    insurance and, II–2.36
    litigation and, II–2.36
    membership, II–2.23, II–2.45
    non-consent clauses, II–2.50–3
    overview, II–2.45–91
    pass mark, II–2.47, II–2.50–1
    powers, II–2.46
    sole risk clauses, II–2.50–3
    subcommittees, II–2.45
    supervision, II–2.26
    ultimate responsibility, II–2.45
    voting rights, II–2.45, II–2.47, II–9.26
        asset sales, II–9.66
joint procurement contracts, II–11.28,
    II–11.35
joint sales contracts, II–11.28, II–11.36–45
joint ventures
    agreements see joint operating
        agreements
    benefits, II–2.01
    competition law issues, II–11.30–3
        federal contracts, II–11.33
        information sharing, II–11.32
    incorporated JVs, II–2.07
    Macondo field, II–2.33
    meaning, II–2.02, II–2.07, II–2.18
    nature, II–2.07–22
    rationale, II–2.01
    tax advantages, II–2.02
    termination, II–2.09
    terminology, II–2.07
    unincorporated JVs, II–2.02, II–2.07–8,
    II–2.97

unlocking value, I–5.53
Wood Review and, I–5.17
Jones, L, I–10.84
Jones, P, II–3.32
junior loans, II–10.09

Kasim, A S, I–2.05
Kemp. A G, I–2.05
Key Programme 3, I–10.55–8, I–10.81,
    I–10.83, I–10.90, I–10.92
Key Programme 4, I–10.91, I–10.92
knock for knock indemnities see mutual
    indemnities
know-how, II–12.26–32
Korea, I–3.11

Ladbury, R, II–2.20
Laggan field, I–3.47, I–3.48
land law   ·
    abandonment, II–13.11
    leases see leases
    licence to occupy, I–4.04
    no separate code for oil/gas,
        II–13.03–5
    ownership, II–13.10–11
    pipelines and, II–13.01
    registration of interests, II–13.15
    servitudes, II–13.25
    servitudes see servitudes
    statutory wayleaves, II–13.31
Land Register of Scotland, II–13.15,
    II–13.25
Lands Tribunal for Scotland, II–13.20
landward licences
    see also onshore regulation
    categories, I–4.03, I–4.79
    consent to drill, I–9.12–28
    licensing authority, I–4.80, I–9.09
    licensing rounds, I–4.80
    overview, I–4.79–81
    PEDLs see petroleum, exploration and
        development licences
    production rate, I–4.81
    proprietary licences, I–4.04
law of the sea
    continental shelf boundaries see
        continental shelf boundaries
    decommissioning and, I–12.07–18,
        I–12.36
    development, I–8.01–2
    interference with other sea users,
        I–11.73–80
    UNCLOS see UNCLOS
Leadsom, Andrea, I–3.51

leases
    assignation, II–13.13
        covenants, II–13.24
    commercial leases, II–13.13
    exclusive possession, II–13.12
    leasehold servitudes, II–13.01, II–13.23
    pipeline routes, II–13.29
    pre-emption rights, II–13.13
    real rights, II–13.06, II–13.12–13
    rent, II–13.13
    statutory regulation, II–13.13
LetterOne Group, II–9.35
letters of credit, I–13.18, II–2.96, II–8.06,
    II–8.53, II–9.28
Lewis, G, II–2.20
liability caps
    access to infrastructure, II–6.85
    importance, II–6.86
    overview, II–6.83–6
    third parties, II–6.84
    transport agreements, II–7.43
    unfair terms, II–6.84
Libecap, G, II–3.05
Liberia, I–3.56
licences
    see also licensing regime; specific
        licences
    assets, II–9.58
    assignment see assignation/assignment
    breach of licence, I–4.84
    categories, I–4.03
    consideration, I–4.32–4
    contractual nature, I–4.12–14, I–4.82,
        I–10.05
    core rights, II–2.06
    cross-licences, II–3.09
    exclusivity, I–4.05
    FEPA licences, I–11.49
    forms, I–4.04
    JOAs and, II–2.06, II–2.09
    landward licences see landward licences
    non-exclusive licences, I–4.05
    online information, II–9.24, II–9.30
    retrospective amendment, I–5.13
    revocation, I–4.51, II–9.36, II–10.99,
        II–10.120
    royalties, I–4.33, I–7.27
    sea deposits, I–11.49
    seaward licences see seaward licences
    securities over, II–10.70–5
    terms and conditions, I–4.15–17
    third party rights and, I–4.06
    transfer
        see also asset trading

consents, II–9.30–6, II–9.45
JOA consent, II–9.56
Open Permission, II–2.12, II–10.70–5
share sale and, II–9.34–5
licensing applications
competing applications, I–4.35–7,
I–4.50
fees, I–4.34
notices, criteria, I–4.37
out-of-rounds applications, I–4.22–3,
I–4.80, II–2.05, II–03.08, II–3.12
promote licences, I–4.72
licensing regime
see also **licences**; specific licences
air pollution and, I–11.62–4
applications *see* **licensing applications**
collaboration, I–5.33
discretionary system, I–4.14, I–4.32
environmental issues, I–4.24–31
EU law, I–4.11
flexibility, I–4.86
grid system, I–4.18–19
intellectual property *see* **intellectual property**
issues, I–4.18–38, I–4.82–6
legal basis, I–4.07–11
legal concept, I–4.04–6
light-touch regulation, I–5.06, I–5.11
replacement, II–1.01
MER Strategy, I–5.42
models, I–4.01–2
national security and, I–4.37
onshore petroleum, I–9.07–11
petroleum licensing, I–4.01–86
pre-Wood philosophy, I–5.05–6
Scottish devolution, I–9.125–6
strict concession regime, I–7.18, I–7.27
territorial rights, I–4.01
transition, I–4.85
transparency, I–4.38
UK approach, I–4.01, I–4.03
UK hybrid system, I–4.12–14
UK law, I–4.07–10
Wood Review and, I–4.82–3
licensing rounds
1st round, I–8.07, II–2.03
4th round, I–3.08
5th round, II–2.03, II–5.13
6th round, II–2.03
14th round, I–9.09, I–9.10
20th round, I–4.52, I–4.63, I–4.74,
II–2.03, II–2.04, II–2.13
21st round, I–4.69, I–4.74
24th round, I–4.27

25th round, II–5.13
26th round, I–4.63, I–4.64, I–4.65
28th round, I–4.43, I–4.68
29th round, I–4.03, I–4.21, I–4.41
30th round, I–4.15, I–4.41
landward licences, I–4.80
numbers, I–4.21, I–4.80
procedure, I–4.20–1
Wood Review and, I–5.16
liens, II–10.08, II–10.09, II–10.69n,
II–10.85n
lifting operations, I–10.45
limited companies, joint ventures, II–2.02
Lindley and Banks on Partnership,
II–2.18, II–2.21
Lindsey Oil Refinery, I–10.84
liquefied natural gas *see* **LNG**
liquidation
disclaiming onerous property,
II–10.112–13
overview, II–10.110–18
Scotland, II–10.110, II–10.114–15
liquidity risks, II–10.08
literary works, II–12.17, II–12.18,
II–12.23
litigation
alternative to arbitration, II–15.70–1
choice of court, II–15.72
civil procedure *see* **civil procedure**
enforcement, II–15.78–81
jurisdiction, II–15.65
operators, II–2.36
overview, II–15.72–7
Technology and Construction Court,
II–15.77
warranties, II–9.71
LNG
competition law, relevant market,
II–11.27
imports, I–3.13
infrastructure, I–3.55
sales agreements *see* **gas/LNG sales agreements**
UK imports, I–3.55–6
loan agreements, II–10.40–6
loan guarantees, II–10.25
loan life cover ratio (LLCR), II–10.57
Loan Market Association (LMA),
II–10.40–1
loan notes, II–10.09
logging while drilling, II–4.14
LOGIC
establishment, II–5.16
fallow blocks and, I–A.14

standard contracts *see* **LOGIC standard contracts**
LOGIC standard contracts
  allocation of risk, II–5.43
  amendments, II–5.21, II–5.56–65
  anti-bribery and corruption, II–5.65, II–5.67
  appendices, II–5.31
  automation, II–5.84
  call-off contracts, II–5.57, II–5.59
  case law, II–5.22–4
  changing terms, II–5.21
  consistency, II–5.25
  contractor-contractor liability, II–5.62–3
  definition of affiliates, II–5.58–9
  disadvantages, II–5.25
  dispute resolution clauses, II–15.31
  drafting, II–5.20, II–5.25
  drilling contract, II–5.22–4
  form, II–5.28–33
  future, II–5.78–84
  general terms, II–5.29–30
  governing law, II–5.29
  guidance notes, II–5.34–5, II–5.66
  IMHH *see* **Industry Mutual Hold Harmless Deed**
  indemnities, II–5.41–55, II–6.05
  intellectual property, II–12.51–61
  key provisions, II–5.40–55
  models, II–5.16–17, II–5.26
  MODU 97, II–4.52–74, II–4.86–91, II–4.97, II–4.114, II–4.116–18, II–5.22–4
  mutual hold harmless indemnities
    amendments, II–5.61–3
    contractor-contractor liability, II–5.62–3
    exceptions, II–5.46–55
    IMHH Scheme *see* **Industry Mutual Hold Harmless Deed**
    key provisions, II–5.41–55
    mobile drilling rigs, II–5.52–5
    well services, II–5.46–51
  negotiation of disputes, II–15.31
  new contracts, II–5.82–3
  On-and Offshore Services, II–15.31
  Purchase Order Terms and Conditions, II–15.31
  special terms, II–5.32–3
  Standard Contracts Committee, II–5.40
  tendering process, II–5.36–9
  updating, II–5.21, II–5.25, II–5.66–8
  variations clauses, I–5.64
  warranties, II–5.60

London Court of International Arbitration (LCIA), II–15.66
Low Carbon Transition Plan, I–3.40, I–3.60–2, I–3.69, I–3.72, I–3.73
loyalty, fiduciary duties, II–2.40

Macondo disaster *see* **Deepwater Horizon disaster**
Magnus field, I–7.58
Major, John, I–12.26
Major Hazard Reports, I–10.62, I–11.95
mandated lead arrangers (MLAs), II–10.42
Mandelson, Peter, I–3.48
Marine and Coastguard Agency (MCA), I–10.61, I–10.90, I–11.26, I–11.67, I–11.69, I–11.85
Marine Management Organisation (MMO), I–4.30
Marine Scotland, I–4.30
marine spatial planning, I–4.28–31
market definition, II–11.26–7, II–11.46
market share, II–11.14, II–11.58
market-sharing agreements, II–11.11
Markham field, II–3.53–7
Marriage, P, I–4.09
Martin, A T, II–5.25
Master Deed, assignation, II–2.13, II–9.32, II–9.43–5, II–9.87, II–9.89–93
maternity leave, II–14.02
Mature Province Initiatives
  Fallow Areas Initiative, I–A.7–34
  overview, I–A.1–51
  Stewardship Initiative, I–A.35–51
  Wood Review and, I–A.1
med-arb, II–15.57
mediation
  agreements, II–15.36
  confidential information, II–15.37, II–15.53–4
  contract clauses, II–15.32, II–15.36
  deal-building and, II–15.20
  evaluative method, II–15.42
  facilitative method, II–15.41
  lawyers and, II–15.40
  med-arb, II–15.57
  mediators' role, II–15.38
  mediators' skills, II–15.39
  methods, II–15.41–2
  mini-trials, II–15.56
  Netherlands, II–15.45
  Norway, II–15.46
  overview, II–15.36–42
  process, II–15.37

SEAM, II–15.56
training, II–15.38, II–15.49
transformative method, II–15.42
United States, II–15.45, II–15.46,
    II–15.73, II–15.79
MER Strategy
abandonment plans and, 1–12.46,
    1–12.47
asset trading and, I–12.45, II–9.03
Central Obligation, I–5.40, I–5.41,
    I–5.44
    infrastructure access and, I–6.50,
    I–6.61
    unitisation and, II–3.11
collaboration, I–5.46–57, II–11.05
    deal-building, II–15.20
commencement, II–9.03
competition law and, II–11.05,
    II–11.58–60, II–11.63–7
conceptual confusion, I–5.28–32
decommissioning and, I–5.42, I–12.47,
    I–12.57, I–12.63, I–12.75
    transport agreements, II–7.113
dispute resolution and, II–15.11
effect, I–1.08
enforceability, I–5.32
frontier licences and, I–4.67
implementation, I–4.17, I–5.02, I–5.25,
    I–5.28–70
infrastructure access, I–6.01, I–6.49–63,
    I–6.50, I–6.54, I–6.59–60, I–6.105,
    I–6.115, I–6.120
legal obligation, I–5.33–40
licensing regime and, I–4.84
LOGIC standard contracts and,
    II–5.78–84
OGA and, I–5.44, I–10.03, II–5.78
    cost switch and, II–7.78
purpose, II–9.03
reinforcing, I–2.52–6
required actions, I–5.45, I–6.51
safeguards, I–5.45, I–5.49, I–5.58–62
    competition law and, II–11.63
    infrastructure access, I–6.50, I–6.54,
    I–6.59–60
sector strategies, I–5.63–70
send or pay clauses and, II–7.83
Supporting Obligations, I–5.40,
    I–5.42–4, I–5.67
    infrastructure access, I–6.51, I–6.54,
    I–6.58
technology and, II–12.04, II–12.64
unitisation and, II–04–5
Wood Review *see* **Wood Review**

merchant shipping *see* shipping
mergers *see* takeovers and mergers
mezzanine loans, II–10.06, II–10.08,
    II–10.09, II–10.47
mini-trials, II–15.56
Mitsui, II–2.33
modelling *see* economic modelling
Monte Carlo technique, I–2.11, I–2.13,
    I–2.41, I–2.45
Monti, Mario, II–11.42
Moodys, II–2.96
MOOIP (moveable oil originally in place),
    II–3.23, II–3.26
mortgages, II–2.11, II–10.69, II–10.77,
    II–10.85, II–10.87, II–10.89
Murchison Field Agreement, II–3.59,
    II–3.63
mutual hold harmless indemnities
    see also indemnities
    back-to-back provisions, II–6.21–3,
    II–6.57, II–7.107
    concept, II–6.06–12
    contractor-contractor liability, II–5.62–3
    ICoP, I–6.93
    LOGIC standard contracts
        amendments, II–5.61–3
        contractor-contractor liability,
        II–5.62–3
        exceptions, II–5.46–55
    IMHH Scheme *see* **Industry Mutual
        Hold Harmless Deed**
    key provisions, II–5.41–55
    mobile drilling rigs, II–5.52–5
    well services, II–5.46–51
    London Bridge, II–6.13, II–6.18–20,
    II–6.28–9, II–6.41–2, II–6.45, II–6.50
    multi-party issues, II–6.48–9,
    II–6.55–76
    mutual indemnities and, II–6.06–12
    oil and gas contracts, II–6.13–20
    qualified indemnities, II–6.23
    transport agreements, II–7.97–9
mutual indemnities
    see also **mutual hold harmless
        indemnities**
    concept, II–6.06
    mutual hold harmless indemnities and,
    II–6.06–12
**Mutual Indemnity and Hold Harmless
    Deed** *see* **Industry Mutual Hold
    Harmless Deed**

National Balancing Point (NBP), II–8.22,
    II–8.30, II–8.42

national emissions ceilings, I–11.64
National Emissions Reduction Plan,
 I–3.37
National Environmental Research Council
 (NERC), I–12.29
National Insurance, II–14.32, II–14.37
national oil companies, I–7.19
national security, licensing and, I–4.37
National Security Strategy, I–3.03, I–3.74
National Transmission System (NTS),
 I–6.11, II–7.10–11, II–7.12–13,
 II–8.22, II–8.30
Natura 2000, I–11.09, I–11.14
Natural England, I–11.19
Natural Environment Research Council
 (NERC), I–11.28
NBP 2015, II–8.22
negligence
 gross negligence, II–2.28, II–2.33,
  II–6.24
 indemnity drafting and, II–6.38–45
negotiations
 clauses, II–15.29–34
 collaborative deal-building, II–15.20
 dispute resolution, II–15.04,
  II–15.21–34
 foreign jurisdictions, II–15.25
 lawyers and, II–15.26–7
 practices, II–15.21–3
 teams, II–15.28
 tiered escalation clauses, II–15.31–4
 training, II–15.24–5
net present value (NPV), II–9.60, II–10.56,
 II–10.57, II–10.58
net profit interest (NPI), II–9.07, II–9.59,
 II–10.08, II–10.16–18
Netherlands
 Brent Spar controversy and, I–12.32
 cross-border fields, II–3.49
 dispute resolution, II–15.25
 exploration strategy, I–5.15
 gas development, I–5.20
 Groningen gas field, I–4.09, I–8.02
 Markham field agreement, II–3.53–7
 mediation, II–15.45
 North Sea continental shelf boundaries,
  I–8.07, I–8.08, I–8.09–12, II–3.52–7
 UK oil exports to, I–3.15
 UK/Netherlands delimitation agreement
  (1965), I–8.08, II–3.52–7
neutral evaluation, II–15.35, II–15.55,
 II–15.64, II–15.79
Nigeria, I–1.08, I–3.56, II–3.27
night work, II–14.56–9, II–14.66

Ninian field, I–10.16n
non-compete clauses, II–11.58
non-consent clauses, II–2.50–3
non-discrimination law, I–4.11, I–4.36,
 II–14.22, II–14.26, II–14.43
non-operators, II–2.23, II–2.24, II–2.42,
 II–2.44
North Sea
 continental shelf boundaries, I–8.03–12
 deep water, I–12.04
Norway
 Brent Spar decommissioning and,
  I–12.26, I–12.32
 cross-border fields, II–3.49
 decommissioning liabilities, I–12.79–80
 development strategy, I–5.06
 Ekofisk Bravo accident (1977), I–10.23
 Frigg Agreement, II–3.59–61, II–3–62,
  II–3.63
 gas exports to UK, I–3.54, I–3.64,
  I–3.70
 Gas Negotiation Committee, II–11.43
 gas transport, I–6.11
 health and safety offshore, I–10.59
 joint gas sales, II–11.43, II–11.45
 mediation, II–15.46
 Murchison Field Agreement, II–3.59,
  II–3.63
 North Sea continental shelf boundaries,
  I–8.08, I–8.09
 oil field size, I–7.17
 oil pollution liability, I–11.103
 oil trade, I–3.12–13
 Petroleum Safety Authority, I–11.98
 pipelines, I–3.13
 production decline, I–5.07
 sovereignty over continental shelf,
  I–8.02n
 standard contracts, II–5.70–2
 Statfjord Agreement, II–3.59, II–3.62,
  II–3.63
 taxation and exploration, I–7.42
 transport tariffs, II–7.75, II–7.76
 UK-Norway Agreement (1965), I–8.08,
  II–3.58–65
 UK-Norway Framework Treaty (2005),
  II–3.63–5
notional interest, II–9.63
nuclear power, I–3.51, I–9.119, I–9.121
numerus clausus principle, II–13.03
Nuttal, W, I–9.119

Occidental Petroleum, I–10.32
OECD, I–3.16, I–3.22, I–3.67

Office of Carbon Capture and Storage, I–5.22
Offshore Fisheries Legacy Trust, I–13.23
Offshore Industry Advisory Committee, I–10.27
Offshore Pollution Liability Agreement (OPOL), I–11.104, I–11.107, II–2.36
offshore work
  accommodation, II–14.77
  dismissal at third party requests, II–14.83–93
  meaning, II–14.44–5
  UK employment regulation, II–14.31–43
  working time and, II–14.44–82
offtake agreements, II–10.76
oil allowance, I–7.09
Oil and Gas Authority (OGA)
  see also licensing
  access to infrastructure and, I–6.28–63, I–6.96–118
    guidance, II–11.52
  Activity Survey (2014), I–5.07
  appeals against decisions, II–15.09
  asset stewardship and, I–5.68–70
  Asset Stewardship Expectations, II–5.78, II–12.64
  asset trading and
    consents, II–9.30–6, II–9.45, II–9.66
    farm-outs, II–9.08
  collaboration and, I–5.45, I–5.54–7, II–1.01
    competition and, II–11.62, II–11.64–6
    deal-building, II–15.20
  creation, I–5.02, I–5.27, I–10.63, I–11.100
  DEAL, I–11.15
  decommissioning and, I–12.42, I–12.46, I–12.47, I–12.48, I–12.63
    Decommissioning Insight 2016, I–12.01
    MER Strategy, II–7.113
    Model DSA, I–13.11, I–13.26–7, I–13.31–3, I–13.38, II–2.03, II–2.92, II–2.95, II–2.96
    role, I–12.75–6, II–7.111
    strategy, II–5.78
  dispute resolution recommendations, II–15.10, II–15.11
  effect, I–5.03
  ICoP and, I–6.74, II–11.52
  innovate licences and, I–4.77, I–4.78, I–4.86
  interventionism, I–4.01, I–4.85, II–1.01
  landward licences, I–4.80

launch, II–5.78
licensing obligations, I–4.13
licensing powers, I–4.14
  discretionary licensing, I–4.17, I–4.32–4
  reasonableness, I–5.59
  revocation, I–4.51, II–9.36, II–10.99, II–10.120
  using, I–5.71
MER Strategy and, I–5.44, I–10.03, II–5.78
  balancing interests, II–9.03
  cost switch and, II–7.78
  decommissioning and, II–7.113
  hub development, II–3.11
on national emissions ceilings, I–11.64
notifications to, surveys, I–11.13
Offshore Fisheries Legacy Trust, I–13.23
online licence information, II–9.24, II–9.30
operators and, I–4.60
programmes, I–5.02
second-tier participants, I–13.27, I–13.32
securities, consent, II–10.70, II–10.74–5
send or pay clauses and, II–7.83
standard contracts, II–5.13–14, II–5.20
  Model JOA see joint operating agreements
Supply Chain Strategy, II–5.01, II–5.18, II–5.78–81
transport and infrastructure role, II–7.76
unitisation and, II–3.13–18
work programmes and, I–4.49, I–4.51
Oil and Gas Independents' Association (OGIA), I–3.50
Oil and Gas Industry Task Force (OGITF), I–A.5, II–5.16
Oil and Gas Technology Centre, II–12.63
Oil and Gas UK, I–3.48, I–5.24, II–12.51
oil crisis (1973), I–3.16, I–3.22, I–3.28
oil crisis (1979–81), I–3.20, I–3.22
Oil Industry Taskforce, I–4.38
oil pollution
  civil liability, I–11.102–8
    EU law, I–11.105–6, I–11.108
    international law, I–11.104
    OPOL, I–11.104, I–11.107, II–2.26
  Deepwater Horizon, I–11.03, I–11.81, I–11.83, I–11.92, I–11.94, I–11.99, I–11.101, I–11.108, II–2.32
  emergency control, I–11.90–3
  insurance, II–2.36

international law, I–11.81
merchant shipping, I–11.82–5
Offshore Safety Directive, I–11.94–8
petroleum activities, I–11.86–9
UK Regulations, I–11.81
oil prices
    crash, I–1.08, I–2.01, I–10.02
        asset trading and, II–9.01
        decommissioning and, I–12.01,
            I–13.02
        insolvencies, II–2.55–6, II–10.93,
            II–10.122
    current trend, II–2.55
    economic modelling and, I–2.20
    high price period, I–7.22
    new normal, II–5.02
    SPAs, II–8.18
    volatility, II–8.01
ombudsmen, II–15.09, II–15.11
Ong, D, II–3.02, II–3.51
onshore allowance, I–7.34
onshore regulation
    consent to drill, I–9.12–28
    environmental impact, I–9.13, I–9.25–8
    fracking see shale gas/fracking
    guidance, I–9.21–3
    health and safety, I–9.13, I–9.15–18
    landward licences see landward licences
    legislative framework, I–9.12–28
    overview, I–9.01–128
    planning see planning
    regulator, I–9.16
    Review, I–9.02
    UK approaches, I–9.07–28
    well integrity, I–9.19–24
opcoms see joint operating committees
Open Permission, II–2.12, II–10.70–5
operating areas, definition, II–4.68–70
operators
    agency, II–2.37–8
    appointment
        ministerial approval, II–2.24
        OGA consent, I–4.60
    choice, II–2.24
    contractor-operator relationship
        dismissal issues, II–14.83–93
        drilling contracts, II–4.04–24
    contractor-operators, II–2.25
    contractual structure, II–5.04–11
    default notices, II–2.54
    definition, I–10.52, I–10.64
    drilling contractor relationship
        commercial, II–04.05–12
        invitations to tender, II–4.49–50

operational, II–4.13–24
oversight, II–4.49
overview, II–4.04–24
duties, II–2.26–31
duty of care, II–2.27
fiduciary duties, II–2.39–43
good oilfield practice, II–2.27, II–2.29
IMHH Agreement and, II–6.73
information duties, II–2.48
insurance, II–2.36
JOAs, II–2.23–43
liabilities, II–2.27–31
litigation, II–2.36
one-stop shop approach, II–12.08
removal, II–2.24
resignation, II–2.24
role, II–2.24, II–2.26
sole risk, II–2.50–3, II–3.43
transfer, asset trading, II–9.15
UUOAs, II–3.17
wilful misconduct, II–2.28–33
OPOL, I–11.104, I–11.107, II–2.36
OSPAR
    Brent Spar and, I–12.38, I–12.76
    chemicals and, I–11.45, I–11.46,
        I–11.47
    Decision 98/3, I–12.37–40, I–12.58,
        I–12.63, I–12.65, I–12.71–3, I–12.76,
        I–12.77, I–12.83, II–2.93
    decommissioning and, I–5.23,
        I–12.13–18, II–93–4
    Recommendation 2001/1, I–11.36
OSPAR Commission, I–11.36, I–12.39,
    I–12.40
out-of-rounds applications, I–4.22–3,
    I–4.80, II–2.05, II–03.08, II–3.12
overseas investment insurance, II–10.25
Owen, David, I–10.29
ownership
    co-ownership and partnership, II–2.20
    Crown see Crown
    doctrine of tenure, I–9.08
    intellectual property
        employee v contractor, II–12.33–7
        joint ownership in collaboration,
            II–12.43–5
        LOGIC contracts, II–12.52–9
    land law see land law

parent company guarantees, II–2.96,
    II–8.52, II–8.53, II–9.77
Park, J, II–5.25
Parliamentary Ombudsman, II–15.09
part blocks, I–4.18

partnerships
  agency, II–2.37
  definition, II–2.20
  joint ventures, II–2.02, II–2.07,
    II–2.15–22, II–7.12–15
passing of title
  gas/LNG sale agreements, II–8.30
  petroleum sales agreements, II–8.12–17
Patent Office, II–11.13, II–11.16
patents
  assignation, II–12.14
  EU law, II–12.15
  European patents, II–11.15
  exploitation options, II–12.14
  international law, II–12.15
  licensing, II–12.14, II–12.48
  oil companies, II–12.16
  overview, II–12.12–16
  requirements, II–12.13
  rights, II–12.12
paternity leave, II–14.02
Peel, E, II–6.43
penalty clauses, forfeiture clauses as,
    II–2.58–82
pendulum procedure, II–3.37
permit to work system (PTW), I–10.32,
    I–10.42
personal protective equipment, I–10.43
personal servitudes, II–13.31
personnel
  definition, II–6.53, II–6.61
  IMHH, II–6.66
petroleum, exploration and development
    licences (PEDLs)
  blocks, II–03.02
  introduction, I–4.79
  landward licences, I–4.03, I–9.125
  licensing authority, I–4.80
  licensing rounds, I–4.80, I–9.09
  model clauses, I–9.11
  numbers, I–9.09
  overview, I–4.79–81
  rights, I–9.09
  shale gas, I–9.09–10
  terms, I–4.79, I–9.10
petroleum licences see licences; licensing
    regime; specific licences
Petroleum Operation Notices (PONs),
    I–11.15, I–11.40, I–11.42
petroleum revenue tax (PRT)
  asset trading, II–9.60
  capital allowances, I–7.33
  decommissioning relief, I–7.47, I–7.50,
    I–13.37

levy, I–7.05
  new fields and, I–7.10
  payments, I–7.04
  rate changes, I–3.45, I–7.09, I–7.13,
    I–7.53–4
  removal, I–7.62
petroleum sales agreements
  collateral support, II–8.52–3
  crude oil see crude oil sales agreements
  drafting, II–8.55–6
  freedom of contract, II–8.02
  gas/LNG see gas/LNG sales
    agreements
  implied terms, II–8.02
  overview, II–8.01–56
  practice, II–8.54
PILOT
  Brownfields Initiative, I–12.81,
    I–A.35–51
  Brownfields Studies Report, I–A.37–40
  Brownfields Work Group, I–A.35,
    I–A.40
  Fallow Areas Initiative, I–A.7–34
  initiatives, I–A.5–6
  PPWG, I–4.52, I–A.7, I–A.8, I–A.9,
    I–A.11, I–A.16
  predecessor, II–5.16
pipelines
  see also infrastructure; transport;
    transport agreements
  abandonment, II–13.14, II–13.18
  ageing pipelines, II–7.25
  agreements, II–7.24
  basic agreements, II–7.55–8
  capacity see infrastructure access
  chemicals, I–11.40
  crossing agreements, II–7.42–7
    standard contracts, II–5.14, II–7.43
  EU law, I–6.66
  land law and see land law
  leasehold rights, II–13.12, II–13.29
  meaning, I–6.22–3
  Norway, I–3.13
  overview, II–7.22–6
  owners' consent, II–13.03
  ownership, II–13.10–11
  Pipeline Work Authorisations, I–11.79
  proximity agreements, II–7.42–7
    standard contracts, II–5.14, II–7.43
  real rights, II–13.04, II–13.07
  regulator, I–9.16
  replacement, II–13.18
  section 29 notices, I–12.59
  servitudes see servitudes

smaller pipelines, II–7.23
statistics, I–12.02
statutory regulation, II–13.09
transparency of terms, II–7.22
trespass, II–13.03
UK-Norway Agreement, II–3.63
Piper 25 Conference (2013), I–10.92
Piper Alpha disaster (1988)
    see also Cullen Report
    casualties, I–9.16
    catastrophic scale, II–4.138
    effect on production, I–3.09
    health and safety and, I–10.31–9,
        I–11.94
    infrastructure and, I–10.57
    legal response, I–9.16
    London Bridge, II–6.41–2, II–6.45,
        II–6.50, II–6.79, II–6.80
    negligence, II–6.38–9
    Orbit Valve, II–6.38–9, II–6.45
    overview, I–10.31–9
    prescriptive approach and, I–11.94
Piper field, I–10.16n
planning
    appeals, I–9.45–6
    applications
        fracking, I–9.35, I–9.55, I–9.58–61,
            I–9.90–110
        material considerations, I–9.56–69
        process, I–9.47–56
        public concern, I–9.70–82
        statutory consultees, I–9.53
    conditions, I–9.43
    development control, I–9.36–42
    England
        appeals, I–9.45
        legislation, I–9.33
        local plans, I–9.42
        National Planning Policy Framework,
            I–9.42, I–9.57
    fracking and, I–3.53, I–9.35, I–9.55,
        I–9.58–61, I–9.90–110
    local development plans, I–9.31,
        I–9.41–2, I–9.49
    onshore wells and, I–9.13, I–9.25,
        I–9.29–110
    politics and law, I–9.83–110
    Scotland
        appeals, I–9.46
        considerations, I–9.56
        legislation, I–9.34
        local development plans, I–9.41
    terminology, I–9.29
    UK development, I–9.30–1

pledges, II–10.65, II–10.69n, II–10.85n,
    II–13.08
political risks, II–10.08
polluter pays principle, I–12.15, I–12.37,
    II–2.94
pollution see oil pollution
portfolio management, II–9.02
posted workers, II–14.36, II–14.42
precautionary principle, I–11.15, I–12.15,
    I–12.22, I–12.37
pre-emption rights
    affiliate route and, II–9.48–54
    assignment and, II–9.21, II–9.27,
        II–9.37–56
    JOAs, II–2.13, II–9.11, II–9.43–54
    leases, II–13.13
    Master Deed, II–9.43–5
    package deals, II–9.47
    unmatchable deals, II–9.46
    waivers, II–9.66
Preese Hall incident, I–9.02, I–9.–05
prepayment facilities, II–10.20
Preston New Road Action Group, I–9.92,
    I–9.99
price-fixing, I–6.68, II–11.11, II–11.15,
    II–11.58
prices
    see also consideration
    gas/LNG sales agreements, II–8.42–8
    oil see oil prices
private equity, II–9.02, II–9.30, II–10.09,
    II–10.12, II–10.14, II–10.121
privatisations, I–5.05
produced water, I–11.02, I–11.34–7,
    I–11.100
production
    competition law, relevant market,
        II–11.27
    decline, I–2.06, I–2.08, I–5.07, II–12.01
    definition, II–11.48
    efficiency, I–2.09–10, I–5.18, I–5.22
    environmental regulation, I–11.16–21,
        I–11.30–3
        chemicals, I–11.38–42
    future estimates, II–11.67
    landward areas, I–4.81
    licences see production licences
    maturity, I–5.07
    modelling see economic modelling
    onshore, I–9.01
    overlifting, II–9.62
    projections, I–2.07–10
    recent experience, I–2.05–10
    statistics, I–7.23

tax incentives and new field
developments, I–2.22–40
UK as energy producer, I–3.43–53
UK exports, I–3.09–10
UK ranking, II–11.64
underlifting, II–9.62
unit costs, I–2.04
production licences
adjustments, I–5.01
air pollution and, I–11.62–4
area rental payments, I–4.45
bespoke licences, I–4.03, I–4.74
blocks, II–3.02
categories, I–4.03
contractual character, I–4.05, I–4.82
early licences, I–A.7
exclusivity, I–4.05, I–6.03
frontier licences see frontier licences
geological surveys and, I–11.13
information rights, I–4.60
interventions, I–4.59–60
JOAs and, II–2.06, II–2.09
model clauses, I–4.17, I–4.56, I–5.01,
I–A.9
numbers, I–4.68
operational control, I–4.59–60
overview, I–4.41–68
promote licences see promote licences
proprietary rights, I–4.04
revocation, II–10.99
seaward licences, I–4.41–68
standard licences, I–4.43–60
terms and relinquishments, I–4.52–8
traditional licences, I–4.43–60
unitisation see unitisation
work programmes, I–4.47–51
products, definition, II–11.48
project finance, II–10.23–4
project management drilling contracts,
II–4.21
project/field life cover ratio (PCLR/
FLCR), II–10.57
promote licences
annual rental fees, I–4.71
applications, I–4.72
concessions, I–4.73
consultation, I–4.69
flexibility, I–4.86
introduction, I–4.69
overview, I–4.69–73
terms, I–4.70
Wood Review and, I–5.15
work programmes, I–4.70
property, definition, II–6.54, II–6.61

proximity agreements, II–5.14, II–7.42–7
PRT see petroleum revenue tax
public concern, planning and, I–9.70–82
public policy
ADR and, II–15.43
energy see energy security
light-touch regulation, I–5.06, I–5.11
replacement, II–1.01
pre-Wood philosophy, I–5.05–6
pumping stations, II–13.12, II–13.23,
II–13.27

Qatar, I–3.13, I–3.55
Quadrant 22, I–4.18
qualified indemnities, II–6.23–5
qualifying combustion installations
(QCIs), I–11.58–9
quantified risk assessments (QRAs),
I–10.35, I–10.38, I–10.53

radioactivity, II–5.54
railway lines, II–13.11
RBL facilities, II–10.06, II–10.09,
II–10.14, II–10.22, II–10.23,
II–10.35–7, II–10.40, II–10.45–6,
II–10.50–3, II–10.78–9, II–10.81,
II–10.121
real burdens, II–13.30
real rights
categories, II–13.06
chain of rights, II–13.07
enforceability, II–13.05
finance and, II–13.08
leases see leases
overview, II–13.06–9
ownership, II–13.06, II–13.10–11
pipelines, II–13.04
servitudes see servitudes
reciprocal indemnities see mutual
indemnities
redundancy, II–14.02, II–14.40
refineries, II–13.10, II–13.12, II–13.22,
II–13.23
regional development strategy, I–5.19–20
renewable energy
Crown ownership and, I–12.82
EU Law, I–3.30
Scottish policy, I–9.121
UK, I–3.63, I–3.73
reputation equity, II–15.18
res nullius, I–4.09
research and development
agreements, II–11.17, II–11.64
encouragement, I–3.51, I–4.36, I–A.14

reserves
  borrowing base mechanism,
    II–10.54–60
  RBL facilities and, II–10.35–6
  recoverable reserves, I–2.21, I–A.47,
    II–2.92, II–3.23, II–3.25
  warranties, II–9.72
reservoirs
  UK-Norway Framework Treaty (2005),
    II–3.63
  warranties, II–9.72
Reynolds, T, II–3.51
Ridley, Nicholas, I–10.13
rig-sharing, II–11.35
ring fence corporation tax (RFCT)
  capital allowances, I–3.44, I–7.28–33
  corporation tax and, I–7.07
  decommissioning relief, I–7.47, I–13.37
  levy, I–7.05
  new field developments, I–2.22
  rate, I–7.07
  technical features, I–7.28–40
Ring Fence Expenditure Supplement
    (RFES), I–2.22, I–7.31, I–7.62
risk allocation
  contractual methods, II–6.02
  indemnities see indemnities
  LOGIC standard contracts, II–5.40–55
  oil and gas contracts, II–6.01–86
  supply chain agreements, II–5.12,
    II–5.40–55
    exceptions, II–5.46–55
risk assessments
  chemicals, I–11.38
  drilling, I–9.14
  EU law, I–10.62, I–11.95
  goal-setting regulations, I–10.43
  Monte Carlo technique, I–2.11
  night work, II–14.57
  offshore installations, I–10.53
  OSPAR, I–11.36
  quantified risk assessments, I–10.35,
    I–10.38, I–10.53
  safety case, I–10.66, I–10.71
  well integrity, I–9.20
Robens Report (1972), I–10.18–22,
    I–10.24, I–10.30, I–10.33
Robinson, J Rowan, I–11.06
Rockall, I–8.13–33
Roggenkamp, M, II–3.57
Roman law, II–13.21
Rough platform, I–3.58
Royal Academy of Engineering, I–9.02
Royal Society, I–9.02

Royal Town Planing Institute, I–9.60,
    I–9.83
royalties, I–4.33, I–7.27, II–9.25, II–10.19,
    II–12.14
Rudd, Amber, I–3.01, I–3.02, I–5.50
rule of capture, II–3.04, II–3.05, II–3.07,
    II–3.10
Russia
  annexation of Crimea, II–9.35
  Energy Charter and, I–3.23–4, I–3.68
  EU energy dependence on, I–3.74
  gas production, II–11.64
  oil price and, I–1.08
  Yukos arbitration, I–3.23

safeguard allowance, I–7.09
safety and environmental management
    systems, I–10.66, I–10.73
safety case see health and safety offshore
safety management systems, I–10.34,
    I–10.37, I–10.41, I–10.58, I–10.66
sale of goods
  implied terms, II–8.02, II–8.08, II–8.12
  international law, II–8.02
  passing of title, II–8.30
sales agreements see crude oil sales
    agreements; gas/LNG sales
    agreements
salvage principle, II–10.104–5
São Tomé and Príncipe, I–3.56
Sasine Register, II–13.15
sceptic tanks, II–13.22
Scilly Isles, I–8.12
Scotland
  civil procedure, II–15.44, II–15.75–6
  devolution see Scottish devolution
  energy policy, I–9.119–22
  independent kingdom, I–9.112
  irritancy, II–10.98
  land law see land law
  liquidation, II–10.110, II–10.114–15
  mixed jurisdiction, II–13.02, II–13.21
  National Marine Plan, I–4.30
  planning
    appeals, I–9.46
    considerations, I–9.56
    fracking case, I–9.101–8
    legislation, I–9.34
    local development plans, I–9.41
  property law, I–9.08, II–10.88
  real burdens, II–13.30
  securities, II–10.69
    over shares, II–10.88
    standard securities, II–10.89

Scott, R, I–3.16, I–3.17, I–3.20
Scottish Constitutional Convention,
  I–9.112, I–9.117
Scottish devolution
  1997 referendum, I–9.112
  2014 independence referendum, I–1.04,
    I–5.29n, I–9.114–16
  energy competence, I–9.117
  energy policy, I–9.119–22
  fracking moratorium, I–9.108,
    I–9.120–2, I–9.126, I–9.128
  further devolution, I–9.116, I–9.123–6
  history, I–9.112
  licensing regime and, I–9.125–6
  overview, I–9.111–26
  recent developments, I–1.04
  settlement, I–9.113
  Smith Commission, I–9.116, I–9.123,
    I–9.125, I–9.126
Scottish Environmental Protection Agency
  (SEPA), I–9.25, I–9.53, I–10.61,
  I–11.19
Scottish Home Rule Association, I–9.112
Scottish Natural Heritage (SNH), I–9.53,
  I–11.19
Scottish Sentencing Council, I–10.83
Scottish Water, I–9.53
sea, law of the sea see law of the sea
sea deposits, I–11.49, I–11.72
Sea Gem Inquiry, I–10.06–11, I–10.12,
  I–10.13, I–10.16, I–10.19, I–10.23–5,
  I–10.32, I–10.35–6, I–10.38
seasonal workers, II–14.30
seaward licences
  area rental payments, I–4.45–7
  bespoke licences, I–4.03, I–4.74
  categories, I–4.03
  exploration licences see exploration
    licences
  frontier licences see frontier licences
  information requirements, I–4.60
  innovate licences see innovate licences
  legal nature, I–4.04, I–4.09
  model clauses, I–4.17, I–5.01, I–A.9
  operational control, I–4.59–60
  overview, I–4.39–42
  production licences see production
    licences
  promote licences see promote licences
  revocation, II–9.36, II–10.99, II–10.120
    work programmes and, I–4.51
  terms and relinquishments, I–4.52–8
  work programmes, I–4.47–51
second lien debts, II–10.08, II–10.09

section 29 notices
  see also decommissioning plans
  asset sale and purchase agreements and,
    II–9.77
  facility information requests, I–12.59
  Guidance Notes, I–12.59–61
  initiation, I–12.42
  ministerial powers, I–12.42, I–13.04
  penalties, I–12.45
  recall, I–12.45, I–12.60
  scope, I–13.09
  service, I–12.43–5, I–13.04, I–13.08
  warning letters, I–12.59
  withdrawal, I–12.45, I–12.60, I–13.05
securities
  decommissioning see decommissioning
    security agreements
  documents, II–10.50
  foreign assets, II–10.92
  forms, II–10.65
  over accounts, II–10.49, II–10.82–4
  over collateral arrangements, II–10.90–1
  over contractual rights, II–10.76–80
  over insurance, II–10.76, II–10.81
  over licences, II–10.70–5
  over real estate, II–10.89
  over shares, II–10.85–8
  overview, II–10.65–92
  registration, II–10.68, II–10.90
securitisation, II–13.08
seismic events, I–9.02
seismic imaging technology, II–12.03
self-employed
  categories, II–14.26–7
  employment status, II–14.03
  workers or, II–14.23–30
  working time rights, II–14.46
semi-sunmersible drilling units, II–4.25,
  II–4.26, II–4.35–40, II–4.112–13,
  II–5.22
send or pay clauses, II–7.80–3
Senior Executive Appraisal Mediation
  (SEAM), II–15.56
sentencing guidelines, corporate
  manslaughter, I–10.83, I–10.84
service contracts, II–5.05–8
servitudes
  access, II–13.23, II–13.29
  conditions, II–13.24
  discharge, II–13.20
  dominant tenements
    pipelines, II–13.24–30
    principles, II–13.23
  form of deeds, II–13.15

leasehold servitudes, II–13.01, II–13.23
owners' exclusion, II–13.22
personal servitudes, II–13.31
pipelines, II–13.07, II–13.14–31
  benefits, II–13.16–20
  dominant tenements, II–13.24–30
  identification of land, II–13.25
  indemnities, II–13.14, II–13.24
  limitations, II–13.21, II–13.26
  negative restraints, II–13.30
  terms, II–13.14
  third parties, II–13.28
praedial servitudes, II–13.06, II–13.31
praedial utility, II–13.27
real rights, II–13.06, II–13.31
registration, II–13.25
Roman origins, II–13.21
storage, II–13.22
utilitas, II–13.27
variation, II–13.20
vicinity requirement, II–13.27
sewage
  definition, I–11.66
  offshore regulation, I–11.65–9
Sewel Convention, I–9.113
shale gas/fracking
  development, I–9.127–8
  energy policy, I–3.53
  environmental impact, I–9.25–8
  guidelines, I–9.12, I–9.22
  health and safety, I–9.15–24
  moratorium, I–9.02, I–9.108, I–9.120–2,
    I–9.126
  onshore exploration, I–9.02
  PEDLs, I–9.09–10
  planning and, I–3.53, I–9.35, I–9.55,
    I–9.58–61, I–9.90–110
  potential, I–3.52
  Preese Hall incident, I–9.02, I–9.05
  revolution, I–4.81
  safety, I–9.15–28
  Scottish policy, I–9.108, I–9.120–2,
    I–9.128
  Task Force on Shale Gas, I–9.03
  UK developments, I–9.01–5
  US exports to UK, I–3.56
shallow drilling, I–11.13–15, I–11.72
share fishermen, II–2.21
shares
  equitable mortgages, II–10.69, II–10.85
  issues, II–10.09, II–10.10
  securities over, II–10.85–8
    English companies, II–10.85–6
    pledges, II–13.08

Scottish companies, II–10.88
transfer
  asset purchase or, II–9.10–18
  consideration, II–9.65
  CVAs, II–10.120
  due diligence, II–9.16
  liabilities, II–9.17
  licensing regulation and, II–9.34–5
  taxation, II–9.18, II–9.86
  warranties, II–9.74
Sharp, D, II–6.18
Shell
  asset trading, I–7.58
  on boundary agreements, I–8.06
  Brent field, I–12.01
  Brent Spar see Brent Spar
  Clair field, II–3.15
  smaller assets, II–2.55
  SPAs, II–08.06
  Stakeholder Dialogue, I–12.06,
    I–12.27–35, I–12.36, II–15.07n
sheriff court procedure, II–15.75
shift work, II–14.53–5
shipping
  air pollution, I–11.51–7
  industry, II–6.04
  oil pollution, I–11.82–5
  sewage and garbage, I–11.65–72
shut-in rights, II–7.84–8
sidetracks, II–4.107
Sierra Leone, I–3.56
small field allowance, I–7.37
small fields, I–2.32–3, I–2.52, I–4.81,
    I–6.05, I–6.10, I–6.90, I–6.103,
    II–7.01–2, II–7.78
Smith, John, I–A.28n
Smith Commission, I–9.116, I–9.123,
    I–9.125, I–9.126
software
  copyright, II–12.23–5
  track participation, II–3.39
sole risk clauses, II–2.50–3, II–3.43,
    II–9.71
South Africa, servitudes, II–13.23
sovereign wealth funds, II–10.09
Soviet Union, I–3.22, I–3.68
special areas of conservation (SACs),
    I–11.09
special protection areas (SPAs), I–11.09,
    I–13.29
specialisation agreements, II–11.17,
    II–11.48
Stakeholder Dialogue, I–12.06,
    I–12.27–35, I–12.36, II–15.07n

stakeholders, II–15.17–18
stamp duty, II–9.86
stamp duty land tax, II–9.86
Standard and Poors, II–2.96
standard contracts
    AIPN *see* Association of International
        Petroleum Negotiators
    automation, II–5.84
    benefits, II–5.21
    Canada, II–5.74
    contents, II–5.12
    contractual structure, II–5.04–11
    CRINE, II–5.15, II–11.35, II–12.51
    field agreements, II–5.14
    form, II–5.28–33
    guidance notes, II–5.34–5
    IADC, II–5.73
    key provisions, II–5.40–55
    LOGIC *see* LOGIC standard contracts
    mistakes, II–5.25
    mutual hold harmless indemnities,
        II–5.41–55, II–5.61–3
    new contracts, II–5.82–3
    Norway, II–5.70–2
    OGA, II–5.13–14, II–5.20
    overview, II–5.012–17
    rationale, II–5.01–3, II–5.18–25
    risk allocation, II–5.12
    supply chain agreements, II–5.11–13
    tendering process, II–5.36–9
    United States, II–5.73
    updating, II–5.25
    variations clauses, II–5.64
standard production licences *see*
    production licences
standard securities, II–10.89
Statfjord Agreement, II–3.59, II–3.62,
    II–3.63
Statfjord B Platform, I–11.02
statutory adjudication, II–15.62–4
statutory wayleaves, II–13.31
Steele-Nicholson, A, II–3.37
step-in agreements, II–10.24
stewardship
    assessment, I–A.49–51
    asset stewardship strategy, I–5.06,
        I–5.09n, I–5.18, I–5.66–70
    asset transfer and, I–A.38
    benchmarking, I–5.68
    Brownfields Initiative, I–5.06, I–12.81,
        I–A.35–51
    Brownfields Studies Report, I–A.37–40
    collaboration, I–5.52–3
    concept, I–A.39

expectations, I–5.67
MER Strategy, I–5.42, I–5.63,
    I–5.66–70, I–6.54
ministerial powers, I–A.41–2
model clauses, I–A.47
OGA Asset Stewardship Expectations,
    II–5.78, II–12.64
pre-Wood state control, I–5.06
process, I–A.41–8
reviews, I–5.70
sanctions, I–A.48
stages, I–A.44
Stewardship Initiative, I–5.06, I–12.81,
    I–A.35–51
UKCS Stewardship Survey, I–5.68
Wood Review and, I–5.18
STOOIP (stock tank oil originally in
    place), II–3.23–6
storage
    carbon capture and storage, I–3.12,
        I–5.22, I–12.45
    FPSO units, II–7.19
    gas, I–3.57–9
    servitudes, II–13.22
strategic environmental assessments,
    I–4.24, I–4.27
Stratfjord B Platform, I–11.02
streaming, II–10.21–2
strict liability, I–11.103, I–11.107,
    I–12.79, II–4.144
study agreements
    allocation of costs, II–7.36
    scope, II–7.34
    site to shore transport, II–7.09
    standard contracts, II–5.14, II–7.33
    timelines, II–7.35
    transport agreements, II–7.33–7
subordinated loans, II–10.09, II–10.76
summary trials, II–15.76
supplementary charge
    capital allowances, I–7.33
    cluster area allowances (CAAs), I–7.40
    decommissioning relief, I–7.47, I–13.37
    energy production policy, I–3.44
    field allowance, I–7.34, I–7.37
    investment allowances, I–7.34–40,
        I–7.62
    levy, I–7.05, I–7.08
    new field developments, I–2.22, I–2.24,
        I–2.28, I–2.30
    rate changes, I–7.08, I–7.50
    rationale, I–7.11
supplementary seismic survey licences,
    I–4.03

supply chain, meaning, II–5.05
supply chain agreements
  allocation of risk, II–5.43
  contents, II–5.12
  contractual structure, II–5.05
  function, II–5.04
  mutual hold harmless indemnities,
    II–5.40–55
  parties, II–5.11
  risk allocation, II–5.12, II–5.40–55
  standard contracts, II–5.11–13,
    II–5.18–25
  trend, II–5.09
Supply Chain Strategy, II–5.01, II–5.18,
    II–5.78–81
Supply Chain Taskforce, II–5.79–80
sustainable development, I–12.57
swap agreements, II–2.90
Sweden, North Sea continental shelf
    boundaries, I–8.08
syndicated loans, II–10.37–8
syndication risk, II–10.08

takeovers and mergers
  asset purchase or, II–9.10–18
  competition and, II–11.11, II–11.27,
    II–11.49, II–11.51
  consideration, II–9.65
  CVAs, II–10.120
  due diligence, II–9.16
  liabilities, II–9.17
  licensing regulation and, II–9.34–5
  taxation, II–9.18, II–9.86
  warranties, II–9.74
Task Force (1999), I–2.08
Task Force on Shale Gas, I–9.03
Taverne, B, II–3.30, II–3.45, II–3.62
taxation
  see also specific taxes
  alternative regimes, I–7.19
  asset transfers, II–9.18, II–9.86
  capital allowances, I–3.44, I–7.28–33,
    I–7.62
  cash flow regime, I–7.04
  categories, I–7.05
  cross-border agreements and, UK/
    Netherlands, II–3.56
  decommissioning and, I–7.45–54,
    I–13.02, I–13.34–8, II–9.78
  energy production policy, I–3.44–5
  evaluation, I–7.62–4
  exploration and, I–7.41–2
  field allowances, I–3.47, I–7.12, I–7.34,
    I–7.37

Fiscal Review, I–1.03
  gas/LNG sales agreements, II–8.51
  infrastructure, I–7.55–7
  investment and, I–9.126
    new driving investment blueprint,
      I–7.24–7, I–7.33
  joint ventures, II–2.02
  LOGIC standard contracts and, II–5.67
  losses, I–7.43–8
    loss relief, I–7.09, I–7.43–4
  mature assets trading, I–7.58–61
  neutral system, I–7.14
  new field developments and, I–2.22–40
  oil allowance, I–7.09
  oil revenues, I–4.34
  overview, I–7.01–64
  payments, I–7.04
  principles of good oil and gas practice,
    I–7.20–3
  recent changes, I–2.53, I–7.06–14
  safeguard allowance, I–7.09
  share purchase, II–9.18
  specific hydrocarbon regime, I–7.15–19
  strict concession regime, I–7.18, I–7.27
  territorial scope, II–14.37
  Wood Review, I–3.45, I–5.08, I–7.01
Taylor, M, II–3.32, II–3.37, II–3.46
technology
  intellectual property see intellectual
    property
  MER Strategy, I–5.42
  Oil and Gas Technology Centre,
    II–12.63
  Wood Review, I–5.22, II–12.62
Technology and Construction Court,
    II–15.77
technology transfer agreements, block
    exemption, II–11.64, II–12.27
tenders
  drilling contracts, II–4.49–50
  LOGIC standard contracts, II–5.36–9
Thatcher, Margaret, I–5.05
title
  see also passing of title
  intellectual property contracts,
    II–12.52–9
  investigation by asset purchasers,
    II–9.24
    insolvency cases, II–10.117–18
  warranties, II–9.71
Tormore field, I–3.47, I–3.48
torts
  agents, II–2.38
  compensatory damages, II–6.30

fault, II–6.30
negligence, II–2.28, II–2.33, II–6.24,
 II–6.38–45
 wilful misconduct, II–2.28–33, II–5.61,
 II–6.24
Total, I–10.84
track participation, II–3.19–34
trade associations, II–11.11, II–11.12,
 II–11.55
trade secrets, II–12.26–32
trade unions, I–9.39, I–10.27, I–10.39,
 I–10.47, II–14.19, II–14.87
traditional production licences *see*
 production licences
training materials, II–12.20
transfer of undertakings, II–10.117,
 II–14.02
transport
 agreements *see* transport agreements
 arrangements, II–7.04, II–7.06–16
 FPSO units, II–7.18–21
 hub-based structure, II–7.02, II–7.04
 multi-field arrangements, II–7.02
 offtake options, II–7.17
 overview, II–7.01–116
 pipelines *see* pipelines
 site to sale, II–7.06, II–7.12–16
 site to shore, II–7.06, II–7.08–11
transport agreements
 basic agreements, II–7.55–8
 capacity booking, II–7.66–70
 collective capacity, II–7.100–3
 competition issues, II–7.65, II–11.46–54
 confidentiality agreements, II–7.29–32
 construction and tie-in agreements,
 II–7.48–53
 cost switch, II–7.77–9
 cost-sharing agreements, II–7.38–40
 crossing agreements, II–5.14, II–7.42–7
 decommmisioning, II–7.110–13
 end dates, II–7.89–91
 form, II–7.27–106
 governing law, II–7.05
 indemnities, II–7.94, II–7.97–9
 crossing and proximity agreements,
 II–7.43–5
 CTIAs, II–7.51–2
 CULAs, II–7.97, II–7.106, II–7.107–9
 liability caps, II–7.43
 new entrants, II–7.106
 initial assessments, II–7.28
 joint supply/purchase of capacity,
 II–11.46–54
 negotiation issues, II–7.64–106

new entrants, II–7.104–6
novation, II–9.84
pipelines, II–7.24
pre-development stage, II–7.41–53
production stage, II–7.54–63
proximity agreements, II–5.14,
 II–7.42–7
send or pay clauses, II–7.80–3
shut-in rights, II–7.84–8
specification, II–7.92–6
study agreements, II–7.33–7
tariffs, II–7.71–6
transport, processing and operations
 services agreements, II–7.60,
 II–7.62–3
transport and processing agreements,
 II–7.59–61
Transport Scotland, I–9.53
trespass, II–13.03
Trinidad and Tobago, I–3.55
trust funds, I–7.45, I–12.81, I–13.18,
 I–13.21–2, I–13.24–5, I–13.31,
 I–13.38, II–2.95–6
turnkey drilling contracts, II–4.17–20
Tuscan Energy, II–2.56
tying, II–11.20, II–11.56

UK Hydrographic Office, I–12.74, I–12.78
UK Onshore Operators Group (UKOGG),
 shale gas wells guidance, I–9.22–3
UKAPP Certificates, I–11.53
UKCS
 boundaries *see* continental shelf
 boundaries
 decline, II–12.01
 diversity, I–1.01–2
 expensive area, II–12.02
 exploration *see* exploration
 grid system, I–4.18–19
 hostile environment, I–10.01
 installation numbers, I–12.02
 legal ownership, I–4.09
 macro-conditions, II–01–2
 maturity, I–1.01, I–2.01–4, I–2.05,
 I–5.07, I–7.58–61, I–10.02, I–10.51,
 II–1.01, II–5.01
 production *see* production
 recent experience, I–2.05–10
UK-Nigeria Energy Working Group,
 I–3.56
UKOOA
 ICoP *see* Infrastructure Code of Practice
 (ICoP)
 infrastructure access and, I–6.73

model JOAs, II–2.03
standard contracts, II–5.13
ullage, I–7.55
Ultramar Exploration, II–3.53
unbundling, I–6.72, I–6.89
UNCITRAL, II–15.66
UNCLOS
    continental shelf boundaries, I–8.15–33
    decommissioning and, I–12.08–10,
        I–12.40, I–12.82
    sovereignty, I–12.82
    unitisation and, II–3.51
unfair contract terms, II–6.05, II–6.28–9,
    II–6.84
unfair dismissal
    dismissal at third party requests,
        II–14.83–93
    employment status and, II–14.02,
        II–14.10
    territorial jurisdiction, II–14.33–43
unfair preferences, II–2.57, II–2.83–91
Uniform Network Code, II–8.30
unit costs, I–2.04
United Nations
    Gulf War (1991) and, I–3.20
    UNCITRAL, II–15.66
    UNCLOS see UNCLOS
    Vienna Convention, II–8.02
United States
    ADR, II–15.45
        mediation, II–15.46, II–15.54,
            II–15.73, II–15.79
    collaboration with stakeholders,
        II–15.18
    dispute resolution, II–15.23, II–15.25
    export control, II–4.69
    indemnities, II–6.28
    JOAs, II–2.02
        fiduciary duties, II–2.43
        model forms, II–2.03, II–2.04, II–2.27
    Macondo disaster see Deepwater
        Horizon disaster
    oil imports, I–3.15
    pendulum procedure, II–3.37
    rule of capture, II–3.04
    shale gas, I–3.52, I–3.56, I–4.81, I–9.02
    standard contracts, II–5.73
    unitisation agreements, II–3.17, II–3.27
unitisation
    alternatives, II–07–9
    case for, II–04–5
    common issues, II–3.19–46
    cross-border unitisation, II–3.47–65
        UK/Netherlands, II–3.52–7, II–3.64–5

UK/Norway, II–3.58–65
    cross-licence agreements, II–3.09
    fixed interest agreements, II–3.09
    geological factors, II–01–3
    international law, II–3.47–65
    legislation, II–3.10–11
    meaning, II–05
    MER Strategy and, II–3.11
    ministerial powers, II–3.12
    MOOIP, II–3.23, II–3.26
    overview, II–3.01–67
    practice, II–3.13–18
    pre-unitisation agreements, II–3.17
    rule of capture, II–3.04, II–3.05,
        II–3.07, II–3.10
    STOOIP, II–3.23–6
    track participation
        computer programs, II–3.39
        determination, 44, II–3, II–3.21–7
        expert dispute resolution, II–3.35–40
        re-determination, II–3.28–34
UUOAs see unitisation and unit
    operating agreements
unitisation and unit operating agreements
    (UUOAs)
    clauses, II–3.17
    competition law issues, II–11.34
    cross-border agreements, II–3.49–65
        dispute resolution, II–3.52, II–3.55,
            II–3.64
        Frigg Agreement, II–3.59–61, II–3.62,
            II–3.63
        governing law, II–3.54
        Markham Agreement, II–3.53–7
        Murchison Agreement, II–3.59
        Statfjord Agreement, II–3.59, II–3.62,
            II–3.63
    decision-making, II–3.44
    default and forfeiture, II–3.45–6
    dispute resolution, II–3.35–40
        cross-border agreements, II–3.52,
            II–3.55, II–3.64
    JOAs and, II–3.17–18, II–3.46
    non-unit operations and, II–3.41–2
    OGA approval, II–3.14
    operators, II–3.17
    overview, II–3.16–17
    pre-emption rights, II–9.37
    pre-unitisation agreements, II–3.17
    purpose, II–06
    sole risk operations, II–3.43
    tract participation, II–3.19–34
    unit operating committees, II–3.17
    voting rights, II–3.44

value allowances, I–7.39
vendor finance, II–10.29–32
Venezuela, I–1.08
Verma, Baroness, I–5.29, I–5.30, I–5.31
vertical agreements, II–11.29, II–11.56–67
Vision for 2010, I–A.5, I–A.35, I–A.37
volatile organic compounds, I–11.56,
    I–11.64
volcanic ash, II–4.132

warranties
    asset transfer agreements, II–9.70–4
    gas/LNG sales agreements, II–8.51
    LOGIC standard contracts, II–5.60
    reserves, II–9.72
    share sale agreements, II–9.74
    time limits, II–9.73
    title, II–9.71
waste
    see also decommissioning
    categories, I–12.11
    London Dumping Convention,
        I–12.11–12
    offshore regulation, I–11.65–72
water
    see also law of the sea; oil pollution
    drinking water, I–10.42
    fracking and groundwater, I–9.27,
        I–9.61
    pipes, II–13.22
    produced water, I–11.02, I–11.34–7,
        I–11.100
wayleaves, II–13.31
Weaver, J, II–3.61
well services
    catastrophic loss, II–5.49
    corrosion by well effluent, II–5.48
    downhole equipment, II–5.47
    LOGIC standard contracts, indemnities,
        II–5.46–51
    pollution, II–5.50
    property lost overboard, II–5.51
wells
    definition, II–4.71–6, II–4.107
    services see well services
    sidetracks, II–4.107
West of Shetland Task Force, I–3.46
whistle-blowing, II–11.03
Wicks Report, 1–3.10, I–3.05, I–3.24,
    I–3.39, I–3.54
Wiggins, D, II–3.05
wilful misconduct, II–2.28–33, II–5.61,
    II–6.24
Willoughby, G, I–4.04

withering interest forfeiture clauses,
    II–2.63, II–2.65
Wood Mackenzie, II–11.67
Wood Review
    asset stewardship, I–5.18
    asset trading, I–7.58
    collaboration, I–2.54, I–5.23, I–5.46,
        II–04.01
    criticisms, I–5.08
    decommissioning, I–5.23, I–12.03
    effect, I–5.71–2
    enhanced regulatory powers, I–5.12–13
    exploration strategy, I–5.15–17
    final report, I–5.08
    on FPSO units, II–7.20
    government response, I–3.45–6, I–5.24
    health and safety, I–10.63
    implementation, I–4.83, I–5.25–72
        MER Strategy, I–4.83, I–5.25,
            I–5.28–70
        OGA creation, I–5.27
        sector strategies, I–5.63–70
    infrastructure strategy, I–2.54, I–5.21
    interim report, I–5.08
    licensing regime, I–4.01, I–4.17, I–4.79,
        I–4.82–3
        retrospective amendment of licences,
            I–5.13
    on LOGIC, II–5.16
    Mature Province Initiatives and, I–A.1
    MER Strategy
        asset stewardship, I–5.18
        implementation, I–4.83, I–5.25,
            I–5.28–70
        legacy, II–1.01
        licensing and, I–4.79
        OGA powers, I–4.51
        regulatory powers, I–5.12–13
        technology and, II–12.04
        tripartite approach, I–5.09
    new regulator, I–4.51, I–5.10–11,
        I–5.19, I–5.27, I–11.100
    objectives, I–4.86
    origins, I–5.07, I–10.63
    overview, I–5.07–72
    pre-Wood philosophy, I–5.05–6
    production diagnosis, I–2.08
    radical change, I–5.02, I–5.24, II–04.01
    reception, I–5.24
    recommendations, I–5.09–23
    regional development strategy,
        I–5.19–20
    sector strategies, I–5.14–23, I–5.63–70
    significance, I–1.03

taxation, I–3.45, I–5.08, I–7.01
technology strategy, I–5.22, II–12.62
tripartite approach, I–5.09
Woolf reforms, II–15.43
Woolfson, C, I–10.47
work programmes
  appropriate programmes, I–4.51
  competing applications, I–4.38, I–4.50
  delivering, I–4.50
  drill or drop, I–4.49
  firm commitments, I–4.49
  initial programmes, I–4.47–8
  meaning, I–4.47
  non-compliance, I–4.49
  OGA requests for, I–4.51
  seaward licences, I–4.47–51
  terms, I–4.49
workers
  definition, II–14.03, II–14.19–30
  dependence, II–14.25
  mutual obligations, II–14.28–30
  self-employed workers, II–14.03
  statutory definitions, II–14.19–20
working time
  amending regulation, II–14.81–2
  annual leave, II–14.62–4
  application to offshore work
    exclusions, II–14.66
    overview, II–14.44–82
  average working time, II–14.47–9
  breach of contract, II–14.68, II–14.72
  case law, II–14.73–80
  collective agreements, II–14.67

  contracting out, II–14.65
  covered workers, II–14.46
  daily rest, II–14.50, II–14.66
  definition, II–14.49, II–14.73
  employment tribunal claims, II–14.68,
    II–14.71
  enforcement, II–14.68–72
  EU law, II–14.44–5, II–14.81–2
  exclusions, II–14.65–7
  health and safety and, II–14.74,
    II–14.80
  holiday pay, II–14.23, II–14.63–4
  H&SE enforcement, II–14.68–70
  night work, II–14.56–9, II–14.66
  on-call time, II–14.74–80, II–14.81
  patterns of work, II–14.60
  purposive interpretation, II–14.80
  record-keeping, II–14.61
  rest breaks, II–14.52, II–14.66,
    II–14.78–9, II–14.80
  shift work, II–14.53–5
  weekly rest, II–14.51, II–14.66
worksites, definition, II–4.68–70
World Bank, II–15.09, II–15.48
World Trade Organization (WTO),
  II–15.48
Worthington, P, II–3.37
wrongful trading, II–10.116
Wytch Farm field, I–4.81, I–9.01

Yergin, D, I–5.24

zero-hour contracts, II–14.81